Salt Effect

Second Arheoinvest Symposium:

From the ethnoarchaeology to the anthropology of salt

20–21 April 2012, 'Al. I. Cuza' University, Iaşi, Romania

Edited by

Marius Alexianu

Roxana-Gabriela Curcă

Vasile Cotiugă

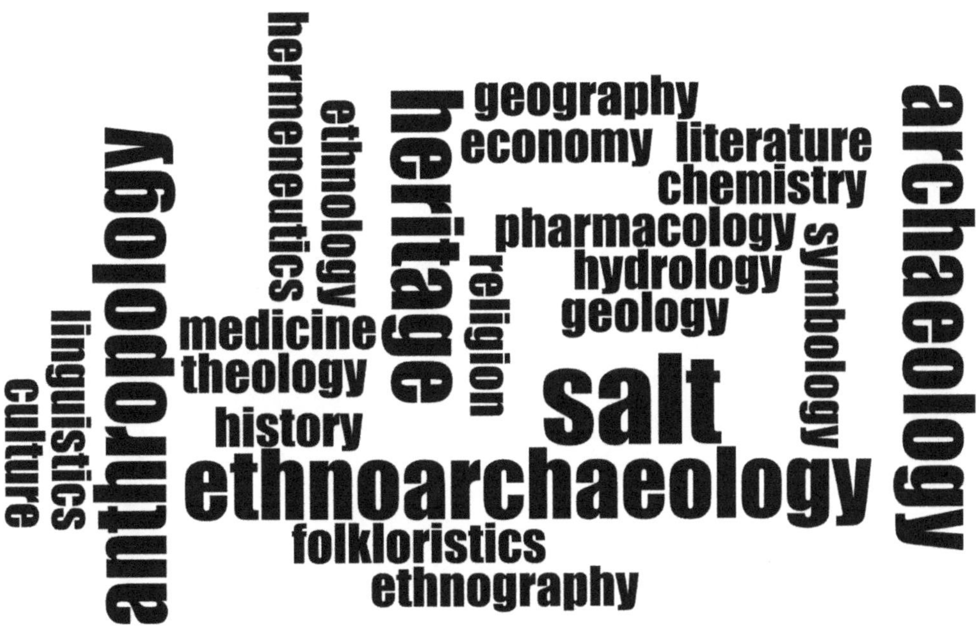

BAR International Series 2760

2015

Published in 2016 by
BAR Publishing, Oxford

BAR International Series 2760

Salt Effect

ISBN 978 1 4073 1422 8

This volume was prepared in the framework of the research project 'The ethno-archaeology of the salt springs and salt mountains from the extra-Carpathian areas of Romania' (code PN-II-ID-PCE-2011-3-0825, 219/5.10.2011), financed by the National Research Council (CNCS), 2011–2016, Romania.
W: http://ethnosalro.uaic.ro / E: ethnosalro@gmail.com

BAR Publishing is the trading name of British Archaeological Reports (Oxford) Ltd. British Archaeological Reports was first incorporated in 1974 to publish the BAR Series, International and British. In 1992 Hadrian Books Ltd became part of the BAR group. This volume was originally published by Archaeopress in conjunction with British Archaeological Reports (Oxford) Ltd / Hadrian Books Ltd, the Series principal publisher, in 2015. This present volume is published by BAR Publishing, 2016.

Printed in England

BAR
PUBLISHING

BAR titles are available from:

BAR Publishing
122 Banbury Rd, Oxford, OX2 7BP, UK
EMAIL info@barpublishing.com
PHONE +44 (0)1865 310431
FAX +44 (0)1865 316916
www.barpublishing.com

Table of Contents

Editors' note

The present volume gathers most of the papers delivered at the Second Arheoinvest Symposium, *From the ethnoarchaeology to the anthropology of salt*, organised in 2012 at the "Alexandru Ioan Cuza" University of Iaşi, within the framework of the project *EthnosalRo — The ethno-archaeology of the salt springs and salt mountains from the extra-Carpathian areas of Romania*. It further includes several studies of established or emerging researchers from Romania or abroad, who have not attended the symposium but who found it opportune, as we did too, to publish their work in this volume.

With respect to the structure of the materials, we've chosen a less common criterion, namely a section dedicated to the anthropology of salt on the international level, grouping authors from Austria, France and Romania, and another section on the anthropology of salt in Romania, grouping authors from Romania and France. In what concerns the order of the studies of a suggestive diversity from the latter section, we followed, in general, the thematic rule and, subordinate to it, the chronological one.

We thank Dr Michael Vickers for the careful proofreading of most of the studies in this volume. The editors thank their colleagues Ştefan Caliniuc, Andrei Asăndulesei, Mihaela Asăndulesei and Felix-Adrian Tencariu for tremendous help in editing this book.

M. Alexianu, R.-G. Curcă, V. Cotiugă

FOREWORD

Eastern Romania, or more specifically the pre-Carpathian and Carpathian area, came almost abruptly, at the end of the eight and the ninth decades of the last century, to the attention of prehistorians, particularly from Europe, on account of the discovery of some of the oldest traces of exploitation of salt springs in Europe and perhaps worldwide. Thus, the discoveries from Solca–*Slatina Mare* (Suceava County) made in 1968 but published only in 1977, followed by those from Lunca–*Poiana Slatinei* (Neamţ County), first published in 1987, have entered relatively quick into the international scientific circuit. These discoveries also elicited reactions in Romania too, most notably from the late archaeologist Dan Monah, who emphasized the role played by the salt springs in the development of the Eneolithic communities from Poduri (Bacău County), even though the salt springs from around this site were devoid of traces of ancient exploitation.

The area with salt springs from Eastern Romania is distinguished from other similar areas in Europe by the continuation, at an unexpected degree of intensity for an EU-member country, of traditional practices of exploitation of brine from salt springs. Save for some generic considerations present in a series of treaties, on the exploitation of salt by rural communities, Romanian ethnographic literature lacked a dedicated work on this living legacy. Having this in mind, the undersigned alongside archaeologists Dan Monah and Gheorghe Dumitroaia, and for a brief period ethnographer Elena Florescu, took upon themselves the task of carrying out ethnographic investigations and surveying the saliferous areas of Neamţ and Bacău counties in search of new salt springs. The 30 salt springs identified on this occasion, and the few ethnographic inquiries of a rather empirical character, served as the foundation for the first ethnoarchaeological study on the Romanian salt springs (1992). Despite a favourable welcome in Romania and abroad, the ethnoarchaeological research on salt ceased for more than a decade after the publication of the study.

The decisive moment for the renewal of these researches was the involvement in a French-Romanian project of archaeologist and ethnologist Olivier Weller, who acted as a genuine catalyst of a professional level for animating such researches in Romania. Very important was the elaboration (by O. Weller, M. Alexianu and L. Nuninger) of ethnological questionnaires from an archaeological perspective, work tools that proved their efficiency starting with the first field campaign conducted in 2003. For several years, the French-Romanian team carried out field ethnological investigations and the interaction between the subjects of the ethnographic inquiries and the research lead to the enlargement of the aspects focused on in the questionnaires, by diversifying and thoroughly addressing the topics of concern.

Decisive impulses for continuing and intensifying such researches on a regional level came from two important Romanian research grants, with French participation (O. Weller and R. Brigand), namely Ethnosal (2007–2010) and EthnosalRo (2011–2016). The more than 300 ethnological investigations conducted up to this moment have firmly established that salt generated in the Romanian rural world a true intangible heritage of a magnitude hard to equal around the globe. From among a far from exhaustive list, are the first-hand informations, never recorded at least in Europe, on the exploitation of salt springs and salt outcrops during the historical present (systems for capturing the salt springs, nonprofessional exploiters and users of the salt resources, production of recrystallized salt from brine from salt springs, the transportation of salt water, ignigenous salt and salt boulders, etc.), and on the present-day uses of the natural brines and salt (human consumption, food preservation, traditional halotherapy, artisanal uses, hunting, ritual uses of salt, barter, trade, gift, etc.).

These revelations drastically changed our initial research stance, as we realised that by limiting ourselves to a purely ethnoarchaeological approach would mean that countless facets of this very rich and varied Romanian intangible heritage on salt will forever be lost, on account of the imminent disappearance of the most aged generations whose (direct or indirect) testimonies quite often harked back to the early 20th century.

The lapse of 20 years since the publication of the first Romanian article on the ethnoarchaeology of salt occasioned in 2012 the national symposium entitled *From the ethnoarchaeology to the anthropology of salt*. The meeting, attended from the most part by Romanian researchers, was meant to test if on the national level there are possibilities to set in motion the anthropology of salt, a discipline established by the author of these lines.

The almost 30 papers in this volume represent an entire range of sciences and disciplines: archaeology, history, ethnography, ethnoarchaeology, cultural anthropology, literary hermeneutics, cultural studies, medicine, geography, geology, toponymy.

The results of this event went much beyond the initial expectations, reason for which, no matter how audacious the idea of organising an international congress on the anthropology of salt seemed even to us, we proceeded with its organisation.

The First International Congress on the Anthropology of Salt, planned to take place in August 2015 in Iaşi, will surely constitute a first challenge at the international level regarding the future of this new discipline.

Marius Alexianu

Anthropology of Salt: a First Conceptual Approach

Marius Alexianu
Universitatea „Alexandru Ioan Cuza" din Iaşi

Abstract
Starting from the diachronic impact of salt on humanity's numerous activities and spiritual reflexes, the author calls for establishing a new humanist discipline: the anthropology of salt. This first exertion lists the themes developed around salt and the sciences/disciplines involved primarily or sectorially in researching this mineral. The anthropology of salt is a discipline of the future, which will gradually become autonomous as the inter- and trans-disciplinary approaches to common salt will prevail over the mono- or multi-disciplinary ones.

Keywords
salt, cultural anthropology, archaeology, history, linguistics, interdisciplinarity, transdisciplinarity

The field researches conducted in Eastern Romania within a multiannual Franco–Romanian project starting in 2003 and within two Romanian projects with French attendance (Ethnosal and EthnosalRo) underlined that there is an entire rural universe generated by salt. This universe is not completely included in the ethnoarchaeological endeavour, but it reflects the structural positions of the man towards the only mineral edible by the animal world. We have realized, gradually, that one should not sacrifice the various dimensions of salt in the Romanian rural world for the (otherwise) natural demands of a discipline and that, after all, the current research should save and valorise this entire universe holistically, and this is possible through cultural anthropology.

In the cultural history of the world, the first who illustrated this model is the Spanish erudite Bernardino Gomez Miedes (1515-1589), the author of a magnificent Renaissance book published—in the final editorial variant—with the title *Commentariorum de sale libri quinque*, Valentiae, 1579. A recent three-volume critical edition (Gomez Miedes 2003) has the remarkable merit of having offered the occasion for the contemporary public to confront a forgotten integrative paradigm. Just for a glimpse of the vision and amplitude of this work, we will mention here the titles of the five books: I. *De sale physico sive philosophico*; II *De sale medico sive empirico*; III *De loco, tempore et modo salis in mensa*; IV *De sale geniali sive iocoso*; V *De sale mystico sive theologico*.

With the emergence of the various sciences and disciplines starting with the modern era, the approaches on salt have specialized progressively, mostly within the past century. Sectorially, the knowledge has reached unthinkable performances, but the researches have atomized and they have become so autonomous, that they tend towards insularisation. Actually, there is not a strong and sufficient communication between certain scientific communities. The perception of the entire salt-generated universe has been threatened; its primordial importance for the whole evolution of humanity has been diminished progressively, especially during the industrialization, refrigeration and globalization eras. For advanced societies, salt is a trivial, insignificant element of daily life. Salt has lost completely the sacred character that Homer talked about at the beginnings of the European culture.

The study object—NaCl (common salt)—has two forms, usually: natural salt water (the water in the seas and oceans, in the salt lakes, in the salt springs, in the salt rivers or salt rivulets) and salt rock. Nonetheless, there other presences, too, natural (naturally-recrystallized salt from salt water, salt lands) or artificial (salt from the evaporation of salt water in special devices, ignigenous salt, artificially salted water) that should be given proper attention.

From the diatopic and diachronic perspective, common salt—with all its natural or artificial metamorphoses—has influenced the humanity in the most diverse aspects. This is why, within a brief enumeration, the salt-related research themes are intriguingly various: natural, tangible and intangible heritage, explorations (hunting for salt), exploitation techniques, techniques to obtain different products, exploitation and use tools, transport and storage containers, human and animal feeding, conservation (meat, bacon, cheese, vegetables, green goods, fruits). The themes also include manufacture-related uses (including the construction of salt houses), mythology, religion, cult, rituals, beliefs, superstitions, mentalities, secret societies, magic, vows, curses, prohibitions, popular medicine, sexuality, economy, hide working, population, alchemical procedures, scientific and cultural representations, treatment of the deceased, barter, commerce, contraband, robbery. On the other hand, the themes also include human and animal mobility, the attraction exerted on savage beasts, symbolic uses, folk literature (stories, tales, and proverbs) and cult literature, the control of salt resources, conflicts, strategic value, geographic perceptions, professions related to salt exploitation and uses, economic, legal and administrative regulations, vocabulary, toponymy, anthroponymy and, of course, the list can go on. All these themes already constitute a study object for an impressive number of sciences, disciplines, or sub-disciplines, such as archaeology, history, ethnography, ethnoarchaeology, heritage studies, economic anthropology, food sciences, statistics, sociology, geology, mineralogy, geography, hydrology, botany, chemistry, medicine, pharmacology, ethology, theology, agronomy, symbology, linguistics, folklore studies, cultural studies, literary studies, hermeneutics, legal sciences, etc.

Obviously, some themes must be approached only in an interdisciplinary vision. On principle, salt anthropology refers to all the periods in human evolution. Nonetheless, currently we believe that the focus should be on the pre-industrial or non-industrial civilizations within contemporary history, though we are aware that even the most advanced current technologies represent a cultural act, a human endeavour. Fortunately, the last are well known.

Our innovating idea is that of putting together, under the sign of cultural anthropology, the (so) diverse approaches on this essential reference of human life. In order to get salt its lost dignity back, *principium individuationis* cannot be fully performing but within a *universa disciplina*. Only such a perspective will make any major research and any detail shows their true significations; outside this perspective, there is no sense in any approach, regardless of its technical sophistication degree.

Several years ago I stressed the difficulty to separate the archaeological approaches from the historical, ethnological and even linguistic or literary ones (Alexianu *et al.* 2011, vii), noting that the international meetings focused on salt organised for instance in Europe during the last two decades had an increasingly anthropological character (Alexianu *et al.* 2011, vii). The necessity to widen and deepen this natural tendency was, for that matter, underlined by a critical review published in a prestigious journal, which extolled the innovative spirit behind the elaboration of the aforementioned volume: "The volume contains a wealth of information for all those interested in the 'white gold' derived from land or sea. More importantly, this book marks a turning point in salt study in the humanities. Unlike previous conferences over the last twenty years, which have been purely historical or archaeological in scope, it brings together archaeologists, historians, philologists and linguistics under the same umbrella" (Galanidou 2012, 936).

But, as it can also be seen from the studies composing this volume, what still predominate are foremost studies of a monodisciplinary character, followed by multidisciplinary ones; rarer still are the interdisciplinary approaches, and the papers rather have an implicit anthropological character than an explicit one. Reaching the parameters demanded by a new discipline such as the anthropology of salt, is a task that can only be reached in a future to come after decades rather than years. Many stages must be completed until a genuine anthropological discourse on salt can emerge, one that will surpass the encyclopedic level, and in which the interdisciplinary and, particularly, the transdisciplinary character will prevail.

The quite reasonable objection of any researcher towards this truly holistic approach to salt is that the concept of anthropology is excessively dilated, one could say unallowably so. But in my opinion this pushing of anthropology's conventional boundaries is fully justified if we agree with a self-evident truth: salt is the only mineral that left an extraordinarily strong mark on almost all aspects of human life, from the structures of the everyday life, to the sciences and the refined symbolistics.

Paraphrasing a famous Latin adagium of Terentius, one could say that "Sal sum et nihil humani a me alienum puto" (I am salt, and I think nothing human is alien to me).

Acknowledgement
This work was supported by a grant of the Romanian National Authority for Scientific Research, CNCS–UEFISCDI, project number PN-II-ID-PCE-2011-3-0825, 219/5.10.2011, *The ethno-archaeology of the salt springs and salt mountains from the extra-Carpathian areas of Romania* — ethnosalro.uaic.ro.

References
Alexianu, M., Weller, O. and Curcă, R.-G. (eds.), *Archaeology and anthropology of salt: a diachronic approach (Proceedings of the International Colloquium, 1-5 October 2008. Al. I. Cuza University, Iași, Romania)*, BAR International Series 2198, viii + 226 pages, 175 b & w & colour illustrations, 2011. Oxford, BAR Publishing.

Galanidou, N. 2012. Review of M. Alexianu, O. Weller and R.-G. Curcă (eds.), *Archaeology and anthropology of salt: a diachronic approach (Proceedings of the International Colloquium, 1-5 October 2008. Al. I. Cuza University, Iași, Romania)*, BAR International Series 2198, viii+226 pages, 175 b & w & colour illustrations, 2011. Oxford, BAR Publishing, 978-1-4073-0754-1 paperback £ 65, Antiquity, vol. 86, No 333, 935-937.

Gomez Miedes, B. 2003. *Comentarios sobre la sal*. Introduction, critical edition, annotated translation and indices by S.I. Ramos Maldonado. Prologue by A. Malpica Cuello. Madrid, Alcaniz.

The Two Faces of Salt in Mesopotamia

Sebastian Fink
Universität Innsbruck

Abstract
This article tries to give an overview of our knowledge about salt in ancient Mesopotamia. A discussion of some cuneiform-texts demonstrates that salt was needed for a lot of activities, especially cooking, but on the other hand salinization of agricultural land caused heavy problems.

Keywords
Mesopotamia, cuneiform, salinization, Sumerian, Akkadian

Cuneiform literature ranges from the third millennium BCE to the beginning of the first millennium CE. Tablets with cuneiform texts are found from Iran to Egypt and from Turkey to Saudi Arabia. Despite this extreme extent of cuneiform script in time and space, salt is in fact not a very prominent theme in the cuneiform texts, maybe because literary texts want to discuss something exceptional – and salt was nothing exceptional. It was an essential part of human civilization, everyone had it and everyone used it. As an additional factor, salt was easily available in Mesopotamia. It was cheap and therefore economic texts do not mention it too often, because for small, private business transactions no business documents were needed. Salt is more or less the only natural resource that exists in abundance in southern Mesopotamia. But this abundance also caused heavy problems. The irrigation system that made the development of Mesopotamian culture possible was responsible for bringing huge amounts of salt into the fields and as a fact the salt-tolerance of agricultural seeds is limited. If there is too much salt in the soil, the field cannot be cultivated any more. Therefore, salt was both a blessing and a curse.

Sources of salt in Mesopotamia

As already mentioned, Mesopotamia has rich sources for salt. Mainly because the water of the two rivers Tigris and Euphrates has a high concentration of dissolved salt, it is even possible to extract salt from fluvial water. In his study "On Salt and Salt Gathering in Ancient Mesopotamia" Daniel Potts has collected the evidence for salt-gathering and has compared it to the situation in the 19th century when the Ottoman Public Debt Administration was charged with the administration of the state salt monopoly. In order to maximize the revenues of the monopoly they were "exploiting even those sources which were inferior in an attempt to deny private individuals or groups any opportunity of taking salt freely, their knowledge of the location of salines within their domain can be assumed to have been very nearly complete." (Potts 1984, 236). The list given by Potts exhibits that there were plenty salines in Mesopotamia, which could be exploited easily. In some places a lot of pits were used to bring the saline water to the surface. The saline water was stored in basins, where the water evaporated and the remaining crystalline salt was collected (Potts 1984, 240-241). Other salines which were repeatedly exploited in summer consisted of the remains of dried up lakes or salt swamps (Potts 1983 discusses the salt gathering activities of Bedouins in Mesopotamia and the North Arabian Desert). There the salt was extracted by cutting blocks (bricks) (Potts 1984, 242. For the situation in Mari (todays Syria) see Durand 1987, Durand 1990 and Guichard 1997). The quality of the salt depended on the saline. Some salines were famous for the fine quality of their salt, while others produced only salt of mediocre quality that was used for livestock, and the worst quality was used for tanning. Besides these sources there are also salt domes, the largest near Basra, where, due to tectonic activity, ancient layers of salt came to the surface and can be exploited. There is also evidence that the ashes of some plants (which are still unidentified) were used as a special salt (Butz 1984, 285-286).

Assuming that the climate and the degree of salt in Euphrates and Tigris have not changed dramatically since ancient times, we can rely on this rich Ottoman information as a base for the system of salt gathering in ancient Mesopotamia. Maybe due to cheapness and the easy availability of salt there is not much evidence for the mining of rock salt and the large scale evaporation of sea water in the Ancient Near East (Potts 1984, 248-249).

Words for salt

Sumerian and Akkadian are the two main languages of cuneiform texts from Mesopotamia. Sumerian, which is not related to any other known language-family, was the language of larger parts of the population in southern Mesopotamia in the 3rd millennium BCE. But around 2.000 BCE Sumerian became a dead language and was no longer the mother-tongue of anyone. Akkadian, which belongs to the big and well-known family of Semitic languages (like Hebrew or Arabic), had taken over its role as an everyday language in southern Mesopotamia. But, like Latin and Greek in Europe, Sumerian was not forgotten completely in Mesopotamia. Scholars and priests still learned Sumerian in school and they continued to use Sumerian, especially for religious purposes. It seems that in a cultic context, where all changes are supposed to be dangerous, the old tradition of Sumerian prayers and incantations were maintained until the very end of cuneiform culture. With the end of cuneiform-script the knowledge of this ancient language also passed away (For a discussion of the linguistic situation in Mesopotamia from the third to the first millennium B.C. see Fink and Lang 2012 with further references). The Akkadian scholars have left behind plenty of bilingual texts (among them there are the so-called lexical lists, a kind of ancient dictionary in list-format. The elaborated versions contained one column with a syllabic rendering of a Sumerian word, the next column gave the Sumerian logogram (word-sign) and sometimes there was also a column for the sign-name and finally we had up to two columns for translations into Akkadian or other languages (For an introduction to the topic see Cavignaux 1980. A digital edition of a considerable part of the lexical list is presented at the Digital Corpus of Cuneiform Lexical texts (DCCLT-*Digital Corpus of Cuneiform Lexical Texts*). Only due to these texts we were able to decipher the Sumerian language. The bilingual lists are a main source for different words for salt in Mesopotamia. Therefore we will discuss salt both in Sumerian and in Akkadian texts, because due to the widespread bilingualism we are hardly able to distinguish Sumerian and Akkadian culture through the means of language. Even a distinction of Sumerian and Akkadian texts is not very useful because we are going to discuss bilingual lists.

The Sumerian word for salt is "mun". This word is not only used as a substantive but also as a verb. As a verb it mostly indicates that something is salty or brackish. The Akkadian word is *ṭābtu* and it derives from the verb *ṭābu* which means "to be good / sweet". Besides these basic words we find a much broader range of designations for salt in the lexical lists. I do not want to discuss every detail because this was already done by Kilian Butz in 1984, and not much new lexical material concerning salt was discovered until then (Butz 1984, 272-316).

The most important Sumerian terms for different forms of salt according to lexical lists are:

- mun.a.ma.nim: red salt
- mun.e$_3$.a: crystal-salt
- mun.ku$_3$.pad: a rough type of salt
- mun.eme.sal.la: table salt ("salt of the fine tongue")
- mun.dur$_5$: brine
- mun.nu.luh.ha: uncleaned salt ("salt not washed")
- mun.HI.HI: Kristall-Salz
- mun.ab: marine salt
- mun.u$_2$.kur: salt made from plants ("salt of the mountain-plant(s)")
- na$_4$.mun: rock-salt ("stone-salt")
- sig$_4$.mun: salt brick ("brick-salt") (Butz 1984, 296-297)

In Akkadian we also have a broad variation of designations for salt:

- *#ābtu šadî*: "rock salt" (CAD 2006, s.v. *ṭābtu*, 10-15)
- *idrānu / idru:* salt found on saline fields (Butz 1984, 310-313)
- *emesallu*: "fine taste". Sumerian loanword. When it is used to designate a certain quality of salt it means table salt, salt of high quality (CAD E 1958, s.v. *emesallu*, 148).
- *mesallu*: "fine taste" (for salt only). This term is only a variant to *emesallu* (CAD M II 1977, s.v. *mesallu*, 28).
- *kuddimmu*: following Butz *kudimmu* designates rock salt (Butz 1984, 297), following the CAD it designates "a kind of salt or lye obtained from a plant." (CAD K, s.v. *kuddimmu,* 493).
- *amānu*: "red salt". The identification of this salt is uncertain, but it is always described as red (Potts 1990 discusses the possibility that the "bacteria Haematococcus found in many parts of the world" is responsible for the reddening of the salt). Maybe this blood-like color was the reason for its use in medicine (see CAD A II 1968, s.v. *amānu*, 2 for further references).

Salt in Sumerian literature

Although there is a broad variation of different words for salt in the lexical lists, we have to take into consideration that these lists are mainly used for the education of scholars who had to work in the administration. There it was necessary to distinguish the different types of salt and to pay the right price for the different qualities. In Sumerian literature we only have attestations of the simplest word for salt, i.e. "mun" without any further specifications. Let us have a look at the texts: (I will not discuss occurrences of mun "salt" in broken context. As a base for this analysis I use the texts in the electronic text corpus of the Sumerian language (ETCSL-*The Electronic Text Corpus of Sumerian Literature*) which offers a searchable database of the most important texts of Sumerian literature).

1. mun kug i$_3$.me.a nanga kug nu.me.[a] When holy salt existed, holy potash did not exist. (Marriage of Martu, Line 4) (For an edition and translation see ETCSL-*The Electronic Text Corpus of Sumerian Literature*)

2. mun gazi gu$_2$ a.šag$_4$.ga.ka ğal$_2$.la bal u$_3$.mu.ak. When salt and mustard which are located at the side of the field are removed. (The debate between cooper and silver, Line 75) (See ETCSL-*The Electronic Text Corpus of Sumerian Literature*)

3. ukur$_3$ ha.ba.ug$_7$ nam.ba.da.til$_3$.i ninda i$_3$.pad$_3$ mun nu.pad$_3$ mun i$_3$.pad$_3$ ninda nu.pad$_3$ uzu i$_3$.pad$_3$ gazi nu.pad$_3$ gazi i$_3$.pad$_3$ uzu nu.pad$_3$ (Variants add: i$_3$ i$_3$.pad šangan nu.pad$_3$ šangan i$_3$.pad$_3$ i$_3$ nu.pad$_3$). The poor man should die, he should not live. When he finds bread, he does not find salt. When finds salt, he does not find bread. When he finds meat, he does not find mustard. When he finds mustard, he does not find meat. When he finds oil, he does not find a flask. When he finds a flask, he does not find oil. (Proverb collection 1, 1.55) (Alster 1997, 346 explains this proverb as follows „When someone possesses something, yet lacks something else which belongs with

it, it may result in his seeming ridiculous. There are many international type parallels.").

4. nitah zig_3.ga.am_3 mun al.gu_7.e munus zig_3.ga.am_3 im.ma al.gid_2.gid_2.de_3 A grown-up man is eating salt. A grown up woman is dragged in the mud (?). (Proverb collection 1, 1.156) (Alster 1997, 355: „The exact meaning is uncertain, but it seems to involve the idea that a man faced with difficulties is in a favorable position compared to a woman faced with the same problems).

5. $utul_2$ du.bu.ul nu.zu mun(?).ta al.si_3 sug.sur.ra nu.zu a.ta. a.si_3 mi_2.us_2.sa_2 nig_2.DI nu.un.zu.a du_{14}.mu_2.mu_2 al.si_3 A kneading (?) trough not (previously) known is put (to the test) by means of salt. A mixing-jar(?) not (previously known) is put (tot the test) by means of water. A son-in-law whose ... is unknown is put (to the text) in quarrels. (Proverb collection 1, 1.196) (For text and variants see Alster 1997, 38- 39)

6. an.na.dug_3.ga mun he_2.gu_7.e nu.un.na.dug_3.ga ^{u2}munzer he_2.gu_7.e
If it is too sweet for him, let him eat salt.
If it is not too sweet for him, he should eat licorice. (Proverb collection 3, 3.131) (Alster 1997, 102 translates the plant in question „bitter plants". In his commentary he explains: "I.e., there is a remedy for every excessive taste." Alster 1997, 389. In my opinion the pair salt – licorice fits much better than salt – "bitter plants", therefore I follow ETCSL's translation at this point (ETCSL-*The Electronic Text Corpus of Sumerian Literature*).

These texts do not give us much direct information on the use of salt in Ancient Mesopotamia, but they show that salt was something of extraordinary importance.

The mythological introduction of the Marriage of Martu – **Text number 1.)** – mentions "holy salt" (the art of cooking) besides the holy crown (the symbol of kingdom), the holy herb (food and medicine), intercourse and kissing and giving birth – the basics of every civilization.

Text number 2.) is not very clear. It mentions salt besides mustard (the identification of this word is not absolutely certain, it could also be something else) – which is a pest plant. This seems to hint to the rather negative aspect of salt, the salinization of the soil.

The next **text – number 3.)** – shows us that salt was a basic need. The poor man, who is always missing something, has bread but no salt – therefore bread and salt go together in Mesopotamia. We are not used to have our bread with salt (maybe because the salt is already used in the baking-process) but when we have a look at Limets description of Mesopotamian bread "it was a rather coarse food, very plain and certainly tasteless" (Limet 1987, 133) we can understand that something spicy was needed to eat it with some pleasure.

It is quite hard to understand **text number 4.)**, but it seems to be a piece of misogyny.

Our **text number 5.)** has some uncertain readings, but Alster points to the possibility that this proverb mentions salt as a cleaning agent („The interpretation is very uncertain and hampered by the fact that the precise function of the jars mentioned is unknown. If correctly understood, just as some unused jars must be washed with salt and water before they can become useful, so must a son-in-law undergo a test of verbal disputes, before he can become truly acquainted with a household." Alster 1997, 358).

Text number 6.) puts salt into contrast with licorice and shows us that for the Mesopotamians salt and licorice marked two extremes of taste, salty and sweet. For us it is surprising that salt is compared to licorice rather than honey but in fact bee honey was not very common in Mesopotamia until Neo-Assyrian times (Ellison 1984, 94).

The danger of salinization

Before having a closer look at the use of salt as a metaphor for destruction in Mesopotamian literature we should explain how salt was linked with destruction. As I have already mentioned agriculture was one of the backbones of the Mesopotamian cultures. To enlarge agricultural land in southern Mesopotamia irrigation was needed. Without irrigation only small pieces of land besides the rivers would have been suitable for agriculture (Charles 1988, 1-39 gives a detailed description of the conditions for agriculture in southern Mesopotamia). As Charles points out the salt level of the rivers Euphrates and Tigris constantly rises downstream the rivers and the evaporation in the irrigation-channels is responsible for another rise in the salt-concentration of the irrigation water. From data collected by scientists in the 1950s it is known that some channels had a concentration of 900 ppm salt in summer. This means that if one cubic meter of water (= 1.000 l) is used per square meter and year 900 grams of salt are brought into a square meter of the field every year. The evaporation and therefore the amount of solved salt in the water rises with the length of the channel. So the area of irrigation is somewhat limited by this factor (Charles 1988, 8-9).

Thorkild Jacobsen even tried to explain the history of Mesopotamia with salinization. He was convinced that the shift of power from southern to northern Mesopotamia was linked with the decline of the southern agriculture due to salinization (Jacobsen and Adams 1958, 1251-1258. See also Jacobsen 1982). This theory would give us an explanation for the decline of the Ur-III-state at the end of the 3rd millennium (Nissen 1998, 67-68), but Jacobsen hypothesis is not generally accepted anymore (See Powell 1985 for arguments against the „theory of progressive salinization". Nützel 1992 argues in favour of this theory). The main argument against this theory is that already the Sumerians had knowledge about salinization, our main-sources for salinization are Sumerian texts and so it is quite likely that they understood this problem and that they found a solution for it (Powell 1985, 38). But despite this it seems to be a fact that the ancient Mesopotamians faced serious problems with salinization. We simply do not know to which extent salinization made agriculture impossible or rendered it ineffective in a certain period of Mesopotamian history because it is nearly impossible to find archaeological evidence for this.

But in fact the Mesopotamians developed strategies against the salinization of the soil, because in the 1st millennium BCE, and later on, until the devastation of the Mesopotamian irrigation systems by the Mongols in the 13th century CE, we have evidence for a flourishing agriculture in Mesopotamia. If they had done nothing, agriculture would not have been possible anymore because the percentage of salt in the soil is constantly rising over time when no measures are taken. Streck lists the main strategies against salinization. The most important techniques are:
-the reduction of the amount of water (therefore reduction of the amount of salt brought into the field);
-the use of a drainage-system to reduce the level of the groundwater and
-leaching. If there is already too much salt in the soil, large quantities of water are used to dissolve the salt in the soil and the water is conducted away from the field somewhat later (Streck 2006, 597-598).

Salt as a metaphor for destruction in Mesopotamian literature

The most important proofs of the awareness for the danger of salinization are literary texts. It is quite a common motive in curse-formulas that salt should be in the field of someone (For references see Streck 2006, 598). The purpose of this curse is quite clear – plants should not grow in the field of the cursed person anymore. Besides curses on Kudurrus (inscribed border-stones) (See Streck 2006, 598 for further literature) we have a very prominent curse in the myth of Atra-Hasis. In the beginning of this text Enlil, the chief god of the Mesopotamian pantheon in the 2nd millennium BCE, is disturbed by the noise of mankind. So he decides to reduce its number to find sleep again. At first he tries to diminish mankind with diseases, but Enki, the clever god and friend of mankind, shows Atra-Hasis, the hero of the story, a way to trick the gods and to put an end to the diseases. After this, Enlil tries to reduce the number of people by a famine:

pur-sa-ma a-na ni-še-e ti-ta i-na kar-ši-ši-na li-me-ṣu šam-mu e-liš ^{d}adad zu-un-na-šu lu-šá-qir li-sa-kír šap-liš ia iš-šá-a me-lu i-na na-aq-bi li šur eqlu iš-pi-ke-e-šu li-né-'irta-šá ^{d}nisaba ṣalmuti^{meš} lip-ṣu-ú ugaru EDIN pal-ku-ú lu-li-id id-ra-nu

Cut off the food supplies from the peoples, Let plant life be in short supply in their stomachs, Let Adad above make his rain scare, Below (let the river) be blocked up and let it not raise the flood from the Abyss. Let the fields diminish their yields, Let Nisaba turn aside her breast, Let the black fields become white, Let the broad plain produce salt (42-48) (Lambert and Millard 1969, 108).

Here we can see that Enlil tried to kill mankind by causing a great famine. First he reduces or stops rainfall, then he blocks the underground-waters in order to make irrigation impossible and finally he turns the black fertile field into infertile white fields covered with salt (For a discussion of this episode see Groneberg 1991, 403-408).

Salt in Akkadian Incantations

One of the most important Akkadian anti-witchcraft incantations is *Maqlû* (= burning) (For an edition of *Maqlû*, see Meier 1937. Tzvi Abusch is preparing a new edition. For a German translation see Abusch and Schwemmer 2008). This incantation is used to help someone who thinks that he is under the spell of a "witch". To free the person from the spell, magical procedures are carried out and the gods are asked for help. In one part of the incantation (Tablet VI) we find a mentioning of salt, which shows us that salt was also used in magic. Tablet VI starts with an incantation concerning Enlil (23 lines), the object of the second incantation is the *kukuru*-plant (44 lines), the third incantation addresses sulfur (ca. 45 lines) and, after a small break, our salt incantation (9 lines) starts. After the salt incantation the text is quite general, it does not address any materia magica anymore. If we have a look at the lines dedicated to the single items, we can see that the *kukuru*-plant and sulfur were very important in undoing magic, in purifying the victim of the witch's magic. In contrast to these two prominent purifying agents salt was only treated shortly in the following paragraph:

šiptu attī ṭābtu ša ina ašri elli ibbanû ana mākālê ilī rabûti išīmki Enlil ina balīki ul iššakkan naptan ina Ekur ina balīki ilu šarru bēlu u rubû ul iṣṣinū qutrinnu anāku annanna apil annanna ša kišpī ṣubbutū'inni upšāšê le'bū'inni puṭrī kišpīya ṭābtu pušširī ruḫêya upšāšê muḫrînnī-ma kīma ili bānīya lultammarki (Maqlû, VI, 111-119) (Text after Stackert 2010, 239. For a philological discussion of the text see Stackert 2011, 191-194. Stackert's line-enumeration differs somewhat from the newest translation in Abusch and Schwemmer 2008, 169).

Incantation: You are salt, that was made in a pure place; For the food of the great gods Enlil appointed you. Without you, the royal banquet is not set in the Ekur; Without you, god, king, lord, and prince do not smell incense.
As for me, so and so[1], son of so and so, whom spells are seizing, whom magical intrigues are afflicting – Release my spell, O salt! Disperse my sorcery! Take from me the magical intrigues so that, like the god who made me,

[1] *Maqlû* is a kind of handbook for the exorcists. Therefore there are no individual names in these texts and the room for the cursed person's name is inscribed for placeholders. This also informs as that these rituals were performed orally.

I will praise you without cease (Translation after Stackert 2010, 239 with slight revisions).

In the second line we read that Enlil, the god who decrees the destinies, appointed salt as the food of the great gods. Salt was used at the royal banquet in the Ekur. The Ekur is situated in Nippur and it is one of the most prominent temples in Mesopotamian tradition. It is known as the place of the assembly of the gods (For references see George 1993, 116-117). That means that salt was used in offerings for the gods. The next line informs us that salt was also used for the burning of incense.

This text raises the question whether salt was deified in Mesopotamia. In fact we have some evidence for this from Mari and Terqa (Guichard 2003, 7 and Stackert 2010, 235) but we do not have to overrate this because a lot of things were described as holy and were written with the DIĞIR-sign (which means god and is used as a marker for divine beings) by Mesopotamian scribes just because they were very important (On divine objects in Mesopotamia see Selz 1997).

The use of salt in Mesopotamia

As everywhere in the world, salt was used as a spice in Mesopotamia and there is also evidence for the use of salt as a necessary element of the nutrition of cattle (The evidence is collected by Streck 2006, 596). We do not have too much evidence for the Sumerian cuisine, but almost every known recipe contains salt (Some evidence is collected by Limet 1987. The Garšana-Archive which was published in the last few years brought some new evidence on the Sumerian cuisine. For a detailed discussion of the information from Sumerian business-documents see Brunke 2010 and Brunke 2008). For example we have a receipt for a kind of soup which is spiced with salt (Brunke 2010, 391. For an edition of Sumerian recipes see Bottéro 1995. For an overview of the usage of salt in the Sumerian cuisine see Bottéro 1995, 221). Besides that salt was used for tanning (Potts 1984, 232-233), for the production of blue glaze (Levey 1958, 340), maybe in metallurgy (Pots considers the possibility that salt was used to separate gold and silver) (Potts 1984, 234) and for the conservation of foodstuff. One proverb points to the possibility that salt was used as cleaning agent (For a discussion of the text see Alster 1997, 358). Especially fish and meat were treated with salt to extend their storage life (See Salonen 1970, 259 and 262 for some details. Salonen also mentions that the Mesopotamians preferred to buy living fish and that the worst quality of fish was "stinking fish" – Salonen 1970 257-258. Due to the heat and the missing possibilities of refrigeration a dead fish starts stinking after some hours without treatment. A lot of evidence for salted meat (mainly from lexical lists) is collected by Streck 2006, 596).

Maybe these techniques of conservation helped southern Mesopotamia to reach the high cultural level which is documented in textual and archaeological evidence, because southern Mesopotamia is alluvial land and therefore extremely poor of stones, metals and – due to the hot climate and the salty soil – also of high-quality wood. All the raw materials for metal-working, stone-working and even the beams, which were necessary to build great halls in palaces or temples, had to be imported. And it is not very convincing that all those products were acquired through raids by the inhabitants of southern Mesopotamia. So we have to look for goods which the Mesopotamians could use to barter. Southern Mesopotamia had a flourishing textile-industry already in the third millennium and Mesopotamia was famous for its richness of grain, so we have two products that were certainly exported from Mesopotamia. This is also reflected in a Sumerian epic text called "Enmerkar and the Lord of Aratta". Enmerkar wants to have precious stones from Aratta, which is situated somewhere in the Iranian highlands, to decorate the temple of his favorite goddess Inanna. At this point we have a conflict between Enmerkar and the Lord of Aratta because Enmerkar wants Aratta's stones and he has no idea of trade – so the only possibility to make the Lord of Aratta obey is to show him his supremacy, in the worst case by means of war. But in the end of the epic Enmerkar invents writing, the basis of book-keeping and Inanna resolves this situation by inventing trade (For an edition and discussion of this text see Vanstiphout 2003, 49-96).

But besides that we also know that huge amounts of fish were caught in the rivers and the sea (For a discussion of fishing in Mesopotamia see Salonen 1970, Englund 1990, Sarhage 1999, Kleber 2004 and Potts 2012) and that these fish were used to produce salt-fish and other fish-products (We also have evidence for fish oil, dried fish and smoked fish. Salonen lists ten methods of fish-conservation that were practiced in Ancient Mesopotamia. See Salonen 259-260 and Sahrhage 1999, 147-151). The most famous regions for stones and metals were situated in the Iranian highlands, where fish, grain, and, at least in some places, also salt could be exchanged for precious goods. Unfortunately a lot of this remains speculation due to missing documentation, because in general we do not have much evidence for Mesopotamian exports. Most of our cuneiform texts only mention the import of precious goods, often as a fabulous deed of the king, and we have no documentation from their business partners (For a discussion of „Mesopotamia's Invisible Exports" see Crawford 1973). But Buccellati has brought up some new evidence concerning salt trade by reinterpreting a certain kind of pottery – the so called beveled-rim bowls – as containers for salt (See Bucellati 1990 for some details. The connection of this bowls with salt was first proposed by Potts 1984, 259-264). Maybe a future detailed study of the distribution and origin of this bowls will give us further hints to the outlines of salt trade in Mesopotamia and its adjacent regions.

Finally I want to point to two rather obscure cases of the usage of salt in Mesopotamia. One letter about terrestrial omens to the Assyrian king mentions an anomaly with eight feet and two tails (which is a very good omen for the king – the king will seize the kingship of the world) (For an edition and discussion of the Mesopotamian omen-texts concerning anomalies, see Leichty 1970). Tamdanu, the owner of the sow which gave birth to the anomaly pickled it in salt in order to prove the occurrence of this omen – maybe in order to be rewarded by the king (Hunger 1992, 287. See also Leichty 1970, 11 for a discussion of this text). Nabû-bēl-šumāte, a rebellious leader of the Chaldeans in the time of the Assyrian king Assurbanipal (628-668 B.C.E), was killed by the Elamites, pickled in salt and sent to Assyria after Assurbanipal offered "the rebel's weight in gold to whoever caught or killed him" (Millard 1998, 30. For other examples for the conservation of dead enemies see Rollinger and Wiesehöfer 2012).

Acknowledgements
I want to thank Robert Rollinger and Martin Lang for their comments on earlier versions of this article and for their hints to literature.

References

Alster, B. 1997. *Proverbs of Ancient Sumer: The world's Earliest Proverb Collections* (2 Volumes). Bethesda, Maryland, CDL Press.

Bottéro, J. 1995. *Textes culinaires mesopotamiens.* Mesopotamian civilizations 6. Winona Lake, Ind., Eisenbrauns.

Brunke, H. 2008. *Essen im Sumer – Metrologie, Herstellung und Terminologie nach Zeugnis der Ur III-zeitlichen Wirtschaftsurkunden.* München, Herbert Utz Verlag.

Brunke, H. 2010. Zur Rekonstruktion von Speisen in Sumer anhand administrativer Urkunden, in G. Selz and K. Wagensonner (eds.), *The Empirical Dimension of Ancient Near Eastern Studies.* Wiener Offene Orientalistik 6, 375-399. Wien, LIT-Verlag.

Buccellati, G. 1990. Salt at the Dawn of History: The Case of the Bevelled-Rim Bowls, in P. Matthiae, M. van Loon and H. Weiss, *Resurrecting the past: a joint tribute to Adnan Bounni*, 17-40. Leiden, Nederlands Instituut voor het Nabije Oosten.

Butz, K. 1984. On Salt Again ... Lexikalische Randbemerkungen. *Journal of the Economic and Social History of the Orient* 27, 272-316.

Cavignaux, A. 1980. Lexikalische Listen. *Reallexikon der Assyriologie und vorderasiatischer Archäologie* 6, Berlin / New York, 1980-1983, 609-641.

Charles, M.P. 1988. Irrigation in Lowland Mesopotamia. *Bulletin of Sumerian Agriculture* IV.1, 1-39.

Crawford, H.E.W. 1973. Mesopotamia's Invisible Exports in the Third Millennium B.C. *World Archaeology* 5, 232-241.

Durand, J.-M. 1987. Villes fantômes de Syrie et autres Lieux. *Mari Annales de Recherches Interdisciplinaires* 5, 199-234.

Durand, J.-M. 1990. Le Sel à Mari (II): Les Salines sur les Bord du Habur. *Mari Annales de Recherches Interdisciplinaires* 6, 629-634.

Ellison, R. 1984. Methods of Food Preparation in Mesopotamia. *Journal of the Economic and Social History of the Orient* 27, 89-98.

Englund, R.K. 1990. *Organisation und Verwaltung der Ur III-Fischerei.* Berliner Beiträge zum vorderen Orient). Berlin, Reimer.

Fink, S. and Lang, M. 2012. Bilingualität im Alten Orient – Räume und Akteure, in C. Ulf and E.-M. Hochhauser (eds.), *Kulturelle Akteure.* Cultural Encounters and Transfers 1, 177-197. Würzburg, Verlag Koenigshausen Neumann.

George, A. 1993. *House Most High: The Temples of Ancient Mesopotamia.* Mesopotamian Civilizations 5. Winona Lake, Indiana, Eisenbrauns.

Groneberg, B. 1991. Atramhasis, Tafel II iv-v, in D. Charpin and F. Joannès (eds.), *Marchands, Diplomates et Empereurs – Etudes sur la civilisation mésopotamienne offertes à Paul Garelli*, 398-410. Paris, Éditions Recherche sur les Civilisations.

Guichard, M. 1997. Le Sel à Mari (III). Les Lieux du Sel. *Florilegium marianum III.* Mémoires de Nouvelles Assyriologiques Brèves et Utilitaires 4, 167-200.

Guichard, M. 2003. Divinité des salines mentionnée à Terqa. *Nouvelles Assyriologiques Brèves et Utilitaires* 7.

Hunger, H. 1992. *Astrological Report to the Assyrian Kings.* State Archives of Assyria 8. Helsinki, Helsinki University Press.

Jacobsen, T. and Adams, R.M. 1958. Salt and Silt in Ancient Mesopotamian Agriculture. *Science* 128, 1251-1258.

Jacobsen, T. 1982. *Salinity and irrigation agriculture in antiquity.* Bibliotheca Mesopotamica 14. Malibu, Undena.

Kleber, K. 2004. Die Fischerei in der spätbabylonischen Zeit. *Wiener Zeitschrift für die Kunde des Morgenlandes* 94, 133-165.

Lambert, W.G. and Millard, A.R. 1969. *Atra-Hasis The Babylonian Story of the Flood.* Oxford, Oxford University Press.

Leichty, E. 1970. *The Omen Series šumma izbu.* Texts from Cuneiform Sources 4. Locust Valley, NY, J.J. Augustin.

Levey, M. 1958. Salt and Soda in Ancient Mesopotamian Chemical Technology. *Isis* 49, 336-342.

Limet, H. 1987. The Cuisine of Ancient Sumer. *The Biblical Archaeologist* 50, 132-147.

Millard, A.R. 1998. Nabû-bēl-šumāte. *Reallexikon der Assyriologie und vorderasiatischer Archäologie* 9, Berlin / New York, 1998-2001, 30-31.

Nissen, H.J. 1998. *Geschichte Alt-Vorderasiens.* Oldenburger Grundriss der Geschichte 25. München, Oldenbourg.

Nützel, W. 1992. Die Bodenversalzung als mögliche Ursache für die Schwerpunktverlagerung von Südmesopotamien über Babylonien nach Assyrien. *Mitteilungen der Deutschen Orient-Gesellschaft*, 79-86.

Potts, D. 1983. Salt of the Earth: The Role of a Non-Pastoral Resource in a Pastoral Economy. *Oriens Antiquus* 22, 206-215.

Potts, D. 1984. On Salt and Salt Gathering in Ancient Mesopotamia. *Journal of the Economic and Social History of the Orient* 27, 225-271.

Potts, D. 2006. Salz, Versalzung (salt, salinization). B. Archäologisch. *Reallexikon der Assyriologie und vorderasiatischer Archäologie* 11, Berlin / New York, 2006-2008, 599-600.

Potts, D. 2012. Fish and Fishing. In D.T. Potts (ed.), *A Companion to the Archaeology of the Ancient Near East*, 220-235. Chichester, West Sussex, Wiley-Blackwell Publishing Ltd.

Powell, M.A. 1985. Salt, Seed and Yields in Sumerian Agriculture. A Critique of the Theory of Progressive Salinization. *Zeitschrift für Assyriologie und Vorderasiatische Archäologie* 75, 7-38.

Rollinger, R. and Wiesehöfer, J. 2012. Kaiser Valerian und Ilu-bi'di von Haman. Über das Schicksal besiegter Feinde, persische Grausamkeit und die Persistenz altorientalischer Traditionen, in H. Baker, K. Kaniuth and A. Otto (eds.), *Stories of long ago – Festschrift für Michael D. Roaf*, Alter Orient und Altes Testament 397, 497-515. Münster, Ugarit Verlag.

Salonen, A. 1970. *Die Fischerei im Alten Mesopotamien.* Helsinki, Snomalainen Tiedeakatemia.

Sarhage, D. 1999. *Fischfang und Fischkult im alten Mesopotamien.* Frankfurt a. M., Peter Lang-Verlag.

Selz, G. 1997. 'The Holy Drum, the Spear, and the Harp' Towards an Understanding of the Problems of Deification in Third Millenium Mesopotamia, in M. Geller and I. Finkel (eds.), *Sumerian Gods and Their Representations*, 149-194. Groningen, STYX.

Stackert, J. 2010. The Variety of Ritual Applications for Salt and the Maqlû Salt Incantation, in J. Stackert, B.N. Porter and D.P. Wright (eds.), *Gazing on the Deep: Ancient Near Eastern, Biblical, and Jewish Studies in Honor of Tzvi Abusch*, 235-252. Bethesda, CDL.

Stackert, J. 2011. An Incantation-Prayer to the Cultic Agent Salt. In A. Lenzi (ed.), *Reading Akkadian Prayers & Hyms*, 189-194. Atlanta, Society of Biblical Literature ancient Near East monographs.

Streck, M.P. 2006. Salz, Versalzung. A. Nach Schriftquellen. *Reallexikon der Assyriologie und vorderasiatischer Archäologie* 11, Berlin / New York, 2006-2008, 592-599.

Vanstiphout, H. 2003. *Epics of Sumerian Kings*, Writings form the Ancient World 20. Atlanta, Society of Biblical Literature.

Abbreviations

CAD = Chicago Assyrian Dictionary, Chicago, 1956-2011.

Online sources

ETCSL-*The Electronic Text Corpus of Sumerian Literature*,
http://etcsl.orinst.ox.ac.uk/cgibin/etcsl.cgi?text=t.6.1.03
http://etcsl.orinst.ox.ac.uk/cgibin/etcsl.cgi?text=c.1.7.1
http://etcsl.orinst.ox.ac.uk/
http://etcsl.orinst.ox.ac.uk/cgibin/etcsl.cgi?text=c.5.3.6

DCCLT-*Digital Corpus of Cuneiform Lexical Texts*
http://oracc.museum.upenn.edu/dcclt

Salt History or Salt in History?

Bernard Moinier
Paris

Abstract
Salt has presumably no history as such but its role in history is of a paramount importance. Since Prehistory it has participated in human activities, and its ever increasing contribution is attested by a tremendous growth rate in terms of outlets and requested quantities. Either mined (rock salt, salt in brine) or obtained by solar or artificial evaporation (marine salt, pan salt, vacuum salt) this mineral substance is known to influence politics (commerce, taxation/smuggling), religion (beliefs, rituals) and health (nutrition, sexuality). Generally speaking, salt is rather well distributed all over the globe, and is exceptionally lacking in areas of noticeable size. Ancient societies appear to have had access to the resource within the framework of cultural attainments, first divine, and practical when cities became better organized. Nowadays, industrialized societies express salt requirements which are by far exceeding the needs of unacculturated ones. In other words, a number of cultures have developed thanks to adequate salt supply. Its production is a source of revenue to governments. It contributes to shape the landscape. Simultaneously, environmental problems arise such as, in the past, lead-poisoning pans, deforestation related to wood-fuelled salt works, and still existing subsidence where over-mined cavities collapse. Linked to Aphrodite's birth, salt crystals have many facets. Mythology abounds in stories and proverbs which are reflected in ancient and modern literature. Salt-related matters are evoked in diaries and note-books, especially those drafted by foreigners travelling throughout the continents and sailing the seas. Salt extraction or harvesting deserves their attention as regards techniques, uses and folklore. A number of examples collected within the framework of a diachronic approach at the world scale (Europe, China, Japan, Americas, and Africa) pave the way to a global elucidation of the extent to which, from the earliest times, salt contributed in making it obvious that life depends on it.

Keywords
salt, salt works, solar salt, salt mining, rock salt, pan salt, brine, geology, landscape, climate, halotoponyms, technology, metrication, nutrition, salt intake, food preservation, animal feeding, population, social forces, civilization, taxation, gabelle, history, literature, caravans, diaries, mythology, rituals, salt cellars

'The second year after this, the Veientes having again broken the truce they had made with Marcius and demanding to get back the salt works which they had surrendered by treaty in the reign of Romulus, he fought a second battle with them, one more important than the first, near the salt works; and having easily won it, he continued from that time forth in undisputed possession of the salt works'. (Dionysius of Halicarnassus, *Roman Antiquities*, III, 41, 3).

During the history of mankind, alliances and civilities have depended on salt. Why not the reverse? Since prehistory, this mineral substance has participated in pastoral community wealth (animal feeding, cheese processing, hide salting), and it was at stake in a number of struggles. Ancient salt facilities similar to Ostia salt works changed hands several times. Both mined and obtained by solar and artificial evaporation in ponds and pans, salt became the cause for waging many a war. Tacit recorded a battle between two German tribes for salt control close to the Werra River, *flumen gignendo sale* (*Annals*, XIII, 57). In the Middle Ages and later, salt trade was a privileged source of revenue. In Northern Europe, salt fleets sailed the Ocean for 'Bay salt'. The Hanseatic merchants suffered from considerable competition from English and Dutch rivals. In the Mediterranean Sea, conflicting interests opposed Genoa and Venice, especially regarding salt. The fortune of Venice depended on it to a wide extent (Hocquet 1978-1979) and, in Central Europe, Florence, Montpellier and other places, merchant adventurers, dealt profitably with it. It is still a basic strategic commodity.

Salt has not only influenced politics, taxation, economy, and commerce but also nutrition, feeding, health, and religion throughout the ages. Specific human activities have been influenced by salt including mythology, religious beliefs and rituals (Latham 1982), travelling, art (Bisaccia 1997), literature (Duché-Gavet and Lapacherie 2005). Besides salt Covenant, this substance is recognized as a symbol for friendship, hospitality, qualities depending on eating at least one bushel together. In the Bible, it is mentioned that Lot's wife turned into a pillar of salt, a mere reference to Sodom rock salt. The origins of salt gathering and later, production, and the development of its uses are paving the way to original developments in human history. Salt production enables population growth, opens regions to discovery and settlement, and contributes to modern State emergency. No industrial countries are known to have developed without adequate access to salt resources which, by chance, are widespread.

To write a thirty pages comprehensive and well balanced history of salt is a forbidding task. Who would not agree that any approach of salt in history is multidimensional? Who would any longer get it in but a political (legislation, institutions, tax levies) description or a rather technical analysis (geology, production methods, tools)? Politics, for example, are inseparable from the physical and intellectual structures of societies.

They cannot be explained unless reference is made to geography or demography, allusion to the individual achievement of religion, and art. Salt history as such has not a significance which can be ascertained indeed. The interrelationship of socio-cultural elements suggests that the role played by salt in history is more pregnant than salt history which would be too ambitious and totally disappointing in its results.

As a matter of fact, many historians of ancient science and technology began this exercise by trying to persuade the readers that this or that theme is indispensable for a better understanding of the universal role played by this mineral substance in every day's life since the very beginning of human adventure. Regardless the option they had chosen, these repute experts aimed at a streamlined integration of social, cultural, physical environment to pare down the political developments, in order to covering many topics organised by date and by region and providing a kind of historical guide concerning '*what you should know about salt in the past*'. Bibliographical compilation is doomed to produce a patchwork, and within this contribution as well. The many isolated studies, which could be of assistance for further reading, aimed at local descriptions and findings regarding salt in the past centuries can neither be compared nor synthesised.

A difficult proceeding

Thus, conceptualising the history of salt is a difficult proceeding. The number of specific studies cannot be adequately synthesised in terms of space and time. On the one hand, a few recent attempts have deserved a dramatic amount of criticism because their compilers had tried to compare materials and data associated with anthropological structures and socio-economic patterns which were not compatible. Moreover these materials and data presentation often depended on personal options uneasily linkable to one single modulus to get them classified and interpreted. On the other hand, one consideration detracts from the value of ancient writings on salt and related customs and duties. The authors were prone to follow approved models. Taking them any longer into account would lead to transfer old clichés to new readers. Both ultra-specialisation and popularisation obfuscate the sight of structural coherence.

Anecdotists are popular as regards salt in the past but history suffers from their turn for trifles. An illustration is given by the popular story that the ground of Carthage was seeded with salt in 140 BC. However, this Roman feat is not evidenced in any ancient sources. It does emerge in none of the Greek and Latin writers. The fact was presumably invented in the nineteenth century. The sowing of the ruins with salt by the Romans was proposed as a symbol of total annihilation. It seems to be the result of a mere contamination by a biblical reminder. After the capture of Sichem, Abimelech pulled the walls down and sowed the site with salt. However this mention in *Judges,* 9: 45 is an *apax*. In spite of explaining the use of salt for such a purpose by alleged purification or elimination ritual, it seems more advisable to excise it from the pseudo-historical tradition (Ridley 1996).

Literature surveying pinpoints stereotypes of the same kind.

-People from around the Mediterranean basin were basically using solar salt and its collection was a common feature of the related civilisations. Sic.

-The high degree of salt taxation is a powerful factor explaining riots and revolutions: the French revolution would be the result of gabelle constraints and inequalities. One consideration detracts from the value of ancient writings on salt and related duties and taxes.

-Dry mining of salt is so painful and detrimental to human health that political prisoners working in salt mines were unable to survive. Why sanatoria have been established in former salt mines?

Lack of correspondence between poorly juxtaposed information based on faulty assumptions and hazardous expectations appears nevertheless to be the major tune of the failure of this approach applied to the history of salt. Aggregates of time spans and special issues involved in how human groups initiated and developed social life based on salt do not bring together the factors in relation to which salt can be regarded as a touchstone for evaluating evolution. It seems more appropriate to get it ensured from brute elements to economical consistency within a given milieu according to a reliable set of identifying dimensions. Merely describing available materials and data is presumed to provide a sufficient understanding of certain mechanisms supposed to rule human behaviour with regard to salt gathering and processing, indeed.

Such dimensions require a special attention for their flexibility to interact or intertwine not only for their several levels of operation but also for their several levels of meaning (from technology to mythology). The challenge is to concentrate to the above-mentioned conditions far from the traditional narrative. Among these identifying dimensions, it seems advisable to focus especially on landscape (climate and eustatic sea-level), population (numbering, growing rate), nutrition, health, technology (experiencing methods and tools), taxation, travels, and interferences between salt making and social structuring.

Dating and other caveats

These identifying dimensions have, as many others, a chronological aspect. Dating the past remains a matter of concern, even of contest when proneness to refinements in investigation techniques opposes to needless to aiming at greater precision dating. Statistics are the core of archaeological and historical studies, even if information is generally incomplete and unreliable. Nothing can substitute what is next to nothing as regards salt. As soon as it is traced and used, salt disappears. That is another argument for a chronological anchorage. Salt boiling and mining are evidenced to go back to Neolithic times but no relevant element of quantification until the Middle Ages… To get the better of archaeology, statistical trials should be in proportion to their worth.

Accurate dating does not matter in the history of salt making, indeed. It would be misleading to suggest a plain chronology by which salt search evolved from uncertain know-how to dispassionate classification into thousands of outlets. Such a concern has little relevance to the various situations that human quest for salt was facing too. The environment for efficient salt production remained dramatically uncertain over centuries and the driving force often unknown when one is tracing back a process to its origin. Archaeological dating (C14) is rather sketchy, and reliable mostly for the larger time scopes of the distant past.

In archaeological dating applied to salt sites and related techniques to get it, and more especially salt boiling, the main questions to be raised are as follows:
- How old is this ceramic artefact supposed to have been used for getting salt evaporated?
- How long the identified process has been going on?
- How fast it has changed (vacuum versus open pan, pump versus bucket)?

Besides that, for which time spans are data series available and comparable to giving the aforesaid dimensions their full significance as regards salt consumption?

A question is making more sense than 'where', 'when', and 'how', it is the question 'why'. "Where" and "when" depend upon history, and "how" on archaeology. And as the answer arises a long time after the question emerged, this exercise seems somewhat frustrating unless done with a pinch of salt. Who would be keen to get accurate and relevant information about salt should rather rely on geology than on history which is a post hoc evaluation of a remote lore. One fond of paradox would add that archaeology is more reliable than history. Both compete to get the proper answer tabled to the questions raised. The most complicated to be elucidated remains the question "why?

Terminology and metrology pave the way to extra difficulties for comparative history applied to salt. As regards salt in Antiquity, words such as '*halopegion*' and '*salina*' are not always correctly interpreted. This is contributing to ambiguity in salt works description by Strabo (*Geography*) or Pliny the Elder (*Natural History*). Moreover, certain experts have neglected to use geological references to get out of trouble. The salt mine of Utique (monograph on salt by Besnier who refers to Aristotle) is an example of such a lack of lateral checking. Ambiguity is still prevailing today in Romania with '*slatina*' a word which means mine and saline as well. Sea salt evaporation is operated in '*salinas*' which seems more suitable for its related engineering than 'salt marshes' or 'solar salt ponds' or 'salt gardens' (Korovessis 1999). Although the *Domesday Book* gives full details about the numbers, values and dues of the salt pans, there are few allusions to the actual process of salt making. Therefore the word '*salina*' may have referred to the very salt pans or the place which the brine was boiled or possibly it even represented shares of salt measured out to those who had the rights of salt making or supply. Similar trouble is experienced for 'patina'. Translated by 'pan' the word can be related to an operative process involving heat or the right to obtain a certain quantity of salt.

The sociological significance attached to salt extraction (mines, springs, lakes) is attested by the fact that there is hardly a place where salt making took place to having not expressed this relationship in its name (Schleiden 1875). Toponyms and hydronyms echoed the presence of the resource as halite or in solution. The influence of salt production is attested by the denomination of English towns with the suffix *–wich* (a place where salt is made), French towns with the prefix *sal-*, *saul- seil-*, German towns with the prefix *hal-* or *Salz-*, *Sulz-*, towns founded in the former Ottoman empire with *tuz*. No wonder that the map of Romania shows a number of toponyms based on salt terminology: *sare* (salt), *slat-*, *slan-*, *sal-* reflecting community contacts in relation to salt exploitation. The salt works of Thessalonike explain the origin of Solun that the Slavs gave to this city. *Sol* means salt. The most ancient salt place in Russia was operated at Usolye, close to the Danilovo Lake according to archaeological evidence. The name of many Russian cities echoes their salt past: Soligalich, Soligorski, Solikamsk... In Japan, *shioyama* is a mountain the trees of which were used as fuel for making salt in the 8th century. Djebel Mellah (Saharan Atlas) reflects the presence of diapiric salt deposits. *Mellah* (salt) is also incorporated in the name of various *sebkha* and lagoons in North Africa and elsewhere. By extension, the *mellah* was a place located to a corner of the city, totally walled with its own gates like medieval salt works adjacent to their city.

Philology should echo technology and the reverse too. Studies are currently carried out in several European countries having the same semantic heritage. Within recent colloquia presentations regarding salt terminology were devoted to Spain, based on lexical units like *salina*, *salar*, *salado* and its evolution between the 12th and 17th centuries (Garcia-Cervigon); halotoponyms and halohydronyms in medieval Romania (Curcă), impact of inter-ethnic contacts in various regions of the country (Poruciuc); and in Portugal where arose a specific terminology related to salt over centuries (Amorim). See *Las salinas y la sal de interior en la historia/Inland salt works and salt history*, 2007.

Produced by Bernardino Gomez Miedes, *Commentarii de sale*, 1572, were issued as the first encyclopaedia about salt in the modern period. Written in Latin, this popular treatise focussed upon the physical and chemical characteristics of this mineral substance within the framework of *argumentationes* based on the best sources from the Antiquity and the 16th century. Its author was also keen to describe the various types of salt and the sites he had paid a visit.

Serious studies about salt in the more or less recent past include detailed appendices regarding measure units which are extremely difficult to keep under reliable equivalents when converting them into SI units. The transposition from volume to weight makes often historians feeling uncomfortable with the results. A *modius* used in Rome for grain and the like represents 8.75 litres. How much in kg as regards salt? It depends on salt granularity and moisture which depend on production process... The same uncertainty is observed for the *muid* the mass of which differed to a wide extent for units deriving from volumetric measurement based on a container (Hocquet 1995). Salt in ancient China requires a number of equivalence tables in as much as values fluctuate. Shi weight moved from 66.4 litres under the Song to 94.9 litres under the Yuan. The question of weights and measures is pregnant in a number of studies. Reservations are made for the attention of the aspiring quantifier (Whatley 1987) besides explanatory notes and conversion tables.

Whereas metrication (SI) is now in force, one would feel somewhat confused with mass expression in hundredweight – *cwt* – which increases evaluation difficulties (Ashead 1992).Together with salt taxation, and deforestation for the only sake of the salt works, the incoherence of multiple measure units entered in France among the major determinants of the Revolution as echoed in the so-called *Cahiers de doléances*. Any conversion in SI units requires taking into account all what characterized it use somewhere and sometimes ago. '*Les hommes ont plus volontiers changé de mesure qu'ils n'ont changé les mesures*' (Hocquet 1995).

Nutrition

Salt has a long tradition as a highly-valued ingredient which makes food more palatable. Over centuries, salt has served many technological purposes beyond seasoning. One of salt's most recognized uses, dating back to early centuries, has been in preserving foods, including meat, fish, vegetables, and even fruit. Salting foods prevented spoiling by drawing water out of the food, depriving bacteria of the moisture needed to thrive. Without salt, the food supply would have been considerably less plentiful and less safe whereas the requirements kept pace with population growth.

A number of cuneiform tablets relate to fish and meat cured with salt. Three tablets propose various recipes which detail the ingredients and their operation including seasoning. They generally specify: 'Salt, on a rough estimate' or 'Salt, as much as needed'. Food preservation is also widely based on salt as attested by verbs like '*madâlu*' (to salt), '*šapâhu*' (to sprinkle with salt). A mixture called '*siqqu*' seems to be equivalent to *garum*. Fish and salt are significantly echoed in Man history. Although Greeks and Romans consumed fresh fish, they often processed fish meat into dried, smoked, and salted form. Fish innards went into producing fish sauce and *garum*. References to salt fish begin in the 5th century, a production with possible origin in the Black Sea from where the process developed westward: Iberia, Mauretania, Narbonensis, Sicily, etc. Large-scale salting facilities have left considerable remains which need cautious interpretation, especially as regards evaluating salt quantities (Etienne and Mayet 2002). The amount of salt requested for food processing is dictated by both technology (wet and dry salting) and taste (stronger flavour for lesser quality). Studies of salt fish and garum amphorae found in shipwrecks or inland sites are also useful to get a clear-cut picture of the importance of the related trade.

Columella described how to salt pork, by laying down alternating layers of meat and salt, the last salt layer being covered by heavy stones to allow salt to permeate meat adequately. He states that the process is the same for salt fish (*Res Rusticae*, XII, 55, 4). In the 2nd century, Athenaeus (*Deipnosophistae*, III 116A-127D) recorded the name of varieties of salted fish products and specialties in relation to the processing process: '*since we are on the subject of salt fish, I shall proceed to tell what I know about it, with full details of the trade*'. Comments on fish cooking are available from Achestratus, reminded by the former. This Sicilian cook recommended that fish be roasted on a grill, gently sprinkled with salt and brushed with oil. Mention is made by Polybius of the export of salt fish from the Euxin.

Cato the Censor complained that Roman snobs would purchase a jar of fish sauce for the equivalent of three hundred drachmas. This fish sauce is presumably garum the price of which reached a thousand sesterces for two conges as observed by Pliny the Elder. '*The Bithynians make garum in the following manner. They use sprats, large or small, which are the best to use if available. If sprats are not available, they use anchovies... They add two Italian sextarii of salt to each modius of fish and stir well so that the fish and salt are thoroughly mixed.*' (*Geoponica,* 20.46.1-5). The taste for seasoning stimulated the development of various recipes such as those collected by Apicius (*De re coquinaria*) where garum competes with salt.

Another important momentum for European salt producers took place in the Middle Ages with the bullish development of herring and cod trade. The Church recommendation to abstaining from meat and the number of fast-days favoured fish and its preservation by salting. When fish salting on board of the herring fleet neared its peak, it amounted to 3 billion units requiring 123 000 tons of salt (Schleiden 1875). And even after the Reformation neither the Dutch nor the Scandinavian people lost their appetite for herrings. The consumer depended upon different fisheries: in Baltic Sea since the early 11th century, in the North Sea from the mid-13th century, and in the Dogger Bank later. Salt fish business was very profitable for the Hansa and Dutch ship-owners and merchants. The Dutch technological superiority, depending on the sea-based method of pickling and curing fish at sea initiated by Wilhelm Benkelszoon (c.1350) helped them to meet an ever increasing demand. Perfected in the sixteenth century, it was modified slightly by the shift shore-based industry. It became popular as the 'Scotch cure' in the eighteenth century, and Adam Smith argued that the duties on salts used for curing herrings for the domestic market had a detrimental impact (*An inquiry into the nature and causes of the wealth of nations*, 1776). Cod fishing around Newfoundland offered new opportunities for Atlantic salt. Before leaving home port, bankers loaded huge quantities of salt for stock-fish preservation. In France, a Royal decree, 1740, enacted that the equivalent of salt 120 kg was the right amount for a thousand cods.

Offering salt and bread is a traditional gesture of hospitality. 'First of all, without salt, there would be nothing eatable which mixed with flour seasons bread also. Neptune and Ceres had both the same temple' (Plutarch's *Symposiacs*). In Roman Egypt, Celsas aks Cronides for salt which is necessary for bread baking in relation to legion subsistence (*Ostraca*, 21). Its poor quality questions its conservation. The introduction of salted bread in the kingdom of Naples was vindicated by Vicinanza *(De salis natura ac sale cum panibus commiscendo commentaries,* 1585). Salt is fit for food preservation. Poor quality bread is not palatable. It will not keep. Besides, salt is a very effective ingredient. For Plutarch, there is sufficient evidence that salt is the most desirable of all relishes. For as colours need light, so tastes require salt. Universal appetite for salt is obvious because 'it makes food, which is necessary for life, to be relishing and pleasant'. Palissy considered that 'salt in the diet revives humans' (*Discours admirables,* 1580). Salt restriction makes food unpalatable and might be detrimental to health in the population at large.

The amount of salt effectively ingested in Europe at the end of the Middle Ages varied slightly from region to region. It did not presumably exceed 15 g/d although it could have been casually higher according to certain sources. Braudel evokes 20 g/d. Exaggerated figures reflect a long prevailing confusion between salt sales (released in account books) and effective absorption which could not be calculated at that time (sodium urinary excretion / 24 hour).

Population and taxation are basically intertwined. Whereas the French population living within the limits of the Grand Party alias *pays de grandes gabelles* were doomed to buy 1 *minot* a year every 14 individuals for cooking, it is easy to calculate which amount of salt had been considered to be dietary salt intake based on discretionary salt at the end of the 17th century. This ration per capita is equivalent to 2.9 kg a year or 8 g/d. Elsewhere the ration averaged 10-12 g/d. It seems it makes sense to focus upon this quantity for the same amount is observed in France and other developed countries. The only major difference depends upon the sources. Nowadays more than 70% relates to sodium content of the food products supplied by the industry. The same situation prevails in almost all of the developed countries.

Taking dietary salt intake into consideration allows making the balance sheet of the actual sodium intake, coming from different sources (Na 17 mmol = NaCl 1 g). By convenience, it is expressed in salt equivalents (g/d). Presently, it would be useful to make the distinction among the following:

The naturally present sodium in certain raw aliments expressed in salt 1-2 g/d

The salt which remains from cooking or use at table 2-3 g/d

The salt which remains from the industrial or artisanal preparations 3-4 g/d

The sodium present in the water and sodium salts besides the chloride c. 1 g/d

The phrase "salt which remains" means not taking into account a certain amount of salt discarded before the ingestion. This could represent currently more than 50%. This explains the difference between food use and dietary intake (salt actually ingested).

Health

Salt—sodium chloride (NaCl)—is essential to the health of every living creature. Man's innate appetite for salt may be related to his evolution from predominantly vegetarian anthropoids, and it is noteworthy that those people who live mainly on protein and milk or who drink salty water do not generally salt their food, whereas those who live mainly on vegetables, rice and cereals use much more salt. Salt was in systemic use long before history. The necessity of getting this substance relates to both innate and acquired appetite for it.

A Chinese treatise on pharmacology (*Pen ts'ao Kang-mu*, 1596) mentions various kinds of salt, including the description of two methods for extracting salt. Gomez Miedes evoked the salt and health issue (*Commentarii de Sale*, 1572) to remind the reader of its bodily functions including generation. It favours fertility according to more ancient authors. In the evolution, the emergence of mechanisms (renin-angiotensin system, aldosterone secretion) which specifically maintained a normal body content of sodium, and thus a normal circulation of blood and tissue fluids, was of cardinal importance (Denton 1984).

Evolution has been shaped by the need for salt. Certain medical experts deny the validity of observations in line with this statement. According to anti-salt zealots: '*The Yanomamo Indians still lead a life very similar to the last million or so years of our evolution, and like primitive man eat a diet that is very low in salt and saturated fat and high in fruit, vegetables and roots. The Yanomamo Indians are not overweight, do not smoke and are very fit. Their blood pressure does not rise with age although they spend much of their time fighting and are under great stress. This tribe does not develop vascular disease, although many die of infection. However, when they migrate to a Venezuelan or Brazilian town and adopt a western lifestyle, they, like native Americans, become overweight and develop diabetes and premature vascular disease. They appear therefore, to be a group which, though predisposed to vascular disease, is protected by the way they live.*'

It has been estimated that the hunter-gatherer diet contained only 16.5 mg/100 g in vegetable food, and 59 mg/100 g in animal food. The alleged sodium 600 to 770 mg estimation by Lev-Ran raises two questions:
-Was it sufficient at that time to meet sodium requirements?
-To what extent had sodium deficiency adverse effects on both reproductive functions and food preservation on which human survival depends?

The breakdown between meat and plant food is totally uncertain for humans living 10 thousand years ago, and varies widely according to models: 35% versus 65%, or 55% versus 45% respectively. Therefore, there is no consensus among paleoanthropologists as to the animal to plant subsistence ratios for pre-agricultural societies. Although sodium need is fundamental for their survival, we suppose that these societies differed in the access they have had to salt, and in their attitude toward this substance (medium, foodstuff).

Anyway, we cannot assume that the human body is biologically fixed for ever. The most that we can do is to collect and interpret archaeological artefacts and historical documents. There is no evidence that dietary salt intake was as low as 1-2 g/d ten thousand years ago. Moreover the allegation that unacculturated populations showing such a low level nowadays provides a point from which we can engage with the salt and health issue in the past is still under question. The salt intakes are between 7 and 10 g/d for most of the population. The big consumers are defined by salt intakes > 12 g/d. We should mention that the kidneys and the heart regulate them somewhere between 4 and 18 g/d, while the minimum requested by the organism is 3-4 g/d. The sodium proportions excreted in the urines per 24 h means an average of 8 g/d. From these statistical elements we should retain the difference which exists between the dietary use of salt and its intake.

As regards the low sodium/high potassium regime prevailing in Prehistory, chemical analysis of the salts from the ashes of the vegetal salt obtained in Colombian Amazon or New Guinea Papua revealed that the bulk of the inorganic component was potassium, calcium and chloride. A high potassium/sodium ratio would have made such salt dangerous if consumed in large quantities. Sometimes, the salt springs are also sulphurous; their contamination would have made the extracted salt toxic. It is difficult to assess the health hazard faced in the past by unacculturated tribes. It was presumed that salt extraction would be actually associated with sexuality in indigenous thought, and confirmed in Uitoto Indians (Echeverri).

In Plutarch' *Symposiacs*, attention is drawn on the fact that 'Salt encourages remarkably generation'. Is not this substance at the origin of Aphrodite birth? From the crystallisation of Uranos sperm would have ensued salt, present in the sea from where she emerged. Close to the myth survived rites the meaning of which was less and less accessible as evidenced in late writings. Nevertheless their reminiscence was echoed across societies and centuries in relation to the doublet 'fecundity-fecundation', and salt remained the favourite medium. Salt was also present in remedies to impotence. The *Shamita* of Sushruta recommended in 8th century that '*Powders of sesame, Masha pulse, and S'ali rice should be mixed with Saindhava salt*'.

Engravers from the Renaissance deplored all forms of idleness, and vindicated salt for restoring hope in women. In his *Traité des superstitions*, 1679, in which Thiers evaluated prescriptions and interdictions valid for all circumstances of human activities, a number of practices relate to salt and fecundity. Nowadays, it is acknowledged that salt restriction might weaken sexual desire. Reduced dietary salt intake would debilitate reproduction capacity. In human subjects, the 'chronic fatigue syndrome' is associated with sodium deprivation. Increased dietary sodium prevents libido failure further to drastic restriction.

Sweating can generate a loss of more than 1 litre/hour under extreme conditions. Prolonged sweating without adequate restoration of saline capital leads to exhaustion characterised by muscle cramps, nausea, and fatigue. Heat stroke can be avoided by increased salt intake. This is not new. In the middle of the 15th century, Alvise Ca da Mosto described how the salt caravans went via Timbuktu to Mali where it was promptly sold. He enquired as to what the Negroes did with salt. He was answered that salt consumption was an absolute necessity for not getting ill in response to hot temperatures (*Travels,* 1455-56). Low salt would affect negatively hydro-mineral balance. Sodium appetite and thirst are recognized complementary motivations that drive to compensate the depletion in both salt and water.

The development of cities about 3000 years ago changed life conditions, more especially as regard the diet within which salt played a major role as a preservative. Salting improved the security of food, and its preservation. With this respect salt also contributed to the development of explorative travels and the implementation of trade routes. Huge rainfalls and floods were observed to compromise both water supply of the cities and coastal drainage systems including that of salt marshes. The contamination of sea salt facilities by urban drainage disturbance has been evoked in relation to grey salt.

Taxation

Salt is a universal necessity, and in the days before refrigeration it was even more necessary, to preserve perishable food, than it is today. In ancient times this precious commodity was actually known as "white gold". It was so important a commodity that it was used as currency, in compensatory payments; in Africa Ibn Battuta observed: '*The Blacks exchange the salt as money as one would exchange gold and silver. They cut it up and trade with it in pieces*. Due to its widespread consumption, salt was taxed by governments in ancient China, as witnessed by Marco Polo. Once introduced in Burgundy, the *gabelle* averaged 25% in 1370 (Dubois 1976). In the United States, the State of New York financed the Erie Canal with its salt tax.

Domesday Book, 1086, describes English society under Norman management, in minute statistical detail. Foreign lords had taken over, but little else had yet changed. The chief landholders and those who held from them are named, and the rest of the population was counted. The King wanted to know what he had, and who held it. The Commissioners therefore listed lands in dispute, for *Domesday Book* was not only a tax-assessment. Thanks to this census, a plenty of information is available as regards salt places in England at the end of the 11th century. Activities based on salt exploitation were presumably well established in a number of sites before the conquest. Thus we can better understand the organization of the salt places, the ownership of the pans, and the modalities of taxation. Most of the Domesday *salinae* (salterns) were located on the coast where salt was obtained by the evaporation of sea water. There were however two inland groups of salt pans in Cheshire and in Worcestershire. A total of 305 were recorded for the latter where no less than 160 were entered for Droitwich, 'the chief salt town of England' (Whitelock). The mention of the Cheshire salt works was centered at the '*wiches*' fo Middlewich, Nantwich and Northwich. As they had suffered from recent disturbances (1070), they were waste when the Earl Hugh of Chester received them.

Extended to many parts of France (with a few exceptions) the salt tax – *gabelle* — became necessary to the government. Richelieu said that it was as vital to France as American silver was to Spain. The repeal of the salt tax was a major claim in 1789. Suppressed it was soon restored by Napoleon, and it continued until 1945. During the French ancient regime salt was particularly indispensable as a source of revenue to the crown, but there was no monopoly. The gabelle was a highly unpopular tax, and salt became a hateful symbol of royal absolutism. In his *Rules for the reduction of a great empire*, 1773, Benjamin Franklin condemned once for all indirect taxes levied in France, with a teasing stance for the gabelle. '*In laying these taxes, never regard the heavy burdens those remote people already undergo...Forget the restraints you lay on their trade for your own benefit and the advantage a monopoly of this trade gives your exacting merchants... To make your taxes more odious, and more likely to procure resistance, send from the capital a board of officers to superintend the collection, composed of the most indiscreet, ill-bred and insolent you can find*'.

This system is characterized by a variety of taxation levels and category exemptions, and by a multiplicity of intermediates. The so-called *greniers* were not only warehouses but also jurisdictions within which certain matters such as smuggling were not dealt with. These cases belonged to the *Cour des Aides*. Although salt supply was depending on the *greniers*, the collection of the salt tax was farmed out to financiers or *traitants* who leased the right to levy the gabelle and other indirect taxes against a lump sum, payable in advance and usually held for a nine-year term before renegotiation. The company of General Farmers was the largest employer in the kingdom of France due to the number of guards, collectors, book-keepers, inspectors and other controllers with a hierarchy of their own. The *Conseil des Finances* was qualified to adjudge disputed claims. Under Louis XVI, the whole company amounted to 28 500 civil servants (Azimi 1987).

The Farm privileges included the right of search over without court warrant. This resulted in the exacting management of a system of unfathomable complexity. It naturally gave birth to various legends such as the shameful number of men put in the galleys and deceitful restrictions regarding animal feeding. The system had two major advantages to the monarchy. First, it provided large funds in advance without having the unwieldy job of collection. Second, it shifted the unpopular consequences of this rabid taxation from the government to private individuals (Hufton 1980). Nowadays, it is admitted that the Farm paved the way to modern administration. Moreover, listings and reports concerning salt tax are very useful for population studies as exemplified by one of the best completed (Dupâquier 1979).

A constant need to improve the efficiency of gabelle collection and other consumption taxes led the authorities and individuals to stick to demographical status and to take a census of the population under different pretexts. Thus, Lazare Ducrot was in a position to specify the number of warehouses (*greniers*) where salt was stockpiled and its quantities delivered per annum in his *Traité des Aydes, Tailles et Gabelles*, 1625. An *Atlas des Gabelles* was mapped by Sanson Jr. in 1685; figures concerning salt deliveries were included. An inquiry related to salt tax collection in 1725 and 1726 made available other statistics about salt sales. It was then feasible to corroborate information valid for the *elections* and the *greniers*, both considered as administrative circumscriptions with their own competences.

This side contribution to the proper estimate of French population deserves some consideration. Salt tax listings or salt toll registers are useful to evaluate ancient population. Demographical reconstructions related to certain regions or certain periods could participate in a rough evaluation of salt requirements. It would then be possible to check to which extent the identified sites have met or not these requirements. Attempts are underway as regards the Greek cities and the Roman Empire. Documents established for the Customs offer other calculation opportunities in relation to salt trade. Although partly defective to a statistical point of view, Sound registers offer a chance to analyse in depth the salt market (Hocquet 1985).

In the sixteenth century, the Hansa east-west trade gained importance. While there were fewer than

1000 vessels passing through the Sound each year between 1500 and 1540, in the 1560s some 3280 vessels, and in the years around 1590 some 5038 vessels passed through it annually. It is estimated that about 80 per cent of this flow was headed towards the Hansa towns situated on the northern German and the eastern Baltic coast. This was reflected in the size of the Hansa fleet. At the end of the sixteenth century, it had about 1000 ships with a total carrying capacity of 45,000 lasts (90,000 tons). Compared to the situation 100 years earlier, this was an increase of 50 per cent.

The King of Denmark established a toll which was to be levied on merchant ships passing the Sound straits. This toll remained in force from 1426 to 1857. From 1497 onwards, the ships were registered. The Sound has been one of the busiest waterways in the world. Ships from many nations and trading cities made more than a million and a half passages through the Sound from the late fifteenth to the nineteenth century, leaving information on their cargoes in the Toll records. Most of these registers are a valuable source of information regarding various commodities like wine and salt loaded on the Atlantic coast. This collection is now available for research and is of incomparable importance for the history of trade and shipping and more generally for the economic history of western, northern and Eastern Europe.

Thanks to the intervention of Charles V, Holland was exempted from the Sound tolls in 1544. This constituted a major step towards Holland's domination of the Baltic trading system for the next two centuries. Free access to the Baltic markets, combined with low transaction costs and a still increasing share of the combined salt and grain trade, paved the way for the establishment of Amsterdam as a major partner. Dutch factors settled in many Baltic ports to organize their business. In 1631, Christian VI suppressed the duties on salt permitting both Danes and Norwegians to buy it under more attractive conditions. The treaties signed by English merchants were founded on those the Dutch had obtained. The treaty negotiated by Charles V for the Dutch provinces remained in force till they became independent from the Habsburg crown. The Dutch Republic was cut off from Setubal salt. In 1625, the entire herring fleet was caught by Spanish men of war off the Hebrides. The list of retaliatory measures taken by the various interested parties would be too long.

Despite the impressive number of ships passing through the Sound in the sixteenth century, both the volume and the value of the Dutch Baltic trade remained modest. The range of products transported to the east consisted mainly of salt, herring, cloth and wines. The majority of the ships had a capacity of less than 50 lasts, and unless laden with salt, vessels often sailed eastward on ballast. The Dutch traders were active in the shipping of both colonial wares and wine and salt from the Iberian Peninsula and France.

Taxation of salt has occurred in India since the earliest times. However, this tax was substantially increased when the British East India Company began to establish its rule over provinces in India. In 1835, special taxes were levied on Indian salt to facilitate import generating an attractive price differential. When the Crown took over the administration of India from the Company in 1858, the taxes were not repealed. The stringent salt taxes imposed by the British were vehemently condemned by the Indian public. In 1885, at the first session of the Indian National Congress in Bombay, the salt tax issue was raised. There were further protests throughout the late 19th and early 20th centuries culminating in Mahatma Gandhi's Salt Satyagraha in 1930, an act of non-violent challenge.

Colonial India was forced to buy salt even from England. Although its proportion decreased, in the early 20th century, total imports amounted to 511 000 tons, 33% being imports from UK (Calvert 1915). At that time the British Commonwealth controlled the trade of salt among most of the world. Therefore the production or the distribution of salt by anyone but the British authorities was not permitted. Natives were forced to buy their salt under British control even though salt was readily available to coastal area dwellers. As the tax had an impact on the entire country of India, Ghandi knew that his decision would gain general sympathetic support. At the end of his 23 day walk to the sea, on April 5, 1930, Ghandi picked up a handful of salty mud and proclaimed: 'With this, I am shaking the foundations of the British Empire.' He proceeded to boil some seawater, illegally producing the controversial substance. He then implored his fellows to start making salt all along the coast.

This contest was followed by other *sathyagrahas* in other parts of the country. A number were arrested, courted and jailed. The authorities invited Mahatma Gandhi to attend the Second Round Table Conference in England. The salt tax, however, continued to remain in force and was repealed only when Jawaharlal Nehru became the President of the Interim Government in 1946. The same year, the salt tax was suppressed in France. It remained in force in Germany until 1993.

Population

Population growth is conditioned by food preservation and climate evolution among other things. The Holocene began about 11 500 years ago. During this period, temperature variations can be considered of relatively modest amplitude as a whole. However, there have been significant changes over the past 9 000 years. Generally speaking, the wettest and the driest episodes occurred slightly before the Holocene Climate Optimum. In both hemispheres, catastrophic floods happened that were then four to ten times greater than those experienced nowadays. Between 7000 and 2000 years ago, the temperate latitude became wetter while the tropics dried out. Lake sediments record the balance between sun and wind driven evaporation and precipitations. On the end of this period to which correspond milder temperatures in the Mediterranean basin, the so-called agricultural revolution took place in the Middle East. Since people were in a good position to grow more crops than they needed, and to husband larger flocks, thanks to food and hide salting, food preservation and product diversification contributed to a significant population growth. They also paved the way to barter trade between communities. Their social structuring is witnessed by the establishment of towns. Elsewhere other Neolithic complexes developed. According to Biraben estimates, the world population would have increased from 7 million inhabitants to more than 30 million between 6000 and 4000 BC. Neolithic people improved relatively their life conditions with handmade pottery. Thus, vessels for making cheese or salt (including food preservation) have been found in various sites. Later mining and metallurgy impacted trade and war not always separable from the fortunes of trade.

On the past 2000 years, an evaluation of hydrological regimes is feasible. For regions ruled by Chinese and European dynasties, information about precipitation and temperature is available in written records. Thus, in feudal Europe, meteorological events were reported with their dramatic consequences on agriculture and saliculture. Pest-related mortality has been observed in Chioggia at various times along the 15th century. About two thirds of the population was eliminated in 1477-78, in 1485, and in 1491. The maintenance of salt marshes suffered from manpower shortage. This state of things increased the difficulties faced by the salt men i.e. milieu alterations due to climatic change, and market restrictions imposed by Venice (Hocquet 1991).

Temperatures collapsed, first after 1250 and again after 1600. Such adverse climatic conditions led to crop failure, food shortage, and starvation. The adverse weather conditions of 1315-17. Torrential summer rains ruined harvest and the resulting famine killed 10% of the European population. Exceptional rains and floods, associated with early cooling seem to have halved the Chinese population with comprehensive repercussions on salt production and consumption. The Black Plague and the Thirty Years War were very detrimental to population standards. Demographic crisis with cyclical rise and fall of urban and rural communities are negatively impacting the evolution of the population, and symmetrically salt production and salt consumption. The general trend concerning the world population estimated by Biraben and the related salt consumption based on 7 kg a year per capita deserve a number of reservations. The same exercise should be done by regions focusing on year spans characterized by meteorological, health and political events. An extensive number of ups and downs would be perceptible. According to statistics, world salt consumption amounted to 13500 thousand tons in 1900 showing that the yearly average exceeds 8 kg from now. The devil holds the figures for spilled salt.

Controversies over ancient population numbering should not prevent experts from considering available estimates for the Mediterranean basin in Antiquity, irrespective of perennial debates about the size of metropolis like Alexandria and Rome or the number of residents in classical Athens. Generally speaking, such debates would be superseded by more important questions about the gross population of core regions concerning the ancient world or the new one (Schleiden 2006). The fact that we want to have an order of magnitude concerning the corresponding quantities of salt is not a mere gadget. For example, at the end of Augustus' reign (who died in 14 AD), Italy would have represented a market of about 50 000 tonnes and it is very probable that the supplying could not have been ensured only by the salt marshes of Ostia... We would also examine to what extent, and with which reservations, it would be possible to make up for the lack of statistics to orient more positively some of the commentaries inspiriting the study of salt in Antiquity.

Different forms of evidence are available for the demographic study of various ancient populations. As regards Rome for which the debate has centred upon the census figures, archaeological evidence and others have shown the weakness of any argument in favour of the low count. Regardless its level, a revisited estimation should pave the way to further calculation of salt consumption and its evolution. Corresponding quantities would have to be connected with production capacity and changes concerning a number of sites and facilities associated with its outlets, exceeding by far culinary needs and fish salting. Although one considers that, thanks to Strabo and Pliny the Elder, we are reasonably well informed about how salt was produced in the Roman empire (Adshead 1992), we do not know how much it was produced and where. Which percentage of the hypothesised production level can be attributed to the sites mentioned by ancient sources is another challenge. Around one hundred sites are not reflecting the precise number of solar salt works, rock salt mines and brine springs. Several thousand places generally with low capacities would have been necessary to meet the requirements of the Roman world.

Further to an estimation of the capacities susceptible of producing salt in Italy during the third century BC, we obtain a theoretical capacity of 20,700 tonnes which corresponds, taking into consideration the variations, to a range of 16–24,000 tonnes. Another one shows that the needs of the Italian confederation give us 18,700 tonnes upon the basis of 1 *modius* × 2.7 million inhabitants. At the beginning of the first century, the population of the Empire counted at least 55 million inhabitants. Thus, the gross salt needs would be somewhere between 397,600 and 401,800 tonnes, calculated on the basis of 7 kg per person and per year. After adjusting it by taking into consideration the number of women and children (6 and 5 kg *per capita*), we have between 340,400 and 351,800 tonnes. This gives us an average of 373,000 tonnes.

Experts feeling concerned with the lack of statistics for salt requirements in the past prefer to adjust their calculation to 7.5 kg a year as far as consumption is concerned. Estimates made to compile the statistical appendix to Neptune's Gift are based on an assumed 'household consumption of 7.5 kg (Multhauf 1978).

Both Schleiden (1875) and Verhille (Mollat 1968) retain the same basis. What deserves attention is a rational attempt to include other uses than food which are barely taken into account. Even in the Prehistory and in the Antiquity, animal feeding, hide treatment were playing a significant role besides food uses (seasoning, fish salting, meat curing). It seems appropriate to segregate, where calculation is feasible, the quantity which participates in dietary salt intake which is not subjected to wide variations over several centuries. Kaufman (1960) sets the figure at 3.8 kg a year. A sensitive point is the universal and constant increase in salt consumption which is not only a matter of population but also a matter of innovation: modern salt uses are innumerable. Cooking salt is but a small percentage of total requirements. This objection is valid above all for the 20th century. A plenty of statistics is at hands for this time span. There is no longer need for estimates.

Even if none can be certain about population number, calculation methods are of interest for acceptable estimates are irreplaceable to evaluate economic performance in the past with regard to salt, and to facilitate a comparison to conditions prevailing in more recent periods. Experts calculated population estimates for the Southern Maya lowlands in order to determine how much salt was needed by the Classic period population living in this area and in some other places. The size and density of ancient Maya population is documented from the archaeological settlement remains of masonry and earthen residential platforms and from the superstructures of varying size and elaboration in plaza groups or in isolation (McKillop 2004). Moreover, allegations about resource deficiency were withdrawn with recent evidence that needed goods such as salt were available locally. The discovery and excavation of salt works at Punta Ycacos Lagoon, Belize, achieved in the broader context of Late Classic Maya civilization shows that the Maya produced salt by boiling brine in pots over fires at specialized workshops. The existence of the coastal salt works illustrates that important production efforts is providing new clues to the Maya sea trade (McKillop 2002).

Salt was long supposed to be scarce in the tropical rainforests of Belize and Guatemala, where the Classic Maya civilization thrived between AD 300 and 900. The prevailing interpretation has been that salt was imported from the north coast of the Yucatan. However, the underwater discovery and excavation of salt works in Punta Ycacos Lagoon demonstrate that the Maya produced salt by boiling brine in pots over fires at specialized workshops on the Belizean coast. These salt works are clear evidence that craft specialization took place in a nondomestic setting and that production occurred away from the economic and political power of the urban Maya rulers, thus providing new clues to the Maya economy and sea trade. Recent research presents a strong case that salt-making in the Port Honduras zone of the Belize coast involved specialized production for trade into the heavily populated interior of the southern Maya lowlands during its Late Classic (AD 600-900) apogee. McKillop identified four specialized salt workshops on the basis of high concentrations of the kinds of traditional salt-making implements - large unclipped jars, vessel supports, charcoal lenses.

Social structuring

How we go about understanding and explaining social forces and the evolution of societies? Is the social theory fit to guarantee a better interpretation of what happened in the past? As regards salt, which data are available and valid to establish a system and link salt to civilisation? The mere observation of animals licking rock salt outcrop is presumably not an element strong enough to determine a group to immediately associate around systematic identification, excavation and extraction of this mineral substance. There is a dramatic absence of reliable notions of the way people behaved socially in the past in relation to the presence of a resource, and the role played by salt in the means to provide structuring is not crystal-clear. In unacculturated societies, the salt places can display characteristics of organised settlement structure. Long-term salt making implies well defendable sites in relation to wealth cumulative barter. Is salt a major determinant? Salt never goes alone in structuring groups, i.e. their capacity to get organised close to a resource. With reservation regarding the physical link: Odysseus spreads salt in ploughed sand whereas he walks along the beach. Feigning to do that is a stratagem to divert. He is reluctant to let people know how and how much he can get a plenty of salt... A similar behaviour has been observed in Dani, a Papuan tribe, as regards vegetal salt making (Weller 1996).

We would recommend relying upon 'society' rather than upon 'civilisation'. Civilisation implies a certain sense of values which integrates later sociological change and experience in other cultures. It can be reasonably assumed that prehistoric societies had their own trajectories rather than being submitted to waves of influence from 'higher' cultures (Johnson 2004). Whereas salt was used not only for its intrinsic value in relation to human diet, animal feeding, leather treatment and other outlets but also as means of exchange and a medium, a few historians considered with favour that the concept of 'salt civilisation' was fairly acceptable (Adshead 1992).

This interpretation is echoed in literature where the ambivalence 'mineral/symbol' deserves writer's attention and is developed without epistemological restriction. '*Le sel, un mot obsédant et fondamental, formé de trois lettres comme le blé, le mil, le riz, le vin, le thé, nourritures de base chargées de symboles et définissant autant de civilisations différentes. Mais s'il existe une civilisation du blé, du mil ou du riz, peut-on imaginer une civilisation du sel* ' (Tournier 1980). Salt presents a duality of structure: incorruptibility, solubility which paves sometimes the way to romantic explanation of certain social systems. This is not incompatible to elucidate some of the trends of ancient and present spirituality. Salt is not the whole of the things. Spirituality includes many variables. An exploration of changes in mentality would be more suitable to ensure the historicity of cultural meanings.

Coming back to salt and civilization, the question of the cultural status of this mineral substance – due to its capacity to carry meaning, even to mediate – was answered by ancient Greeks who would have been reluctant to imagine that somebody could ignore salt use: *'When you get home you will take your revenge on these suitors; and after you have killed them by force or fraud in your own house, you must take a well-made oar and carry it on and on, till you come to a country where the people have never heard of the sea and do not even mix salt with their food, nor do they know anything about ships, and oars that are as the wings of a ship* (Homer, *Odyssey*).

This remark has been taken for granted within the framework of the Mediterranean basin. It can be adjusted extremely well as regards other geographical areas (Le Roux and Ivanoff 1993). One of the most striking features concerning salt is its universal character combined with its ahistorical essence, indeed. Therefore, we are doomed to apply too much consistence to the various uses of salt throughout history and not enough significance to myths, beliefs and rites within which its meaning is involved. Aristotle recommends we should eat a measure of salt together in order to fasten friendship and Cicero too.

Understanding the internal dynamics of salt socialization and the triggers for cultural change with this respect requires a global approach. The correlation between the material and the social parameters is far from being easy to determine. In the case of salt, is the material construction a reflection of the social organization? Once again, interaction seems the most acceptable solution. Salt mines and surface salt facilities including salt ponds are confined and isolated for technical and security reasons. They are provided with structures and regulations of their own. As regards dry mining, underground extraction requires specific measures and activity control which prevail now. Over centuries, subsidence and deforestation were two major negative factors impacting social life.

Although little evidence remains, it is admitted that salt springs at Northwich were used in pre-Roman times. At Upwich (Droitwich) a possible Iron Age brine boiling site was excavated and probable remains of salt pits were found nearby. The importance of Droitwich as a salt producer in Roman times is indicated by its Roman name of Salinae, a name also applied to Middlewich and used for Saltzburg in Austria. Droitwich has been recognized as the *Salinae* noted in Claudius Ptolemy's *Geography* compiled in ca. 140-150 AD. The Roman remains showed a well-organized salt industry with a wood-lined well and a probable brine lifting structure. It contributed to social structuring as shown by the excavations at Hengistbury which reveal a commercial network with continental ramifications (Cunliffe).

The evolution of continental salt works is especially lightening the debate where and whenever the sites have been operated. Salt was obtained by boiling brine from salt springs, and many towns and cities in Europe located near salted sources. A number of towns developed with an industrial enclave dedicated to salt production. Such settlements developed around either the castle or salt works. Huge firewood stockpiles close to open pans where salt is evaporated and timber structures constituting the salt works required a sustainable stone separation from the town which could be devastated in case of accidental fire. Moreover, the wall surrounding the whole installation prevents robbery and smuggling.

The aforesaid interaction has been studied in relation to the alterations of the milieu. The long term evolution of the Adriatic coastline modified drastically the conditions for solar salt gathering in the Italian peninsula. Solar salt production was directly affected by eustatic moves and erosion, and indirectly via the worsening of coastal settlements (malaria) and the conversion from agricultural to pastoral activities in the hinterland (Gravina 2005). Likewise, when considering the sea level eustatic moves, McKillop enters the environmental-versus-cultural debate over the Classic Maya collapse by evaluating the factors that led to the abandonment of the Punta Ycacos salt works at the end of the Classic Period, synonymous with the abandonment of inland Maya cities.

An issue to be further investigated relates to causes and consequences of the abandon of salt works in connection with the withdrawal of a human settlement. How does it happen? On the one hand, one will evoke the intentions of the members belonging to the community susceptible of variation for political or psychological reasons. On the other hand, the question will be whether the resource, i.e. salt, is less accessible for geological or technical reasons. Due to the perfect conservation in rock salt a great number of organic materials have been found, among mostly wooden artefacts. Chronology building started at the Dachstein plateau, in the vicinity of Hallstatt, where tree-trunks were discovered in the Schwarzer See, and after recovery a spruce-larch chronology was compiled that dates back to 1475 BC.

The Hallstatt era is divided into four phases by modern day archaeologists who expect to explain the emergence social organization and its later dismantling. During the Hallstatt C fortified settlements on hilltops north of the Alps seem to be established with greater frequency. Consequently, many burial mounds mark the graves of the rising noble classes. Increased trade volumes seem to have contributed to their fortune including salt. In Hallstatt D, the richest graves are more concentrated in the west than previously. Why not to be closer to other trade routes connected with the Mediterranean basin? In the fifth century BC began the rapid extinction of the rich chiefdoms of the Hallstatt D. Hill forts all over Central Europe were dismantled, and rich burials ceased. The discovery of various salt production centres in Bulgaria and Romania opens new possibilities for research on role of salt in the structuring of the earliest European societies.

A Bulgarian team made extremely interesting discoveries at an archaeological site located near Provadia, called Solnitsata, whose shape and dimensions were until quite recently not very well defined. The archaeological excavations, which have been published (Nikolov 2009) have confirmed the hypothesis and produced evidence for the earliest salt extraction in Europe. It happened ca. 5400 BC, in the late Neolithic, when a group of people from Thrace crossed the Stara Planina Mountains and settled down at the salt springs near the present-day town of Provadia. Evidence for brine boiling technique are the thousands shreds of ceramic bowls, 40-45 cm in diameter. The final product was a lump with standard dimensions, which was ready for exchange or trade.

The importance of the discovery at Provadia is related to the necessary daily intake of salt for the normal functioning of the human body as high as 12-18 g. The animals, especially the domestic ones, have the same needs. Until quite recently there was no data on the salt production in the Balkans. However, the simple calculations reveal that the early farmers (6th millennium BC) of Thrace only, the area where the settlers of Provadia have come from, needed at least 500 tons of salt per year (including the needs of the domestic animals). Such estimates remain somewhat disputable. The salt producing "colony" apparently traded the salt and supplied the population of Thrace with salt and probably got in return food and other necessary products.

In Transylvania, the newly excavated sites reveal that both dry mining and brine evaporation were in use between 1430 and 1010 BC in this region of Romania. Salt mining was attested in the sites Băile Figa, Sasarm and Caila while evidence for artificial evaporation was coming from Olteni and Zoltan. As regards Băile Figa, the site is placed above the underlying deposit (1.5 to 4 m from the surface). There were three main kinds of evidence: wooden structures and wooden tools, cavities and mounds of earth resulting from bore holes made in different periods for salt extraction, and isolated artefacts. In Sasarm, various timber structures were alleged corresponding to mining galleries.

In 1984, excavations were undertaken at the Poiana Slatinei site in Lunca, Vânători-Neamţ. A Franco-Romanian archaeological mission resumed the excavations in 2004 (Weller and Dumitroaia 2005; Weller 2008), to study the earliest salt production in Europe, an issue which has been the subject of significant debate linked to human sedentarisation, the development of complex economies and the emergence of hierarchical societies. This site is unique in Europe for the preservation of its 60 by 25 m stratified conglomerate of ash, charcoal and red-coloured burnt soil. Located in the vicinity of a still-used salt spring, it contains large amounts of combustion soils. The successions of combustion and refuse areas have formed a sizeable mound of almost 3 m high. This exploitation was facilitated by the strong concentration of brine from this salt spring: 160 g/l. Painted ware from the different trenches (Starčevo-Criş culture IIIb-IVa) shows that most of this deposit can be dated to the Early Neolithic. The examination of this deposit has also revealed Precucuteni, Komarov-Costişa, Noua and Hallstattian exploitations. Nine C14 AMS dates are available. These calibrated dates are consistent and confirm the early dating for European salt exploitation as being the very beginning of the sixth millennium BC. They evidence that salt exploitation emerged in Eastern Europe during the earliest Neolithic (6050-5500 BC) with exploitation continuing into the 5th and 4th millennia and even beyond.

Technology

Defining technology is not so easy. Technology is such a major determinant of human progress that all accessible sources of information, methods of investigation, and means to explaining how it works should be exploited to improve our understanding of the role of salt in the long run. History of technology combines technical studies about tools and devices, and anthropological observations regarding the context of innovations. Dry mining has only a few specificities as regards salt. They relate basically to its plasticity and its hygroscopicity. Generally speaking, the working practices have changed dramatically in modern times because of the use of mathematical calculations, detection techniques and the introduction machines, conveyor belts and other tools and interchangeable parts. Anticipating solution mining, brine boiling offers a distinct field of investigation with various routes. Both archaeologists and historians implement research to evaluate the economic and social background of human groups attracted by salt making. Answers to the question *why* should no longer be discriminated by those to the question *how*. It is much more for the former than identifying and classifying ceramic artefacts (briquetage). Moreover, what deserve consideration is the will to conform no longer to some Western technological framework. There is no abstract evolutionary model applicable in ancient salt search.

Although it seems acceptable to consider that the history of world salt production parallels its consumption by Man, one would be permitted to refuse that the quest of this substance became more significant with the 'advance of civilization'. Industrialized societies express salt requirements exceeding by far salt needs of unacculturated societies. An inroad of salt in chemical processes is a matter of progress. May be... This 'hidden salt' has certainly a less cultural significance than table salt (use of standing salt in medieval societies). The purpose of this remark is not to deny any importance to technical improvements. Throughout the evolution of salt making, there has been a constant search for ways and means to make techniques more efficient and less costly.

The first important and early uses of salt seem to be food preservation rather than seasoning, and animal feeding. However, its preservative properties contribute to get this substance adopted in tanning and leather treatment where salt is added to animal hides to inhibit microbial activity on the underside of the hides and to attract moisture back into the hides. Salt chemical properties have facilitated its non-food uses. For instance, addition of salt to the silver-gold alloy is attested in Antiquity. The sodium chloride attacks the silver and forms silver chloride which is absorbed by the walls of the crucible. The facilitation of cupellation and cementation by the salt process is mentioned by Pliny the Elder (*Hist. Nat.* XXXI).

From a geological point of view, salt is available in almost every country in the world. It is made available by solar evaporation of the sea water or the salty water of continental lakes where the coastline and the climate are suitable. Inland saline springs keep pace with a number of halite deposits. Rock salt can be mined by dry mining or by solution mining, natural brine from the springs being first evaporated in ceramic containers (boiling vessels), and later, in metal open pans. Artificial evaporation is now carried out in vacuum evaporators. Direct boiling of salty water produces tiny quantities; filtering and concentration yield much higher quantities. Tree and plant derived ashes either in combination with salt springs or not have been used to obtain vegetal salt in Amazonian areas or in Papua New Guinea. Inhabitants of the Polar Regions have even used removing water by freezing (separation of dehydrate) since prehistoric times for getting salt (Kaufman 1978).

Reconstruction of prehistoric salt-making techniques has been attempted by reference to anthropological fieldwork in such areas as New Guinea (Weller 2000) where techniques are to combine mineral and vegetal, and Niger (Gouletquer and Kleinmann 1984) or Central Africa (Lovejoy 1986) where salt is still manufactured by traditional pottery–using technologies in the 20th century. Early salt making techniques are difficult to be traced in archaeological records and the scale of production as well.

The Mediterranean climate is ideal for salt making. According to ancient writers, salt was widely produced in coastal sites; the Mediterranean basin and various salt lakes (Tatta, Mantiane, Sirbonis, Maeotis, Thopitis) offered the same chance to get salt thanks to solar evaporation. In certain places salt crystallized naturally. In others bowls were cut out by hand in the rocky coast. Elsewhere, rudimentary facilities were managed on a limited surface with channels and ponds (Carusi 2008). Seawater concentrated and once evaporated, salt was collected and stockpiled (Manilius, *Astronomica* IV, 5, 680). Although there was a need to improve both quality (white salt) and yield, technology continued with little change, irrespective of more accurate views about water circulation and concentration. Aristotle was anxious to understand fully the causes of its salinity (*Meterologica*). The question is still open of who were the first to regulate production via developed pans, linked by channels and protected by dykes: Phoenicians, Etruscans? The addition of mentions from ancient literature does not reflect at all the number of operated salt places, including salt mines and springs where salt was evaporated artificially by boiling or filtering ashes further to throwing salted water on fire. One has high concentrated brine: evaporation and crystallisation are speeded up. Elsewhere the difficulties of achieving concentrated brine from sea water or peat ashes or from salt impregnated soil are slowing down the processing of salt the quality of which is rather poor (Brisay and Evans 1974).

Approximately 5 000 years ago, Chinese people were boiling sea water to produce salt. The hinterland was subjected progressively to high density human settlement and it relied increasingly on farming. Thus, critical to human survival for its constitutive ions (sodium and chloride) salt became a matter of political significance and its production from inland salt springs was boosted accordingly. The primitive exploitation of natural brine in the Ganjing River valley impacted both geographical and cultural landscapes. The first pit was dug in Sichuan, around 2 250 years ago. This was the first time solution mining was applied successfully to the exploitation of salt resources and marked the beginning of Sichuan's salt drilling industry. About 2 000 years ago this technology evolved within a change of scale. The inhabitants began to dig wells with percussive drilling systems instead of digging them by hand with shovels. By the beginning of the third century AD, wells were being drilled up to 140 m deep. Throughout the history of Imperial China, from 221 BC to AD 1911, the salt monopoly was a major source of state revenue. Textual sources on salt administration are plentiful and have been a major topic of research. By contrast, the material remains of early salt manufacture have been little studied until recently-unlike other areas of the world such as Central Europe, Japan, and the Americas, where "salt archaeology" has a long existence.

The finds from Zhongba have pushed back the dates for the origins of salt production in this part of China by over 2000 years, to the third millennium BC. In spite of a bunch of historians who had traditionally disregarded this area, the enormous volume of salt-making pottery shows that, during much of prehistoric times, salt production went on at a quasi-industrial scale, involving sophisticated division of labour. Clearly, salt was not merely produced for local consumption, but for trade, probably down the Yangzi River to the salt-poor areas of Chu in the present-day provinces of Hubei and Hunan. In spite of being located outside the purview of the early Chinese dynasties, this area obviously was a major economic centre. Archaeological evidence based on reliable methodology is to change the degree of investigations by enlarging their field to human structuring (Flad and Shu 2005).

Thus the Chinese had the understanding that, besides sea water evaporation, salt could be operated deep under earth' surface. Drilling offered the opportunity to extract salt either by dry mining or by solution mining (brine being pumped to the surface and boiled. 'In this country there are salt springs from which they manufacture salt by boiling it in small pans. When the water has boiled it becomes a kind of slurry which is formed into cakes' (*Il Milione* or Kingdoms of the East, 1307). As witnessed by Marco Polo during his stay in the Khan empire, the categories of salt produced in the Song and Yuan period are as follows: sea salt (*haiyan*), lake salt (*chiyan*), well salt (*jingyan*), earth salt (*tuyan*), and rock salt (*yayan*).

The importance of studies related to the salt monopoly system prevailing in ancient China has been emphasized several times; more especially under the Yuan dynasty (1271-1368). The *Aobo tu* was completed and published by Chen Chun in 1334. Further to the issue of a Japanese edition, an English translation was worked out recently with appropriate comments (Tora and Vogel 1993). By combining research on various aspects of the salt industry during the Song (960-1279) and Yuan (1271-1368) periods, it provides a better understanding of the fiscal and economic importance of salt.

Moreover the *Aobo tu* is basically an illustrated description of the Xiasha saline which was located along the coast. It is aimed at assisting Empire officials in the reorganization of the salt works. A former controller of the saline was featured to have a deep knowledge of the methods related to sea water boiling, of the differences in soil and wind conditions, and of the timing for the subsequent steps. According to its title (*aobo*=boiling of sea water), this treatise refrains from giving a general description of salt making processes. In the reverse, the various steps for brine boiling are explained with the support of sketches:

1. Construction of hearth houses and leaching basins
2. Changeling the tide water
3. Construction of the spreading place and levelling of its surface
4. Spreading and leaching the ashes for the production of brine
5. Transport of the brine
6. Fuel to be used for brine boiling
7. Designing and casting the iron pans
8. Recovery of salt
9. Collecting the loose salt (*sanyan*)

Although the salt produced by the Xiasha saline constituted only part of the output of the Huai and Zhe region which participated for more than 50% in the total salt production quota assigned within the existing monopoly to the Chinese salt works, it was presumably considered as a suitable standard for maritime salt making conditions. Improvements were noticeable for the Xiasha saline which was liable to some 93 000 Song shi in 1127-1162 plummeted to more than 260 000 shi — after translation in Song shi — by about 1335-1368. Thus, solar evaporation was not a standard for getting salt from sea water. The Mediterranean way of salt processing has no universal value irrespective of its generalization in Classic Antiquity around this highly salted basin. The Japanese salt industry offers another example which is insofar valuable than the Nippon islands are often alleged not to be in capacity to produce salt.

Since prehistoric times, salt has been obtained from sea water. Solar facilities were not feasible for climatic and pedological factors. Therefore a two stage process was developed combining sea water concentration into brine and the latter crystallization into salt. From the 8th century, people managed salt fields (*agehama*) to concentrate sea water. It was manually transported from the sea; brine was extracted from sand, and then evaporated into open pans fuelled with wood. In the middle of the 17th century, large scale salt fields appeared on the Eastern coast of Seto inland sea and Northern Kyushu taking advantage of the tidal range and larger salt making units. The salt place was divided into rectangular plots by a network of channels designed to get concentrated brine from sea water. Most of these new facilities (*irihama*) were located near Kyoto and Osaka. In the region called Jisshu Enden, they provided almost all the salt produced in Japan. In the late 18th century, their production amounted to 372 000 metric tons, exceeding 80% of the national production (456 000 metric tons). It is noteworthy that a graduation device (*shijoka*) was introduced to increase the productivity in modern salt fields (*ryuka*) with less energy and reduced labor force. After the Meiji Restoration, 1868, the government implemented drastic measures in favour of the Japanese industry. Quality considerations stimulated efforts to improve technology by reference to American (Kanawha-type evaporator, ion-exchange membrane) or European innovations (vacuum evaporator, thermo-compression). When the third reorganization of the salt industry took place in 1959–60 combining vapor compression and vacuum evaporation, the capacity reached 900 000 metric tons (Ohno 1993).

Archaeologists working in prehistoric periods focus upon innovation which seems easier to identify than utilization. Medieval archaeology is in a position to connect both of them for material evidence and writings can be combined. *De re metallica* (Agricola, 1556) was the first systematic treatise on mining and metallurgy, and one of the first technological books of modern times. It embraces everything connected with mining including salt extraction and brine boiling in its book XII. '*There should be as many salt pits dug as the circumstances of the place permits, but there should not be more than can be used.*' It is a landmark in technological developments which '*salinists*' have continuously forwarded and described (Fürer 1905; Kaufman 1978). The salinists are experts in science applied to salt mining and salt boiling.

At the end of the 16th century, Heinrich Schickhardt (1558–1635) was in charge of improving the yield of Württemberg salt works (incl. Sulz and Schwäbisch Hall, and Saulnot in the comté de Montbéliard), and visiting several salt facilities located in the neighbour countries: Hall in Tirol, Porto Cesenatico (between Rimini and Ravenna), Salins in Franche-Comté,

Salonnes and Rosières in Lorraine, Schönfelden in Palatinat. He also documented the following salines: Orba, Salzkotten, Ollendorf in Hesse, Halle in Saxony, Frankenhausen in Thuringen, Brunswick, Lauenburg, and the Polish salt mines: Bochnia and Wieliczka. Later on, the technological evolution of these continental salt works has been duly documented either from a technical point of view (Emons and Walter 1988) or from a legal and social one (Palme and Hocquet 1991).

He was accordingly in a good position to observe the specificities for each of them; he provided a number of practical suggestions, followed by further experiments in order to restore efficiency in spite of the decline of boiling houses, and to allow a revival of salt production hampered by low salinity and huge wood requirements for getting brine properly evaporated. Graduation and other processes aimed at high salt concentration (pre-heating arrangements) were introduced. Thus, at Sulz, graduation has saved nearly 50% of the wood previously used to fuel the pans as mentioned in the description of the saline. The invention of coal-pits close to certain places paved the way to substitute coal to wood. The most determined efforts were obfuscated by technical (iron pan resistance) and political (Thirty Years War) difficulties. The use of boring in prospecting for coal developed nevertheless during both 17th and 18th centuries. Simultaneously, engineers initiated geological investigations related to salt layers to be operated either by dry mining or by drilling and brine evaporation in traditional pans (Bouvard 1983). Amongst the eminent writers addicted to innovative manufacturing practices were Langsdorf Bros. Johann-Wilhelm published '*Ausfuhrlichere Abhandlung von Anlegung, Verbesserung und zweckmässigen Verwaltung derer Salewerke*, 1781, which develops a previous introduction to salt technology. For salt works operated with weak brine, the major problem to solve was preconcentration of water either thanks a combination of heated vats (the grainer perfected later the device) or in building graduation systems such as that in operation at Schwäbish Hall from 1739. Wood consumption was reduced there to 30% of the volume required in the late 17th century.

In Germany, the situation of salt works remained almost unaltered in the 16th-17th centuries by comparison with the Middle Ages. Innovation was nevertheless a matter of research based on benchmarking. The increase of the demand once war and pest were over, and the struggle for efficiency (less energy for more salt) led to check pragmatically the different production factors in relation to brine boiling (pan material, pan resistance, pan size, brine concentration, brine purity, insolubles and additives, fuel material). Graduation which depends on contingencies like meteorology was often the last resource. The list of 86 salinists drafted together with the salt works where they have been active evidences the weight of boring and heating elements (Piasecki 1987). In continental places, it is obvious that fuel drives the innovative attempts for better evaporation (Multhauf, 1978), and the high cost of energy remains currently a matter of concern for evaporated salt. He who wants to produce some report about the former salt works is doomed to be an expert in geology, technology and economy. He has also to take into consideration the regional and international circumstances under which salt making have reached its critical size. Within the framework of the Holy Roman Empire every sovereign wanted to own, at least, one saline. A Bill of Charles IV, 1356, states that '*the kings of Bohemia, and all the electoral princes, ecclesiastical and secular, shall hold and possess with full rights, all mines... and all salt works, both those already discovered and those which shall be discovered in the future, situated within their lands, domains, and dependencies*'. Their capacity varied widely, from a few hundred kg at Thale up to 60 000 tons at Schönebeck. A number of tables and figures contribute in that case to a well-balanced overview of the specificities of the salt sector (Emons and Walter 1988). Such a *haligraphia* is manifolds; studying juridical aspects contributes to a better understanding of the reasons why salt production developed (Palme 1983) and which interest got involved in the related revenue.

Studies about ancient technology could be an opportunity for quantification essays. Once again, a majority opposes the casual results. They pinpoint that it is impossible to evaluate the global incidence of historical events and to discriminate their relative impact. Moreover, the scarcity of data makes any post hoc evaluation vague and void in the absence of an exhaustive appreciation of existing documentation (Piasecki 1987). Its continual and universal increase in its consumption reflects both population evolution and scientific innovation which generates new uses especially in chemistry (Multhauf 1973). At the end of the 18th century, the Le Blanc process was developed to produce synthetic soda ash from marine salt. It was even one of the first patented. A more efficient one was introduced in 1862 by Solvay. Chlorine was first studied by Scheele in 1774 and used to bleaching textiles from 1785. In the latter half of 19th century, the adoption of electrolytic methods led to a large scale production. By passing a current of electricity through a sodium chloride solution the salt is decomposed into chlorine at the anode and sodium at the cathode. But the latter decomposes at once a molecule of water of the solution, forming caustic soda and setting free hydrogen.

Landscape and environment

Contemporary understandings of landscape are set within a disputable conception of the world based on a segregation of Man from its environment as if its creation was a gift to his reason. In other words, Man became the administrator of nature instead of God further to acquisition of knowledge. Information available in the outside world is internalised to get nature ruled according to human standards. Even if the existence of God is denied, Man is part in nature. Filtering the reality through an individual interpretation within which science is juxtaposed to fancy generates particular understandings of salt places and their respective milieu as it was prior to human activities, as it evolved over centuries before being that you can see just now.

Salt production also helped shape the landscape. Historical developments, in the long term, are influenced much more by environmental and evolutionary processes than by conscious human choices; but human activities can affect the environment in important ways, for instance by selecting salt production techniques. Solar evaporation is more sensitive to ecological requirement than dry mining or solution mining which results in environmental damages such as subsidence. In Cheshire, around Medieval salt extraction sites which is indicated by the local names ending in "*wich*", and in the 1870s, the pumping out the brine from the flooded mines, and using it to supply the salt works had catastrophic effects. Houses, shops and offices disappeared without warning. Brine extraction in these areas accentuated salt karst development, causing severe subsidence problems. The Great Subsidence of 1880 caused a great upheaval. In 1891, a Brine Subsidence Compensation Act was introduced.

The importance of ecology with regard to salt gathering relates to climate, sea level, erosion or alluvium inroads. Environmental changes as a result of both natural phenomena and human activities are impacting salt production with regional differences, amplified either by agricultural decline or the use of land for developing animal management strategies. An issue to be further investigated is the causes and consequences of deteriorating material infrastructures of a salt place. The withdrawal of salt mining in the former kingdom of Noricum or the abandon of salt boiling in the Seille valley at the end of La Tène deserves serious attention. Explanations would be favoured as regards the Danube area or the former Amerindian empires. The lack of technical means to renovate a failing production system due to flooding or some other natural accident is generally interconnected with community inertia or political uncertainty. The degree of sustainability is another aspect which has to be investigated as regard salt supply. The *salinae* of Ostia have produced salt from their primitive establishment up to 1874. Those of Salapia are still in operation although the morphological evolution of the coastal plain where they are located was characterized by environmental crisis (seismic events, climatic change) impacting settlement conditions and anthropic influence. When the local environment became more suitable for human inhabitation, some settlements developed around the lagoon of the Murgian coast. During the Roman period, the corresponding basin was divided into smaller lakes: Salpi and Salso. The former hosted salt marshes which are still operated although the site moved along the gulf of Manfredonia (Di Vittorio 1981). The major cause of the closure of continental salt works was the lowering of water salinity and the failure of graduation equipment to restore it. More recently, the rising cost of energy and increasing flexibility of capital enhanced research aptitude at a form of social analysis. The challenge in steering a field of research towards new perspectives is complicated by current interpretation based on politically correct views (microstructures) or bare journalistic lucubration (salt caravans).

Continuous transformation of the landscape over various periods of time is a matter of fact because of natural phenomena and human activities. Flood-prone marshes were managed along the sea from the protohistoric age either to manufacture ignigenous salt bricks or to gather salt in bulk further to natural evaporation. No landscape is more artificial than the ponds designed for solar evaporation (Hocquet *et al.* 2001). Landscape archaeology generally follows a well-integrated approach combining a number of factors critically interpreted with regard to the physical transformation of an exceptional site in relation to human activities which have evolved or disappeared. The green movement leads currently to biased discussion concerning salt sites. When considering on the *longue durée* the specific landscape which is formed by salt gardens, we try to elucidate how the humanisation of the milieu with ups and downs complexified its evolution which is neither natural nor artificial (Reault-Mille 2003). The cultural value of the Guerande salt works has been exaggerated in the sense of salvation of something which no longer exists. Concerning sea salt production, human impact inscribed in the long run is modelled by short term interventions e.g. every year, and casually altered by natural perturbations. One of the most relevant questions is how people cope with in adverse conditions. Experts should focus upon regions where the significance and the survival of salt making is less controversial and seems adequately related to landscape conservation.

From this point of view, two more convincing examples deserve attention: 'the salt Gardens of Kibiro' (Uganda) and 'the Salt Route of Niger' (Rössler and Souama-Ferero 2000). In the first case, eight areas of salt production are associated with hot springs on the border of the western Rift. All the activities concerning salt are carried out by women, and their know-how transmitted in the full respect of the vivid tradition. The trans-Saharan itineraries have relayed Black and White people over centuries. The Salt Road of Niger is organised around two major crossroads: Agadez and the oasis of Kaouar, a traditional centre for salt production too (salt works of Bilma and Fachi). One might consider that the identity of this itinerary offers a synergy between natural and cultural elements which can be put in a perspective of sustainable management.

Environmental problems have arisen in the past such as lead-poisoning pans, salt-related deforestation and subsidence. Lead ore was another vital commodity to the Romans. It was used extensively for the fabrication of containers (for example, coffins) and other vessels for many purpose (pans) as well as the making of pipes to supply clean water to their houses. Celtic Britain was a major supplier of lead to the Roman Empire and control of the mines of Somerset and North Wales was an important economic reason for the occupation of Britain. The crude lead also contained silver, and separation of silver from the lead was an important stage in the lead making process. Lead ingots and lead pipe with dateable AD 70's inscriptions have been found at Chester and lead brine-evaporation pans for making salt have been found at Northwich, Middlewich and Nantwich. Worcestershire Domesday lists 318 salt houses and refers to the existence of lead pans, a lead smithy, and furnaces. The need for firewood is obvious.

Salt pan supply with firewood contributed to deforestation at large in areas where salt springs were exploited to get white salt for huge quantities were used to fuel the various types of pans. Wood use for salt making was prodigal in old England and in Central Europe as well. One of the furnaces operated in Nantwich consumed alone 6 000 cartloads a year. At the end of the Middle Ages, fuel reserves in the Royal Forests were widely depleted whereas control of timber felling tightened. Ore mining and salt boiling in Hall area (Tyrol) would have used around 1 million m^3 during the late 15th to early 16th centuries. Visiting the salt works of Hall in October 1580, Montaigne was impressed by the huge quantity of wood stockpiles for fuelling the several round pans in operation. In a Memorandum regarding Lorraine salt works, 1791, mention is made of the equivalent of 169 000 m^3 to supply the 56 pans; restrictions for other uses than salt making were so prejudicial that State arbitrary seemed to plummet with them for the population (Husson 1991). The same situation occurred in Japan. In the middle of the 17th century, large deforestation and climatic change (2°C lower) resulted in slowing down production and increasing price. The introduction of coal encountered strong opposition from the farmers and the feudal lords. The Meiji Restoration paved the way to a drastic modernisation of the salt industry.

The traditional method for extracting fine white salt from brine was to boil it up in immense pans until the water evaporated. During the eighteenth century, the efficiency of the boiling brine process was improved by using coal instead of wood as fuel. This contributed to late salvation of forests although in many places the damage was irremediable. Huge iron pans the size of swimming pools were heated by coal – it took about one ton of coal to make two tons of salt. Northwich and other towns suffered with great clouds of smoke and steam, as antiquarian John Leland noted in the 16th century: '*Northwich is a pratie market town but fowle, and by the Salters houses be great stakes of smaul cloven wood, to seethe the salt water that thei make white salt of.*' Air pollution was a problem in many industrial countries. The relationship between salt and coal needs examination. For example, both industries were intimately linked in Scotland where the availability of coal was one of the factors determining how much salt could be produced as long as collieries and pans remained profitable. The larger salt works became over years (1570-1850) the appendages of collieries (Whatley 1987).

Table 1. World population (million units) and salt consumption (thousand metric tons)

400	152	1064
0	250	1750
500	205	1435
1000	257	1799
1300	429	3003
1400	374	2618
1500	458	3206
1650	493	3451
1700	682	4474
1800	969	6776
1900	1613	11291

At the end of the 20th century, salt production exceeded 210 million tons whereas a century ago it totalled hardly 10 million tons. A consistent series gives the following figures for the United States in short tons: 1800 = 1 400, 1900 = 2 921 708, 2000 = 45 600 000. Based on these data and various environmental studies, there do not appear to be negative impacts to the ecosystem in relation to the material flow of salt (Kostick 1993).

Outside salt production, the options to secure ecological sustainability and economic viability of arid, semi-arid and sub-humid regions have varied considerably due to soil salinity. In Sumer, an extensive canal system was developed for watering fields in response to soil salinization during the third millennium. Due to seepage, flooding and over-irrigation, an ever-increasing rise in saline groundwater levels occurred. Wheat was gradually abandoned in favour of the more salt-tolerant barley until the lands were abandoned, ultimately caked white.

Globally, about 10 M ha of agricultural land is lost annually due to salinization, of which about 1.5 M ha is in irrigated areas. Water logging and salinization affect 23% of the irrigated land in China and significantly reduce production on an estimated 15% of China's irrigated land. Although sea- water desalination (the process of removing the salt from salt water to make it potable) offers some hope for the long term, the worst-case scenarios could now reveal probable since the resource has been spoiled over centuries, impacting on landscape and climate.

Travellers and caravans

Salt making and distribution are evoked in diaries and journals, notably those of foreigners who travel throughout the continents and sail the seas. Within what is worth to be seen salt is a topic deserving attention. Digressions regarding salt observed in various writings left by travellers whose journeys took place in the remote centuries constitute another approach. Far from their mother country, they merely expressed rapid consideration about production and consumption of a mineral substance, known at least as a foodstuff which is often subjected to taxation. It is still today a fascinating story for globe trotters (Ritter, 1981; Denhez 2005) and photographs (Valli and Summers 1994; Dortes 2005): the salt caravans that cross the Sahara, the

Himalaya, the Antiplano, for transporting salt from continental salt lakes to traditional markets. Salt was used to keep food from spoiling as well as for the hydro-mineral balance as a dietary supplement in hot, dry climates.

Herodotus vindicated the large rivers falling into the Black Sea including the Borystenes. At its mouth a large mass of salt is formed of itself in natural salt works which were said to yield inexhaustible quantities. They were available for fish salting and, presumably, hide tanning due to the presence of numerous flocks (*Histories*, IV, 53). Dio of Prusa evokes the same facilities: 'And it is here also that we find the vast number of salt-works from which most of the barbarians buy their salt as do also those Greeks and Scythians who occupy the Tauric Chersonese' (*Discourses*, XXXVI, 3). Leaving Soldaia (Crimea) on June 1253, Guillaume de Rubroek paid attention to several lagoons and springs from which salt is extracted. Mongol rulers are said to get important revenues from that salt for Russian people are prone to buy it. In 1672, Jean Chardin described elements of salt trade at Caffa (Kertch area) where more than two hundred ships used to load it. 'This salt comes from salt facilities located eastward of the harbour and called Dousla (Tuzla). See his *Voyage de Paris à Ispahan*, 1686).

The Japanese monk Ennin journeyed in China where the salt monopoly contributed substantially to the revenue of the Tang dynasty (618-907). In 838, moving from the coast to Yang Chow on one of the lateral tributaries of the Grand Canal, he mentioned the numerous salt works which had been built or repaired from the San Kuo to the Sui, facilitating a fruitful business. The annual revenue amounted to ten million strings of cash. Ennin described a train of forty boats, most of them lashed two or three abeam, was slowly pulled by two buffaloes. When reaching the Grand Canal, such trains followed one another, continuously (Diary). Marco Polo emphasized the importance of salt taxation in the Grand Khan's empire. As he has been entrusted several times to survey the fiscal management of salt revenues, he was in a good position to comment about. Russia was repute for its salt resources, one of the most important areas being located at Staraya Russa (Novgorod). Its salt works (brine boiling) were mentioned by Herberstein (Journey 1549) who observed judiciously that the salt houses were built within a separate district, although close to the town on which they depended.

According to Leo Africanus, the habits, regarding table salt, are different in Africa. 'The people of Ethiopia do not put it in salt cellars set on a table, but when eating bread they hold a piece of salt in the hand and lick it so as not to use too much salt' (*Descrizione dell'Africa*, 1550). Considering the behaviour of Brazilian natives with regard to their diet, especially how they use salt, Jean de Lery related that '*they do not salt the meat or the fish or other food before raising it to their mouth as we do. They take each piece separately and accompany the mouthful by dipping two of their fingers into jonquet*,' a mixture of pepper and salt they use as seasoning (*Histoire d'un voyage en la terre de Bresil*, 1578). Jean de Léry's account of a year spent living among the Tupinamba Indians of Brazil ranks among the masterpieces of early modern ethnography. In Europe salting has been subjected to rules and rituals which cannot be transgressed by a gentle man, for centuries. The delineation of the cultural significance of salt remains difficult because it is a matter of belief, and the ways to sapience and superstition have not the same crossings. According to Robert Knox, the salt facilities established on the west coast of Ceylon have enough capacity to ensure the whole island supply. Provided that they get rice and salt, the inhabitants feel comfortable with their diet (*An historical relation of Island Ceylon*, 1681).

During his journey in several Andean regions Jean-Baptiste Boussingault observed that the salt collected in *salares* contained iodine as did water. Consuming iodized salt would prevent a goitre which is frequent in subjects suffering from iodine deficiency when they eat grey salt from the coast (*Mémoire sur les salines iodifères des Andes,* 1830). His observations have resulted in salt iodization programmes supported by both WHO and UNICEF in order to eradicate iodine deficiency disorders throughout the world.

Being both a human necessity and source of commerce, salt has been in high demand in Central Sudan since the 12th century when it was first found in the sand dunes of the desert. The king of ancient Ghana controlled trading through taxation, especially the export of gold from Ghana and the import of salt from north of the Sahara. The prosperity of the sub-Saharan empire of Borno was closely related to salt business. For centuries, West Africa got most of its salt from Taghaza, a place in the desert where salt was mined by slaves or subjects from other groups of people than the conquerors. Later, camel caravans and the trackers that lead them traversed the Sahara desert in search of the salt of Taudenni (*azalai*) and brought back salt slabs. From there, from the salt works of Bilma eastwards, and from the sebkha Ijil westwards, many thousand camels loaded with salt were travelling across the windswept sand dunes, 800 km north of Timbuktu (Arbaumont 1994).

In 1468, ownership of Timbuktu transferred to the Songhai Empire. Under their leadership the city flowered into the commercial and religious Mecca, which would stir the passions of European adventurers. During its golden age Timbuktu was also a repute centre for Islamic scholarship. Libraries full of hand-written documents were treasured (and well-preserved in the dry desert climate). Morocco had conquered the city in 1591 but had little success in holding firm. Several smaller raids by such groups as the Berbers, Fulani and Bambara slowly chipped away at Timbuktu's glory. Little by little traders moved elsewhere (Ritter 1981). The published descriptions of Leo Africanus, Ibn Battua and other Muslim travellers sparked the imagination of Europeans in the late 18th and 19th centuries. They believed Timbuktu to be a fabled city with roofs made of gold. Exploration societies held competitions aimed at seeing Timbuktu. A number tried, few succeeded. The only entry was by camel or via the Niger River. Most camel caravans were reluctant to take explorers on it. Dehydration and dysentery provided more hurdles for adventurers.

When René Caillé set out to discover Timbuktu (Caillé 1985), he already had a solid knowledge of Africa. By the age of twenty he had explored regions around Senegal. Before setting off for Timbuktu, Caillé learned Arabic and studied Islam. Posing as an Arab trader en route to Egypt, he left Senegal in April 1827. He arrived in Timbuktu one year later. Disappointing! By early 19th century, the splendid city has sharply declined. He described it as grey and dingy. The gold-topped roofs imagined by Europeans were actually flat mud slabs atop adobe structures. The sands dunes surrounding the city seemed to be drowning a lot of poor houses. He was most impressed by the recurrent aspects of salt trade of Taoudeni and its importance in the daily life of the tribes he visited. The Kounta nomads still travel across the desert to reach the salt mines and make money with the slabs quarried by hand. Once bought to the workers, salt is carried to Timbuktu (Puigaudeau 2004).

Within the context of traditional salt making in Central Africa, any effort aimed at quantifying the production and the volume of trade in the pre-colonial periods must be undertaken with caution because of an obvious absence of references for the population and its activities as well. Scientific studies are available: can they supplement the scattered statistics (Lovejoy 1986)? Tonnages are computed from the seasonal caravans and the number of loaded camels. Calculations vary to a wide extent due to the means of establishing preliminary estimates on the scale of salt-making based on geological conditions and political events. The generate levels in as much difficult to appreciate as they are unusual. Although opinions admit that, at least, data regarding the early years of the 20th century reflect the presumable capacity of production in the past centuries, the matter does appear so easy to tackle (no demographical anchorage). Total output for Soudan is estimated within a wide span from 20 to 65 000 tons. Where is the truth? How to decide that this span was valid for the 17th century or earlier? What do we know regarding trade trends between tribes and kingdoms? When, in 1353, Ibn Battuta visited Azelik in the Teguida region, the major trade concerned gold and copper. What can we say regarding the history of transhumant migrations, and other elements of the demand? As regards extraction feasibility on which the supply depends, who were the workers? Under which status they were doomed to dig here and to boil there?

The technological backwardness and the apparent inability to integrate modernization and innovation are showing that the different salt places evaluated in Central Africa are remains of the past without being the mirror of the past. The salt of the desert is an emotional mirage. Its caravans continue to attract comments and pictures. The *kantous* (cakes) of Bilma are now transported by truck to Agadez. One trailer carries 5 tons, the same weight as 150 camels (Dortes 2005). Where it was appropriate, modern facilities have been juxtaposed to traditional salt making as observed in Djibouti where sea salt has superseded rock salt from Lake Assal as long as the international market was buoyant (Dubois 2003).

Since pots have been broken to extracting salt, the resultant mounds of shards of salt boiling vessels can make this type of production visible on site. In East Africa, salt extraction was undertaken in a large scale in a number of locations like Kibiro using this technology (3 m of deposits). Kibiro, on the shores of Lake Albert in western Uganda, is now a quiet and remote village. Historical sources from the late 19th century allude to the scale of the salt trade in the pre-colonial kingdom of Bunyoro, one of the two dominant inter-lacustrine states at that time. In Kibiro, women have been making salt for a thousand years. Archaeology reflects an active history of the long and deliberate development of a sophisticated and unique salt-making process, with no indication that it was learned from other salt production sites in Africa (Connah 1996). In the course of technological development Kibiro salt makers discovered that leaching the salt-impregnated soil, rather than boiling the saline spring water, produces higher concentrations of higher quality salt, consumes less fuel: boiling the brine at night, inside kitchens, protects the fires from fuel-consuming wind.

This research underlines the importance of archaeology to the reconstruction of African history, although it fails to escape dating illusions. When? The contents of one layer of the dig suggest a pause in human activity about 1860, and in fact that the village of Kibiro was completely destroyed twice in the final decades of the nineteenth century. But the dating of the origin of salt production is confusing. The findings are not enough to allow Connah to declare that salt production began simultaneously with human settlement. Anyway, no effort to maintain Kibiro salt making is observed when salt imports coming from British dependences and heavy taxation affect such a traditional activity. In 1910, Kibiro only produced £ 800 worth of salt.

In a number of areas, people obtain salt from sebkha or salt lakes. In Sahel, salt is extracted from salted hearth, using the evaporation method. In most salt places, there are two round decantation basins and many small evaporation basins. Salt exploitation can be made only during dry season, from November to April. Men and women carry water (which is also salted) in leather bags from the springs to the decantation basins where it is mixed with salted hearth. The water, of which salt concentration increases, is left to clarify and deposed in the small basins where, under the solar heat, it evaporates. The resulting salt is traded by traditional caravans. The point is to ascertain that tradition means history. It is impossible to make sure that salt was extracted in Central Africa or in the Andean Belt with unchanged techniques over centuries, without disruption. The exact impact of climate cycles when going further back in Man history is not determinable and the extremes pose a problem in climate modelling.

There are also salt caravans that travel from Lake Assal into the Ethiopian Highlands. Lake Assal (Djibouti) is Africa's lowest geographic point; it lies 155m below sea level, saltier than the Dead Sea. The Afar, a nomadic tribe, has been cutting slabs of salt from this lake for centuries and continues to do so today. Slabs are shaped into rectangles weighing around 7 kg are hoisted onto the backs of camels. A similar method is employed with llamas in the Bolivian Altiplano, a high dry plateau in the Andes. But many primitive populations had not easy access to salt before the contact with the Europeans. Speckled with lakes, marshes, salt flats, and geysers, crowned by volcanoes, the Altiplano is Chile's link with the great civilizations of the Central Andes. An immense smooth surface of crystallized salt culminates at 3640 meters above sea level. From 40 000 to 25 000 years ago, Lago Minchin occupied the Southwest part of Bolivia. After evaporation of its water, the area laid dry for 14 000 years before the emergence of Lago Tanca. When the latter dried up, it left two major salt concentrations, the *salares* of Uyuni and Coipasa.

The *salar* of Uyuni (Tunupa) was created by the alternation of rainy and dry seasons. Located at the lowest point of the Altiplano, it offers a natural spillway for the rivers of the region that drained before disappearing quantities of minerals, coming from the surroundings basins. The *salar* is formed of a complex where alternate salt layers and sedimentary layers (sedimentary deposits left when the *salar* was under water, then mineral deposits when the water evaporated). With an extension approximately of 12.000 km², it is said to having a depth of 120 meters. The first 6 meters layer is of pure salt and the rest contains among the elements of salt, boron, magnesium, sodium, potassium and others. The most valuable element is lithium, a strategic mineral. Northwest of the salar of Uyuni is located the salar of Coipasa. The village where the salt miners live occupies an island in the middle of the lake. It is constructed mainly with salt blocs, like Thagaza described by Ibn Battuta (*Journeys – Black Africa*): '*amongst its curiosities, the construction of its houses and its mosques is of rock salt with camel skin roofing and there are no trees in it, the soil is just sand. In it is a salt mine. It is dug out of the ground and is found there in huge slabs, one on top of another as if it had been carved and put under the ground.*'

Further to intensive works related to the traditional exploitation of salt springs in the Highlands of Irian Jaya, long distance trade and migratory movements offer an opportunity to understand the stakes of barter agreements within which Moni and Dani salt cakes appear, like the polished axes, to participate in social regulation (Weller 1996). These remarks were replicated for re-evaluating the ethno-archaeological patterns of salt making or mining in Pre-historical Spain (Figuls 2005). Salt mining at Cardona would have to be considered in the Solsones economics as a factor of complexation integrating added value of highly appreciated items (Morere 2007). Similar investigations should be initiated elsewhere to delineate the role of salt in positive migration.

Worth of their salt

The pre-eminence of salt as food and as symbol is enlightened by the various masterpieces produced by famous goldsmiths. Besides gilded and silver salts, pewter and ceramics have played their part in households as standing salts and cellars; and wood boxes too, in the kitchen. The ceremonial salt produced by Benvenuto Cellini represents two figures personifying the Ocean and the Earth who both offer huge amounts of salt. Between them, is a beautiful ship designed to carry it. In 1570, King Charles IX offered this chef d'oeuvre to Archduke Ferdinand who made no bones about accepting. It is the reason why this splendid salt is now in Vienna.

Neolithic salt cellars are placed on display in the Archaeological Museum of Naples and Syracuse. Pythagoras said that salt was the emblem of justice; for as it preserves all things and prevents corruption. He therefore directed that a salt cellar should be placed upon the table at every meal, in order to remind men of respecting equity. The Romans adopted salt as an ingredient in the sacred confection of the *mola salsa*. A shell (*concha salis puri*) served as a receptacle for salt on the table of peasants, but the wealthy citizen possessed an handsome piece of silver, which was usually handed down from generation to generation (*paternum salinum*), as attested by Horace: the inherited salt cellar shines on his table. The so-called treasure of Boscoreale includes some Roman salt cellars.

Salt was a precious foodstuff in traditional England, valued both as a preservative and as flavouring. So important was salt at the table that it was often enshrined in a masterpiece of architectural form adorned with the figure of a female supporting the cellar. In a standing salt of the early Renaissance (1542) the shallow receptacle of the salt is enclosed within a cup-like section reminiscent of a medieval hour-glass. By the end of the reign of Queen Elizabeth I, the ceremonial standing salt became rather simpler than before, like the Gibbon salt made in 1576. Four ionic columns enclose a central cylinder of crystal in which is enclosed a figure of Neptune. When the diarist Samuel Pepys went to Portsmouth in 1662 as part of the delegation meeting Queen Catherine of Braganza, he was shown the splendid spool-shaped salt to be given to her by Portsmouth Corporation. Four eagles with wings displayed and four hounds decorate the flat top with the well for the salt. The central part is of rock crystal, while the stepped base stands on eight lions couchant (*Journal*).

From the late medieval period a large ceremonial covered salt (great salt), was placed on the high table at the host's side. This distinguished the status of the diners, who sat either 'above' or 'below' the salt. This type of salt cellar is known as a standing salt. Smaller salts were arranged around the tables, next to the trenchers, or plates. These are known as trencher salts. Great salts were common by the mid-16th century, but were still an important part of household silver, valued as high status objects. Among the rarest pieces of old silver and pewter is the ceremonial salt, the principal article of domestic plate in early English houses of whatever degree.

The salt cellars known as trencher salts, were used at the lower end of the table and by the side of the trencher. These were small in size and made in various forms of pewter and of silver. In early 18th century small salts were frequently made in sets of two or eight. The finest ones were cast and the most popular shape was the octagonal with incurved sides and an oval well for the salt. The circular salts with three feet and the salts with the glass lining belong to the Georgian period. The period saw changes in social habits and silver salts developed. More recently, small so-called trencher salts, placed beside each plate, are circular, octagonal or hexagonal in outline. The Victoria and Albert Museum in London owns a rare Tudor trencher salt of triangular form on three cast mask feet, the upper surface embossed and chased around the central circular well.

Many of the customs surrounding its usage emanate from the medieval manners of the European continent The size of a number of standing salts and the arranging of the guests « above » or « below the salt » together with uses and taboos suggest questions relating to salt as a determinant of social status. Everybody understands the phrase « below the salt ». There was a time when the whole English household ate together at one table, the master and his family at one end above the salt cellars and the servants at the other end below them. The French fashion dictated that separate tables or more often separate rooms should be more appropriate. By the middle of the 17th century the position of the standing salt lost its significance. Anthropologists are somewhat divided. But there is no doubt that white salt played a significant role in the social evolution of mankind. To spill salt is considered unlucky. Not to be « worth his (her) salt » is even worse by reference to the word salary (*salarius*). Overturned salt cellars bring bad luck. When that happens, counteract the ill effect by tossing a pinch of salt over the left shoulder into the eyes of the devil…

The instruction regarding salt remains simple over centuries: Never use fingers for salt. The *Babees Book* (1475) says: '*Enfant, garde qu'en la saliere / Tu ne mettes point tes morceaux / Pour les saler ou tu deffaulx / Car c'est deshonneste maniere*' and the *Young Children's Book* (1500) contains a similar passage: '*It was not graceful to take the salt except with the clene knyfe; far less to dip your meat into the salt-cellar*'. A no less scrupulous rule is proposed by Erasmus in *De Civilitate morum puerilium*, 1530: 'Three fingers stamped in salt constitute, as people say, the canting arms of the villain. One must pick up salt with the sharp end of his knife'. English servants resisted this custom sturdily and Jonathan Swift revenged them forever in giving negative directions to their fellow-workers: '*When any salt is split on the table, do not let it be lost, but when dinner is over, fold up the table-cloth with salt in it, then shake the salt out into the salt cellar, to serve next day*'... '*That the salt may lie smooth in the salt-cellar, press it down with your moist palm*'... '*Do not forget with your thumb and two fingers to put salt one the side of the plate... Only remember to lick your thumb and fingers clean, before you offer to touch the salt*' (*Instructions to Servants*, 1745). The pre-eminence of salt as a foodstuff is set at naught definitely while taking the precedence of guests becomes obsolete.

Can these bits and pieces of salt inheritance constitute a coherent whole once put together? Indeed, it may be judged thoroughly unsound by some academic experts. Within the reigning diversity, the comments about salt in history from the recent literature might deserve a special attention provided that they relate not only to salt but to its uses too. Many areas remain obscure. By chance, archaeology stimulates research and further criticism contributes to improve the evaluation of salt role in human bondage. Evaluation of excavation results and depositing of new documents have boosted contribution to recent symposia and colloquia (Weller 2002, Figuls and Weller 2005; Morere Molinero, 2007; Weller 2008). The participation of Eastern Europe (Monah 2007; Alexianu and Weller 2011) and Middle East emphasizes how unfinished remains the debate, and deserves special congratulations for investing in new fields of investigation and requesting for open and enlarged discussion aimed at further recycling of the specialist debates. Thanks to these new findings, we can expect the issue of opera updating and optimizing what we should know about salt in history. '*Il sale nel mondo greco (VI BC–III AD)*' is an excellent presentation concerning salt production and trade in the Mediterranean basin during the Antiquity (Carusi 2008). A more extensive survey is currently on the stocks. Once again, we may conclude with Pliny the Elder 'that the higher enjoyments of life could not exist without salt'.

References

Adshead, S.A.M. 1992. *Salt and Civilization.* London, Macmillan.

Agricola (alias Bauer Georg). 1556. *De re metallica.* Basilea, Frobenius.

Alexianu, M., Weller, O. and Curcă, R.-G. 2011. *Archaeology and Anthropology of Salt: A Diachronic Approach – Proceedings of the International Colloquium, 1-5 October 2008 (Iaşi, Romania)*, BAR International Series 2198. Oxford, BAR Publishing.

Azimi, V. 1987. *Un modèle administratif de l'ancien régime : les commis de la Ferme Générale et de la régie générale des aides.* Paris, Editions du CNRS.

Bérenger, J. 1975. *Finances et absolutisme autrichien dans la seconde moitié du XVIIe siècle.* Paris, Publications de la Sorbonne.

Bergier, J.-F. 1982. *Une histoire du sel*. Fribourg, Office du Livre.

Bintliff, J. (ed). 2004. *A Companion to Archaeology*. Oxford, Blackwell Publishing.

Bisaccia, C., De Santo, R.M., De Santo, L.S. *et al.* 1997. The symbolism of salt in paintings. *American Journal of Nephrology* 17(3-4), 318-339.

Bouvard, A. 1983. Heinrich Schickhardt, technicien des salines – Les techniques de fabrication du sel vers 1600 – Les salines de Salins et de Saulnot à la fin du XVIe siècle. *Bulletin de la Société d'Emulation de Montbéliard* 106, 55-115.

Bryant, E. 1997. *Climate, process and change*. Cambridge, Cambridge University Press.

Cabourdin, G. (ed.) 1981. *Le sel et son histoire. Actes du colloque interuniversitaire de Nancy, 1979.* Nancy, Publications Université Nancy 2.

Caillé, R. 1985. *Voyage à Tombouctou*. Paris, La Découverte.

Calvert, A.F. 1915. *Salt in Cheshire*. London, Spon Ltd.

Carusi, C. 2008. *Il sale nel mondo Greco (VI AC-III DC) – Luoghi di produzione, circulazione commerciale, regimi di sfruttamento nel contesto del Mediterraneo antico*. Bari, Edipuglia.

Connah, G. 1996. *Kibiro: the salt of Bunyoro, past and present*. London, British Institute of Eastern Africa.

D'Arbaumont, J. 1994. Le sel en Mauritanie et au Soudan. *Les Cahiers du sel* 7, 29-39.

D'Arienzo V. 1996. *L'Arrendamento del sale dei Quatro Fondaci – Struttura, organizzazione, consume (1649-1724).* Salerno, Elea Press.

De Brisay, K.W. and Evans, K.A. (eds.) 1975. *Salt – The study of an ancient industry. Report of a conference held at the University of Essex, 1974.* Colchester, Colchester Archaeological Group.

Du Puigaudeau, O. 2001. *Le sel du désert*. Paris, Phébus.

Denhez, F. 2005. *Les voyages du sel*. Paris, Kubik.

Denton, D. 1984. *The hunger for salt – An anthropological, physiological and medical analysis*. Berlin, Springer Verlag.

Di Vittorio, A. (ed.) 1981. *Sale e saline nell'adriatico (secc. XV-XX) – Atti del convegno internazionale di Bari*. Napoli, Giannini.

Dortes, M. 2005. *La caravane du sel*. Collection Mémoires des peuples. Paris, Vilo.

Dubois, C. 2003. *L'or blanc de Djibouti – salines et sauniers (XIXe-XXe siècles)*. Paris, Karthala.

Dubois, H. 1976. *Les foires de Chalon et le commerce dans la vallée de la Saône à la fin du Moyen Age*. Paris, Imprimerie Nationale, Publications de la Sorbonne.

Duché-Gavet, V. and Lapacherie, J.-G. (eds.) 2005. *Du sel. Actes de la journée d'études « Le sel dans la littérature française », 2003*. Biarritz, Atlantica.

Dupâquier, J. 1979. *La population rurale du bassin parisien à l'époque de Louis XIV*. Paris, Editions de l'EHESS.

Emons, H.-H., Walter, H.-H. 1988. *Alte Salinen in Mitteleuropa*. Leipzig, VEB Deutscher Verlag für Grundstoffindustrie.

Etienne, R. and Mayet, F. 2002. *Salaisons et sauces de poisson hispaniques*. Paris, De Boccard.

Figuls, A. and Weller, O. (eds.) 2005. *La Trobada internacional d'arqueologia envers l'explotació de la sal a la prehistòria i protohistòria – Colloque de Cardona, 2003.* Barcelona, IREC.

Flad, R., Zhu, J., Wang, C. *et al.* 2005. Archaeological and chemical evidence for early salt production in China. *Proceedings of the National Academy of Sciences of the United States of America* 102(35), 12618-12622.

Forbes, R.J. (ed.) 1968. *Het zout der aarde*. Hengelo, De Koninklijke Nederlandse Zoutindustrie.

Fürer, F.A. 1900. *Salzbergbau- und Salinenkunde*. Braunschweig, Friedrich Vieweg & Sohn.

Gouletquer, P. and Kleinmann, D. 1984. Les salines du Manga. *Techniques et culture* 3, 1-42.

Hocquet, J.-C. 1978. *Le sel et la fortune de Venise – Production et monopole*. Lille, Publications Université de Lille.

Hocquet, J.-C. 1979. *Le sel et la fortune de Venise – Voiliers et commerce en Méditerranée 1200-1650*. Lille, Publications Université de Lille.

Hocquet, J.-C. 1985. *Le sel et le pouvoir – De l'An mil à la Révolution française*. Paris, Albin Michel.

Hocquet, J.-C. (ed.) 1987. *Le roi, le marchand et le sel – Actes de la Table ronde sur l'impôt du sel en Europe XIIIe-XVIIIe siècle, Arc-et-Senans, 1986*. Lille, Presses Universitaires de Lille.

Hocquet, J.-C. 1989. *Le sel de la terre*. Paris, Du May.

Hocquet, J.-C. 1991. *Chioggia, capitale del sale nel Medioevo*. Sottomarina, Il Leggio Led.

Hocquet, J.-C. 1995. *La métrologie historique*. Paris, Presses Universitaires de France.

Hocquet, J.-C., Malpica, C.A. and Weller, O. 2001. *Hommes et paysages du sel.* Paris, Actes Sud.

Hocquet, J.-C. and Sarrazin, J.-L. (eds.) 2006. *Le sel de la Baie – Histoire, archéologie, ethnologie des sels atlantiques*. Rennes, Presses Universitaires de Rennes.

Hufton, O. 1980. *Europe : privilège and proteste 1730-1789*. London, Fontana Press.

Husson, J.P. 1991. La consommation de bois des salines lorraines à la fin du XVIIIe siècle. *Le Pays Lorrain* 1, 3-12.

Just, R. and Meisner, U. (eds.) 1996. *Das Leben in der Saline - Arbeiter und Unternehmer. Internationale Salzgeschichtetagung*, Halle: technisches Halloren- und Salinemuseum. Schriften und Quellen zur Kulturgeschichte des Salzes 3.

Kaufmann, D.W. (ed.) 1978. *Sodium chloride – The production and properties of salt and brine.* ACS Monograph 145. New York, Hafner Publ. Company.

Komarek, A. 1998. *Österreich mit eine Prise Salz – Ein Mineral macht Geschichte.* Wien, Kremayr & Scheriau.

Korovessis, N.A. and Lekkas, T.D. (eds.) 2000. Saltworks: preserving saline coastal ecosystems – Satellite Symposium Samos 1999. Athens, Global Nest.

Kostick, D. 1993. *The material flow of salt.* Washington D.C., U.S. Department of Interior, Bureau of Mines.

Lamschus, C. (ed.) 1989. *Salz – Arbeit – Technik – Produktion und Distribution in Mittelalter und Früher Neuzeit.* Lüneburg, Deutsches Salzmuseum.

Latham, J.E. 1982. *The religious symbolism of salt.* Théologie Historique 64. Paris, Beauchesne.

Le Foll, N. and Moral (del), J.-M. 1997. *Le sel.* Paris, Editions du Chêne.

Le Roux, P. and Ivanoff, J. (eds.) 1993. *Le sel de la vie en Asie du Sud-Est.* Patani, Prince of Songkla University („Grand Sud", 4).

Lev-Ran, A. and Porta, M. 2005. A Salt and hypertension: a phylogenic perspective. *Diabetes/Metabolism Research and Reviews* 21, 118-131.

Lovejoy, P.E. 1986. *Salt of the desert sun – A history of salt production and trade in the Central Sudan.* African Studies Series 46. Cambridge, Cambridge University Press.

Mac Gregor, G.A. and de Wardener, H.E. 1998. Salt, Diet and Health: Neptune's Poisoned Chalice: the Origins of High Blood Pressure - The uses and abuses of salt throughout human history. Cambridge, Cambridge University Press.

Mager, J., Just, R. and Meissner, U. 1993. *Salz- Gott erhalt's.Kulturgeschichte alter Salzstädte und Meisterwerke deutscher Salinentechnik.* Halle, Technische Halloren- und Salinemuseum.

Malpica, C.A. and Gonzalez Alcantud, J.A. (eds.) 1995. *La sal : del gusto alimentario al arrendamiento de salinas – III Congreso internacional del CIHS, Granada, 7-11 Septiembre, 1995.* Granada, Diputación Provincial de Granada.

Marion, M. 1989. *Dictionnaire des institutions de la France XVIIe-XVIIIe S.* Paris, Picard.

McKillop, H.I. 2002. *Salt: White gold of the ancient Maya.* Gainesville, University Press of Florida.

McKillop, H.I. 2004. *The ancient Maya: new perspectives.* Santa Barbara, California, ABC-CLIO.

Moinier, B. 1993. Salt in Europe: Historical aspects and economic outlook. *Seventh International Salt Symposium.* Elsevier 1, 29-45.

Moinier, B. 1997. *Sel et société – Une affaire de métier.* Paris, Nathan.

Mollat, M. (ed.) 1968. *Le rôle du sel dans l'histoire.* Paris, Presses Universitaires de France.

Monah, D., Dumitroaia, G., Weller O., Chapman, J. (eds.) 2007. *L'exploitation du sel à travers le temps – Colloque de Piatra Neamt, 2004.* Piatra-Neamţ, Editura Constantin Matasă.

Morère Molinero, N. (ed.) 2007. *Las salinas y la sal de interior en la historia : economia, medio ambiente y sociedad.* Madrid, Dykinson.

Multhauf R.P. 1978. *Neptune's Gift – A history of common salt.* Baltimore, The John Hopkins University Press.

Ohno, M. 1993. Technical progress of salt production in Japan. *Seventh International Symposium on Salt.* Elsevier 1, 13-19.

Palme, R. 1983. *Rechts-, Wirtschafts-und Sozialgeschichte der inneralpinen Salzwerke bis zu deren Monopolisierung*, Rechtshistorische Reihe 25. Frankfurt am Main, Peter Lang.

Palme, R. and Hocquet, J.-C. (eds.) 1991. *Das Salz in der Rechts- und Handelgeschichte* Salzgeschichtekongress. Hall in Tirol, Berenkamp.

Pétrequin, P., Weller, O., Gauthier, E., Dufraisse, A. 2001. Salt spring exploitations without pottery during Prehistory – From New Guinea to the French Jura, in S.

Beyrie, P. Pétrequin (eds.) *Ethnoarchaeology and its transfer. Ve Meeting of The European Association of Archaeologists, Bournemouth, September 1999*, 37-65, BAR International Series 983. Oxford, BAR Publishing.

Piasecki, P. 1987. *Das deutsche Salinenwesen 1550-1650 – Invention, Innovation, Diffusion.* Idstein, Schulz-Kirchner.

Pira, S. (ed.) 1997. *Storia del commercio del sale tra Mediterraneo e Atlantico. Atti del Convegno internazionale, Cagliari, 1996.* Caligari, AM&D.

Réault-Mille, S. 2003. *Les marais charentais – Géohistoire des paysages du sel.* Rennes, Presses Universitaires de Rennes.

Ricoeur, P. 1955. *Histoire et vérité.* Paris, Seuil.

Ridley, R.T. 1996. To be taken with a pinch of salt: the destruction of Carthage. *Classical Philology* 81(2), 140-146.

Ritter, H. 1981. *Caravanes du sel.* (Traduit de l'allemand par Arminjon Thérèse). Paris, Arthaud.

Rössler, M. and Souama-Ferero, G. (eds.) 2000. *La convention du patrimoine mondial et les paysages culturels en Afrique.* Working session at Tiwi, Kenya, 1999. UNESCO.

Schleiden, M.J. 1875. *Das Salz – Seine Geschichte, seine Symbolik und seine Bedeutung im Menschenleben.* Leipzig, Engelmann.

Swift, J. 1973. *Directions to servants and miscellaneous pieces – 1733-1742.* Oxford, Basil Blackwell.

Tora, Y. 1993. *Salt production techniques in ancient China – The Aobo Tu* (Translated and revised by H.-U. Vogel). Leiden, New York, Köln, E.J. Brill.

Tournier, M. 1980. *Gaspard, Melchior et Balthazar.* Paris, Gallimard.

Treml, M. (ed.) 1995. *Salz macht Geschichte - Katalog.* Augsburg, Haus der Bayerischen Geschichte.

Valli, E. and Summers, D. 1994. *Les voyageurs du sel* (Dolpo). Paris, La Martinière.

Volk, O. 1984. *Salzproduktion und Salzhandel mittelalterlicher Zisterzienserklöster.* Sigmaringen, Thorbecke.

Weller, O., Pétrequin, P., Pétrequin, A.-M. *et al.* 1996. Du sel pour les échanges sociaux – L'exploitation des sources salées en Nouvelle-Guinée (Irian Taya). *Journal de la Société des Océanistes* 102(1), 3-30.

Weller, O. (ed.) 2002. *Archéologie du sel. Techniques et sociétés dans la Pré- et Protohistoire européenne. Actes du colloque international, XIVe congrès UISPP, Liège (Belgique), sept 2001 et de la table ronde du Comité des Salines de France, Paris, mai 1998, Internationale Archäologie, ASTK, 3.* Rahden, Verlag Marie Leidorf GmbH.

Weller, O. and Dumitroaia, G. 2005. The earliest salt production in the World: an early Neolithic exploitation in Poiana Slatinei, Lunca, Romania. *Antiquity* 79, 306. www.antiquity.ac.uk/ProjGall/weller/index.html

Weller, O., Dufraisse, A. and Pétrequin, P. (eds.) 2008. *Sel, eau et forêt – D'hier à aujourd'hui.* Besançon, Presses Universitaires de Franche Comté.

Whatley, C.A. 1987. *The Scottish salt industry 1570-1850 – An economic and social history.* Aberdeen, Aberdeen University Press.

*** *Compilation de l'ordonnance de Louis XIV donnée au mois de mai 1680 sur le fait des gabelles.* Besongne, 1727.

Salts in the Passage to the After Life in Ancient and Recent Times

Alexandra Comşa
Institutul de Arheologie „Vasile Pârvan" Bucureşti

Abstract
The paper refers to the salt and its use in the treatment of the people, either as purifying rituals, or for preservation of the bodies. There are also described some embalming methods, as well as rituals used for facilitating the travel of the soul between the world of the living and that of the dead.

Keywords
salt, treatments of the deceased, purifying rituals, past and present

The desire and sometimes the necessity to preserve a corpse, either of a human or animal being, has been an important challenge, since very old times. During thousands of years, the specific techniques were improved gradually, thus becoming a real art, mostly appreciated in certain communities. Such concerns continue, even today.

In order to go deeper into the use of salts for different practices, connected with the afterlife and especially with the protection or preservation of the deceased's soul, we will briefly present here some of the most interesting utilities of salt and generally of salts. If we look for its Biblical meanings, we would not be surprised to find a reference in the New Testament, Matthew 15, 3, when Jesus addressing his disciples told them so: "You are the salt of the earth!"

The most frequent interpretation of this sentence is that he wanted to emphasize how valuable they were. On the other hand, some opinions connect the salt with the physical body, so that the conclusion inferred in this case was that they should pay much more attention to their spirit, instead of their physical body.

Meanings of salt

We should not forget that especially common salt had not only an important physiological role, but also a religious one. For many centuries, it was used in activities connected with the protecting, blessing and purifying rituals. Of course, one of its most persistent significances was as a ward against bad energies or evil spirits.

Very old traditions and beliefs are still maintained and carefully preserved, at least by people of Europe, of various religions. Such cases we find in the Orthodox, but also in the Catholic world, the salt still keeping its ancient meaning as the owner of the protective powers.

The chemical composition of salt

In fact, salt is a mineral composed of two main chemical elements: sodium and chloride, in a proportion of 60%, respectively 40%.

When unrefined, it also contains 82 other minerals, enclosed in its crystalline structure. This assemblage is a vital compound for life on Earth (*Himalayan Crystal Salt* 2014). It cannot be produced by the body, which, on the other hand, cannot function without it. This is a vital compound for life on Earth. Of course, we will not insist here upon this fascinating issue, restraining our focus just on the use of salt and generally the salts after the death of a human body.

Interestingly, it is to be pointed out that, after cremation, the human body transforms itself in nothing else but salts. Therefore, we could point out here the transgression of salt from life to death, at least in what concerns the human body. We could say that, on account of its function within the body, it is a compound of the essence of life, as well as one of death (*Why Don't Bones Burn When a Body Is Cremated?* 2014).

Embalming the body

Embalming is a process that is meant to preserve the corpse for a longer time, with different purposes: to expose it for the community or certain groups of people, to adjust to certain rituals, to help the soul of the deceased to find and recognize his own body.

The best known people who were preservers of the human bodies, should be considered the Egyptians. They used natural chemicals, but also took advantage of the naturally dry environment that, working alongside the substances, accelerated the dehydration of the tissues, thus assuring the preservation of the body for the afterlife.

The embalming methods have evolved in time by trial and error and were faster improved much later, during the Renaissance period as a result of the most intense concerns about research and innovation in medicine.

The re-creation of features, or even complete bodies after being damaged by accident or disease, is called today restorative art, or demisurgery, being a subdivision of the embalming procedure.

In Europe, embalming was brought from Africa and Asia and was used during certain periods, especially later, during the Crusades, when noblemen wished to be embalmed in order to be buried back home, or at least as close as possible to that spot. Descriptions of such embalming methods in Europe, used for about 1200 ears beginning around 500 A.D., have been preserved by the contemporary physicians in their written works.

Other cultures who also practiced embalming processes were ancient people like the Ethiopian tribes, aboriginals of the Canary Islands, Sumerians, Babylonians, Persians, Greeks, Romans, and others.

A variant of embalming is the mummification, practiced by many cultures, but being better known from the Egyptians. The word "mummy" seems to be of Persian origin, meaning "bituminized thing". This term was applied to the dead after mummification, as they had a black appearance and were considered to have been immersed into bitumen.

The Egyptians believed that the preserving the body as a mummy empowered the soul of the deceased owner, which would anyway return to the dead body in a certain moment. In such a case, it was important that the soul could have recognized its own body and this was facilitated by the practice of mummification, in order to preserve the corpses as similar as possible to their life-like form.

But, as we have more clear data, we will render here the steps used for embalming noble Egyptians, as described by Herodotus. The steps are as it follows:

1. The corpse was carried to the mortuary workshop, stripped and placed on a wooden table;

2. The brain was removed through the nostrils using a hooked, or spiralled needle, and the skull was filled with resins;

3. Evisceration, which means the removal of the internal organs, by an incision in the abdomen, made using an obsidian blade. Afterwards, the cavities were washed with palm-wine and spices. The cavities were temporarily filled with dry natron, a mixture of salts which could be found on the dry bottom of some lakes. There were also put inside linen bags containing natron, resin, straw and fibres. All the important organs, excepting the heart and kidneys, would have undergone a treatment with natron, oils, fats, resins and perfumes. Afterwards, they would have been wrapped and placed into the so-called canopic jars. The liver, stomach, intestines and lungs were preserved in a saline fluid. The canopic jars were covered with lids shaped as the four sons of Horus: *Imset* for the liver, *Duamntef* for the stomach, *Quebehsenuf* for the intestines and *Ha'py* for the lungs. All canopy jars would have been placed in the tomb, together with the mummy. The corpse was immersed into a solution that was also based on salt, a procedure that would have lasted between 20 and 70 days. A longer exposure would have been more beneficial.

4. The dehydration took place on a specially inclined table, upon which the corpse was put and covered with dry natron. The liquid that came out from the body was gathered in special jars and buried in the grave.

5. As dehydration was one of the most important goals, some other preparing steps had to be followed. Therefore, the corpse was taken out from the saline solution, was washed, its limbs were properly arranged and afterwards the body was exposed to sunlight, in order to dehydrate as much as possible;

6. The last step consisted in using linen strips with appropriate hieroglyphic inscriptions for wrapping the body, by the help of a natural resin. Sometimes, the resin was poured upon the entire body. This substance was considered to be bitumen (*ar*), that the Persians used to call *moumio.* According to the ancient practice, a heart amulet was placed on the chest of the deceased (Knight 2009).

Some authors speculate that the linen used for wrapping the corpse were the old sheets that the deceased has used before his death and therefore, they should be used further for "sleeping" until the resurrection. The linen ere applied with gum Arabic.

7. Afterwards, the mummy was placed in a sarcophagus and given back to its family.

8. Together with the deceased, there were also buried mummified animals. There were four kinds of such companions: the sacred ones (i.e. the Apis bull), the pets and the victual animal mummies, that serves food for the dead (i.e. ducks, beef, fish). The fourth category of animal mummies was the votive ones, sold to pilgrims for being used in everlasting prayers (ibises for the god Toth and cats for the god Bast) (Knight 2009).

9. The embalming of the poor Egyptians was performed in a more simple way. It was used the cedar oil injected into the corpse and subsequently the immersion of the body into the salt solution. Another procedure comprised the purging of the intestines and again the use of the salt solution.

10. The Egyptians believed that a human being was composed of six important elements: the physical body, the shadow, the name, *ka* (the spirit, or sould), *ba* (the personality) and *akh* (the immortality). A body with all its components was necessary in order to pass towards the afterlife. This is why the mummification was the optimal procedure for assuring the revival from the death (Bakry 1965, 10-16).

11. Other embalming variants, used by the Guanches of the Canary Islands, as well as by Ethiopian tribes, were somehow similar to the methods practiced by the Egyptians. After removing the internal organs, the abdomen was filled with salt, but also with vegetable powders.

In ancient times, the deceased from the Hawaii Islands used to be buried in caves. In some cases, the corpses were placed in extended position, while in others they were flexed with their knees tied and brought close to the head, in order to get a rounded shape. Afterwards, it was wrapped in a cloth of tapa (made from the bark of the paper mulberry tree) and thus was ready for the burial. In some cases, the internal organs were taken out of the body and the remaining cavity was filled with salt before the dead being interred.

The fishermen were thrown into the sea, wrapped in a red cloth, so that they could be easily be eaten by the sharks. By doing that, it was considered that the spirit of the recently deceased would pass into the shark and therefore, the fisherman would be protected from other attacks of those fish (Redmond 2014).

In Tibet, there is still in use the embalming that consists of putting the corpse into an appropriate box and covering it with salt for about three months, thus assuring its mummification (Ezugworie *et al.* 2009).

Old customs have also survived to this day. We could give some examples in this sense.

Thus, in Ireland, during the vigil, a wooden plate containing earth and salt is usually put on the chest of the dead. The earth signifies that the body is to be interred in it, while the salt is considered to mean that the soul doesn't rot (*Weird Death Customs and Burial Traditions* 2011).

In the Hawaii Islands, at the funeral is kept the *pi kai* ritual, meaning "the salt water sprinkling" that is applied those who carry the dead and those in the convoy, all being just men. This purifying is also done in the house or yard, in order to expel the evil spirits. It is used salt water and *olena* root or herb of *mauna kiaki*, all these assuring the return of the deceased's soul into the house as, otherwise, it could come back to do bad things.

Salt water is also used for washing the dead, having a purifying but also conservation role, being also "wai kala", namely "the water of forgiveness" (Green and Warren Beckwith 1926 180).

The Ladakhi are a people from Tibet. Generally, their funerary customs are almost identical to those of the other populations of that region, so that their corpses are usually burned, with some certain exceptions: the people who died by epidemics and the children who died under the age of 8. In that later case, the dead child is put into a box, strewn with salt, and buried on a hillock, or under a stone. Sometimes is even thrown into a river (Brauer 1982, 319).

As we can see, in the past, but also today, humanity has had death rituals connected with the use of salt, with its specific, already mentioned meanings. This paper is just and introduction into these practices, that are very interesting and useful for interpreting the funeral customs.

References

Bakry, H.S.K. 1965. *A Brief Study of Mummies and Mummification*. Cairo, Takkadum Press.

Brauer, M. 1982. *Death Customs in Ladakh*. Kailash, 9(4), 319-332.

Ezugworie, J., Anibeze, C. and Ozoemena, F. 2009. Trends in the Development of Embalming Methods. *Internet Journal of Alternative Medicine* 7(2), Online: ispub.com/IJAM/7/2/3694 [accessed: 11.06.2014].

Green, L.C. and Warren Beckwith, M. 1926. *Hawaiian Customs and Beliefs Related to Sickness and Death. American Anthropologist* 28, 1, 176-208.

Knight, S.C. 2009. *Egyptian Funerary Practices*. Online: www.andrew.cmu.edu/course/98 030/mummies.pdf [accessed: 02.05.2014].

Redmond, J. 2014. *Hawaii Burial Methods*. Online: dying.lovetoknow.com/Hawaii_Burial_Methods [accessed: 10.02.2014].

Online sources

Weird Death Customs and Burial Traditions http://www.nerdygaga.com/4155/weird-funeral-and-burial-practices/, accessed on 14.08.2014
Himalayan Crystal Salt

http://www.himalayancrystalsalt.com/, accessed on 06.11.2014

Why Don't Bones Burn When a Body Is Cremated? http://www.ehow.com/facts_7437682_don_t-bones-burn-body-cremated_.html, accessed on 09.04.2014

Salt in the *Adagia* of Erasmus of Rotterdam

Mihaela Paraschiv
Universitatea „Alexandru Ioan Cuza" din Iaşi

Abstract
Desiderius Erasmus, called "the crowning glory of the Christian humanists" (Scott Latourette 1953, 661), wrote both on ecclesiastic subjects and on topics of general human interest. The *Adagia* – a collection of proverbs – are part of his great humanist works, with numerous reprints and successive annotations (in the *editio princeps* of 1500 there were 820 proverbs, while the last edition while the author was still alive– the 30th, of 1536 – counted 4,151 proverbs). The author had an excessive admiration for the Greco-Latin Antiquity and for the pagan authors, whose spirituality—succinctly expressed in maxims—he wanted to share with his contemporaries; thus, in 1599, the Council of Trent placed the *Adagia* on the *Index librorum prohibitorum*. Hence, it was prohibited in Catholic countries until 1900. Though registering fewer occurrences than in another contemporary book (the monograph of Bernardino Gomez Miedes, *Commentarii de sale*), salt is also present in this *florilegium Erasmicum*. The salt-related proverbs and their commentaries are the subject of the present paper; our intention is also to follow the connection between text and paratext, as configured by the author.

Keywords
salt, proverb, Erasmus of Rotterdam, Greco-Roman Antiquity, textual analysis

The Dutch humanist, known by his Latinized name Desiderius Erasmus Roterodamus[1], was convinced from the start of the possibility of, and a need for, cohabitation between classical and Christian culture, as had been demonstrated almost ten centuries earlier by Augustine and Jerome. Erasmus approached this issue in the dialogue *Antibarbarorum liber*, but he outlined the true dimension of ancient Greco-Roman spirituality in the work considered a genuine "Renaissance sun", namely his collection of famous proverbs and phrases, currently known as the *Adagia*[2]. The *editio princeps* of Erasmus' proverbs was published in 1500 in Paris, at the Jean Philippe Printing House and it included 818 Latin proverbs. They had been edited in collaboration with Publio Fausto Andrelini; the title was *Veterum maximeque insignium paroemiarum, id est adagiorum collectanea*, and Erasmus presented it as the first collection of the kind (*Porro apud Latinos nemo quidem ante nos, quod sciam, huiusmodi negocium tentavit*). Actually, two years before, the collection of the Italian Polydoro Virgilio had been published under the title *Proverbiorum libellus*, reprinted in 1500 as *Adagiorum liber*. In a subsequent edition of his *Adages* (that of 1533), Erasmus admits to have read the Italian's book, but he arrogantly claims the superiority of his own collection, both in quality and quantity. In fact, Erasmus' collection was the first attempt to open a broad perspective upon the ancient world, through proverbs and the carefully selected wisdom-based samples, commented by the humanist scholar. This collection soon became a best-seller of the time, a genuine pagan Bible, and it had numerous reprints (30) and successive annotations (10). This is why the last edition published in his lifetime, that of 1536, contained 4,151 *adagia*.

1508 witnessed an important moment in the publication of the proverbs, when Erasmus, displeased by the fact that the Greek characters had not been printed properly in the 1506 edition, chose the Venetian printing house of Aldus Manutius[3]. This led to the publication of the third edition, *editio Aldina,* with 3,260 proverbs, entitled *Adagiorum chiliades ac centuriae fere totidem*. The increase in the number of proverbs from the first edition to the last produced in his lifetime, attests to his enduring effort to permanently consult classical texts. He of course focused on Latin texts, since Latin was naturally familiar to all humanists; he could access the Greek texts later, starting in 1500, when he began to learn the Greek language. In the *editio aldina*, in a letter addressed to the contemporary reader, Erasmus explains the difficult path of collecting and editing the proverbs, given the lack of Greek writings. This lack makes "wishing to write about poems nothing else but, as Plautus states, wishing to fly without feathers" (*sine qua de proverbiis velle conscribere nihil est aliud quam sine pennis, ut ait Plautus, velle volare*). He considers the first edition rudimentary (*primus ille partus Adagiorum rudior fuit*) and he eulogizes the opportunity provided by Aldus Manutius of accessing the Greek books published by him and to elaborate an "edition of the Proverbs with fewer errors and richer in general and, if I am not mistaken, better" (*et emendatius et locupletius atque omnino, ni fallor, melius Adagiorum opus*).

By stating to his readers in the *Prolegomena* to the book that, as author of the work he has "the duty to inform

[1] In the opinion of Huizinga 2002, 9), the name Geer Geertsz – traditionally ascribed to Erasmus as his original name – is uncertain. The only sure thing is that he received the name Erasmus when he was baptized, and he adopted the second one – Desiderius –in 1506, when the second edition of the *Adagia* was published.
[2] The Latin word *adagium* is a loan from the Greek paroimía, meaning "proverb, maxim".

[3] A literary man himself and a great admirer of Antiquity, Aldus Manutius had founded an *Academia Aldina* in Venice, reuniting notable Hellenists and scholars of the time, refugees from Constantinople, after the Turkish conquest.

them" (*munus docendi*), Erasmus provides a definition for the proverbs and he differentiates them from other similar phrases. The author also mentions how useful proverbs can be how they should be employed. Unlike other proverbial phrases, Erasmus believes, the *adagium* has three characteristics: it is concise, sententious and it has a figurative meaning (*Adest enim simul et brevitas et sententia et figura*). He justifies the necessity of ancient proverbs, just like Aristotle, and he considers them relics of an ancient philosophy, ascribing to them the three important values: 1. **rhetorical-persuasive** (*ad persuadendum conducere proverbia*); 2. **aesthetic** (*ad ornatum conducere paroemiam*); 3. **cognitive** (*ad intelligendos auctores conducere paroemiam*). In the same Aristotelian key, Erasmus draws attention to the way in which proverbs should be used: "we should treat them not as food but as condiments, not for sufficiency but for delight" (*ut illis utamur non tamquam cibis, sed velut condimentis, id est non ad satietatem sed ad gratiam*). Another emblematic aspect for humanist culture is the triple nature of the Erasmian work. The first element was knowledge of the classical authors in their historical context. The second was to underline the idea-related continuity between Antiquity and the Renaissance. The third was to state the actuality of the ancient texts. As he asserted in the introductory letter addressed to his patron, William Blount, 4th Baron Mountjoy, Erasmus offered the collection of proverbs to his contemporaries with a twofold purpose: to enrich and embellish their expression in Latin (*ad locuplendam venustandamque orationem conducere*). He wanted to make known to everyone – not just to clergymen – the simple and elegant Latin of the *Adages*, with the commendable intention of "democratizing" Latin; this, however, would remain but a beautiful utopia. Each referential Latin proverb is followed by its original or by its Greek equivalent, as well as by its provenience, by the authors who used it and by its interpretations. Erasmus adds his own interpretation, which often includes copious digressions. This reveals successive cultural layers, which justifies the assertion of Claudie Balavoine that "an entire view of culture crystallizes around an proverb" (Balavoine 1984). The most frequently cited classical authors are the Latin Cicero, Horace, Virgil, Pliny the Elder, Aulus Gellius, Macrobius and the Greek Homer, Pythagoras, Plato, Sophocles, Aristophanes, Aristotle, Plutarch, Lucian and Diogenes Laertius. Erasmus displays an intellectual honesty, uncommon among authors of the time, for he does not allow the reader to believe he had read the works cited from cover to cover, but he mentions, at times, that he had lifted quotations from collections of ancient Greek and Latin proverbs. There are also quotations from encyclopaedists, ancient and mediaeval paroemiographers, Byzantine scholiasts, who provided important references on Antiquity.

Nevertheless, the Church could not forgive Erasmus for expressing a preference for pagan literature so ostentatiously. Thus, after the Council of Trent (*Concilium Tridentinum*) in 1559, Pope Paul IV promulgated the *Index librorum prohibitorum*, and Erasmus' books were immediately indexed, including the *Adagia*, which was part of the *Index* until 1900 in Catholic countries. Just as with other works by Erasmus, the *Adages* continued to be published and read in both protestant and Catholic Europe, and Jesuits were allowed to research and read them in colleges given their acknowledged pedagogical value, but only with the express permission of a bishop.

Recently, a team of 58 Latinists and Hellenists edited the paroemiological collection of Erasmus disseminated it to scholars of European culture. The leader of the team was Jean-Christophe Saladin, and they published an impressive edition, in five volumes, of 4,151 proverbs, which the editor labelled "the royal access road to the ancient literature"[4]. They were grouped in chiliads (*chilias/-adis* = group of 1,000); the first four volumes comprise a chiliad each, while the fifth contains the remaining 151 proverbs. The collection offers a "state of the art" of Erasmus' book in the last edition of 1536, and each proverb was presented just as the author's contemporaries read it.

Considering the symbolic connotations of salt in Antiquity, of which Erasmus was well aware, his anthology naturally included the principal proverbs that refer to it, taken from ancient Greek and Latin authors. There are not too many of them in the collection: 20 such proverbs were identified. In what follows, we inventory and discuss them, focusing on the connection between text and paratext; in other words, between the ancient proverbial phrase and Erasmus' commentary upon it, meant to illustrate his profound erudition and a mind-set imbued with humanist values. We have used as the principal bibliographical reference the French edition of Saladin, which will explain why we have depended on the order in the five volumes of this edition[5]:

Vol. I (7 proverbs): 2. *Salem apponito*; 12. *Qui circa salem et fabam*; 510. *Salem et mensam ne praetereas*; 537. *Salsuguinosa vicinia*; 679. *Sale emptum mancipium*; 680. *Salis onus unde venerat, illuc abiit*; 681. *Salem vehens dormis*;

Vol. II (2 proverbs): 1014. *Nemini fidas, nisi cum quo prius modium salis absumpseris*; 1251. *Salsitudo non inest illi*;

Vol. III (4 proverbs): 2326. *Sale perunctus hic adjuvabitur*; 2420. *Salem et fabam*; 2525. *Oleum et salem oportet emere*; 2633. *Salem lingere*;

Vol. IV (5 proverbs): 3327. *Lingere salem*; 3487. *Salem et caseum edere*; 3763. *Piscis eget sale*; 3868. *Sale nihil utilius*; 3880. *Sal et mensa*;

---

[4] Érasme de Rotterdame, *Les Adages*, complete bilingual edition in 5 volumes, edited by Jean-Christophe Saladin, collection Les Miroirs des Humanistes. Paris, Les Belles Lettres, 2011.

[5] For the translation into English, we have used *Collective Works of Erasmus, Adages*, published at Toronto University Press.

Vol. V (2 proverbs): 4008. *Salillum animae*; 4016. *Sale et aceto.*

In the first volume, the first three proverbs involving salt are part of the 36 quotations that Erasmus labelled *Pythagorae symbola*, taken from Diogenes Laertius and compared to oracles (*oraculorum instar*), given the deepness of thought and the subtlety of the Greek philosopher's expression.

-The first, ***Salem apponito*** (Put salt on the table), is the Latin version of the Pythagorean phrase Σὸν ἅλα παρατίθεσθαι, drawing attention on the need to always use the spirit of justice and right (*Admonet justitiam et aequitatem omni adhibendam negotio*), "whatever it has penetrated, salt preserves" (*sal enim quicquid occupaverit servat*) "and it consists of the purest of things [like the two virtues], water and sea" (*et ex rebus purissimis constat, aqua et mari*). In the commentary on the third proverb, Erasmus explicitly ascribes to Pythagoras the paternity of this proverb and its meaning: *Idem [Pythagoras] censuit salem potissimum in mensa apponendum, quod aequitatis ac justitiae nos admoneat, ut quod et servet tueaturque quicquid occupat et ex liquidissimis rebus, aqua marique, fiat.*

-The second proverb, ***Qui circa salem et fabam*** (On salt and beans), the Latin equivalent of the Greek phrase Οἱ περὶ ἅλα καὶ κύαμον, is indicated by Erasmus as an allusion to those who pretend to know what they do not know (*In eos dicebatur, qui se simularent scire quod nescirent*). He refers here to the fact that the interpreters of divine will, the so-called keepers of profound secrets used salt and beans in their oracular ceremonies. In a copious commentary on this proverb, Erasmus notes, with real philological accuracy, another version of this Greek phrase, mentioned by Plutarch. In the other version, κύαμον is replaced with κύμινον (lat. *cuminum* "cumin"); this way, the second version would imply – in the opinion of the grammarian Apollophanes – the "supreme familiarity" (*summa familiaritas*) of those who share a frugal diner. Erasmus continues with a eulogy for the virtues of salt, based on the appreciations of people in Antiquity, starting with Homer, who labelled it "divine", an epithet taken over by Plato. Erasmus also mentions the use of salt in Christian ceremonies, especially Baptism.

-The third proverb (Pythagorean), ***Salem et mensam ne praetereas***, is the Latin translation of the Greek exhortation Ἅλα καὶ τράπεζαν μὴ παραβαίνειν. In Erasmus' interpretation, it would recommend not to neglect the relationship between friends and not to forget the laws of friendship (*Ne negligas amiculorum consuetudinem aut ne violes amicitiae iura*), since salt and food were symbolic of feasts (*convivia*) shared by friends. The commentator consolidates the meaning of this proverb by citing fragments of Theocritus, Euripides, of the Old and New Testament (where the communion of the diner, symbolized by salt and the meal, is the proof of sincere friendship).

-The phrase **Salsuginosa vicinia** (A brackish neighbourhood) is the translation of the Greek phrase Ἁλμυρὸν γειτόνημα and it literally designated barren farmlands near the sea, unsuited to cultivation because of their high salt content. Through a semantic transfer, Erasmus notes, it was included in the paroemiological sphere. The author says that, according to Aristides and Plato, it connotes a piece of business that is very toilsome but unproductive (*negotium multi quidem laboris, at parum frugiferum*) or to the inhospitable people who live near the sea, "as though they acquired some unpitying quality from that most pitiless element" (*tamquam immanitatis nonnihil ex immanissimo contrahant elemento*). We believe that the allusion to the "most pitiless element" (*immanissimum elementum*) refers to the sea, not to salt.

-Sale emptum mancipium (A slave exchanged for salt) is the Latin equivalent of the Greek phrase Ἁλώνητον ἀνδράποδον, referring, in Antiquity, as Erasmus finds in Zenodotus, "to any barbarian and worthless creature" (*In barbarum quempiam ac vilem homuncionem olim dicebatur*). In fact, it refers to the merchants of those days who carried salt into lands far from the sea, and who used to return with slaves from the barbarians, and that it was a practice specific to the Thracians to exchange slaves for salt (*Et Thraciae genti peculiaris erat mos mancipia sale commutare*).

-Salis onus unde venerat, illuc abiit (Salt to water whence it came), a translation of the Greek expression Ἁλῶν δὲ φόρτος ἔνθεν ἦλθεν ἔνθ'ἔβη, became the proverbial phrase for one who cannot look after his gains. This was based on a true story: a certain merchant had chartered a ship and loaded it with salt and, because the crew fell asleep, the sea got in, dissolved the salt and sank the ship. The result was that, what had originated in seawater returned to seawater again.

-Salem vehens dormis (You sleep on a cargo of salt) is the translation of another version of the previous proverb, Ἅλας ἄγων καθεύδεις, meant to stigmatize a man who behaves in an idle and careless fashion at a moment of danger (*Quadrabit in eum qui in re periculosa socorditer atque indiligenter agit*).
In the second volume, two proverbs on salt make Erasmus include other erudite incursions into ancient Greek and Latin literature:

-Nemini fidas, nisi cum quo prius modium salis absumpseris (Trust no man, unless you have eaten a peck of salt with him first). According to Erasmus, this proverb was in common circulation among his fellow citizens (*Vulgo apud nostrates circumfertur adagium*), because people could easily understand it in relation to knowledge of the human character (*Significat autem hominis ingenium non posse perspici nisi diuturno convictu longoque commercio*), an eagerly-debated moral issue among humanists. Erasmus mentions that this proverb is clearly derived from Antiquity (*Apparet proverbium hoc ab antiquitate profectum esse*) and ascribes it to Aristotle. Erasmus also reproduces two quotations from the *Ethics* of the Greek philosopher

(*Ethica Nicomachaea*, VIII, 3, 8; *Ethica Eudemica*, VII,2), where the phrase ὁ μέδιμνος τῶν ἁλῶν is found (a peck of salt = approximately 50 litres). In Latin, it reads *modius salis* (approx. 8.75 litres), through a highly different quantification. Erasmus also notes another Latin version of the proverb, cited by Cicero in *De amicitia* (XIX, 67): *Verum illud est, quod vulgo dicitur, multos modios salis simul edendos esse, ut amicitiae munus expletum sit* (But this is expressed in the common saying, that one should eat many a peck of salt with a man in order to fulfil the duty of a friend). This version referred to keeping old friendships, tried by time. We note that there was also a Latin version of the proverb cited by Aristotle, ascribed by some editors to the mimographer Publilius Syrus (1st century BC), the author of a collection of aphorisms (*sententiae*), highly appreciated in Roman education: *Salis absumendus modius, priusquam habeas fidem* (You have to eat a bushel of salt with a man before trusting [him]). However, Erasmus does not mention the version of Syrus in his commentary, though he had edited his aphorisms using a *codex Cantabrigiensis* (the *editio princeps* seems to have been published in 1514, though some later editors ascribe this edition to the year 1516). The proverb ascribed to Syrus is not, however, mentioned in the copious collection edited by Erasmus, where besides the Publilian mimes, there were also the precepts of the moralist Cato the Elder, entitled *Disticha Catonis,* which also included the maxims of the seven wise men of Antiquity (*Septem sapientum illustres sententiae*). One reason might be that it was missing from the rather limited codex he used; another might be that Erasmus was the first to notice that proverbs from other sources circulated as written by Syrus. In editions of Syrus' aphorisms that we consulted, the proverb is noted as his or else it is mentioned at the end, in the category of those falsely ascribed to Publilius Syrus (*Sententiae falso inter Publinianas receptae*)[6]. Similar changes to the original forms of Greek proverbs occurred in the Latin and Romnce worlds because of their having been adopted from the popular reserve of proverbs. There are thus several versions in circulation. **French**: *Pour connaître un homme, il faut avoir mangé un boisseau de sel avec lui.* **Italian**: *Prima di conoscere uno, bisogna consumare un maggio di sale. / Prima di sceglier l'amico bisogna averci mangiato il sale sett'ani.* **Spanish**: *En tu amigo confiaras, cuando hayas comido con el medio fanega de sal.* **Portuguese**: *Não te deves fiar señao daquele com quem j'a comeste un moio de sal.* **Romanian**: *Trebuie să mănânci un car de sare cu cineva ca să-l poţi cunoaşte. / Până nu mănânci cu omul o maje de sare nu-l poţi cunoaşte* (Gheorghe 1986).

-Salsitudo non inest illi (There is no saltiness in him). Erasmus also presents the Greek version of this proverb, without indicating its Greek source. He mentions what the proverb refers to (*In infacetos et fatuos ac stupidos* – It refers to witless, fatuous or stupid people). He motivates the figurative use of *salsitudo* (the concrete sense "brine, salty taste") by referring to Pliny the Elder (*Naturalis historia*, XXXI, 88) and to Quintilian (*Institutio oratoria*, VI, 3) regarding the metaphorical use of *sal* and its derivations. He then reproduces Catullus' well-known epigram on Quintia (*Carmina*, 86: *Quintia formosa est multis; mihi candida, longa,/ Recta est: haec et ego singula confiteor, /Totum illud formosa nego, nam nulla venustas,/ Nulla in tam magno corpore* ***mica salis***). In this sense, he criticizes the interpretation of a certain Fabius (a translator of Catullus at the time) and he offers his own interpretation of *mica salis*. In Erasmus' opinion, Catullus does not mean that there is nothing in Quintia's body that one could laugh at (*non hoc dicit, nihil in corpore eius esse ridiculum*), but that "anything laughable is not 'salty'". Erasmus underlines this because one of the commonly used figurative senses of *sal* and *salsum* was a synonym of *ridiculum* in that time.

When commenting on salt-related proverbs in the third book, Erasmus also appears as a critic of his own edition:

-Sale perunctus hic adjuvabitur (A rub with salt will do this man good). Here, too, he indicates the Greek original of the proverb, as it appears in Aristophanes' comedy *Clouds* (Ἁλσὶν διασμηθεὶς ὄναιτ'ἂν οὑτοσί), as well as the meaning of the proverb (*In eos dicebatur qui vel aetate vel vino delirarent* – It was said of those who were out of their senses because of age or because of wine). Erasmus tries to determine which sphere of current life could have generated the semantic transfer implied by the proverb and he indicates three possibilities: two of them are related to craftsmanship: the leather bottles became softer if washed with salt and earthenware jars were impregnated with salt to contain better the liquid put in them. The third possibility belongs to folk medicine: there was a custom to smear people who were ill through drunkenness with salt and oil.

-Salem et fabam. Erasmus explains why he repeats an abbreviated and much more frequent version of a proverb he had noted in the first chiliad – in some old copies of Greek writers (**Qui circa salem et fabam**) because he was not sure whether or not this was different from the other one.

-Oleum et salem oportet emere (Buy both oil and salt). Erasmus believes it to be the Latin equivalent of the proverb mentioned by Aristotle (*Rhetoric*, II, 23, 1399a), related to the use of contraries in demonstrative enthymemes: τὸ ἔλαιον πρίασθαι καὶ τοὺς ἅλας. In Erasmus' commentaries on this proverb, an ancient medical procedure is used to explain its rhetorically figurative meaning: *opus esse nonnunquam remediis contrariis, nunc acrimonia, nunc lenitudine. Sal enim mordet et siccat, oleum lenit et humectat. Sic medici oleum addunt Sali ne plus erodat et qui sapiunt, admonitionis acrimoniam bonis verbis leniunt*

[6] In the first category, we mention the editions of Zell 1829, where the proverb is listed among Syrus's aphorisms, at no. 810; the editions of Chenu 1835, Baudement 1903, Diaconescu 2003, where the proverb is inventoried as no. 869. In the same categories we include the editions of Woelfflin, 1869; Mureşanu 2002 (after the text of Otto Friedrich, Berlin, 1880).

(sometimes opposite remedies are necessary, now sharp, now mild. For salt is biting and drying; oil soothes and moistens. Thus physicians add oil to salt, lest it be too corrosive, and wise people soften the sharpness of rebuke with kind words). Erasmus indulges in some textual criticism here, stating that he is aware that in the Aldine edition of the *Adages* people read τὸ ἕλκος (wound) instead of τὸ ἔλαιον, but he considers it improper to the logic of the text (*sed quis volet vulnus emere*? – But who normally buys a wound?). Nevertheless, several modern editions of the Aristotelian text[7] present a third variant reading of the word, meaning τὸ ἕλος (swamp). In one of them, the editor explains in a note[8] that "to buy the swamp with the salt" means to take the bad with the good: the good because salt is bought, the bad because the salt is not pure, being mixed with the impurities of the swamp. In the editor's opinion, in this version, the Greek proverb would correspond to the Italian proverb *comprare il miele colle mosche* (buying the honey with the bees).

-Salem lingere (To lick salt) is the translation of the proverbial Greek phrase Ἅλα λείχειν, referring to those who live a very meagre life (*dicuntur qui perquam tenuiter victitant*). Taken up again in Rome by Persius and Plautus, the Greek phrase is ascribed to the cynic Diogenes as a reply to the invitation of a rich citizen named Craterus to enjoy his hospitality. Diogenes would have replied to this Macedonian that he preferred to lick salt in Athens than to enjoy sumptuous meals in his house[9].

In the fourth volume, two of the previous proverbs are repeated in other expressive versions, with several additional facts, which does not neutralize Erasmus' earlier exegeses. We mention in this context **Lingere salem** and **Sal et mensa.** The three new proverbs within this volume are the following:

-Salem et caseum edere (To eat salt and cheese). Erasmus says that Pliny the Elder would have alluded to it[10], by ascribing to Varro the information that people in ancient times proverbially often ate bread and cheese instead of other food. In the context cited, however, Pliny only mentions *salem et panem* (salt and bread), as the food of the ancient inhabitants of Italy[11].

-Piscis eget sale (Fish needs salt). This is the Latin equivalent of the Greek proverb Ἰχθὺς δεῖται ἁλός, which, in Erasmus' interpretation refers to "something absurd and hard to believe. It is not very likely that fish that spend their lives in the sea should need salt" (*De re absurda et incredibili. Nec enim verisimile est pisces in mari agentes opus habere sale*). In *Rhetoric* (II, 23, 1400 a), Aristotle cites this utterance as an example of an incredible and improbable assertion, in the context of a speech delivered by Androcles, Alcibiades's adversary.

-Sale nihil utilius (Nothing is more beneficial than salt). Erasmus states that he found this as a maxim in the works of scholars. This appreciation of salt is more amply formulated by Pliny the Elder and by Isidore of Seville: *Humano corpori nihil esse utilius sale et sole* (Nothing is more useful for the body than salt and sun)[12].

In Saladin's edition, the fifth volume of the *Adages* includes the following two phrases:

-Salillum animae (A little salt-cellarful of salt). This is a metaphor through which a character in Plautus (Philto in *Trinummus,* v.492) designates a human being. This means, or so Erasmus believes, that his life is as short as the grain of salt served to each guest at dinner (*sentiens hominis brevissimam esse vitam, videlicet quantulum salis cuique in convivio apponitur*). He probably quoted the lines of Plautus containing the metaphor (lines 490-494)[13] from memory; we note that two syllables are missing from the last line, of which we believe Erasmus had an inaccurate recollection (the word *censu*, but Erasmus believes it was *Pelops, Cresus* or something similar). The problem is that, in modern editions of the comedy *Trinummus*, besides the variant reading *sal*[14], there are several more: *satillum* (trifle)[15]; *scintillula* (mere spark)[16]; *vatillum* (a pan for holding hot coals)[17]. The diversity of variant readings suggests an uncertain text, for which the authors of the manuscripts proposed conjectures, most of them within the sphere of the diminutives denoting something insignificant.

---

[7] *Aristoteles graece* ex recensione Immanuelis Bekkeri, Berlin, 1831, vol. IV; *Rhétorique d'Aristote*, text and French translation by Norbert Bonafous, Paris, Durand, 1856; Aristote, *Rhétorique*, vol. I-III, bilingual edition, by M. Dufour, A. Wartelle, Paris, Les Belles Lettres, 1989; Aristotel, *Retorica*, bilingual edition, translation, introductory study and index by Maria-Cristina Andrieş, notes and commentaries by Ştefan-Sebastian Maftei.

[8] Norbert Bonafous edition, note 23, p. 431.

[9] Cf. Diogenes Laertios, *Despre vieţile şi doctrinele filosofilor*, VI, 57, translation by C.I. Balmuş, introductory study and commentaries by Aram M. Frenkian, Iaşi, Polirom Publishing House, 1997, p.206.

[10] Plinius Maior, *Naturalis historia*, XXXI, 89.

[11] Cf. Plinius, *Naturalis historia. Enciclopedia cunoştinţelor din Antichitate*, vol. V *Medicină şi farmacologie*, edited, preface, notes and index by Ioana Costa, Iaşi, Polirom, 2004, p.186.

[12] *Ibidem*, XXXI, 102; see also Isidore of Seville, *Ethymologiae*, XVI, 2, 6.

[13] *Dei divites sunt, deos decent di opulentia / Et factiones, verum nos homunculi, / **Salillum animai,** quam cum extemplo amisimus,/ Aequo mendicus atque ille opulentissimus/ Censetur ad Acherontem mortuos.*

[14] Cf. *Comedies of Plautus*, translated by Bonnell Thornton, London 1769; in a footnote at page 34, the translator informs that the variant reading *tantillum* appears in some manuscripts.

[15] *The Trinummus of Plautus*, with introduction and notes by H.R. Fairclough, Toronto, The Macmillan Company, 1910.

[16] Cf. *Ausgewahlte Komodien des T.M.Plautus, Trinummus*, ed. Iulius Brix, Leipzig, Teubner, 1864; *Titus Maccius Plautus*, ed. Calvin Sears Harrington, Harvard University, 1879.

[17] *The Captives and Trinummus of Plautus*, with introduction and notes by E.P. Morris, Boston and London, 1898; in a footnote at page 128, the editor explains *vatillum* as a secondary form of *batillum*, also mentioning that the text is uncertain.

-Sale et aceto (With salt and vinegar). This is the expression denoting, in Plautus' comedy *Rudens* (line 937)[18], a frugal and modest lunch for Grippus the fisherman (*Parcum tenueque prandium significans Grippus piscator in Rudentibus*). Erasmus also mentions other gnomic connotations of vinegar, which the Italians extracted even back then from buttermilk (*Id erat jus e caseo confectum, quod et hodie Itali vocant menestram*). In his opinion, in Plautus and Persius it meant sharpness of the mind (*acumen*), in Varro, with great subtlety (*subtilitas*). In Erasmus' text, we underline the phrase *usurpare in significationem* to designate the figurative use of the word *acetum*.

After listing the proverbs referring to salt symbolism, we stress that Erasmus' selection depended on his personal taste. His interpretation was intended to put back in circulation the Latin and Greek paroemiological reserve and to stimulate the desire of the contemporary reader to read the ancient classics, so often cited throughout the *Adages*. In this way, he operated a beneficial axiological transfer from the spiritual and cultural code of the Antiquity to the informative and formative horizon of Humanism, essentially enfolded in the same perennial values.

References

Andrieş, M.-C. (trad.). 2004. Aristotel, *Retorica*. Bucureşti, IRI.

Balavoine, C. 1984. Les principes de la parémiographie érasmienne, in F. Suard and C. Buridant, *Richesse du proverbe*, vol. II, 9-23. Lille, Presses universitaires de Lille.

Balmuş, C.I. (trad.) 1997. Diogenes, Laertios, *Despre vieţile şi doctrinele filosofilor*, VI, 57, (Introducere şi comentarii de Aram M. Frenkian). Iaşi, Polirom.

Baudement, Th. (ed.) 1903. *Sentences*. Paris, Firmin Didot.

Bekkeri, I. (ed.) 1831. *Aristoteles graece* ex recensione, vol. IV. Berlin, Academia Regia Borussica.

Bonafous, N. (ed.) 1856. *Rhétorique d'Aristote* (Text and French translation). Paris, Durand.

Brix, I. (ed.) 1864. *Ausgewahlte Komodien des T.M.Plautus, Trinummus*. Leipzig, Teubner.

Chenu, J. (trad.) 1835. *Sentences de Publius Syrus*. Paris, C.-L.-F. Panckoucke.

Costa, I. (ed.) 2004. Plinius, *Naturalis historia. Enciclopedia cunoştintelor din Antichitate*, vol. V *Medicină şi farmacologie*. Iaşi, Polirom.

Diaconescu, T. (ed.) 2003. Publilius Syrus, *Sententiae Maxime*. Bucureşti, Saeculum I.O.

Dufour, M. and Wartelle, A. (eds.) 1989. Aristote, *Rhétorique*, vol. I-III. Paris, Les Belles Lettres.

Fairclough, H.R. (ed.) 1910. Plautus, *The Trinummus*. Toronto, The Macmillan Company.

Fay, H.C. (ed.) 1969. Plautus, *Rudens*. Bristol, Bristol Classical Press.

Gheorghe, G. 1986. *Proverbele româneşti şi proverbele lumii romanice*. Bucureşti, Albatros.

Harrington, C.S. (ed.) 1879. *Titus Maccius Plautus*. Harvard, Harvard University Press.

Huizinga, J. 2002. *Erasmo*. Torino, Einaudi.

Morris, E.P. (introduction and notes) 1898. *The Captives and Trinummus of Plautus*. Boston and London.

Mureşanu, C. (trad.) 2002. Publilius Syrus, *Maxime* (Text original, traducere din limba latină şi note). Cluj-Napoca, Editura Cartimpex.

Saladin, J.-C. (ed.) 2011. Érasme de Rotterdame, *Les Adages*. Collection Les Miroirs des Humanistes. Paris, Les Belles Lettres.

Scott Latourette, K. 1953. *A History of Christianity*, New York.

Sears Harrington, C. (ed.) 1879. *Titus Maccius Plautus*. Cambridge, Harvard University.

Serbat, G. (ed.). 1972. Pline l'Ancien, *Histoire naturelle*, livre XXXI. Paris, Les Belles Lettres.

Thornton, B. (transl.) 1769. Plautus, *Comedies*. London, T. Becket and P. A. De Hondt.

Woelfflin, E. (ed.) 1869. *Publilii Syri sententiae* / Ad fidem codicum optimorum primum recensuit Eduardus Woelfflin. Accedit incerti auctoris liber qui vulgo dicitur de moribus. Leipzig, Teubner.

Zell, K. (ed.) 1829. Publilius Syrus, *Sententiae falso inter Publinianas receptae*. Stuttgart.

[18] *Sed hic rex cum* ***aceto*** *pransurust* ***et sale*** *sine bono pulmento*: Plautus, *Rudens*, edited by H. C. Fay, Bristol Classical Press, Bristol, 1969.

A Latino-Hispanic Paroemiological Saline "Feast", by Bernardino Gomez Miedes (*Commentariorum de sale libri V*)

Mihaela Paraschiv
Universitatea „Alexandru Ioan Cuza" din Iaşi

Abstract
Reading the book of the Spanish bishop Bernardino Gomez Miedes (whose Latinized name is *Bernardinus Gomesius Miedes*) is a true revelation of the breadth and thoroughness of the scientific concerns of European humanists. It is a generous monograph on salt, drafted initially in four volumes (as it was published in the *editio princeps* of 1572 in Valencia), to which a fifth volume was added (the *editio altera* of 1572, also published in Valencia). The Spanish humanist's book has recently benefited from a critical bilingual edition in three volumes thanks to classicist Sandra Inés Ramos Maldonado of the University of Cádiz;[1] it was actually her PhD thesis (1995), and won a prize from the Society of Latin Studies in Spain. The partiality of the Spanish cleric for this subject is attributed by an editor to circumstances specific to the geographical, cultural and economic environment of his life: he was from Alcañiz, an important Hispanic saline area; note the importance of salt, a genuine "white gold" for European commerce at the end of the Renaissance and the beginning of the Modern Era. This editor also underlines the interest of theological humanism for any given subject as "the theologian's duty and mission" (*officium ac munus theologi*) of not leaving subjects untreated. In the *Praefatio*, the author states that he has dedicated a long and elaborate labour of polishing (*labor limae*) to salt, since in his opinion the topic had either rarely been treated or actually neglected by other writers, despite its real importance. Following the tradition of a recurrent topos in the prefaces of Cicero's philosophical works, Gomez Miedes confesses to the reader that he had to face hostile reaction from his friends and contemporaries. This came about because they could perceive any useful purpose in a book about salt, when there were so many other more important and potentially fruitful subjects out there. This was all the more of a challenge for him. The first four books reveal the physical, medical, gastronomic and symbolic properties of salt, while the fifth book represents a *translatio ad allegoricam salis intelligentiam*, a metaphorical translation to the ethical or mystical register of the previously mentioned properties of salt. The author suggests, in this last book, an *interpretatio anagogica* of salt, able to "raise" (Greek *anago* - to raise) the soul from the terrestrial to celestial, from the temporal to eternal, from the human to divine. The author asked the reader's permission to become an ancient *epulo* (the organizer of a sacred feast), staging an authentic saline feast, where the most copious "dishes" are the proverbs dedicated to salt from Latin and Hispanic paroemiological reserves. The typology of proverbs is both interesting and original, depending on the ethical virtues symbolized: *prudentia, iustitia, aequitas, fortitudo, sobrietas, parcitas*. The aim of our essay has been to follow the paroemiological inventory of this last volume. We will focus on the author's commentary, which shows in detail the intellectual concerns of 16th-century European humanists.

Keywords
salt, Bernardino Gomez Miedes, paroemiology, Renaissance, *interpretatio anagogica*

To read the book of the Spanish bishop Bernardino Gomez Miedes (whose Latinized name is *Bernardinus Gomesius Miedes*) is a true revelation of the breadth and thoroughness of the scientific concerns of European humanists. It is a generous monograph on salt, drafted initially in four volumes (as it was published in the *editio princeps* of 1572 in Valencia), to which a fifth volume was added (the *editio altera* of 1572, also published in Valencia). The Spanish humanist's book has recently benefited from a critical bilingual edition in three volumes thanks to classicist Sandra Inés Ramos Maldonado of the University of Cádiz;[2] it was actually her PhD thesis (1995), and won a prize from the Society of Latin Studies in Spain.

The partiality of the Spanish theologian for this subject has a twofold inspiration, in the opinion of the editor Antonio Malpica Cuello, a professor of medieval history at the University of Granada and the president of the International Commission for the History of Salt (CIHS). The first inspiration is derived from the geographical, cultural and economic circumstances of his personal background: he was from Alcañiz, an important Hispanic saline area. Salt was of immense importance, a genuine "white gold" for European commerce at the end of the Renaissance and the beginning of the Modern Era.[3] The second inspiration was the interest of theological humanism in any given subject. It was "the theologian's duty and mission" (*officium ac munus theologi*) not to leave subjects untreated. The author dedicated the book to King Philip

[1] Bernardino Gomez Miedes, *Comentarios sobre la sal*, Introduction, critical edition, annotated translation and indices by Sandra Inés Ramos Maldonado, Prologue by Antonio Malpica Cuello, Madrid, Alcañiz, 2003.

[2] Bernardino Gomez Miedes, *Comentarios sobre la sal*, Introduction, critical edition, annotated translation and indices by Sandra Inés Ramos Maldonado, Prologue by Antonio Malpica Cuello, Madrid, Alcañiz, 2003.

[3] Cf. Pores Marijuan, *Sazon de manjares y desazon de contibuyentes. La sal en la corona de Castilla en tiempos de los Austria,* Universidad del Pais Vasco, Bilbao, 2003, p. 17.

II of Spain, by skilfully comparing the virtues of salt with those of kings: "as, in the virtue of the similarity between salt and kings, you alone, the most prominent of all kings, appear as the salt of the world. Because just like salt – through its taste and properties – not only spices meals, but it also cures the human body from its eternal rottenness, the same way you, in the virtue of your justice and equity (as salt represents them both, through its rough and sweet taste), bring happiness [through these virtues] to those whom you rule."[4]

In the *Praefatio*, the author states that he has dedicated a long and elaborate labour of polishing (*labor limae*) to salt, since in his opinion the topic had either rarely been treated or actually neglected by other writers, despite its real importance. Following the tradition of a recurrent topos in the prefaces of Cicero's philosophical works, what we might call the *topos (sive locus) adversitatis*, Gomez Miedes confesses to the reader that he had to face hostile reaction from his friends and contemporaries. This came about because they could perceive any useful purpose in a book about salt, when there were so many other more important and potentially fruitful subjects out there. This was all the more of a challenge for him.

Nevertheless, just before the book came out, some contemporaries did eulogize the novelty of the subject and the author's virtues, also alluding to possible criticism he might face; Palmirano, a friend of Gomez, thus wrote to him in 1564: *"O, happy and lucky are you to dare approach a subject nobody ever had the courage to approach before! Who could criticize those remarkable* Commentaries *of yours on salt, truly admirable for the elegance and charm of the style, for the spirituality of the words, for the abundance of information?"* Another contemporary, Domingo Andreas, dedicated several epigrams to Gomez, among which was one referring to his *Commentarii de sale*: "*De sale fecisti quos, Bernardine, libellos / Te facile ostendunt salsius esse nihil*" ("On salt, you wrote books, Bernardinus, /clearly showing there is nothing 'saltier' than you are"). As regards the novelty of the subject, we must say that the author of the *Commentaries* admits to have been inspired by ancient sources. He states that the main inspiration for the book was a brief allusion in Plato's *Symposium* (177b: "I have met with a philosophical work in which the utility of salt has been made the theme of an eloquent discourse").[5] He included other information from Homer, Hesiod, Aristotle, Theophrastus, Dioscorides, Plutarch, Plautus, Cicero, Horace, Virgil, Quintilian, Pliny the Elder, Martial, Valerius Maximus, Celsus, Galen.

In view of the importance of Arabic culture in Spain, Antonio Malpica believes that a significant source must have also been the great book on pharmacology written by the Andalusian botanist and pharmacologist of the 13th century, Ibn al-Baitar, who also wrote about salt, [6] but whom the author fails to cite. He also overlooks Erasmus of Rotterdam whom, as the editor also notes, Gomez cited copiously (Erasmus was also a competent commentator on the Greek and Latin paroemiological literature, as seen in his work *Adagia*).[7] At times he merely says *quidam dicit, quidam confirmat*. He thus used a variety of sources, some properly acknowledged (mostly ancient sources, sometimes inaccurately quoted), others not (writers more or less contemporary with the author);[8] Tacitus' assertion *Maior e longinquo reverentia* (*Annales*, I, 47) thus applies in this case, too.

According to the division established by Gomez, the first four books of the book *Commentarii de sale* deal with the physical, medical, gastronomic and symbolic properties of salt, while a fifth book would treat a *translatio ad allegoricam salis intelligentiam*, a metaphorical transfer to the ethical or mystical register of the previously mentioned properties of salt. Gomez proposes, in this last book, an *interpretatio anagogica* of salt, able to "raise" (Greek verb *anago*-to raise) the soul from terrestrial to celestial, from temporal to eternal, from human to divine. The author asks the reader's permission to become an ancient *epulo* (the organizer of a sacred feast), staging an authentic saline feast, where the most copious "dishes" are proverbs dedicated to salt from the Latin and Hispanic paroemiological reserves. The typology of proverbs is both interesting and original, depending on the ethical virtues symbolized: *prudentia, iustitia, aequitas, fortitudo, sobrietas, parcitas*.

In what follows, we present several items in the paroemiological inventory elaborated by the author, accompanied by his own commentaries which offer a real insight into the intellectual potentialities of 16th-century scholars. There are 117 Latin and Hispanic proverbs, of which 61 refer to salt.

As the fundamental principle of the entire book, the author states ***Sale et sole nihil corpori utilius*** (Nothing is more useful for the body than salt and sun), quoted

[4] The translations into English are the author's.

[5] Plato, *The Dialogues of Plato,* ed. and trans. by Benjamin Jowett, Brandon Press, Boston, 1996, p. 34.

[6] Cf. Toby Huff, *The Rise of Early Modern Science: Islam, China, and the West*, Cambridge University Press, 2003, p. 218.

[7] This book by Erasmus, published in 1500 with the title *Adagia collectanea*, is a collection of proverbs and quotations from Greek and Latin authors, with commentaries, and which enjoyed great success; it went through 30 editions during the author's lifetime (that 1536 comprised 4,151 *adagia*). An explanation for Gomez's failure to cite the *Adagia* could be that it had been placed on the *Index librorum prohibitorum* by the Council of Trent in 1559. Although some laymen might dare to ignore the prohibition, by keeping and secretly reading the forbidden book, a churchman could not afford such a "luxury"; see Érasme de Rotterdam, *Les Adages*, complete bilingual edition, in 5 volumes, edited by Jean-Christophe Saladin, Paris, Les Belles Lettres, 2011.

[8] Cf. Bernardino Gomez Miedes, *Comentarios sobre la sal*, Introduccion, II, 4, 6, p. CLII: "Una de las caracteristicas de los humanistas es silenciar en sus obras los nombres de autores más o menos contemporáneos".

from Pliny the Elder [9]. In fact, he performed a *translatio ad intellectum*, by turning its connotative into a denotative sense and by presenting it in the hermeneutical version ***Sole et sale nihil humanis animis salubrius*** (Nothing is more beneficial for the human spirit than salt and sun). After explaining this basic Latin adage, Gomez admits that his entire work is an *enodatio proverbiorum*, meaning that its fund of ideas is mostly based on explaining the aphoristic virtues as they relateto wisdom, in accordance with his personal belief. This belief is *Quot sunt de sale proverbia, totidem ex iisdem elici virtutes ut medicamenta ad sanandos animos* (As many proverbs on salt there may be, they all contain virtues used as remedies to cure souls).

The author legitimates his commentaries on the paroemiological symbolism of salt by concluding that "ancient authors came up with numerous symbols arising from salt" (*complura e sale apud antiquos orta symbola*). He begins with the one referring to salt as a "symbol of concord" (***concordiae symbolum***), by correlating it again with the sun, as in the previous proverb. This way, the saying ***Salem et solem offuscare*** (Denigrating salt and sun) becomes the reference point of ultimate disobedience. The author tells a story about two female nonagerians who had an argument, by accusing each other as follows: *posses tu Salem et solem offuscare, ut est in proverbio* (XII, 4), and they only made peace after tasting some salt (*neque vero antea confecta pax fuit neque invicem datae dextrae quam allatum salem ambae gustarunt*). The author's commentary recalls the "salt pact" (*pactum salis*) in the Old Testament (*Numbers*, 18; *Paralipomena*, 13) and the explanation provided by Jerome for this ritual, in *Quaestiones Hebraica.*[10] In direct connection with the symbolism of concord, the author then reproduces and comments upon numerous proverbs where salt is a "symbol of friendship" (***symbolum amicitiae***), among which we will cite two proverbs with rich commentaries by the author.

Salem et mensam ne praetereas (V, 19, 1 – Pass not over salt and the table), an aphorism attributed to Pythagoras. Diogenes Laertius does not mention this *adagium* in the philosopher's biography, but he speaks of the importance of salt in his teachings: "*Of salt he [Pythagoras] said it should be brought to table to remind us of what is right; for salt preserves whatever it finds, and it arises from the purest sources, sun and sea.*"[11] Pythagoras' belief motivates both the association between salt and sun and the recurrence of another Pythagorean adage, recalled by Gomez: ***Salem apponito*** (Put salt on the table). After Gomez's commentary, we deduce that the *sal et mensa* within the first aphorism do not form a hendyadis, as one might be tempted to interpret this maxim (salt on the table). On the contrary, each of them represents a symbol of the two constants of friendship: *consuetudo* (habit) and *iura* (laws, norms); in other words, the existential and moral foundation of a vital relationship, in the philosopher's opinion. We present in what follows the transposition (*translatio ad intellectum*) that the Spanish humanist proposes. *Tam magna siquidem communicati salis et mensae vis inest mortalibus, ut quoties nos eius suavissimae familiaritatis et convictus recordatio subierit, toties in veros amoris igniculos usqueadeo accendimur, ut ab individua et communi mensa ad individuam animorum copulationem servandam transferamur* (V, 19, 2). We propose the following translation. "Indeed, salt and the common table are so important for people that, every time we think of their remarkable intimacy and cohabitation, we are caught in the fire of love, so much so that, starting from the inseparable and common meal [of friends], we interpret it figuratively as the conservation of the inseparable union of spirits".

Sale et cymino tenus amici (V, 64, 1 – Friends are only drawn together by salt and cumin), a proverb taken from Plutarch (*Moralia* 684e) without mentioning Erasmus as the intermediary. Gomez does, however, cite critically the explanation given by a certain grammarian called Apollophanes: *Ii amici proverbio notantur qui usqueadeo sunt familiares, ut vel solo sale et cymino contenti convictarint* (V,64,2: This proverb refers to those friends who are so close, that they do not need anything more to eat than salt and cumin). Gomez finds the grammarian's explanation ridiculous, as he believes that the two spices cannot compose a meal, but can only spice it, though salt alone symbolizes a frugal meal. In his opinion, the proverb emphasizes the qualities added to food by the two spices: *per salem suavia optimeque condita, per cyminum vero sobria atque salutaria esculenta* (V, 64, 3: salt makes the meals pleasant and well-spiced, but the cumin makes them well-balanced and healthy). In the context of this explanation, the adjective *sobrius* is not coincidental since, according to Pliny the Elder's explanation,[12]the cumin offered to those who had drunk too much (*ebrii*) made them look paler, which masked their red cheeks and made them look moderate and sober (*sobrii*). For this reason, Horace uses the metaphor *exsangue cuminum* (the pale cumin). At the end of this commentary, the author compels himself to offer to his friends feasts both pleasant and moderate (*convivia una cum suavitate sobria fiant*); indispensable spices symbolize the two qualities: *sal et cyminum*.

Among the proverbs referring to the quality of salt as a "symbol of prudence" (***symbolum prudentiae***), a virtue categorized by the author as "the head and queen of all the other virtues" (*aliarum omnium virtutum princeps et regina*), we take special note of the one that Gomez

[9] Pliny the Elder, *Naturalis historia*, XXXI, 102; see also Isidore of Seville, *The Etymologies*, XVI, 2, 6.

[10] Jerome, *Quaestiones Hebraicae*: *Pactum salis in quibusdam locis pro lege ponitur, quae omnium condimentum est, hic vero pro domo David, qui universo Israeli fuit condimentum*; for salt as symbol of reconciliation, see also Ovid, *Fasti*, I, lines 337-338: *Ante deos homines quod conciliare valeret/ Far erat et puri lucida mica salis.*

[11] Diogenes Laertius, *Lives of eminent Philosophers*, trans. by Robert Drew Hicks, Harvard University Press, 1970, p. 351.

[12] Pliny the Elder, *op.cit.* XIX, 160.

believed was often invoked at feasts. ***Salem ne aliis porrigito Romae qui versatus non fueris*** (V, 23, 5: Do not give salt to anybody if you did not live in Rome). The author gives the following explanation for the associating the words *sal* and *Roma*: "Since salt means prudence, another hypostasis of wisdom, while Rome is the most significant place, thus a sort of nursery garden for prudence" (V, 23, 6: *Per salem enim prudentiam, alteram sapientiae partem, per Romam vero praecipuum prudentiae locum et quasi seminarium denotari*). The copious commentary on Rome, considered to be "the most famous monument on Earth" (*illustrissimum terrarum monumentum*), "the theatre or school of the world" (*theatrum sive schola mundi)*, is in accordance with the attachment of the European humanists to Antiquity, with an inalienable first place up to that moment. As one who lived ten years in Rome (1548-1558), Gomez can afford to state arrogantly "that the people born or educated in Rome become, beyond doubt, not only wiser, but even more generous and more illustrious" (V, 24, 5: *Nam Romae natos aut in ea educatos homines non solum aliis prudentiores, sed omnibus longe fortiores, quin et magnanimos atque clarissimos evadere necesse est*). Among other proverbs with the same wisdom-based connotation, Gomez mentions one that he states was regularly employed in Italy, France and Spain, with the Latin translation ***Nullo conspersum sale caput*** (V.51,1: No salt in his head). The Hispanic version of this colloquial aphorism, meant to stigmatize the lack of prudence or its infatuated simulation is ***No tiene sal en la mollera***. The author's commentary suggests the possible analogy with the well-known custom of salting the meat of animals to preserve it. "In the same way, all words and deeds are dead and tasteless and they even smell bad, as if rotten, unless they were spiced with the highly spiritual line: Whatever you do, do it with moderation and keep in mind the end" (V, 51, 3: *ita profecto mortua sunt ac fatua, immo quasi corrupta solent putere dicta factaque omnia quae salsissimo illo condita versu non fuerint*: *Quicquid agas, prudenter agas et respice finem*[13].

In another Hispanic proverb, reproduced in its Latin translation, salt is a "symbol of bravery" (***symbolum fortitudinis***): ***Salem cur manu excipis qua virum non occideris***? (V, 60, 2: Why do you take salt with the hand you have not killed anyone with?). The Spanish original is ***¿Por qué coges la sal con la mano con la que no has matado a un hombre?*** The author's interpretation of this interrogative adagio, *quod ad nobilissimam Hispanorum firmitudinem exprimendam spectat* (that aims to externalize the well-known bravery of the Spaniards) begins with a reference to two Latin versions in circulation: an older one, with *virum,* and a more recent one, with *maurum*. The interpretation continues with a short history of Spain, with the proverbial bravery of its inhabitants and with the right to kill an enemy, for civil or religious reasons, that persisted for a long time until the Arab occupation. Gomez believes this is the explanation for the modification of *virum* into *maurum* (V, 61, 5: *At vero posterior proverbii sensus, qui pro viro maurum habet, ab eodem ipso symbolo duci atque ex similibus prope caussis ortum habuisse fertur*). Gomez also lists phrases designating the quantification of salt, such as ***cum grano salis,***[14] ***mica salis,***[15] ***salis modius,***[16] ***saccum salis***. The last is the Latin translation of the phrase *un saco de sal*, of the Hispanic proverb ***Aunque coma un saco de sal, no te podrás fiar de él*** (V, 59, 4: You can eat a bag of salt and you still cannot trust it). Gomez is preoccupied with the moralizing aspect of the phrases cited and he does not give importance to the way in which the amount of salt oscillates between derisory and consistent: grain (*granum*), crumb (*mica*) // bushel (*modius*), bag (*saccus*), depending on the minimizing or hyperbolizing intention in the valorisation of its symbolism in the ancient and contemporary sayings. We might conclude that the insignificant amount of salt is specific to the phrases where it symbolizes the minuscule coefficient of good taste and spirituality, necessary to human speech or behaviour, and the hyperbolically augmented amount appears in the proverbs where salt symbolizes the hard-to-earn trust within human relationships.

An explanation from the perspective of Christian morality is attributed to the old Spanish folk proverb ***Allá vayas mal donde comen el huevo sin sal***, translated into Latin by the author (V,79,4: ***Eo conducare malum quo editur sine sale ovum*** – Go away, trouble, where they eat the egg without salt!). Gomez notices the apotropaic connotation of the imprecation (*ad avertendum impendens sibi malum*) and he attempts to explain it through the popular belief that demons live in Hell, deprived of divine grace. "As, through the egg – seen as the most amazing of foods, as I have shown before – what else does it mean but the existence and life of demons? But what does salt – which they live without or through which they are banished (maybe because it tastes like benevolence) – show but divine benevolence, forever out of their reach?" (V, 79, 8: *per ovum quidem, quod omnium, ut supra probatum est, alimentorum habetur excellentissimum, quid prater quam ipsorum daemonum esse atque vivere denotatur? Per salem autem, sine quo vitam traducunt quove, ut ostendimus (forte quia gratiam sapit ille), fugantur, quam divina ipsis in perpetuum interdicta gratia demonstratur*?).

Hence, this is an authentic anagogic explanation of salt; its symbolism moves from the terrestrial to the celestial world, representing the most desired expression of divinity and as a sign of immortality, which is divine grace.

In direct correlation with the previous explanation

[13] As with most quotations of Latin authors, Gomez again fails to reproduce the original form accurately; it should read: *Quidquid agis, prudenter agas et respice finem !* (*Gesta Romanorum*, 103-104).

[14] Pliny the Elder, *op. cit*., XXIII, 149.

[15] Catullus, *Carmina*, 86,4: *Nulla in tam magno est corpore mica salis.*

[16] Cicero, *De amicitia*, 67: *multos modios salis simul edendos esse ut amicitiae munus expletum sit*; Aristotle, *Ethica Nicomachea*, 1156 b.

regarding the divine character of salt, the author ends his hermeneutic endeavour beginning with an interesting interpretation for the tripartite phonemic structure of the words designating salt in Latin (***sal***) and Ancient Greek (*hals, halos*). Based on the authority of Plato, Gomez starts from the premise expressed by the Greek philosopher in *Kratylos*: that names were not instituted arbitrarily, but depended on the specifics of the reality defined. After noticing the identical graphic structure of the two words, he ascribed a mystical symbolism to each grapheme within the Greek name for salt. According to him, it symbolizes the divine principle, also common to the other two divine persons: the Son and the Holy Ghost. It represents the Son as Divine Word; the two sides of the letter symbolize the propensity of the Son towards the Father and the Holy Ghost. Through the two semicircles, it symbolizes the Son and the Holy Spirit, with God as "co-equal and co-eternal" (*coaequalis et coaeternus*), overwhelming them with His infinite love. The conclusion of this explanation can be seen as the natural finishing line of the entire anagogic interpretation provided by the author in the last volume of the *Commentaries*. "Salt keeps a trace of that sacrosanct and indivisible trinity, which we venerate both in unity and as unity in trinity" (V, 84, 1: *in sale esse impressum vestigium aliquod illius sacrosanctae atque individuae triadis, quam in monade, peraeque ac monadem in triade veneramur*).

Following a profuse digression through the Old and New Testaments taking this assertion as its starting point, the book ends with an exhortation for readers. They should beg Christ: "the supreme salt bearer" (*summus salis lator*) always to provide mortals with the precious spice for every meal, not only to appreciate its taste, but mostly to contemplate the divine mysteries that it symbolizes, "for us to partake in His eternal salt and celestial meal" (V, 107, 12: *caelesti atque perpetuo nos suo sale et mensa frui efficiat*). The paroemiological saline feast, officiated at by the Spanish hermeneutist in the tradition of the ancient *epulo*, had as its primary objective (as attested also by his last interpretation) to intertwine the two symbols of human and divine fraternization: ***sal et mensa***.

References

Bywater, J. (ed.) 1894. *Aristotle's Ethica Nicomachea*. Oxford, Clarendon Press.

Combès, R. (ed.) 1968. Cicéron, *De l'amitié*. Paris, Les Belles Lettres.

Hicks, R.D. (transl.) 1970. Diogenes Laertius, *Lives of eminent Philosophers*. Harvard, Harvard University Press.

Huff, T. 2003. *The Rise of Early Modern Science: Islam, China, and the West*. Cambridge, Cambridge University Press.

Jones, W.H.S. (ed.) 1951. *Pliny. Natural History*, Books 20-23. Harvard, Harvard University Press.

Jones, W.H.S. (ed.) 1963. *Pliny. Natural History*, Books 28-32. Harvard, Harvard University Press.

Jowett, B. (ed. and transl.) 1996. *The Dialogues of Plato*. Boston, Brandon Press.

Lafaye, G. (ed.) 1984. Catulle, *Carmina*. Paris, Les Belles Lettres.

Pores, M. 2003. *Sazon de manjares y desazon de contibuyentes. La sal en la corona de Castilla en tiempos de los Austria*. Bilbao, Universidad del Pais Vasco.

Rackhan, H. (ed.) 1950. *Pliny. Natural History*. Books 17-19. Harvard, Harvard University Press.

Ramos Maldonado, S.I. (transl.) 2003. Bernardino Gomez Miedes, *Comentarios sobre la sal* (Introduction, critical edition, annotated translation and indices by Sandra Inés Ramos Maldonado, Prologue by Antonio Malpica Cuello). Madrid, Alcañiz.

Saladin, J.-C. (ed.) 2011. Érasme de Rotterdame, *Les Adages*. Collection Les Miroirs des Humanistes. Paris, Les Belles Lettres.

Schilling, R. (ed.). 1993. *Ovide, Les Fastes*. Tome I: Livres I-III. Paris, Les Belles Lettres.

Traditional Production of Salt in Chile. The Case of Cáhuil Lagoon

Răzvan Victor Pantelimon
Universitatea Ovidius Constanţa /
Universidad Católica de Valparaíso

Abstract
In this article we'll try to present a case of traditional salt production that is still present in the central part of Chile, in an area known as the Cáhuil Lagoon. We'll start with a brief history of the salt production and economic importance in the Latin American region. Then will follow a short description of the salt production in Chile, with an emphasis on the salt produced in the coastal area. The largest part of our study will deal with the description of the way in which is produced the salt in the area of the Cáhuil Lagoon using the same methods, proceedings and tools as a few hundred years ago.

Keywords
salt, salt pans, Cáhuil Lagoon, Chile, salterns

Salt is a necessary element for human existence, and it is said that physiologically 'Man requires five to six grams of salt per day, which usually is contain in the food itself. In the case of being private of the salt with which can preserves his food, part of humanity would be threatened in his survival' (Lemonier 1996, 656). Salt has also been used for the development of multiple technical activities such as the 'pottery, soap production, glassware, tanning and preserving hides, mummification, etc. (Multhauf 1985, 21).

In the monumental study *The Cambridge History of Latin America* edited by Leslie Bethell in 11 volumes, there are some references at the use of salt, both in the pre-Columbian period as in the colonial one, but there is no systematic analysis of salt production and importance. During the pre-Columbian period, salt, 'a vital resource', was produced by the original peoples of 'Muisca from highland salt springs and by the Tairona and Cenvi from coastal sources. The Quimbaya realm included productive gold fields and salt springs, and Quimbaya craftsmen were highly skilled in weaving and probably also in metallurgy. Once again, salt, textiles and gold and tumbaga objects were exchanged both with neighbouring peoples and with more distant groups' (Helms 1984, p. 42). Salt was considered not only an exchange product, but also was charged with a powerful political-religious symbolism: 'Many of the scarce and valued goods exchanged, not only the emeralds, gold pieces and fine textiles but also such products as salt, dried fish and war captive slaves, were most likely highly charged with political — religious symbolism signifying the sacredness, efficacy and authority of chiefship' (Helms 1984, 43). In other parts of pre-Columbian world, the 'rich salt beds on the southern coast of Puerto Rico may have been associated with the domain of the highest chief of the island' (Helms 1984, 51) which gives us another proof of the relation between the salt sources possession and political and symbolical authority. The same scholar affirms that 'Textiles and salt also probably facilitated the long-distance contacts which yielded rough stone for crafting and finished golden pendants' (Helms 1984, 48). In Andean region, the Inca methods for census *khipu* (*quipu*) recorded some activities related with the exploitation of salt (Murra 1984, 65–66).

After the *Conquista* the salt remains a very important product not only for the colonial powers in the region, Spain and Portugal, but also for other emerging powers, as the United Provinces 'from the end of the sixteenth century Dutch shippers were showing an unhealthy interest both in the Brazil trade and in the Caribbean, to which they turned in search of salt' (Elliott 1984, 328). In that era of large sea voyages, salt was essential in order to preserve the food use on the ships, especially salted fish and pork. The Dutch assured themselves the predominance of the salted fish commerce using the resources of the New World: 'They (*the Dutch*) were attracted to the great salt flats of the coast of Tierra Firme. The salt extracted, combined with the Newfoundland fisheries, enabled them to dominate the European trade in salt codfish and herring, both staples of the everyday diet' (Macledo 1984, 377). In the seventeen century salt played a very important role in the trans-Atlantic commerce salt and also in the relations between metropolis and colonies, especially in the case of Portugal and Brazil: 'Salt played a notable part in the Atlantic trade. It was not only essential for the diet of man and beast, but it was also indispensable for preserving fish and meat. Salt production was important to Portugal. [...] In 1632 the supply of salt to Brazil came under state control, a monopoly which only ended in 1821. In 1665 and again in 1690–1691 attempts were made to prohibit the production of salt in Brazil to prevent competition with salt imported from Portugal' (Mauro 1984, 460).

Salt was one of the important resources of the colonial economies, and it was obtained both from mines and the costal deposits. It was use in diverse activities: obtaining metals through the amalgamation process (Bakewell 1984, 115–119); conserving meat in the meat-salting plants called *saladeros* (Morner 1984, 211). Giving his importance royal monopolies were established over the production and commercialization of salt, monopolies in which contractors would lease the rights to exploit this resource (Schwartz 1984, 453) and also a very high level of taxation which, both being kept also in the Republican period.

In the first half of the 19th century in some new independent republics, the state revenue from the salt monopolies and taxation was the third after revenues from customs and tobacco (Deas 1985, 519). In the second part of the century, the development of the meat packing systems and the apparition of the meat freezing plant destroyed the old *saladeros* system and the importance of salt started to drop (Lewis 1986, 276); nonetheless, revenues from salt remains important, for example, in Columbia at the end of the century, where the salt mine leases were the second source of revenue after the custom duties (Cortés Conde 2006, 213).

In the specific case of Chile, the situation is more or less the same as in the rest of the Latin America. It is clear that salt production precedes the *Conquista*, and here we can quote one of the writings that refer to the extraction of salt in Cáhuil since before the Spanish arrived: '...in Cáhuil seems to have existed since the Spanish times—perhaps in the years of the Inca invasion (1463–1525)—salt-producing activity which needed a strong established base in the place. They are suggestive in this regard, the remains found by an elderly indigenous woman in the beginning of this century, as well as early granted categories of Indian Communities (1777) and the installation of a County Council (1787) which was obtained by this region on the occasion of the establishment of the gubernators regimes' (Polanco and Marcelo 1996, 24–25).

For the Spaniard conquerors which arrived with Pedro de Valdivia, the issue of marine salt production wasn't unknown, because in the south of the Iberian Peninsula, the sea salt exploitation was an ancient tradition in the Salinas de Torrevieja and La Mata in the province of Alicante on the Mediterranean shore, and also a little further west on the Atlantic coast of the province of Andalusia, where stood the salt banks from Cadiz, Jerez de la Frontera, and Sanlucar de Barrameda.

Since the onset of the *Conquista*, the Spaniards sought to get salt. In the 16th century, the institution which held the responsibility to seek, distribute and organize the salt production was the *Cabildo* of Santiago. During the first years of the Conquest, one of the main topics discussed in the *Cabildo* of Santiago was related to the supply of salt. By August 1548, the council stated that any person who wanted to trade salt must previously declare in writing the value of transactions. A few years after, on the 2nd of January 1552, the Royal Envoy Francisco Miñez, proclaimed to the corporation which he convoked, that 'all the salt pans are common, as HM sends him by his royal ordinances, in order that everyone can bring salt and have salt pans for obtained it; any chieftain, could not take the others salterns, but everyone can do it and take salt: because somehow all the shelves of this city lack it and then the people complain that they spend their wealth in order to purchase the said salt, because with having salt one can keep his house, as there are no butchers in this city, and by force they must always have salty meat' (TVH 2012, 139). This valuable document clearly indicates that a few years after the Conquest commenced, the Spanish began to regulate the activity of salt production, as the *conquistadores* were authorized to construct salt pans, 'pits' or pools that allowed them to make and take salt. Is also noteworthy that is permitted the construction of artificial salt pounds or salt pans which by law were common, being able to work the salt without fear of others which can try to take it. We also can observe that because his relative scarcity salt had a high price in that period.

This statement makes it clear to us how important it was salt during the early developing period of Santiago. Our current abundance seems infinite, but during the conquest and colonization the salt extraction and trade issues were of first order of importance and must have been firmly protected. Proof of this is another ordinance of the *Cabildo* in 1566, which banned harvesting salt before March 1st due to the constant plundering of the salternos that supplied Santiago, which besides draining the production, spoiled the curdle process their premature removal ruined their quality. After this date: 'can go and come all and any Indians and Spanish to take salt to their homes, and no person can prevent to any Indian or Spanish to harvest this salt, under the penalty of fifty pesos of gold for the Spanish, and the penalty of two hundred lashes for the Indians, and the loss of salt that has captured.' (TVH 2012, 139).

The earliest reference we have of the existence of salt on the coast of the central region was the work of Geronimo de Bibar (Jerónimo de Vivar) *Crónica y relación copiosa y verdadera de los Reynos de Chile* written in 1558. It tells us that 'there are very good salterns in the lagoon called Topocalma and in Quillota and also 'there are salt pans in other many parts' (Bibar 1558, 133). He describe one of the salt pans, located in what is call 'Topocalma Lagoon', 'one thousand and four hundred feet long and half a league in round, at two shots of stone from sea, in which doesn't enter or exit the sea water because is rounded with some earth walls' (Bibar 1558, 75). The indigenes provide themselves with 'a lot' of salt from the lagoon during the summer months (Bibar 1558, 74–75). The harvested salt 'is as white as crystal, and in summer, as the lagoon has little water, the sun penetrates more the heat on earth, which makes the first salt to be blonde or red, and the one they draw from below of this is whiter' (Bibar 1558, 75).

Another important reference to the salterns can be found in the *Histórica relación del Reyno de Chile* write by Alonso de Ovalle and published in 1646, in which he spoke of the salt pans called Laguna de Rapel. This lagoon, in January when the sun are stronger, is 'closing the mouth from where it continued with the sea, and the water remained inside, so is curdled which produce a crust of more than two or three spans of thickness of white salt of a very good flavour' (Ovalle 1646, 35). Another interesting information from Ovalle speaks about the fact that 'the salt pans are never missing, at least the one they do to hand, which there are like a few small lagoons in which, entering the water during the winter, it turns into salt the one that stays inside and, as it is less material they obtain the curdle effect with less sun' (Ovalle 1646, 35). These 'hand-made' salt pans artificially diminished their depth in order to accelerate the evaporation process of the salted water. We'll further see that this type of strategy has been used till our days in the Region of Cáhuil.

In the middle of the 18th century (1760) appeared the *Historia geografica e hidrografica con derrotero general correlativo al plan del Reino de Chile* which described as functional all the salt evaporation pounds existing today (Quiroz, Poblete and Olivares 1986, 117).

With the introduction of the Republican Government in Chile, it was established a new policy with respect to the salt. Under the rule of Bernardo O'Higgins, the Chilean State assumed the *estanco* of the salt, which meant that the State collected all the salt produced in Rancagua and Curicó, and from there it would sell it, and at the same time ban the import of salt from abroad (Peru mainly). One of the reasons invoked for this measure is to avoid the traffic of salt, but the basic idea was to make the citizens pay a new 'indirect' tax as resulting from the salt price established by the monopolistic State. 'The new State, in order to protect his incipient economy, imposed severe penalties to those who risked smuggling illegal salt: 5 years of forced prison' (Vera 2003, 39). There is also another clause which was acting if some producer was leaving his salt pound 'for one year' this one was declared desert and other persons could appropriated it. This measurement intended to prevent the producers from ceasing production because the low price that would be paid to them' (Vera 2003, 40). But the bureaucratic apparatus which must apply this *estanco* wasn't functioning very good, so in a couple of years the State renounced this practice.

At the end of the 19th century, Enrique Espinoza, described in his work *Geografía descriptiva de la República de Chile*, published for the first time in 1890, some characteristics of the salt pans presented in the estuaries of the Rapel-Mataquito interfluves. He points out that, at the Northern margin of the estuary of Cáhuil is a 'small hamlet', with the same name, 'whose residents exploiting the abundant salt of the estuary, which although is sold at low prices in the towns of the interior, it is, however, a source of income' (Espinoza 1890, 273). Espinoza added that the 'annual production is estimated at 240,000 metric quintals of salt, which if there were easy means of locomotion for the interior of the country, could give it a very profitable result to those who exploit this production' (Espinoza 1890, 281).

During the 20th century there was a very low interest in the issue of salt exploitation and description of salterns. In this period appeared only a few studies concerning this topic: Díaz Gárces (1933); Manríquez (1955), and Quiroz, Poblete and Olivares (1986). In the last fifteen years there has been a greater interest for this theme mainly from the anthropological and ethnographical standpoint, materialized in a few undergraduate theses: Vera Rodríguez (2003), Araya Muñoz (2006) and Fahrenkrog Borghero (2009).

Before we'll start to present the process of salt production we must explain which are the conditions that make possible the creation of salt evaporation pounds in the area of the Cáhuil Lagoon. The climatic and geographical compositions that must occur for the artisanal production of salt are quite particular, and in the case of the Cáhuil they are given by the seasonal connection of the Nilahue estuary with the sea, by the wind and the temperature. During the periods with high rainfall, the flow of the stream grows and is connected to the sea, which permitted the entering of the saltwater with the waves of high tides, 'the height of the tides in this area of the coast is 1.85 meters, which added to the low slope of the lower path of the estuary [...] allows that the tide back up at least seven kilometres till the sector where we can find the salt evaporation pounds used for the extraction of salt' (Andrade and Grau 2005, 61).

When the rains stop in September or October, the flow of the stream drops very much, revealing muddy terraces at the edges, which specifically '... corresponds mainly to metamorphic basement and batholiths of the coast, on which are located, in a jarring position, deposits of Quaternary marine terraces. The rocks of the batholiths and the basement are deeply altered, delivering powerful debris mantels of gravel (grus). This gravel is waterproof, therefore prevents filtering of the freshwater in the saline soils'. Dropping the flow of the stream, this can no longer lead to the sea settling a sandy barrier that disconnects the sea from the estuary, giving origin to the fails to lead to the sea complying a sandy barrier which switches off the sea in the estuary, giving rise to the lagoon of Cáhuil (Andrade and Grau 2005, 62).

The distribution of salinity changes strongly in the summer period in which the mouth is closed and the stream discharge decreases to 0.06 m^3/s on average. In this situation the actual estuary conditions are mooted and there is a pattern of distribution much more homogeneous, in which you can see a general increase of the salinity of the body of water with predominance in salinities of about of 24%, with a slight decrease in water above the lagoon of up to 21.5%, and with a vertical distribution that increases from the surface to the bottom, arising the higher salinities to 5 m deep, in the vicinity of the sealing bar (32.3%).

To resume, the area defined as the *Salinas de Cáhuil* falls from the Gulf of Cáhuil within the coastal profile of the VI-th region of Chile. Formed as an arm of the sea that is interiorized toward the continent several kilometres between two capes made basically by a topographic depression in relation to the level of the sea. The cape interior corresponds to a gorge formed by the estuary and the Nilahue marshes as a source of freshwater. In this arm there is an area of greater influence, where the level of the sea stretches out towards both coasts of the gulf, increasing the distance between these limits (in some points reaching 1.5 km). In these areas close to the shore it was built a system of locks that partitioned the seawater with passive methods from the slopes between the land and the sea level, in a complex network of content sites by the edges of earth. These areas of maritime influence are called "wetlands" or marshy areas located at the end of a river and in this case also act by differences in the height of tides (1.6 m in some periods).

Seawater is the raw material in the production of the costal salt. Currently the entry of saltwater takes place between the months of June and October; for the *salineros* the period in which the estuary remains connected with the sea diminishes day by day, as over the years the sedimentation and the clogging of the lagoon has continuously increased, which has led to an early closure of the mouth of the lagoon and this, prevents more sea water to enter the lagoon. It noteworthy that the 1955 geographical study of the area describes the closing of the mouth of the lagoon as late as February. Currently, the mouth closes in mid or early November. Also, the same study mentions the beginning of a process of sedimentation of the lagoon, which means a decrease in the discharge, which implies the growing need for pumps to fill the pen. This, together with the early closing of the mouth, means that today water pumps are needed not only to raise the water to the salt flats, but it also to drain them, because if the lagoon mouth closes very quickly it is impossible to fully clean the sites, an operation which increases the cost of production (Araya 2009, 111). The situation worsened a lot after the 2010 earthquake, which further raised the mouth of the estuary to the degree that today, according with some sources, the change of water can only be achieved using water pumps (Bustamante 2011, 23).

The process which ends with the production of salt is described more or less in the same manner in all the studies. The salt works are quite complex, structured productive units of very harmonic way. Immediately to the side of the lagoon we find the *corral* or *corralón,* an artificial pond where the waters originated from the lagoon are stored in order to be processed later. This pond is separated from the lagoon by a *fuerte* made buy mad embankment and branches, 1.70 m tall and 0.5 m wide and 50 m long, on average. The salty waters of the lagoon pass into the corral through a gate inserted in the *fuerte*, which allows the manipulation, control and regulation of the flow of salt water. The water is transported from the corral to the salt pans using smaller gates. A site of saline or short site is composed of *sal calles* or simply *calles*. The streets and the *piezas* are separated by small elevations of mud and branches, called *parapetes,* and connected through a system of *compuertas, canales* and a system of communicating vessels. In every *calle* are possible find four different types of *piezas*, according to the paper that they'll have assigned in the production process. The first types of pieza are the *sancochadoras*, followed by the *cocedoras* or *cocederas*, then the *recocedoras* or *recocederas* and, finally, the *cuajadora*, also called *cosechedora* or *cuartel*. The pools (*piezas)* are rectangular in shape, separated by long terraces of earth of approximately 30 cm breadth, parapets on which it is possible to walk. The saline waters are stored in the pools or *piezas*, where the process of evaporation occurs. These dimensions of a *pieza* (pool) are approximately 15 × 30 meters. Through this complex system of ponds, sluices and channels, is distributed the saltwater lagoon in order to obtain, after about 30 to 35 days, the salt harvest (Quiroz, Poblete and Olivares 1986, 105–107).

Due to the increase of the flow of the estuary during winter, the salterns are flooded with water and mud; therefore, when the production cycle starts, the salineros (salt worker) must drain the pools, operation needed to maintain the lagoon connected with the sea, so that a flow takes water from the salt sites. When the lagoon is already unable to connect with the sea, and the distance between the sea and the lagoon makes it impossible for the salineros, by means of their own, to the open mouth of the lagoon, they accordingly see the need to organize with fishermen, boaters and other organizations in order to call the authorities with heavy machinery to open the lagoon mouth. While this helps to drain, it is not enough; the salt workers must resort to pumps to clean the pools of water left by the stream in the winter (Araya 2006, 127).

In general, in the middle of September or early October begins the individual work in the Salinas. The first task after removing the water of pools is to remove the layer of mud which was left by the water of Nilahue, in order to let the soil clayish and waterproof for the processing of salt. In the process of salt production, the *Salineros* used a variety of tools, which are almost entirely manufactured by themselves, which together with their relevant and appropriate handling, represents being in possession of an adaptive technology. To remove the layer of mud pools first uses a rake (*rastrillo*) with which the mud is piled up, then a shovel (*pala)* with which mud is coiled. To clear the channels, through which water runs by transferring of one to another, a the shovel carrilana (*palo carillana)* is used. In addition, the workers should use the pot (*macete)* to tighten and match the parapets. With the trolley (*angarilla*) the salineros transferr the mud pieces towards the outside of the site, walking on the parapets. The *angarilla* is designed to be manipulated by two workers; consisting of a central plate and composed from multiple tables, the tool is attached to both sides of the stringers that serve to grab, and is about 70 cm long by 50 cm wide. Finally, the workers use the *pizon* to match and give consistency to the pieces (Araya 2006, 127–128).

When 'the salt water is delivered' to the salineros, from the pound or directly from the lagoon through the motor pump, the salt workers must carry the water to the first pool called *cocedera*, where the process of crystallization begins. The productive capacity of each piece depends on the terrain or, in words of one salinero, on the 'force of the mud to evaporate the water', a force which according to the salineros is innate in the site, under which the sites are classified according to the productivity of the soil. In addition, the soil to crystallize salt needs to absorb solar radiation, so the soil which receives solar energy is more efficient to evaporate the water. For this, the salineros leave each piece between 1 to 3 days without water, exposed to the sun. After that, the water is moving from *piezas* to *piezas* with the function of go composing it, which mean to increase the condensation of the salt through heat. As the water is taking temperature it is ready to begin to 'walk' on the different parts of the salt evaporation pans. The tour is not random, each piece has its name and function, so that the water first reaches the *cocederas*, which are the pieces located in the first line from the impound, then pass the waters to the rooms called *sancochadores,* where the water temperature is higher, and then go to the space predecessor of the *cuartels* — the *rececedera* (Fahrenkrog 2009, 53).

In the *corcederas* the water stays for one day, in the *sancochadores* for two, while in the *rececedera* the water is kept for approximately three days (Araya 2006, 132). In the beginning, the differences in the 'water' are not easily distinguishable, however, the salineros know only by looking at the waters when they are ready to be moved; for example, one of the signals for recognising the increased salinity of the water is that being denser, the wind does wavelets. The different salt pounds are filled with water from the neighbouring piece that will be the next to rave. The transfer of water is made with wooden handle buckets, and is directed toward the pipe carrying water to another part, otherwise the sharp blow of the water with floor piece turn the muddy water. Whit the stable time of heat (summer), the difference of the water begins to note: the water takes a pink colour, then acquires a deep copper colour, and if you touch is it becomes hot, dense and oily. At that moment they begin a more careful process of water transfer, since a delay in transferring water from the *rececedera* to the *cuartels* causes the salt to begin to curdle in the wrong sector, *rececedera,* resulting in a dirty salt, and lack of water to feed the *cuartel* (Fahrenkrog 2009, 54).

The *cuartel* must remain vacant for ca. one day before receiving the first transfer of water, for the same reason as the *cocedera.* The *cuartels* will receive water from the *rececedor* for 25 days, during which the water will crystallize and lose volume, so it should go filling the permanent water part. In general, the *cuartels* form a kind of emulsion that is floating on the water and prevents that the solar radiation to fall evenly across the water surface. Because of this, the salineros must break the emulsion, so that the crystallization is even; this is done with a rake with which they also move the water to prevent the formation of this emulsion (Araya 2006, 133).

After 26 to 30 days, all the water in the *cuartels* has been transformed into salt ready to be harvested. The salt is taken out using wooden spoons and deposited in the carts; during this part of the work the workers must be especially careful to avoid damaging the floor. Six to eight hours for the workday are dedicated to harvesting the salt of a single *cuartel*, and the rest of the day is used to clean the *cuartels* for the next production. They carry the *angarillas* loaded with salt out of salt—on the edge of the road—leaving the salt into piles that exude for a few days. The places where the piles of salt were left are betrayed by circles of burnt grass of various sizes. Only dry salt is stored in sacks, avoiding the dust of the dirty roads (Fahrenkrog 2009, 54). The extraction period runs from October to March, and during the rest of the months the *cuartels* are in ruin or temporary abandonment condition.

The position of the sea salt harvester has been transmitted from generation to generation orally, being the experiential event that has allowed to maintain the technique over time, there is no record bibliographic capable of realizing the dimensions of the harvest. They were initiated into this work as children, when they accompanied the father. One of the main concerns today it is that children no longer participate in the work alongside the father, since this activity bearing excessive physical effort and as schooling levels have increased overall in the communities throughout the country, children migrate to the city in search of better opportunities that involve less effort and which provide larger incomes than from the sale of salt. The office of Productive Promotion of the Municipality of Pichilemu indicates that they are 86 persons dedicated to this seasonal work, which has a duration brought near of 7 months. On the other hand, women are not involved in the extraction process, even though they are part of the 'social system' of Salinero and plays their role in other complementary areas related to agriculture, crafts, food, marketing, as well as caring for the home and family, and this despite the fact that some women are owners of salt pans.

A survey realized in this area with respect to the age of salt workers shows that 87% of respondents have more than 50 years and there are also people younger than 40 years. This configures one larger human group, which you may have difficulties to adapt to changes in the environment, but on the other hand there is experience and knowledge of the craft, also leaders who currently lead the process of formation of the cooperative are enabled to be partners able to undertake and lead the group that they represent.

In the salt industry, we find the system of work called *mediería* (a form of share the production in half between the holder of the terrain and the person which effectively worked) or the cases in which the wealthier land owner hired day labourers in order produce salt. The system of tenure and exploitation basically considers two indivisible elements, the owners of salterns which for this file they corresponded to 55% of the respondents and the sharecroppers correspond to 45%. The condition of the legal ownership of the streets represents 56% of the streets is regularized, i.e. their owners, they have the title. Of these, 71% have initiation of activities and invoice for their movements; other 19% is done on an informal basis (Sepúlveda 2009, 57–59).

In the course of this investigation, we got deep into the Salinas of Cáhuil Lagoon, knowing their procedure, noting that that is a long process—from seven to eight months—which requires specialized knowledge, and dedication of the salt workers. Also, that this way of producing salt, has been kept almost intact along the time, as well as the technology used — with the exception of the use of pumps and trucks.

As we met the salt industry activity in this research, it seems that it is slowly coming to an end, and that to reverse this situation today requires initiatives—absent today—that can improve the marketing and the working conditions, so that it competes in terms of profitability and required labour with other activities.

References

Andrade, B. and Grau, S. 2005. La laguna de Cáhuil, un ejemplo de estuario estacional en Chile central. *Revista de Geografía Norte Grande*, Julio 33, 59-72.

Araya Muñoz, C.G. 2006. *Salineros de la Laguna de Cahuil. Cristalizadores de Oro Blanco,* Unpublished Ba Thesis, Santiago, Universidad de Chile, Facultad de Ciencias Sociales.

AAVV. 2012. *Tesoros Humanos Vivos*, Consejo Nacional de Cultura y Arte, Santiago.

Bakewell, P. 1985. Mining in colonial Spanish America. In L. Bethell (ed.), *The Cambridge History of Latin America. Volume II Colonial Latin America.* Cambridge, Cambridge University Press.

Bustamante, C. 2011. *Informe, Diagnóstico.* Informe Final Intervención Cuenca Nilahue 2010-11 del Programa Servicio País de la Fundación para la Superación de la Pobreza, Pichilemu, VI Región.

Cortes Conde, R. 2006. Fiscal and Monetary Regimes in V. Bulmer-Thomas, J.H. Coatsworth, R. Cortes Conde (eds.), *The Cambridge Economic History of Latin America, Volumen II The Long Twentieth Century.* Cambridge, Cambridge University Press.

Deas, M. 1985. Venezuela, Colombia and Ecuador: the first half century of independence. In L. Bethell (ed.), *The Cambridge History of Latin America. Volume III From Independence to c. 1870, Colonial Latin America.* Cambridge, Cambridge University Press.

Diaz Garces, C. 1933. *El clorura de sodiu: su industria en Chile.* Santiago, Lagunas y Quevedo.

Eliott, J.H. 1984. The Spanish Conquest and settlement of America. In L. Bethell (ed.), *The Cambridge History of Latin America. Volume I Colonial Latin America.* Cambridge, Cambridge University Press.

Espinoza, E. 1890. *Jeografía descriptiva de la República de Chile.* Santiago, Gutenberg.

Fahrenkrog Borghero, K. 2009. *Salinas de Cahuil. Una etnografía sobre la actividad salinera en Cáhuil, Barrancas y La Villa*, Unpublished Ba Thesis. Santiago, Universidad Academica de Humanismo Cristiano.

Helms, M. 1984. The Indians of the Caribbean and Circum-Caribbean at the end of the fifteenth century. In L. Bethell (ed.), *The Cambridge History of Latin America. Volume I Colonial Latin America.* Cambridge, Cambridge University Press.

Lemonier, P. 1996. *Sal*, in P. Bonte and M. Izard (eds.), *Diccionario de etnología y antropología.* Madrid, Akal.

Lewis, C.M. 1986. Industry in Latin America before 1930. In L. Bethell (ed.), *The Cambridge History of Latin America. Volume IV c. 1870-1939.* Cambridge, Cambridge University Press.

Macledo, M. 1984. Spain and America: the Atlantic trade 1492-1720. In L. Bethell (ed.), *The Cambridge History of Latin America. Volume I Colonial Latin America.* Cambridge, Cambridge University Press.

Manriquez Munoz, F. 1955. *Salinas de Cáhuil: ensayo de geografía cultural,* unpublished Thesis. Santiago, Universidad de Chile.

Mauro, F. 1984. Portugal and Brazil: political and economic structures of empire, 1580-1750. In L. Bethell (ed.), *The Cambridge History of Latin America. Volume I Colonial Latin America.* Cambridge, Cambridge University Press.

Morner, M. 1984. The rural economy and society of colonial Spanish South America. In L. Bethell (ed.) *The Cambridge History of Latin America. Volume II Colonial Latin America.* Cambridge, Cambridge University Press.

Multhauf, R. 1985. *El legado de Neptuno. Historia de la sal común.* Mexico, Fondo de cultura económica.

Murra, J. 1984. Andean Societies before 1532. In L. Bethell (ed.), *The Cambridge History of Latin America. Volume I Colonial Latin America.* Cambridge, Cambridge University Press.

Ovalle (de), A. 1646. *Histórica relación del Reyno de Chile.* Online: /www.memoriachilena.cl/archivos2/pdfs /mc0012104.pdf [accessed: 02.10.2014].

Polanco, M. and Marcelo, J. 1996. *Historia Urbana de Pichilemu: Origen y crecimiento.* Santiago de Chile, Bogavantes.

Quiroz, D., Poblete, P.P. and Olivares, J. 1986. Los Salineros en la costa de Chile Central. *Revista Chilena de Antropología* 5, 103-120.

Schwartz, B.S. 1984. Colonial Brazil, *c.* 1580-*c.* 1750: plantations and peripheries. In L. Bethell (ed.), *The Cambridge History of Latin America. Volume I Colonial Latin America.* Cambridge, Cambridge University Press.

Sepúlveda, C. 2009. *Asistencia técnica y apoyo a captación de negocios asociados a los productores de sal del territorio. Informe Final.* Proyecto Consultora Alta Mar. Chile/ Pichilemu Región de O´Higgins.

Vera, J. 2003. *Las salinas de Boyeruca 1644-2001.* Unpublished Md Thesis, Santiago. Universidad de Chile.

Vivar (de), J. 1558. *Crónica y relación copiosa y verdadera de los Reynos de Chile,* Online: memoriachilena.cl/602/w3-article-8175.html [accessed: 10.10.2014].

Online source

http://www.memoriachilena.cl/602/w3-article-3436.html last visited 02.10.2014, 16.20.

Salt in Magical Procedures

Ileana Oana Macari

Universitatea „Alexandru Ioan Cuza" din Iaşi

Abstract

In Romanian traditional culture, salt holds an important place, and indications for procedures that used it have often survived to the present day mostly in the form of superstitions. Ordinary people knew—and still do—how to avoid the perils brought about by forbidden gestures or deeds. They grew with sets of do's and don'ts, but when any of these were broken, they had 'professionals' to resort to, because not everybody was qualified to perform such acts.

Keywords

salt, Romanian folklore, magic, witchcraft, ethno-mythology

In the Romanian traditional culture, salt holds an important place, and descriptions of procedures that used it have often survived to the present day, mostly in the form of superstitions. Ordinary people knew - and they still do - how to proceed in order to avoid the perils brought about by forbidden gestures or deeds. The analogies in magical practices identified in different cultures provide evidence for the idea that certain beliefs transcend the boundaries of time or nationality exactly because they are so profoundly human.

From an interdisciplinary point of view – by using historical, archaeological, anthropological, ethnological and linguistic or philological data – salt is more than just a mineral; it is a complex phenomenon that has affected the evolution of humanity in the course of time. In the Romanian traditional culture, this chemical compound holds an important place and indications for procedures that used it have often survived to the present day in the collective consciousness, mostly in the form of superstitions, proverbs and rhymes. Consequently, probably anyone in this country, irrespective of age, level of education or social origin, will be able to come up with examples in which salt is the common denominator.

A facile explanation for this may be that, because of its social and political evolution, and due to its rural nature, Romanian culture is a living depository characterized by the continuity of traditional practices to the present day. Another is that, thanks to the conjugated efforts of folklorists, linguists and historians, there are important anthologies and dictionaries that offer records of Romanian folk beliefs that were still in use in the 19th and even the 20th centuries, and the autochthonous education system employs such resources, either directly (as in specialized subject matters) or indirectly (as in literature or cultural studies). Nevertheless, the identification of the underlying reasons is less important than the possibility to reconstruct the essential features of the Romanian traditional culture starting from a synchronic and diachronic analysis of the elements we can resort to, in an endeavor based on analogies that account for old and new salt-related beliefs present in other cultures as well.

In Thomas A. Green 's ***Folklore: an encyclopedia of beliefs, customs, tales, music, and art*** popular belief is defined as a broad genre of folklore that includes expressions and behaviors variously called superstition, popular belief, magic, the supernatural, old wives' tales, folk medicine, folk religion, weather signs, planting signs, conjuration, hoodoo, root work, portents, omens, charms, and taboos. Alan Dundes has formulated a structural definition of folk belief (underlying patterns remain the same even if the content changes) in order to avoid the pejorative connotations of many of the terms just listed (Green 1997, 89).

Indeed, the above-mentioned critical attitude towards traditional magical practices has a long history going back to Antiquity, when the educated "scientific" minds mocked the people who were taken over by the compulsive observance of superstitions seen as lack of courage in front of the supernatural. A famous example is offered by Theophrastus, a 4th century BC scholar and pupil of Aristotle, whose portrait of the superstitious man is quoted in G. Anderson's *Greek and Roman Folklore*: The deisidaimon [...] is the kind who if anything pollutes him will wash his hands, sprinkle himself with holy water, put bay-leaves in his mouth, and walk about like this for the rest of the day. [...] When he goes past the smooth stones at the crossroads he anoints them with oil from his flask, falls on his knees, and reveres them before he can go on his way. [...] And he is adept at frequently purifying his house, claiming that something has happened to attract Hecate. And if he hears the hoot of owls as he walks along, he is in a panic, and exclaims 'Athene is greater' and only then passes on. And he does not like to set foot in a tomb, or go near a corpse or a woman in labour, but says he has to avoid pollution. [...] and when he has a dream, off he goes to the dream-interpreters, the soothsayers, or the interpreters of birds, to ask what god or goddess he should pray to. [...] And I imagine he would be one of those who scrupulously sprinkles himself with salt water on the seashore (Anderson 2006, 35-36).

Following the combination between the establishment of powerful official religions and scientific advance, such attitudes became dominant and got to account for the condescension and even contempt the modern man professed towards folk lore, as well as for the disruption that has occurred in modern societies between the inherited customs and everyday life. The recovering of such "data" builds on the conviction that humankind used to live in a continuum populated with shared symbolic representations and values.

Salt is such a common denominator present inside different communities in similar divination procedures that can be incorporated in halomancy (Ancient Greek ἅλς (*hals* 'salt') + μαντεία (*manteia* 'divination'). Some sources (Mystica...) consider halomancy a branch of pyromancy, which involves throwing salt into the fire, with the forthcoming events announced by the color, speed and direction of the resulting flames. Other authors (Occultopedia...) think that it involved the interpretation of random patterns formed by the scattering of salt over a flat surface, but admit that the exact method remains obscure.

However, in the following examples, remnants of such methods appear to have survived in simplified forms even in England and Germany which, in the late 18th and early 19th centuries, were among the first countries to undergo the industrial revolution that dramatically replaced the rural lifestyle with the machine-powered industry and commerce. School education became available to more people, including women, and whole communities were uprooted from villages and moved to towns to work in factories or coal mines. Surprisingly enough under those new circumstances, the most enduring symbols and beliefs were still present more than a century later in practices that were so obscured and essentialized that most of their practitioners and followers were unaware of original significances. According to Frazer's *Golden Bough*, a fairly down to earth method of salt divination that does not require any complex initiation was still used in the Isle of Man at the beginning of the 20th century, on Halloween: The housewife fills a thimble full of salt for each member of the family and each guest; the contents of the thimblefuls are emptied out in as many neat piles on a plate, and left there over night. Next morning the piles are examined, and if any of them has fallen down, he or she whom it represents will die within the year (Frazer 1922, 244).

An alternative ritual suggestive of pyromancy comes from Scotland; it was performed on Imbolg night, usually celebrated on February 1st or 2nd (about halfway between the Winter Solstice and the Spring Equinox) as a festival marking the beginning of spring.
In Scotland [...], it was a tradition for each member of the family to throw protective salt into the fire and divine their immediate futures by the pops and sparkling it made. This was considered to be a subset of Capnomancy, which usually covered all forms of "throwing something on the fire and figuring out what it meant." (Occultopedia...).

Girls everywhere are eager to know about their future husbands and are ready to devise and employ ways to get a glimpse of their prospects. In Romania, for example, on November 29th, on St. Andrew's Eve – a night that is quite similar to Halloween – among other practices meant to protect the household, young unwed women who want to get married in the following year (still!) knead a very salty wheat dough which they cook and eat before going to bed, and they expect to dream their future husband bringing them water to drink. In a testimony about life back in the 50's and Halloween traditions, J. Murphy recalls her grandfather's *Book of Fortune* which contained several "methods of divination, including Palmistry, Phrenology, Reading the Tea leaves, oracles and omens and other popular superstitions" (Halloween ceremonies...). Among other things, she describes a very similar method under the heading "Ceremonies to see your future husband, The Salted Herring": a salted herring eaten on Halloween Eve just before going to bed will make your future husband "appear in a dream, carrying a glass of water with which to quench your thirst" (Halloween ceremonies...). Similarly, in Greece on the Eve of St. Catherine's day (November 26th) girls feast on salt cakes and drink wine, then go to sleep "with the specific intention of dreaming of their future husbands" (Magic and religion revisited...).

A method belonging to cromniomancy (divination by onions) is still used nowadays in traditional Romanian households on New Year's Eve to predict weather in relation to the agricultural calendar. A detailed description was recorded at the beginning of last century by Elena Niculiță-Voronca in her impressive anthology of Romanian folk beliefs and customs; according to one informant, in order to find out which months were going to be rainier than others, one could make the "onion calendar". That involved selecting 12 fleshy leaves, one for each month, from January to December, placing equal amounts of salt on each and observing the quantity of water resulted, one could identify the rainy and the dry months (Niculiță-Voronca 1998, 107).

Historically performed all across Europe, Africa and northern Asia, cromniomancy is usually done by interpreting either the sprouting behaviour or the thickness of the onion skin, and an example of Southern German extraction is strikingly similar to the above-mentioned Romanian practice:

At Silvester evening you take twelve big, sphere-like pieces of onion and put them on a wooden board in a row. They represent the months from January to December . Then a grain of salt is put into every piece of onion, and the whole thing is left overnight in a room without heating, but without frost. The amount of liquid in every onion piece tells how much rain/snow there will be in the respective month (Cromniomancy...).

Since in Germany Silvester is the common name for December 31st and the accompanying celebrations, it is hard to believe that the similarities in the divination circumstances in the two countries are purely accidental.

The natural properties of salt that were empirically discovered quite early in the evolution of mankind made it an essential ingredient in purification, protective and healing rituals.

Amongst ancient magicians it was commonplace to lay down a pinch of salt in each corner of the room before performing a spell or ritual. This practice carried over into contemporary African-American hoodoo tradition as well. Both Greeks and Romans also mixed salt with their sacrificial cakes (Occultopedia…).

Folk medical practices may differ from culture to culture but they have in common, besides the underlying magical component, the belief that certain ingredients, such as salt, can offer a cure for several common ailments, such as colds, tonsillitis, sprains, skin affections and many others. In "Estonian Folk Medicine, Witches And Healers", Marju Torp-Kõivupuu notes that salt is a remedy that achieved special popularity with Estonians in the practice of so-called 'monks' medicine'. Before giving the healing salt to a patient, the monks chanted prayers to go with it (in Estonian - *puhusid soola peale*, blew on the salt). The synonym in Estonian for a folk or country doctor, *soolapuhuja*, blower on salt, probably dates from that time (Torp-Kõivupuu 2005).

The prayers accompanying the monks' treatments are suggestive of the charms the initiated rural women in Romania incant in order to scare away physical, psychical and sentimental affections by "doing and undoing" their age-old spells. In traditional cultures, ordinary people knew from their predecessors how to proceed in order to avoid the perils brought about by forbidden gestures or deeds. They grew with sets of "do's and don'ts", but when any of these were broken, they had such "professionals" to resort to, because not everybody was qualified to perform such acts. Nevertheless, common knowledge has preserved only the essence, in encrypted texts or in sequences of gestures to be performed on certain occasions, while the original meaning of the procedures is (intentionally) obscured because of their magical character.

The Romanian folklorists Grigore G. Tocilescu and Christea N. Tapu, who published the anthology *Materialuri folcloristice III* in 1900, noted that the few old women who still knew how to do spells were loath to communicate the charms because they believed that the magic might no longer work (Tocilescu 1981, 39). Another belief is that one should learn the charms by stealth, always from an older person, and never by writing anything down. That would be quite expectable, since this type of knowledge is supposed to be transmitted orally from one generation to another, without explanatory notes or instructions. The procedures have to be learned exclusively by imitation and practice. Fortunately specialists managed to find ways in which to extract and record valuable data that otherwise would have been lost today, and the above mentioned anthology includes, besides 380 incantations and 60 spells, the detailed descriptions of the procedures that accompany them and the required ingredients.

Some of them concern, for example, toothache, which was a common health problem in the times when no specialized medical services were available to ordinary people. For them, rural witches did toothache spells in salted water with which the person rinsed his/ her mouth. The spell could also be worked in salted water, salt, saltpeter, garlic cloves or incense, which the patient introduced afterwards in the decayed tooth. Some women did the spells in salt with a hot iron, on Wednesdays and Fridays, as those were fasting days (Tocilescu 1981, 164). According to Hatfield's *Encyclopedia of Folk Medicine: Old World and New World Traditions*, some of these remedies – apparently unaccompanied by the incantation itself – were used in Britain and the United States for the same purpose. For bad breath, "one simple British folk remedy was to drink and gargle with a solution of salt" (Hatfield 2003, 32), "garlic has been widely used to treat toothache" (Hatfield 2003, 170) and "applications to the sore tooth include salt, soda, ashes, a hot raisin and red pepper" (Hatfield 2003, 347).

Remedial procedures with salt shared by different cultures can be identified for several other affections, of which I will mention colds and tonsillitis. People in Ireland used "salt herring applied to the soles of the feet was a remedy for a sore throat" (Hatfield, 2003: 324), while for similar conditions Romanians even nowadays use hot salty water to bathe their feet, apply salted warm polenta, cabbage leaves or warmed salt on their chests when they go to bed, and that is to mention only the first that have come to my mind. Especially with children, tonsillitis could become a serious health hazard, and different incantations are recorded as a treatment for it. The symbols behind the characters that appear in them are nowadays obscured (in some versions there are three old women – one deaf one dumb and the other blind – who guard the tonsils and get eaten by wolves because of their handicaps), but the reasons for the "medical" procedures are obvious:

The charmer massages the big veins in the neck, rubs them with (edible, usually sunflower, *our note*) oil and repeats the words above (the charm, *our note*). Often a soft oiled dough is placed around the neck to soften the inflammation (Tocilescu 1981, 197).

At least equally interesting are the folk medicine treatments for skin affections. The Anglo-Saxon world has the following remedy for warts: "Some folk cures used the idea of flowing water carrying away disease. Warts were rubbed with salt, then washed in running water" (Hatfield 2003, 366). It is discernible that the affliction is fought with the help of two powerful symbols, water and salt, whose significances are too complex to be discussed here. Likewise, in Romania the spells for skin swellings usually require salt, water and, additionally, a special incantation. One procedure goes as follows: Salt is touched to the purulent swelling then dropped down, and the following words are uttered: Just as salt dissolves in water, So I command the pain, the stabs and the swelling Of (the name of the patient) to melt away (Tocilescu 1981, 139). The actual incantation that precedes the activity above is much longer and uses the power of numbers and several imprecation adjectives pointed at the disease.

For both the practitioner and the believer, the power of the words in a charm at least equals that of the material or performative components, even if many phrases have lost their sense or if some of the terms have altered beyond recognition in the course of their voyage through time by word of mouth. Moreover, anthologies and encyclopedias such as the ones I used as sources for the recorded instances provide evidence for the idea that certain enduring beliefs built around the essential elements of life transcend the boundaries of time or nationality exactly because they are so profoundly human.

References

Anderson, G. 2006. *Greek and Roman folklore: a handbook.* London, Greenwood Press.

Elworthy, F.Th. 1895. *The Evil Eye.* London, J. Murray. www.sacred-texts.com/evil/tee/tee03.html

Frazer, J.G. 1922. *The Golden Bough.* London, Wordsworth. www.bartleby.com/196/pages/page244.html

Green, Th.A. (ed.) 1997. *Folklore, An Encyclopedia of Beliefs, Customs, Tales, Music, and Art.* Santa Barbara, ABC-CLIO.

Hatfield, G. 2003 (ed.). *Encyclopedia of Folk Medicine: Old World and New World Traditions.* London, ABC-CLIO.

Holmstrom, L., *Magic and Religion Revisited: Gender and Dream Divination in the Ancient Mediterranean,* Sonoma State University (oral communication).

Mercatante, A.S. and Dow, J.R. 2009. *The Facts On File Encyclopedia of World Mythology and Legend,* Third Edition, The Facts On File. New York, Infobase Publishing.

Niculiţă-Voronca, E. 1998. *Datinele şi credinţele poporului român adunate şi aşezate în ordine mitologică* , vol. I. Iaşi, Polirom.

Noegel, S., Walker, J. and Wheeler, B. (eds). 2003. *Prayer, Magic, and the Stars in the Ancient and Late Antique World.* Pennsylvania, The Pennsylvania State University Press.

Tocilescu, G.G. and Ţapu, C.N. 1981. *Materialuri folcloristice III,* Bucureşti, Minerva.

Online sources

Torp-Kõivupuu, M. 2005. *Estonian Folk Medicine, Witches And Healers* www.estinst.ee/publications/estonianculture/I_MMV/koivupuu.html.

Occultopedia www.occultopedia.com/h/halomancy.htm.

Mystica www.themystica.com/mystica/articles/h/halomancy.html.

Magic and religion revisited www.sonoma.edu/users/h/holmstrl/DIVINATION.html

Salt as a Metaphor

Ludmila Bejenaru
Universitatea „Alexandru Ioan Cuza" din Iași

Abstract
Our metaphorical thinking is reflected in words and expressions; we can therefore say that we think and speak in metaphors; *salt* is one of these because it is the element without which man cannot live. Beyond any chemical, historical, or archaeological research, the aesthetic field has built an entire literature, from biblical models to cultural and literary figures, through the various connotations of *salt*. The stylistic diversity of salt is infinite, if we consider the topic of salt, the attributes of salt, the music of salt, the Salt Road, the Salt Riot, salt painting, salt between sacred and profane, statues of salt, the salt of kissing. Salt is ubiquitous in the reality of the languages of the world. The expressiveness of the Romanian, Slavic and other languages abounds in catchy phrases, proverbs, place names, names of rivers and lakes, in which salt is used as a symbol or metaphor. The overwhelming influence that salt had on the development of human society has turned the commodity into an essential symbol of universal culture, and its multiple meanings have become since ancient times part of the universal heritage of the people of the world.

Keywords
salt, the salt road, salt riot, salt metaphor, salt painting

In their *Dictionary of symbols: myths, dreams, habits, gestures, forms, shapes, colours, numbers*, Jean Chevalier and Alain Gheerbrant describe the symbol, or the concrete images and signs that describe the qualities of phenomena and concepts, as the relation between an image and a set of ideas, beliefs or emotions (Chevalier and Gheerbrant 1998, 438). According to Jung, symbolic communication is specific to ages, cultures and individuals. The concepts are turned into language, while metaphorical thinking is reflected in words and expressions; these issues were the subject of Lakoff and Johnson's cognitive linguistics. According to this science, the metaphor is an essential instrument in the organization of our conceptual system. Our metaphorical thinking is reflected in words and expressions; we can therefore say that we think and speak in metaphors; *salt* is one of these because it is the element without which man cannot live. Beyond the practical utility of this mineral it is also the case that in traditional societies, both in those of the distant past and the present, *salt* is assigned a number of imaginary properties, the symbolism of the most important natural curing agent being extremely varied. Beyond chemical, historical, or archaeological research, through the various connotations of *salt*, the aesthetic field has built an entire literature, from biblical models to cultural and literary figures: Mihail Sadoveanu, Theodor Rapan, Pavel Vasiliev, Ernst Lissner, Nichita Stanescu , Bashir Sultani. The stylistic diversity of salt is infinite, if we consider the topic of salt, the attributes of salt, the music of salt, the Salt Road, the Salt Riot, salt painting, salt between sacred and profane, statues of salt (pillars of salt), the salt of kissing. Our approach will concentrate on this area. Thinking about what surrounds us, we see the world borrowing the attributes of salt. In relation to language, the salt metaphor lies at the very foundation of Christianity, which uses *salt* along with water, fish, and bread, as a central symbol. In this respect, biblical references to the image of this commodity are legion.

Jesus Christ made a startling, resonant, statement about those who believe in Him and follow His teachings: "You are the salt of the earth and the light of the world". The salt metaphor suggests what the Christian is, and the light metaphor shows what the Christian does. As a Christian, his/her involvement in the life and fate of the world is twofold, as *salt* and as *light*.

"Salt and Light" is the metaphor used by Jesus in the Sermon on the Mount, which has parallels in the Gospel of Luke, the Gospel of Matthew, and the Gospel of Thomas. While in the Gospel of Thomas *salt* acts as a metaphor of light, in the Gospel of Matthew the "salt and light" metaphor is divided into two figurative segments that offer us new meanings: "the salt of the earth" and "the light of the world". The term "the light of the world" is amplified and extended by similarity: "the city on the hill cannot be hidden", "those who have lighted a candle, do not hide it under a bushel, but illuminate with it the whole house." In the Gospel of John we can also find the "the light of the world" metaphor, but this time it makes direct reference to Jesus.

"You have salt in yourselves and live in peace with each other" we read at *Mark* 10.50, this being the weightiest and heaviest burden thrown on our shoulders: to live in peace with each other, to love one another, to be effective as "salt and light" to those around us, giving meaning and significance to their lives, and becoming responsible for the confidence granted to us.

In Christianity, *salt* passes on the one hand for an essential substance, and on the other for a pure substance, filled with a special magical power. Just as salt is a basic element of human life—and things do not change over time—Christianity is the essence of humanity. The symbol of a religious elite, the meaning of the expression "the salt of the earth" is hard to define, since it was also a symbol of the covenant with God, of wisdom, and the need to preserve the purity of the world. It was no accident that the Apostles were chosen by Jesus as guardians of the essentials of faith, and were called by Him "the salt of the earth", alluding to their mission. In addition, the preservative qualities of salt are taken into account here in order to show the disciples that they were called in order to preserve society, protecting it from moral decay. According to Christian tradition, "the salt of the earth" consists of chosen, sanctified, people such as the Apostles, the Prophets, and priests.

In literature, the salt metaphor is based on a possible similarity between *salt* as a commodity and some of its new meanings, based on implied comparisons or by actual concepts and representations. The expression "the salt of the earth" is a symbol of the elite of a society or of a social group worthy of admiration; figuratively "the cream, the foam" of society (Mircea and Luiza Seche 2002, 219). "A group of people who represent what is better, more valuable, especially in a community, in a society" (*Dicționar explicativ roman* 1998, 374), "part of a society which holds a top position in the respective social group" (Marcu and Mânecă 1986, 142), "the salt of the earth" represents, in Russian literature, that segment of society that is able to change the world. The character Rahmetov in N.G. Cernisevski's novel *What shall we do?* is overcome by the obsession to search: the hero seeks, walks, struggles. He wants to find the beauty of the world, to discover the way to a better life for his followers. The hero strongly believes that one man alone could not do anything for the world; one must seek one's people, with whom one must go hand in hand, look for followers who can change the world, namely for those who are "the salt of the earth".

"Remember that we [the engineers] are the salt of the earth, the future belongs to us" says A. Kuprin. The author supports the Social Theory of Elites according to which in every society the task of social management is assigned to an elite, to a small group of people with superior education. This elite is able to assume responsibility for the good life of the society, being worthy of the trust that was granted to it by those who held it accountable.

Sadoveanu places his heroes on the opposite side. Through the expression, "the many who speak are the salt of the earth" the author promotes the values of the masses and not those of the elites. The hero of his novel *The Golden Bough*, Father Filaret of Amnia who gives away to the poor his considerable fortune after being prompted "by a sign from God", says: "The many, the poor and the stupid, are the salt of the earth; the world powers oppress them; the seas, the mountains and the rivers cannot give them as much as they deserve" (Sadoveanu 2005, 138).

One of the most interesting aspects of the early history of the Slavs relates to the extraction of salt. In Russia the extraction of rock salt, of coarse salt in the shape of large crystals, is attested between the fifth century and our own day and is related to the proto-Slavic tribes. The beginnings of extraction of salt in the Russian principalities from salt marshes and from underground deposits are attested by written sources of the 11th and 12th centuries. Salt acquisition was widespread in Pomorie, where mountains of salt were formed by the tides of the saltier seas in the region. Through a *gramota* of 1137 from St Sophia's Cathedral in Novgorod, the knez Sveatoslav Olegovici exercised strict control over the price of salt, increasing it in order to get funding for wars. The *gramota* set fees and taxes for other monasteries as well: for Kirilo-Beloziorsk, Solovki, Murom, and Troite-Serghievo Lavra.

The exploitation and marketing of salt deposits became part of world culture, the phenomenon giving birth to the Salt Road. The *Ipatiev Chronicle* of 1170 makes mention of the Russian knezs, who, after losing influence on the road "from the Varangians to the Greeks" across the Polovets, sought other ways of transporting and selling the salt they extracted from the Staro, Krasno and Ciongarsk rivers in the Crimea. In reference to the Crimea region, William of Rubruck writes in his journal of the middle of the 13th century, that "in the north there are many lakes, on the banks of which one can find salt springs; as only their water gets into the lakes, salt forms, which is hard as ice; from these saline deposits, the Batii and Sertah generate high incomes, since wagons from all over Russia go there for salt, and each one filled with salt has to pay a tax": the salt tax (Вильгельм де Рубрук 1993, 86-87). Apart from the Flemish traveller, some Russian scholars (Liubavin, Seredonin) believe that the Salt Road begins in Crimea for Kievan Russia; Seredonin states that some of the Russian knezs used the land route towards the salt mines from the tenth century onwards. This, however, did not cause them to give up their "saline" relationships with Galicia; evidence for which is to be found in the *Pecerska Patericon* edited by Yakovlev. The Salt Road on the River Dnieper passed through Pereprava which were below the estuary of the river, from Perekop to the Dnieper. The oldest crossing was at Krarii where the river was crossed by the Byzantine merchants who travelled by land from Kherson to Kiev.

17th century chronicles provide information concerning *chumaki*. These were Ukrainian townsmen and carters, who between the 16th and 19th centuries were sent to fairs with carriages full of salt and fish, from the Black Sea and the Sea of Azov. Being involved in the delivery of other goods, *chumaki* played an important role in Ukraine's trading relations with other states. The Ukrainian Salt Road, called the road of *chumaki*, passed by Nikitin Rog (Nikita's horn) on the left bank of the Dnieper to the town of Kahovka, where it reached Perekop through the steppe. A branch of this road leading to the Dnieper led to the Crimea. This Salt Road was the shortest of all, and this is why it was the most frequented. The importance of the Nikitin Rog pass was so great that the Tartars, who conquered the southern steppes of Russia in the thirteenth century, established one of their most heavily fortified positions there.

Salt was an especially valuable commodity, and there developed across Europe a trade route along which salt was carried from saline sources to the marketplace. The

Salt Road through Europe (Germany, Austria, Romania) can be reconstructed in part by archaeology and in part through documentary sources. Places located on such a trading route, such as Lübeck, Salzburg, Magdeburg, Leipzig, Prague, Venice and Rome, experienced a period of prosperity. The salt monopoly brought great wealth to Venice. Salzburg, Hallstadt and Hallein, towns located along the Salt Road in the Salzach river valley, in the centre of Austria, took their name from *Salz*, the Germanic root for "salt". Halle is another German "city of salt", and an old "salt route" connected it with Baltic Sea ports.

Salt was also an important commodity in the Adriatic and the Balkans. The Salt Road could not bypass Romania, a country with rich saline resources; the fact that neighbouring countries were poor in salt gave Romanian salt great value. The Salt Road passed through the eastern part of the city of Bacau as it was in 1816. In Bucharest, on the Salt Road, carts would carry blocks of salt from Romanian salt sources to be traded in Turkey. The Mures, Tisa, and Somes rivers provided water transport on both the northern and southern Salt Road.

France was another major producer of salt in both the Mediterranean and the Atlantic, and figures in our general picture thanks to the Salt Road along the Mediterranean coast, to Celtic salt, and also to *la gabelle* or the tax on salt, one of the causes of the French Revolution in 1792. Celtic salt, with the most expensive kind *Fleur de sel* is harvested from the salt water of the marshes of Brittany, Guérande, Noirmoutier and the Camargue. In Brittany, water comes from the Celtic Sea and the salt obtained from it has a very fine salty-sweet taste. It is said that salt can be obtained only when the wind blows from the east. *La gabelle*—the most burdensome tax in France—was introduced in 1286 and remained in force until 1890. Thanks to this tax, the price of salt increased dramatically, a situation that led to emigration, invasion and war.

In Russia, on 1 July 1648 the Salt Riot broke out, one of the largest riots of the middle of the 17th century. It came about on account of the extravagantly high taxes on salt, the price of which rose from five kopecks to 2 hryvnias a measure. The poet Nikolay Nekrasov (1821-1878) depicted these events in his *Salt Riot in Kolomensko* as did the painter Ernst Lissner in his *Salt Riot in Red Square* (1930). The modernist Afghan painter Bashir Sultani uses salt in another way: he uses it instead of colours to create portraits and other images. Instead of canvas he uses a dark wooden board and salt shakers with fine sea salt serve him as paint brushes. The painter's skilful movements, salt, a knife and a few cotton buds shape reality into images that reverberate between the soul of the characters and the world that becomes the object of reflection.

Portraits of Bob Marley, John Lennon, Michel Jackson, or Mahatma Gandhi depicted through the classic contrast between black (the "canvas") and white (the salt) are an expression of the personalities of these notables, organized into a whole, as a product of artistic intuition as well as the intentional achievement of their creator. Life and art mingle in Bashir Sultan's creations, but in a manner quite different from that cultivated by earlier painters. In his artistic creations, salt is the material through which we may reach the magic of art.

In literature, the Russian poet Pavel Vasiliev devotes the poem *Salt Riot* to this topic; it employs literary and artistic images to depict the reality of the Kazakh people in whose bosom he was born. Deeply impressed by their plight, the author expresses the organic connection between the Kazakh people and the land in which they have been rooted for centuries. The poem metaphorically defines the moral profile of the people, who defend their rights, by reference to the ancestral land and to history. The lyrical universe focuses around two poetic terms that define one another: *salt* and *taxes*, and suggests those moments of suffering through which entire peoples had experienced during their troubled history on account of a simple product that was however extremely valuable and cherished at all times. The author's vision is not local, provincial or personal. It is directly related to the realities of the world and was born of great pain—the pain of humiliation.

The Romanian poet Theodor Râpan, in *The Gospel of Silence,* places in new contexts and gives unexpected meanings to constructional elements including *salt,* which acquires stylistic connotations: "the salt of the earth is a sweet wound", "He looked at me sideways, his heart sends back its longing, it is a letter without which I cannot love you, the salt of the earth is a sweet wound." An oxymoronic construction, expressing an initiating process reveals new meanings to the reader: "The bitter taste of happiness beats me! *The salt in the kiss hurts*!" The flavour of the question is even more pleasant, as the association of terms is unexpected: "In this submission of the soul do you still want me, salt sprinkled on the wound of your hunger?" The magic language takes us through a poetics of what Nicoleta Milea has called "existence and knowledge, going beyond a programmatic text, as love and creation equally belong to the existential design" (Milea 2012, Metafora tăcerii la Theodor Râpan).

The symbolic nature of salt has penetrated deeply into popular culture, being still present today in various proverbs and sayings, superstitions and beliefs, even if their origins are unknown by many. Pagan collective rituals which survive only in the works of ethnographers, traditions and customs of many nations, exploit and use the symbols of salt, which is used as a purifying element, as a symbol of spiritual food, or as an offering along with bread in Christianity. For the Greeks, the Hebrews, the Arabs, as well as for Romanians, Russians, or Bulgarians, salt is a symbol of hospitality and friendship. Thus, the Romanian folk tradition, along with those of Bulgaria, Russia, Ukraine, or Belarus, still preserves the custom of honouring illustrious and valued guests "with bread and salt", without any terms of comparison, such as "the salt of the earth".

In earlier times, salt was so hard to find and its value was so great that no one wasted it; hence the popular saying "if salt is poured, a fight will start". Among the symbols of salt there is an image of a salt shaker with salt poured on a table, symbolizing evil and predicting possible disagreement, argument or betrayal that might occur between guests or within a community. In this respect, Leonardo da Vinci's famous *Last Supper* stirred and still sparks controversy among scholars and art critics, giving rise to various interpretations, sometimes generating confusion between the reality of the moment and the artist's vision. To emphasize the drama of Jesus' last meal with the disciples, the painter appears to have immortalized a symbol of the evil that was about to occur. This concrete sign suggesting betrayal is spilled salt placed on the table near Judas's right elbow. It goes almost unnoticed in the large fresco. Judas's face in the fresco, which became the symbol of the traitor and cunning individual, looks at Jesus. The dark face, full of treachery, shows hatred and wicked betrayal, and the right hand which grasps the bag containing the thirty pieces of silver, hits the salt shaker which pours on the table. The symbol of spilled salt also expresses deep feelings of disunion between people.

Salt is ubiquitous in the reality of world languages. People have admirably summarized the importance of the commodity. The expressiveness of the Romanian, Slavic and other languages abounds in epigrams, proverbs, place names, names of rivers or lakes, in which salt is used as a symbol or metaphor. Figuratively an "unsalted" way of speaking is an improper way of speaking, lacking in "good taste" and common sense. In a well-known Romanian fairy tale, a king's daughter tells her father that she loves him as "salt in food." The expressions "dear as salt in the eyes" or to be "like salt in the eyes" for someone, are phrases that describe an obnoxious, objectionable person.

The overwhelming influence that salt had on the development of human society has turned this commodity into an essential symbol of universal culture, and its multiple meanings have become, since ancient times, part of the universal heritage of the people of the world.

References

Chevalier, J. and Gheerbrant, A. 1998. *Dictionarul de simboluri—mituri, vise, obiceiuri, gesturi, forme, figuri, culori, numere.* Iaşi, Polirom.

Marcu, F. and Mânecă, C. 1986, *Dicţionar de neologisme.* Bucureşti, Editura Academiei.

Milea, N. 2012. *Metafora tăcerii la Theodor Râpan.* www.observatorul.com/articles_main.aps?

Sadoveanu, M. 2005. *The Golden Bough.* Bucureşti, Agora.

Seche, M. and Seche, L. 2002. *Dicţionar de simboluri.* Bucureşti, Litera Internaţional..

Вильгельм де Рубрук 1993. *Путешествие в восточные страны*, Alma-Atî, Gîlîm.

*** *Dicţionarul explicativ al limbii române.* 1998. Bucureşti, Univers Enciclopedic.

Salt and Health

Tilman B. Drüeke
INSERM Unit-1018, Team 5, CESP, Villejuif

Bernard Moinier
Paris

Abstract
Sodium chloride (NaCl), commonly called "salt", is added by households and the food industry in meals and processed products to improve their palability (flavour and texture). Urinary sodium excretion is used to estimate daily sodium intake in the majority of which is ingested with the diet. Sodium (Na) and chloride (Cl) are basic elements of the *milieu interieur* contributing to the maintenance of normal body fluid volume distribution. The presence of Na is required for a number of physiological functions, including cardiac and skeletal muscle contraction and nerve conduction velocity. As a whole, salt participates in growth and development of living organisms. A sufficient intake also is essential for sexual performance, reproduction and pregnancy. The minimum daily requirement amounts to 2 g/day. Under usual physical activities it would average 7-9 g/day, with wide variations according to gender, age, body weight, social status, region and climate. The importance of salt appetite is illustrated by its constant search by animals and man since the early ages.

Blood pressure (BP), which is the product of cardiac output of blood and peripheral resistance of arteries, is regulated by a subtle interplay between endogenous and exogenous factors including neurohormonal regulatory systems, physical activity, body mass index, stress, and last not least daily water and nutrient ingestion. Salt uptake (gut) and excretion (kidney) are controlled in such a way that excessive gains and losses are avoided. Accumulation of NaCl can favour arterial hypertension in people who have a variety of other predisposing factors. The objective evaluation of optimal, as opposed to excessive, salt intake is often obfuscated by misconceptions concerning the role of salt in the regulation of blood pressure, morbidity and mortality risk of healthy people. The blood pressure response to dietary salt intake is highly variable from one individual to the other, with no change in the majority ("salt resistance") but an increase in some of them ("salt sensivity"). The relationship between the salt intake and arterial hypertension has been investigated for at least a century and has been the object of numerous epidemiological studies and several intervention trials since the eighties. In contrast, the number of observational studies on a possible association between salt intake and the risk of cardiovascular events and mortality has remained scarce and inclusive at best, if not in favor of a U-curve relationship with higher mortality risk in people with either low or high salt intakes, respectively, as compared to those with usual salt intakes. Sadly, there are no randomized controlled trials on the effects of long-term salt restriction on cardiovascular morbidity and mortality in the population at large. Due to feasibility issues it is probable that such trials will never be done.

Salt and its possible role in other disease states constitute another field of epidemiological studies. The relationship of salt intake with bone mineral density and osteoporosis, if any, remains controversial despite numerous studies devoted to this issue. High salt intake has also been incriminated in the development of stomach cancer but published epidemiological data are unconvincing. As regards asthma, there is no clear-cut evidence for a deleterious role of dietary salt either. The potential role of NaCl remains also a matter of debate in the case of kidney stone disease. In any case, observational studies can only be considered as hypothesis generating and there are no randomized controlled trials in all these other disease states.

Special attention should be paid to salt supplementation with other micronutrients. Common forms of micronutrient malnutrition include iodine and iron deficiencies. Special attention has generally been paid to iodine and iron with salt as an appropriate carrier. Moreover, salt is also used to carry fluorine, and folic acid. Salt iodization and fluoridation both participate in WHO mass prevention recommendations regarding physical and mental retardation on the one hand, and dental decay on the other. Iodine disorders are widespread throughout the world, including industrialized countries. Even moderate levels of iodine deficiency can have detrimental effects on human health. Improvement of oral health depends on several factors which include topic and systemic action of fluorine. Salt fortification is a low-cost, relatively simple strategy that may reach a wide range of people, and can contribute to reducing the high prevalence of micronutrient deficiencies affecting mothers and children, especially in poor countries, with important health and economic consequences.

Despite the impressive number of studies and trials on the potentially adverse impact of excessive salt ingestion and restriction respectively, the debate on optimal salt intake is eagerly going on. What consumers and the health authorities as regulators need to know is whether a population-wide reduction in salt intake, with a concomitant reduction in iodine and fluorine intake, will lead to a better or worse state of health and longer or shorter life expectancy. Challengers of the arbitrary universal salt restriction to less than 6 g/day have repeatedly called for randomized prospective trials to test the validity of this recommendation.

In conclusion, both excessive salt intake and excessive salt restriction are probably harmful. The higher and the lower limits of salt consumption and the possible danger of insufficient iodine intake with generalized salt restriction need to be more clearly identified. Salt sensitivity and tolerance in terms of blood pressure are highly variable from one individual to the other. Cardiovascular morbidity and mortality appears to increase at both high and low salt intakes. Therefore, it may be more appropriate to tailor recommendations on optimal salt consumption based on individual needs and tolerance rather than on simplistic extrapolations from weak, inconsistent associations with blood pressure.

Keywords
sodium, chloride, salt, iodine, fluoride, blood pressure, salt sensitivity, cardiovascular events, kidney, daily intake, diet, food products, development, mental retardation, deficiency, oral health, health policy, epidemiological studies, intervention trials, prospective trials

Introduction

The terms 'sodium' and 'salt' are indifferently used by the public, as if they were interchangeable terms. However, strictly speaking sodium is not salt. Sodium is only one among the numerous elements of the environment. It does not occur naturally in isolated form. It represents one of the two elements necessary to form the most common form of 'salt'. The other element is 'chloride'. When sodium associates with chloride it becomes sodium chloride (approx. 40% sodium, 60% chloride). Chemically speaking, sodium chloride is but one among various types of salt, such as sodium bicarbonate, sodium fluoride, potassium chloride, calcium chloride, calcium carbonate, and magnesium sulfate. Notwithstanding these limitations we, as the majority of other authors, will use the term 'salt' as a surrogate term for sodium chloride, if not specified otherwise.

Sodium chloride is added by the food industry and households as a preservative and an ingredient which provides flavor and texture to many foods. Dietary salt is mainly derived from sodium chloride naturally present in foodstuffs, salt added in food processing, and salt added at the table or in cooking (discretionary salt). Nutrition data suggest that natural sources make up 1 to 1.5 g ingested sodium chloride daily. Discretionary salt provides 2 to 3 g per day. The main amount derives from salt added to food by food manufacturers which represents 3 to 4 g per day. Salt equivalent from other sources is negligible.

After oral ingestion, salt dissociates in the intestinal fluid into its two constituents, to become sodium and chloride ions. These ions are absorbed across the intestinal wall by partly separate mechanisms. Under usual steady state conditions, both sodium and chloride are absorbed entirely, and the amounts absorbed are excreted in the urine. This is the reason why 24-hr urinary sodium excretion is generally considered as a useful surrogate parameter for the estimation of daily sodium intake. Under hot environmental temperatures or sustained physical exercise, however, a large proportion of the absorbed sodium chloride (up to 50% or even more) is eliminated with sweat, i.e. via the cutaneous route. In such circumstances, reliance on urinary sodium excretion alone may lead to a more or less considerable underestimation of total sodium intake. Fecal excretion is usually minimal. It can however increase in presence of diarrhea.

Physiological roles and needs of salt

Physiological roles of sodium and chloride

Sodium and chloride are important elements of the « milieu intérieur » bathing the cells of the organism. They are indispensable for the maintenance of normal volume distribution between the extracellular and intracellular milieu, i.e. the amount of fluid inside and outside the cells of the body, respectively. Na and Cl are the most important ions in the extracellular fluid, whereas K and a variety of negatively charged ions predominate in the intracellular fluid. The extracellular volume comprises the blood space and the interstitial space surrounding the cells. The maintenance of a normal blood volume is essential for maintaining normal blood pressure (see below), in concert with the pump activity of the heart and fine tuning of peripheral resistance, that is the degree of openness (vasoconstriction/vasodilatation) of the peripheral arteries and arterioles. A precise control of sodium and water input into and elimination from the organism, guaranteeing neutral sodium and water balance, is of utmost importance for the maintenance of physiological steady state. If the control mechanisms governing sodium and chloride fluxes across the cell membranes fail to function correctly, disease states will ensue.

Until recently, it has been assumed that sodium accumulation in the body is paralleled by commensurate water retention to maintain normal osmotic pressure (« iso-osmolality ») of body fluids. However, human sodium balance studies and experimental data performed in the last decade have suggested that this is not necessarily the case. The group of Titze and his coworkers have shown that water-free storage of sodium may occur in interstitial tissues (i.e. tissues surrounding the cells), particularly in the skin, and that this is a regulated process involving the lymphatic system, activation of a specific protein called « tonicity-responsive enhancer binding protein » (TonEBP), and activation of mononuclear phagocyte system cells (Machnik *et al.* 2009, 545-552). The discovery of this regulatory system may have implications for the salt-blood pressure relationship. If disturbed, it could contribute to the pathogenesis of arterial hypertension.

The presence of sodium is required for a large number of other physiological functions, including skeletal and cardiac muscle contraction and nerve conduction velocity. As to chloride, it participates in the digestion of meals by contributing to the acid milieu of the stomach in the form of hydrochloric acid (HCl).

Salt also plays an important role in growth and development (2) (Moinier and Drueke 2008, 2154-2161). Thus it has been shown that experimental animals put on low sodium intake do not grow normally. One of several possible mechanisms is the induction of insulin resistance. Moreover, sufficient salt intake is essential for sexuality, reproduction, and pregnancy (Moinier and Drueke 2008, 2154-2161). Salt-induced alterations of neurophysiological functions and sexual behaviour have already been presumed by the ancient Greeks, and it is generally admitted to date that salt restriction can weaken sexual desire. Significant associations have been observed between salt restriction during gestation and low birth weight. Recent literature provides ample information on the effects of an optimal dietary electrolyte intake, including the intake of sodium chloride, on reproductive performance, and the physiological status of various species.

Physiological control of blood pressure

Blood pressure, which is the product of cardiac output (of blood) and total peripheral resistance (of arteries), is controlled by a subtle interplay between endogenous factors including age, gender, the cardiovascular system, the kidneys, the endocrine system, neurohumoral factors, and genetically determined predisposition, and exogenous factors including dietary factors, physical activity, body mass index, and psychosocial stress. Among the exogenous factors, sodium chloride plays a central role. Its uptake by the organism from the gut and its excretion by the kidney are subject to a tight control by these organs and regulatory systems, so excessive gains or losses are avoided. Accumulation of sodium chloride favours arterial hypertension, whereas depletion may lead to hypotension.

The other most important dietary element for the control of blood pressure is potassium. High intakes of this nutrient are protective against, whereas low intakes are permissive for arterial hypertension. Nutrients such as calcium and magnesium also contribute to the regulation of blood pressure, with sufficient to high intake of these elements being probably protective against high blood pressure. The role of diet in blood pressure regulation involves a number of interrelated nutrients and, as food intake varies widely, the arbitrary modification of one or the other among them may be beneficial or detrimental, especially when considering hard outcomes such as cardiovascular events and mortality (see below).

Several other exogenous factors play a role as well, either via an interaction of salt with blood pressure or independently. Thus regular physical activity is protective whereas stress, obesity, and excessive alcohol consumption favor arterial hypertension (Muntzel and Drueke 1992, 1S-42S).

Physiological needs of salt and salt appetite

Since the body has only limited reserves of sodium and chloride, it depends on a regular intake of salt. The minimum daily requirement compatible with life has been estimated to be 1-2 g sodium chloride. However, this minimum amount is not compatible with usual physical activities. Total salt intake estimates derived from 24-hour urine samples usually are in the range of 7 to 9 g per day (Muntzel and Drueke 1992, 1S-42S), with the majority of samples being comprised between 6 and 14 g per day in developed countries (IntersaltCooperativeResearchGroup 1988, 319-328). Reviewing salt intakes in various European countries more than 20 years ago James *et al.* (James, Ralph and Sanchez-Castillo 1987, 426-429) noted that, while average salt intake in Finland and Sweden was in the same range as in the UK, a lower proportion of salt appeared to be added in food processing in these countries. Table 1 shows an update of the amount of salt ingested with the diet in various countries worldwide. The relative importance of assumptions about both discretionary salt used by the cook and salt added by the industry has to be underlined, keeping in mind that a significant proportion of this salt is not ingested, but discarded.

Increased needs of salt intake are related to changes in external conditions and enhanced sodium losses such as hot climate and high levels of physical activity, and also to disease states such as salt losing endocrinopathies and nephropathies and the chronic fatigue syndrome.

Decreased needs of salt intake may be associated with several disease states including salt-sensitive hypertension, chronic heart failure, chronic kidney disease, the nephrotic syndrome, and decompensated liver cirrhosis, that is clinical conditions associated with salt and water retention. A number of randomized clinical trials have however established that salt restriction also stimulates sympathetic nerve activity, increases serum cholesterol and triglyceride levels, activates the renin-angiotensin system, decreases insulin resistance and alters aldosterone secretion (Graudal, Hubeck-Graudal and Jurgens 2011). These stimulations are not necessarily beneficial, they may even be harmful.

The importance of salt appetite is illustrated by the constant search for salt of animals and of man since the early ages. The complexity and sophistication of the central control of sodium appetite offers compelling support for the proposition that vertebrates evolved a mechanism to assure that their physiologic needs for sodium are defended when dietary access to it is limited or when excessive amounts of sodium are lost under conditions of stress such as hemorrhage, sweating, or diarrheal illness. The importance of angiotensin II, aldosterone, and other peripheral signals to these central nervous system circuits activating sodium appetite is consistent with an objective of maintaining optimal extracellular volume status (Geerling and Loewy 2008, 177-209; McCarron *et al.* 2009, 1878-1882).

Table 1. Mean salt intake according to Intersalt, Intermap and various other sources* (n.a., not available; n.r., considered non reliable).

	Intersalt	Intermap	Other sources	Extreme data
Argentina	9.1	n.a.	<12	
Australia	n.a.	8.5	10.3/7.7	12
Austria	n.a.	n.a.		
Belgium	8.5	n.a.	10.4	
Brazil	n.r.	9.9	10.8	1.9
Canada	10.3	n.a.	8.5	
China	12.1	11.9	8.4/7.7	
Denmark	8.2	n.a.	10.6/7.1	
Finland	9.1	8.4		
France	n.a.	n.r.	7.9	
Germany	9.5	n.a.	9.0/6.5	
Hungary	11.7	n.a.		
Iceland	8.1	n.a.	9.3/6.5	
India	10.7	n.a.	9.9/9.0	
Italy	10.3	11.0	11.4	
Japan	11.0	12.6		23.5
Mexico	8.5	n.a.	6.3	
Netherlands (The)	8.8	7.1		15.4
Nigeria	n.a.	6.7	6.6	
Poland	11.1	n.a.	13.5	
Portugal	n.r.	n.a.	12.3	
RSA	n.a.	8.7	7.8/9.5	
Russia	n.a.	n.r.	13.4	
South Korea	12.2	n.a.	11.5	
Spain	8.9	n.r.	11/8	5.5
Switzerland	n.a.	n.a.	9.3	
UK	8.9	**8.9**	**8.8**	
USA	7.6	**9.1**	**9.0**	

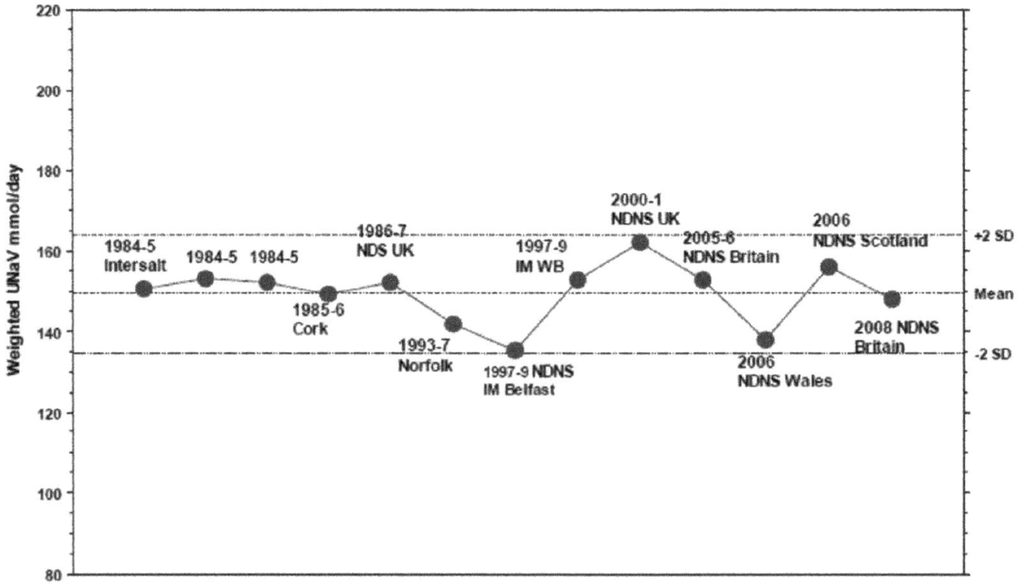

Figure 1. Mean and ± SD 24-h UNaV from 13 published surveys in the UK between 1984 and 2008 with essentially equal representation of women and men ($n = 6343$). Trend line equation $y = -0.097x + 150.4$; $R2 = 0.0026$; UNaV, urinary sodium excretion; UK, United Kingdom; NDS, National Diet Survey; NDNS, National Diet and Nutrition Survey. From McCarron DA *et al.* (McCarron *et al.* 2009, 1878-1882) (reproduced with permission).

* Sources: INTERSALT (4) (Intersalt Cooperative Research Group. 1988, 319-328); INTERMAP (93) (Brown et al. 2009, 791-813); and several additional articles identified in PubMed, including McCarron et al (8) (McCarron et al. 2009, 1878-1882) and (9) Bernstein and Willett (Bernstein, Willett 2010, 1172-1180).

An objective evaluation of salt appetite and needs for optimal development is often obfuscated by present misconceptions of the salt and health issue. Sodium appetite could well be a physiologically set parameter with a relatively narrow range intended to assure optimal function of a myriad of physiologic systems (McCarron *et al.* 2009, 1878-1882). In addition to evidence for this proposition derived from research on the central nervous system and peripheral mechanisms that control vertebrate sodium appetite, there is evidence from precise measurements of salt intake in humans that have accumulated over the past two to three decades. These measurements have shown that dietary sodium intake has varied minimally in the UK over the 25 years, encompassing several surveys from the UK and also the INTERSALT data (IntersaltCooperativeResearchGroup 1988, 319-328), as put together by McCarron *et al.* (McCarron *et al.* 2009, 1878-1882) (Figure 1). Of interest, they do not support recent pronouncements by the British Foods Standards Agency that their national campaign directed at sodium reduction has achieved a significant reduction in the population. Similary, according to a recent report by Bernstein and Willett based on the analysis of 38 studies dating from 1957 to 2003, sodium intake in the US adult population appeared to be well above current guidelines and did not appear to have decreased with time (Bernstein, Willett 2010, 1172-1180).

Disturbances of sodium and water homeostasis

Changes in serum sodium concentration are among the most common electrolyte disorders in man. They are generally associated with disturbances in water balance. A decrease in serum sodium concentration is called hyponatremia and an increase is called hypernatremia. Depending on the severity of the concentration changes and the underlying disease, they can be clinically inconsequential or lead to serious health consequences including death.

The serum sodium concentration is primarily regulated by water homeostasis. The latter depends on water intake and excretion which are adjusted by thirst and a hormone called “antidiuretic hormone” (ADH) or “vasopressin”.

Hyponatremia

The most common cause of hyponatremia is water retention (“dilutional hyponatremia”), due to an insufficient capacity of the kidney to excrete ingested water like for instance in patients with chronic heart failure, liver failure, nephrotic syndrome, or kidney failure. More infrequently, it is due to enhanced renal reabsorption of water like for instance in patients with inappropriate secretion of ADH. Another, relatively uncommon cause is excessive water intake. The majority of causes involved in hyponatremia are acquired although a minority among them is the consequence of hereditary diseases.

The clinical manifestations of hyponatremia are mainly associated with dysfunction of the central nervous system caused by cerebral edema. Nausea, malaise and headache can be followed in more severe cases by lethargy, obtundation, seizures, coma and respiratory arrest.

Hypernatremia

Hypernatremia frequently develops in elderly and/or hospitalized patients with fever and unreplaced water loss due to impaired thirst or water intake. It also can develop in children and adults as a consequence of prolonged vomiting or diarrhea. More infrequent conditions include a large variety of causes including endocrine diseases such as primary hyperaldosteronism, central or nephrogenic diabetes insipidus due to insufficient secretion of ADH by the pituitary gland or renal resistance to the action of this hormone; certain forms of chronic kidney disease; hypercalcaemia; hypokalemia; renal causes of hypotonic fluid loss like in diabetes mellitus or with diuretic drugs; nephrotoxic effects of drugs such as lithium; and ingestion or infusion of hypertonic sodium-containing solutions. Most of these conditions are acquired. Only a minority among them is the consequence of hereditary diseases.

Clinical manifestations of hypernatremia also largely reflect central nervous system dysfunction, with accelerated breathing, restlessness, muscle weakness, insomnia, lethargy, and coma. Both severe hyponatremia and hypernatremia require rapid expert medical intervention.

“Iso-osmotic” retention of sodium and water

In many disease conditions there is retention of both sodium chloride and water, in the absence of marked changes in serum sodium and chloride concentrations and hence of osmotic pressure in the extracellular fluid. The most common diseases involved in this type of “iso-osmotic” salt and fluid retention are chronic heart failure, the nephrotic syndrome, chronic kidney disease, and chronic liver disease. These conditions may lead to edema, pulmonary congestion, ascites and arterial hypertension, respectively. Most of these diseases are acquired although they may be favored by genetic background. Only a minority among them have a predominantly genetic origin.

Several genetically transmitted diseases have been described in the last 20 years which involve primary or secondary disturbances in renal tubular sodium transport. Numerous monogenic forms of hypertension have been identified which however have a very low prevalence in general population. In contrast, large numbers of common allelic variants of candidate genes for hypertension have been identified in relation to salt sensitivity but their relative importance for this phenotype remains uncertain and is probably small. To name but a few among them: Liddle syndrome, Bartter syndrome, glucocorticoid-remediable aldosteronism (GRA), and apparent mineralocorticoid-excess (AME). The pathogenic role of gene-gene interaction has received increasing attention but remains at present in the research domain. The search for quantitative trait loci in the human genome linked to the transmission of

Table 2. Salt sensitivity: variable individual response to changes in dietary salt intake.

Relative salt-sensitivity	* old individuals * black subjects * some hypertensive subjects (30 - 40 %) * baseline level of blood pressure * calcium deficiency * expansion of extracellular volume and increase of atrial natriuretic factor (ANF) * stimulation of sympathetic nervous system * biochemical and genetic markers : – low plasma renin activity – increased plasma ANF – decreased urinary kallikrein – decreased urinary endothelin-1 – haptoglobin 1-1 phenotype – 17a-hydroxylase deficiency – impaired nitric oxide synthase activity – mutations of sodium channel genes – genetic polymorphism of a2-adrenergic receptor
Relative salt-resistance	* young individuals * white subjects * most hypertensive subjects (60 à 70 %) * biochemical and genetic markers : – elevated plasma renin activity – low plasma ANF – haptoglobin 1-2 and 2-2 phenotypes – genetic polymorphism of a2-adrenergic receptor
Variable salt sensitivity during lifetime, potential factors	* age * baseline level of blood pressure * obesity * pregnancy * several disease states * some medications
Certain biochemical and genetic markers may be used to assess salt sensitivity. For example the activities of several plasma enzymes and the amounts of certain proteins in plasma and tissues are related to salt sensitivity.	

salt-sensitive hypertension has so far been less successful (Strazzullo and Galletti 2007, 25-32).

Salt intake, hypertension, cardiovascular disease and mortality

Salt sensitivity and salt resistance

A number of physiological and genetic factors have been shown to determine salt sensitivity in humans. Salt sensitive individuals are by definition prone to exhibit a blood pressure decrease in response to salt restriction, and a blood pressure increase in response to salt overload. There is no commonly accepted method to assess salt sensitivity and thus no precise definition of it. Salt sensitivity can be evaluated by placing subjects respectively on a high and/or low salt diet for several days and to measure blood pressure changes in response to these diets. Roughly 40% of individuals in the general population are generally considered to be "salt-sensitive". It is often forgotten that this means the other 60% are "salt-resistant". Some of them, mostly younger people, may even be "reverse reactors". They increase their blood pressure in response to a marked restriction in salt intake. Figure 2 shows a histogram of salt sensitivity in terms of systolic blood pressure changes in normotensive individuals in response to salt depletion (Weinberger *et al.* 1986, II127-134).

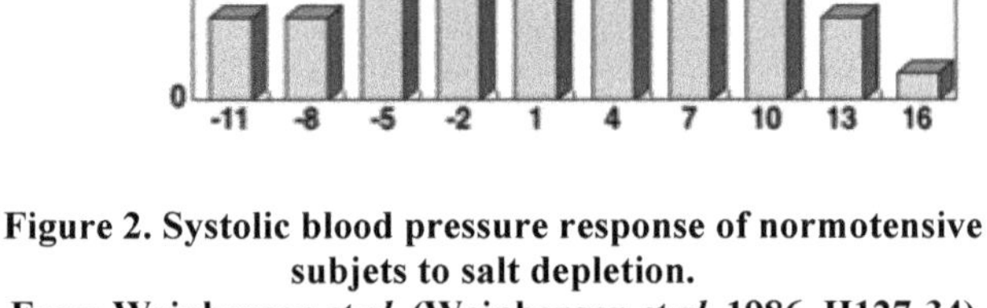

Figure 2. Systolic blood pressure response of normotensive subjets to salt depletion. From Weinberger *et al.* (Weinberger *et al.* 1986, II127-34) (reproduced with permission).

Salt intake and blood pressure

The medical community and the lay public continue to be confused about the role of sodium chloride in the regulation of blood pressure. As pointed out above, several disease states are associated with a reduced capacity of the kidney to excrete sodium and lead to arterial hypertension. Moreover, lowering of salt intake often brings about an amelioration of associated adverse conditions including high blood pressure. Lay people and also numerous physicians take these observations as sufficient proof for the postulate that in healthy subjects a relatively high salt intake must be deleterious as well, and that universal salt restriction should be beneficial.

There is also a historical basis for the assumption of a close relation between salt intake and blood pressure (Muntzel and Drueke 1992, 1S-42S). It relies on the fact that in the early years of the last century practically no measures were available to decrease high blood pressure other than a drastic reduction of salt intake. The concept that dietary salt is a major determinant in raising blood pressure was further developed by Dahl in the fifties of the last century, with a type of laboratory rat that was sensitive to salt. Thereafter, information on the influence of salt on blood pressure has been gathered from human epidemiological studies, such as those performed in Northern and Southern Japan, which suggested an association between salt intake and high blood pressure, although the evidence provided for this hypothesis was extremely weak, if not entirely speculative (Muntzel and Drueke 1992, 1S-42S). Nonetheless the idea progressively developed that salt is a villain for everyman's blood pressure.

Association of salt intake with high blood pressure – epidemiological studies

Since the eighties, well-designed epidemiological studies have been conducted in the attempt to establish definitively the presumed relationship between the level of salt intake and the prevalence of high blood pressure. The largest such effort has been undertaken with the INTERSALT study which showed a weak positive association between urinary sodium excretion (reflecting salt intake) and blood pressure based on 10,079 volunteers in 52 participating centers worldwide (IntersaltCooperativeResearchGroup 1988, 319-328). This suggested that systolic blood pressure augmented by 2.2 mm/Hg for every 100 mmol per day increase in habitual sodium intake (6 g salt per day) (Figure 3). However, the association disappeared when four centers in Brazil, Kenya and New Guinea were excluded (Figure 3). The populations in these specific centres had unusually low salt intakes and blood pressures and differed from the populations of industrialized countries in many respects. Moreover, the generally limited life span of these primitive tribes did not allow one to obtain any information on the possible development of hypertension with age.

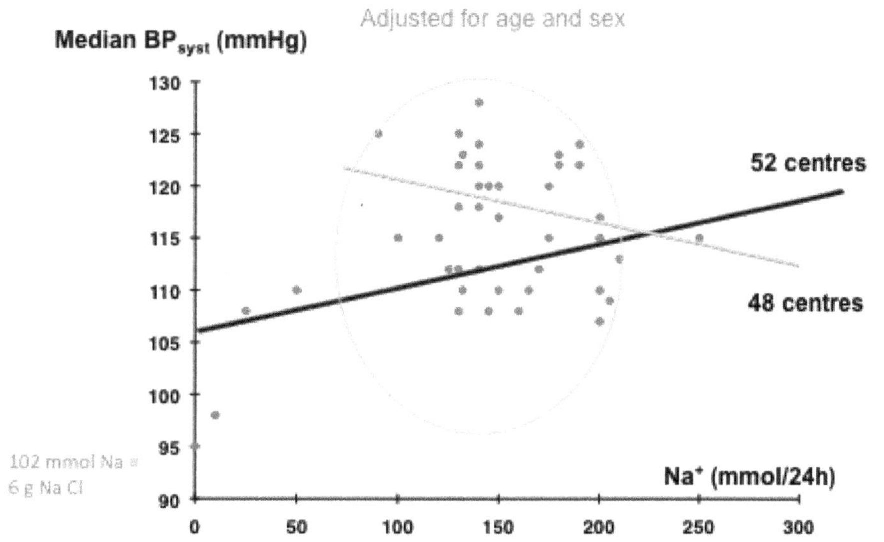

Figure 3. INTERSALT 1988: within population studies (Intersalt Cooperative Research Group. 1988, 319-328) (reproduced with permission).

Of interest, another observational study done in a large population sample in Scotland, which was published at the same time as the INTERSALT study, was unable to identify a relationship between urinary Na excretion and blood pressure (Smith *et al.* 1988, 329-330). An important source of contention in these studies is the effect of allowing for differences in so-called « confounding factors ». When the effects of age, body mass index (BMI), physical activity and alcohol consumption are taken into account, the association between salt and blood pressure becomes weaker or vanishes. Clearly excessive alcohol consumption and obesity are more important lifestyle factors influencing blood pressure than is salt intake. Another important finding is that the correlation between blood pressure and urinary potassium or urinary Na/K ratio is much higher than that found with urinary Na alone (Geleijnse *et al.* 1996, 737-741). Thus the evidence for high potassium intake being protective against hypertension is much stronger and less controversial than the evidence for a deleterious effect of high sodium intake. Reports from the literature on the potential role of calcium are more controversial. Whereas some showed higher intakes of calcium to be associated with lower blood pressure values others could not confirm this observation (Dickinson *et al.* 2006; Griffith *et al.* 1999, 84-92; McCarron 1985, 607-627).

Effects of salt on blood pressure – intervention studies
It is worth mentioning that epidemiological studies can only identify statistical associations between lifestyle factors and disease. To demonstrate that increasing salt intake causes blood pressure to rise or that salt reduction causes blood pressure to fall requires direct experimentation – so-called « intervention studies ». Even with well-done intervention studies, the results have been once more conflicting. Some studies have found significant reductions in normotensives as well as in hypertensives when salt intake was decreased, whereas others have not. Differences between studies included the total numbers of subjects investigated, their age, their initial blood pressure, their willingness to reduce salt intake and the length of the study. In one initial analysis (Law, Frost and Wald 1991, 811-815), salt reduction appeared to have been effective in studies lasting 5 weeks or longer (33 of 78 trials analyzed) but not in those lasting less than 5 weeks (45 of 78 trials analyzed). However, in another analysis (Swales 1991, S42-S6), when the author took care to include only those studies that had an appropriate design, the opposite conclusion was reached. The effect of salt reduction was nil for studies with a duration lasting more than 4 weeks, a conclusion that has been reinforced by three subsequent meta-analyses (Graudal, Galloe and Garred 1998, 1383-1391; Hooper *et al.* 2002, 628-632; Midgley *et al.* 1996, 1590-1597), although meta-analyses carried out by other groups reached the opposite conclusion (Dickinson *et al.* 2006, 215-233; He and MacGregor 2006, 861-869; Strazzullo *et al.* 2009).

The prospective randomized DASH-2 trial published in 2001 (Sacks *et al.* 2001, 3-10) examined this issue again. The authors allocated 412 adult American volunteers to six different treatment groups. The subjects ingested for a one-month period either a typical, high caloric US diet or a diet enriched in vegetables, fruits and low-fat dairy products ("DASH diet"). In addition, they were allocated to a salt intake of either 9 g, 6 g or 3 g per day. The most marked decrease in blood pressure as obtained with the DASH diet. The additional blood pressure reduction induced by concomitant salt restriction was much smaller (Figure 4). Thus the systolic blood pressure decrease with the DASH diet, as compared to the standard US diet, was 5.9 mmHg, whereas the decrease observed in response to a reduction of salt intake from 9 g to 6 g per day was only 1.3 mmHg. A further decrease of 1.7 mmHg could be induced by a reduction of salt intake from 6 g to 3 g per day, but it is well known that such a drastic restriction is practically impossible. We also would like to point out that the DASH study has been performed in a selected cohort of US individuals with high salt sensitivity (i.e. a high percentage of black and overweight subjects), and that the primary criterion of evaluation was blood pressure, i.e. an intermediary endpoint, not a hard endpoint such as cardiovascular morbidity or mortality. Extrapolations from these findings to populations outside the US may therefore not be justified.

Observational studies on the association between salt intake and cardiovascular events and mortality
As recalled by Alderman in 2010 (Alderman 2010, 448-449), at least 13 observational cohort studies examined the relationship between sodium consumption and clinical outcomes, and these studies included more than 100 000 participants and more than 800 morbidity and mortality events. The results have been conflicting, as shown in Table 3. In two studies with high mean daily sodium consumption (Finland, 4600 mg/day; Japan, 5428 mg/day corresponding to 11.6 and 13.7 g of salt per day), the association between salt intake and cardiovascular disease events was positive (Nagata *et al.* 2004, 1543-1547; Tuomilehto *et al.* 2001, 848-851). In the study carried out in Finland, in 1173 men and 1263 women, there was a direct relationship between salt intake (based on daily sodium excretion) and cardiovascular outcomes including all-cause mortality (Tuomilehto *et al.* 2001, 848-851). However, this relationship was significant only in individuals with excess weight, not in those with no overweight, in line with the notion that obesity favors salt sensitivity. However, in this study there was no clear-cut relationship between urinary sodium excretion and blood pressure level. In the other study with a positive relationship between salt intake and cardiovascular events, carried out in Japan, sodium intake was determined in 13 355 men and 15 724 women using a validated food frequency questionnaire (Nagata *et al.* 2004, 1543-1547). The authors reported that, in the men, the highest compared with the lowest tertile of sodium intake was positively associated with death from stroke. They found a similar trend in the women, but the difference was not statistically significant.

Table 3. Observational Studies Linking Dietary Sodium to Cardiovascular Disease Outcomes.

Lower Salt Intake Associated With Cardiovascular Events Worksite Hypertension Study, 1995 National Health and Nutrition Examination Survey I, 1998 II, 2006 III, 2008
Salt Intake Had No Association With Cardiovascular Events Honolulu Heart Study, 1997 Scottish Heart Health Study, 1997 Health Professional Study follow-up, 1997 Multiple Risk Factor Intervention Trial, 2000 Trials of Hypertension Prevention (Cook *et al.* 2007, 885-8B) 1 and 2 follow-up mortality, 2007 Controls, 2009 (30) (Cook *et al.* 2009, 32-40)
Increased Salt Intake Associated With Cardiovascular Events National Health and Nutrition Examination Survey, involving obese patients, 2000 Finnish Heart Study, 2001 Takayama, 2005 Trials of Hypertension Prevention 1 and 2, follow-up morbidity, 2007 (Cook *et al.* 2007, 885-8B)

In eleven other studies with mean sodium intakes between 2070 and 3680 mg/day (corresponding to salt intakes between 5.2 and 9.3 g salt per day), two detected a positive association between salt and cardiovascular disease events, and each was a post hoc subgroup analysis with findings not entirely consistent with the overall study results (Alderman 2010, 448-449). For instance, in the NHANES I study, the association between salt intake and cardiovascular disease was positive in the obese subset of study participants but was inverse in the entire sample (Alderman, Cohen and Madhavan 1998, 781-785). Likewise in the morbidity follow-up of a subset of participants in TOHP I and II, the association between salt and CVD was positive, whereas in the mortality follow-up of the entire population, the association also was positive, but not statistically significant (Cook *et al.* 2009, 32-40). In five of these eleven studies, there was no association between salt intake and clinical outcome. One of these five studies involving follow-up of 2275 TOHP controls had the unique advantage of estimating usual sodium intake (3634 mg/d) by collecting 4 to 5 urine samples over 24 hours during the initial 3-year period (Cook *et al.* 2009, 32-40). In four studies, sodium intake was inversely associated with cardiovascular disease events. Thus, higher salt intake was adversely associated with cardiovascular disease risk in societies with high salt intake, whereas lower salt intake was associated with worse outcomes in some societies with moderate salt intake. A J-shaped curve, in which the most favorable dietary sodium range surrounds 3450 mg/day – with possible risk above and below this level – is a hypothesis that may best accommodate the available data (Alderman 2007, 903-907). In a recent meta-analysis including many of the same observational studies (Strazzullo *et al.* 2009), but without stratification to account for the sharp differences in ambient sodium intakes, the authors detected a significant positive association of sodium intake with stroke. This finding was driven by studies involving populations with high salt intake. However, there was no association between salt intake and CVD events, and total mortality was not reported. Not surprisingly, given different populations, different diets and sodium intakes, different methods and conflicting results, more than one explanatory hypothesis has emerged. Neither more observational studies nor further debate over the available studies is likely to resolve this public health controversy (Alderman 2010, 448-449).

Nevertheless, Bibbins-Domingo et al performed an additional analysis in 2010 based on the above mentioned and other observational data (Bibbins *et al.* 2010, 590-599). They estimated the effects of reduced salt intake by applying previously published data to a computer-simulation model. They hypothesized that reducing dietary salt by 3 g per day might reduce the annual number of new cases of chronic heart disease by 60,000 to 120,000, stroke by 32,000 to 66,000, and myocardial infarction by 54,000 to 99,000 and to reduce the annual number of deaths from any cause by 44,000 to 92,000. All segments of the population would benefit. However, this process relies on numerous unverified assumptions, in particular the one that salt reduction lowers blood pressure and that in turn lowering blood pressure by this means reduces the risk of stroke and chronic heart disease (Appel and Anderson 2010, 650-652).

Of note, several more recent observational studies found either a negative or a U-curve type association between salt intake and mortality. This means that low salt intake was associated with an increased prevalence of cardiovascular events and death. Such findings were made in general population and also in patients with diabetes and patients with heart failure. Concerning the general population, Stolarz-Skrzypek *et al.* examined 3681 participants at two European studies on the effects of genes versus environment on hypertension. They had no cardiovascular disease, 2096 among them were normotensive at baseline and 1499 had blood pressure and urinary sodium excretion measured at baseline and last follow-up (2005-2008), with a median 7.9 years. Cardiovascular disease deaths decreased across increasing tertiles of 24-hour sodium excretion, from 50 deaths in the low (mean, 107 mmol), 24 in the medium (mean, 168 mmol), and 10 in the high excretion group (mean, 260 mmol; $P<0.001$), resulting in respective death rates of 4.1%, 1.9%, and 0.8%. Baseline sodium excretion predicted neither total mortality nor fatal combined with nonfatal cardiovascular events. Of note, the risk of hypertension did not increase across increasing tertiles. However, in multivariable-adjusted analyses, a 100-mmol increase in sodium excretion was associated with 1.71 mm Hg increase in systolic blood pressure ($P<0.001$) but no change in diastolic BP. Thus in this population-based cohort, systolic blood pressure, but not diastolic pressure, changes over time aligned with change in sodium excretion, but this association did not translate into a higher risk of hypertension or cardiovascular complications. Most importantly, lower sodium excretion was associated with higher cardiovascular mortality.

O'Donnell *et al.* performed a post hoc analysis of two populations enrolled in drug trials (O'Donnell *et al.* 2011, 2229-2238). They reported that both high and low sodium intakes, based on urinary sodium spot samples, were associated with increased cardiovascular events in a J-shaped curve. As compared with participants who had a baseline sodium excretion of 4 to 6 g per day (10 to 15 g per day of sodium chloride), participants who excreted more than 6 g of sodium (15 g of sodium chloride) per day and those who excreted less than 4 g of sodium (10 g of sodium chloride) per day in that study showed an increase in cardiovascular deaths, strokes, or heart attacks.

Two other recent studies examined persons with type 1 (Strazzullo *et al.* 2009) and type 2 diabetes (Sacks *et al.* 2001, 3-10). Ekinci *et al.* evaluated prospectively 638 patients with type 2 diabetes. They measured 24-hr urinary sodium excretion at a single visit, and the patients were followed for 10 years. Baseline urinary sodium excretion was 184 ± 73 mmol/24-hr (10.7 g sodium chloride/day). After 10 years of follow-up, 175 deaths were observed, 43% of which were of a cardiovascular cause. The main finding of this study was that all-cause mortality was lower by 28% for every 24-h sodium excretion higher by 100 mmol. Cardiovascular mortality also significantly and inversely correlated to sodium excretion. This finding has been interpreted as evidence that lowering salt intake may cause adverse effects. However, as pointed out by Lambers Heerspink *et al.* (Lambers *et al.* 2012, 3435-3442), patient characteristics across the range of urinary sodium excretion were not balanced since the low salt group had a longer duration of diabetes, lower body mass index, lower glomerular filtration rate, higher frequency of macrovascular disease, and received less renin-angiotensin-aldosterone blockade and insulin treatment more frequently.

The other recent report on the role of salt intake in diabetes outcomes was done in 2807 adult patients with type 1 diabetes (FinnDiane Study) (Thomas *et al.* 2011, 861-866). Thomas et al followed these patients for 10 years and found that urinary sodium excretion was non-linearly associated with all-cause mortality: both low and high urinary sodium excretion were associated with higher mortality. The nadir of mortality was at a consumption of 100 mmol sodium /24-hr (5.8 g sodium chloride per day). The association was independent of confounders such as age, gender, duration of diabetes, glomerular filtration rate or albuminuria. Also for this study, Lambers Heerspink *et al.* (Lambers *et al.* 2012, 3435-3442) pointed out that in the low salt cohort there were more females, individuals with lower BMI, less anti-hypertensive medication and more macrovascular disease, all findings being consistent with a higher cardiovascular risk.

In a recent report, the Institute of Medicine (IOM) in the US reexamined the consequences of sodium reduction in populations, including the most recent, new evidence mentioned above (Brian, Strom and Oria 2013). The committee states that « the evidence from the current literature is inconsistent with regard to associations with sodium intakes below 2,300 mg per day, with results ranging from lower, similar, or higher risk of CVD, stroke, or mortality, including all-cause mortality. All of the studies identified have limitations of different types. The evidence in some cases is suggestive, however, of associations between lower sodium intake (below 2,300 mg per day) and potential increased risk of adverse health outcomes, though reverse causation, confounding and systematic measurement error cannot be ruled out. »

As pointed out by Alderman already in 2010 (Alderman 2010, 448-449) surrogate markers such as high blood pressure, are not clinical events, although they are usually associated with the incidence of subsequent stroke, myocardial infarction, kidney dysfunction, or heart failure. However, as evoked above, a reduction of sodium intake sufficient to lower blood pressure also is a powerful stimulus for several neuro-humoral systems such as the sympathetic nervous system and the renin-angiotensin-aldosterone axis with potential negative consequences for the organism. The health effects of sodium reduction will be the net result of these conflicting effects. The four new observational studies above clearly support the notion that simple extrapolation from observed changes in blood pressure to hard outcomes, such as the one recently proposed by Bibbins Domingo *et al.* (Bibbins *et al.* 2010, 590-599), is premature according to Kotchen *et al.* (Kotchen, Cowley and Frohlich 2013, 1229-1237), if not erroneous according to McCarron *et al.* (McCarron, Drueke and Stricker 2010, 1005-1006).

Randomized controlled trials with morbidity and mortality endpoints are the standard with which health and medical interventions are tested. The only such trial that has compared different sodium intakes involves patients with heart failure (Paterna *et al.* 2008, 221-230). Of note, randomization to a restricted sodium intake led to a significant increase, not decrease, in mortality and hospitalization compared with those randomized to higher sodium intake. Strictly speaking, however, the results of this study are applicable only to patients with heart failure.

There is thus no convincing study from controlled trials on the effects of salt restriction on cardiovascular outcomes and mortality in the general population. Certainly, the Trials of Hypertension Prevention (TOHP) studies, which were initially randomized controlled trials, examined the effects of salt restriction on cardiovascular events and mortality in individuals without known cardiovascular disease (Cook *et al.* 2007, 885-888B]. However, the final conclusion reached by the authors was based on an observational follow-up, using an intent-to-treat analysis. They claimed that the finding of a small reduction in mortality in this extension study was proof for the efficacy of salt restriction. However, the problem with this observational follow-up study was that the association between urinary sodium excretion and mortality was not significant.

Feasibility of a large scale, marked reduction in salt intake?

Over 75% of salt intake in industrialized diets is likely to come from processed foods. Therefore national authorities, supported by the World Health Organization, have been encouraging food manufacturers in the last decade to reduce the sodium content of their products. As mentioned above, the feasibility of such recommendations is however questionable since recent surveys from the United Kingdom (UK) and USA showed that salt consumption has essentially remained unchanged during the past decades (McCarron *et al.* 2009, 1878-1882). It is probably premature to discount the apparently paradoxical cardiovascular outcomes associated with low salt intake, particularly in specific clinical conditions such as type 1 or type 2 diabetes and congestive heart failure (Kotchen, Cowley and Frohlich 2013, 1229-1237).

Thus a marked, generalized reduction could be harmful. One must also no forget that salt plays an important role in food technology and taste, and salt is needed in a number of food products for preservation and safety. Currently chosen approaches have resulted in salt reductions between 20-30%. Further reduction, if effectively implemented, will require new technologies and careful evaluation of potential risks.

Salt and other diseases

Osteoporosis

The relationship of salt intake to bone mineral density (BMD) and osteoporosis remains controversial despite numerous studies devoted to this issue. Since calcium is the most important mineral element in bone, and since calcium and sodium share a common transport mechanism in the kidney the idea came up many years ago that high urinary sodium excretion linked to high sodium intake should go along with high urinary calcium losses. As an example, a relatively recent study again examined the relationship of sodium intake to calcium excretion, the factors that may modify the relationship of sodium excretion to calcium excretion, and the relationship of calcium and sodium excretion to BMD of the total hip (Carbone *et al.* 2003, 415-420). However, there was no significant relationship between urinary sodium loss (based on 24 h urinary Na excretion) and BMD of the total hip in the study population. This suggested that dietary sodium intake in the range of 1236-6516 mg/day did not have a significant impact on bone mineral content. As another example, a study ancillary to the DASH-2 trial (see above) showed that urinary calcium excretion decreased in proportion to dietary sodium restriction (Lin *et al.* 2003, 3130-3136). When sodium intake was reduced from the higher to the lower level, urinary calcium decreased effectively by 22 mg/g urinary creatinine in the control, American type diet group, but only by 12 mg/g creatinine in the subjects on the DASH diet. Moreover, the difference between the two groups was not significant. This observation is consistent with the finding of another study showing that a 90 mmol/day potassium supplement (similar to the additional potassium provided with the DASH diet) prevented sodium induced increases in urinary calcium excretion. Although the DASH diet led to a significant decrease in circulating markers of bone formation and resorption this appeared to be largely independent of salt since in the control group, the reduction of salt intake from 9 g/day down to 3 g/day did not modify bone turnover markers consistently. Thus the short-term effects of sodium intake on bone resorption remain unclear, in line with long-term studies aimed at investigating the relationship between dietary sodium and BMD or osteoporosis which has produced conflicting results (Teucher and Fairweather-Tait 2003, 859-866; Cohen and Roe 2000, 237-253). It is possible that deleterious effects of excessive salt intakes on bone occur at low calcium intakes. With calcium intakes at equal or above currently recommended levels, there appear to be no deleterious effects of prevailing salt intakes on bone or the calcium economy, mainly because adaptive increases in calcium absorption offset the increased urinary loss (Heaney 2006, 271S-276S). However, even this attractive view as a compromise of contradictory literature reports is unable to integrate all of the findings of a recent calcium balance study in postmenopausal women where salt was found to be responsible for a negative bone calcium balance when consumed as part of a high calcium diet (Teucher *et al.* 2008, 1477-1485). However, with a low calcium intake the bone calcium balance was found to be negative on both high and low salt diets. Clearly more studies are needed, especially those examining the effect of salt intake on fractures as hard outcomes, to solve this issue.

Stomach cancer

Stomach cancer remains the fourth most common cancer and the second most common cause of cancer death in the world (Brenner, Rothenbacher and Arndt 2009, 467-477). Therefore, continued search for factors involved in its pathogenesis is warranted. Helicobacter pylori infection, the by far strongest established risk factor, and also male sex, a family history of stomach cancer, and smoking are among the established factors. Dietary factors including high salt intake have long been incriminated to also favour its occurrence, including a very recent study by D'Elia *et al.* (2012, 489-498). While some factors related to diet and food preservation, such as high intake of salt-preserved foods and dietary nitrite or low intake of fruit and vegetables, are likely to increase the risk of stomach cancer, the quantitative impact of many dietary factors remains uncertain, partly due to limitations of exposure assessment and control for confounding factors (48) (Brenner, Rothenbacher and Arndt 2009, 467-477).

Thus the increased risk attributed to salt, as observed in some studies, may be attributable to food constituents other than salt. An interesting finding is a possible interaction between salt and *H. pylori* infection. High salt concentrations could for instance alter gene expression in *H. pylori* and render the bacteria more aggressive, by this or other pathways, with respect to inflammatory changes in the gastric mucosa, as suggested by some animal experiments (Kato *et al.* 2006, 1558-1566), but not by others (Rogers *et al.* 2005, 10709-10715).

Taken together, at present the available literature does not provide convincing evidence in favor of a direct association between salt consumption and stomach cancer (Dias-Neto *et al.* 2010, 133-147).

Asthma

A large number of reports have dealt with the issue of the relationship between salt intake and asthma. As for the other salt-related health issues addressed above, there is no clear-cut evidence for a role of salt. Most of the studies are observational in nature. They can therefore only be considered as « hypothesis-generating ».

A Cochrane search has been performed in 2004 to identify randomized controlled trials aimed at assessing the effects of dietary salt restriction on symptoms and signs of asthma. A total number of only six such trials could be included in the review (Ram and Ardern 2004).

All studies were of small size and short duration. Data from only four studies could be pooled. Low sodium diet was associated with a significantly lower urine sodium excretion than normal or high salt diets. There were no significant differences in any asthma outcome between low salt and normal or high salt diets. However, the confidence intervals were wide. The authors concluded that based on currently available evidence it is not possible to say that dietary salt reduction has any place in the treatment or management of asthma. The results of the review did indicate an improvement in pulmonary function with low salt diet, but further large scales trials were required before any firm conclusions could be reached.

At least one other randomized controlled trial has subsequently been performed and reported in 2008. The authors tested the hypothesis that a low sodium diet would improve asthma control over a 6-week period. Participants had a diagnosis of asthma and measurable bronchial reactivity to inhaled methacholine, which induces bronchoconstriction. All of them adopted a low sodium diet and were randomized to receive either 80 mmol/day (i.e. 4.7 g salt/day) of oral sodium supplements (normal sodium intake) or matched placebo (low sodium intake) for 6 weeks. The primary outcome was change in bronchial reactivity to methacholine. Secondary outcomes were other changes in lung function. A total of 199 completed the study protocol. In the low sodium-intake group, mean daily urinary sodium excretion decreased by 20 mmol and in the normal-sodium-intake group it increased by 28 mmol. There were no differences between the two groups in the primary or secondary outcome measures. The authors concluded that the use of a low sodium diet as an adjunctive therapy to normal treatment has no additional therapeutic benefit in adults with asthma and bronchial reactivity to methacholine.

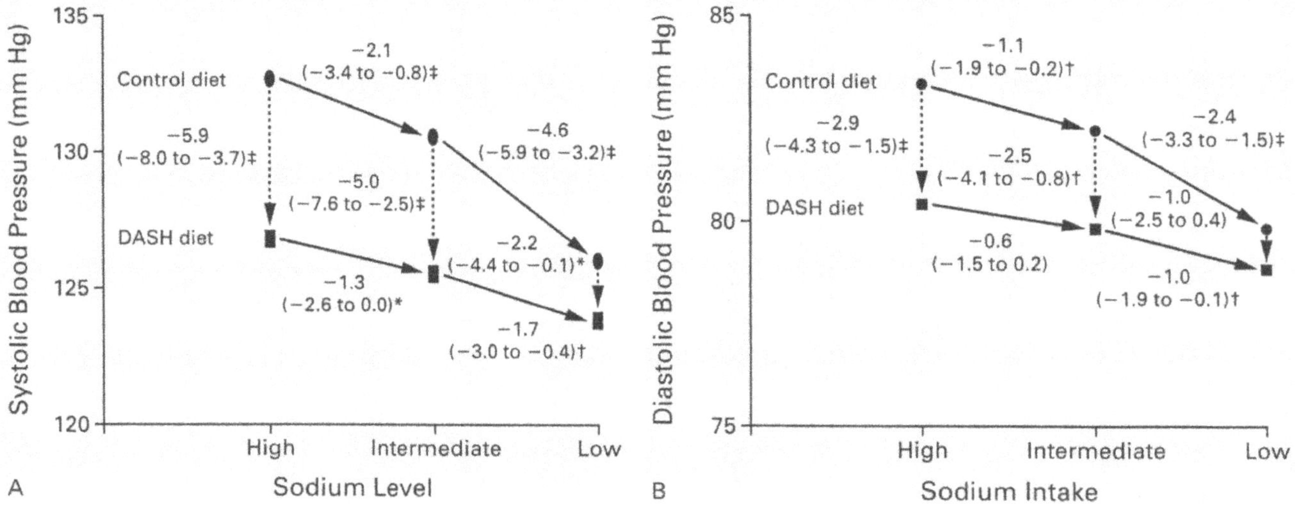

Figure 4. Sodium Reduction, the DASH Diet, and Changes in Systolic Blood Pressure. Beneficial effects of the DASH diet and reduced intake of sodium on systolic and diastolic blood pressure in 412 patients with mild hypertension, age > 45 years. The participants were randomly assigned to follow a DASH diet or a typical U.S. diet for 30 days. During that period, each group consumed, in a crossover design, three versions of the diet adjusted for daily sodium content : high, 8.8 g NaCl/day: intermediate, 5.8 NaCl/day; low, 3.0 NaCl/day. Body weight was held constant. The two downward-sloping arrows on the left depict the effect of intermediate sodium intake as compared with higher sodium intake, and the two downward-sloping arrows on the right depict the effect of lower sodium intake as compared with intermediate sodium intake. The dotted lines show the effect of the DASH diet as compared with the typical U.S. diet at each level of dietary sodium. Numbers shown represent the mean changes with 95% confidence intervals. (From Sacks *et al.* (Sacks *et al.* 2001, 3-10) (reproduced with permission).

Kidney stone disease

This disease, also called nephrolithiasis, is characterized by the formation of stones in the urinary tract, from the kidneys down to the bladder. It affects a large number of patients worldwide. It can be extremely painful but is rarely life-threatening. Its causes are numerous. A variety of endogenous and exogenous factors have been identified. Among the latter, dietary factors clearly play a role, the most important among them being fluid intake, pH, calcium, phosphate, oxalate, citrate, phytate, urate and vitamins. The role of sodium have remained a matter of debate (Borghi *et al.* 2006, 301-312).

Numerous trials have been done to evaluate the efficacy of diet, fluid, or supplement interventions for secondary prevention of nephrolithiasis. In contrast, few have been included in systematic reviews or referenced in recent nephrolithiasis management guidelines. Fink *et al.* performed a systematic review and meta-analysis of trials published between January 1950 and March 2008 (Fink *et al.* 2009, 72-80). Eligible trials included adults with a history of nephrolithiasis, compared diet, fluids, or supplements with control, compared relevant outcomes between randomized groups (eg, stone recurrence), and had 3 or more months of follow-up. Eight trials were eligible, with a total number of 1855 participants. The quality of the trials selected for analysis was considered to be mixed. We would like to underline from the outset that the authors failed to identify any trial that examined the independent effect of altering dietary sodium. Similarly, there was no prospective randomized trial to examine independent effects of altered intakes of calcium, animal protein, fruit, fiber, purine, oxalate, or potassium. In two trials, water intake of more than 2 liters per day or fluids to achieve urine output exceeding 2.5 liters per day significantly reduced stone recurrence. In one trial, fewer high soft drink consumers assigned to reduced soft drink intake had renal colic than controls. Content and results of multicomponent dietary interventions were heterogeneous. In one trial, fewer participants assigned increased dietary calcium, low animal protein, and low sodium had stone recurrence versus controls (20% vs 38%, p=0.03), while in another trial, more participants assigned to diets that included low animal protein, high fruit and fiber, and low purine had recurrent stones than controls (30% vs 4%, p=0.004). Two trials showed no benefit of supplements over control treatment. Adverse event reporting was poor. The authors thus confirmed that fluid intake decreased risk of recurrent renal stone disease. Moreover, reduced soft drink intake lowered risk in patients with high baseline soft drink consumption. Data for other dietary interventions were inconclusive, although limited data suggest possible benefit from dietary calcium.

Taken together, there is no convincing evidence allowing the recommendation to lower salt intake to prevent the occurrence or recurrence of kidney stone disease.

Salt as a carrier of other substances

The most common forms of micronutrient malnutrition relate to iron, iodine, and vitamin A. Special attention is paid hereunder to iodine and iron with salt as an appropriate carrier. Moreover, salt is also used to carry fluorine, and folic acid. Salt iodization and fluoridation both participate in WHO mass prevention recommendations regarding physical and mental retardation on the one hand, dental decay on the other hand. Iodine disorders are widespread throughout the world, including industrialized countries. Even moderate levels of deficiency can have detrimental effects on human health. Improvement of oral health depends on several factors which include topic and systemic action of fluorine. Salt fortification is a low-cost, relatively simple strategy that may reach a wide range of people, and contribute to reducing the high prevalence of micronutrient deficiencies affecting children, especially in poor countries with important health and economic consequences.

General debate on the pros and cons of fortified foods

Fortified foods are products whose original composition has been modified through addition of essential nutrients to meet specific population needs. The fortification of foods will have a positive impact on nutrition status, provided that micronutrient bioavailability is ensured. Diverse factors affect bioavailability such as dietary factors, micronutrient interactions, illnesses, and chemical characteristics of the compound used for fortification. In countries such as Chile, Venezuela and Mexico, striking effects have been obtained in reducing iron deficiency anemia in children. In less than a decade, the salt iodization programme has also proven its effectiveness. Iodized salt was introduced in Switzerland in 1922 and in the USA in 1924 to reduce severe iodine deficiency. In 1938 voluntary enrichment of flours and breads with Fe was initiated to reduce the incidence of iron deficiency anemia. The Canton of Zurich was the first political unit world-wide to authorize the sale of salt enriched with fluoride in 1955.

The General Principles for the Addition of Essential Nutrients to Food issued by the Codex Alimentarius Commission (General Principles for the Addition of Essential Nutrients to Food CAC/GL 09-1987 (amended 1989, 1991) [database on the Internet] 1991. Codex alimentarius) defines fortification as '*the addition of one or more essential nutrients to a food whether or not it is normally contained in the food, for the purpose of preventing and correcting a demonstrated deficiency of one or more nutrients in the population or specific population groups.*' Micronutrient intakes appear to be generally adequate for most nutrients in various countries. However, a number of population subgroups are at risk of suboptimal intakes (below the lower reference nutrient intake) for some micronutrients, e.g. iodine, iron, and folate: children, adolescents and young women are concerned. Dietary surveys indicate that fortified foods play a role in mitigating such risks for such nutrients. Food suited to fortification is limited by several factors, including technological properties (moisture, pH and chemical complexation), leading to unacceptable taste and appearance, far from consumer expectations. This is not the case with salt as evidenced in a subsequent chapter.

In terms of salt based mass prevention practice, it appears essential to build links among academic researchers, politicians, food manufacturers and consumers for effective and sustainable food fortification. A number of countries have successfully experienced combined iodine and fluorine fortification of salt. By recommending both iodization and fluoridation, the authorities were in a position to maximizing their related public health initiatives. Technological issues concerning salt combined fortification have been fully resolved by the stakeholders with regard to the appropriate levels of micronutrients, their respective interaction and the acceptability of the end-product by the consumer. The major challenges remain the approximation of legal provisions at national level, and the access of fortified salt to both household and food industry whereas dietary salt intake depends currently on processed food within a span of 60 to 75 %.

Salt iodization

Food fortification can take several forms. Mass prevention such as Universal Salt Iodization (USI) requires well designed legal provisions for being sustainable. USI promotes the iodization of salt for all human consumption (households, food industries), and livestock consumption. Most of the industrialized countries prefer voluntary acceptation by the consumer. In developing countries, iodine mandatory fortification is more likely to succeed, especially when health authorities seek industry cooperation, and take restrictive measures to prevent the distribution of coarse salt whether imported or locally produced. In countries in which voluntary fortification is widely practiced, micronutrient intakes are considerably below tolerable upper safe limits. Concerns about safety are addressed in relation to the potentially increased level or proportion of fortified foods (e.g. following exuberant EU legislation), for nutrients with relatively low tolerable upper intake levels and where the potential benefits and risks differ in subpopulations.

Iodine deficiency is a perpetual threat for persons living in areas with iodine-deficient soils or eating iodine deficient food. For such areas, USI is the consensus strategy to improve iodine nutrition because fortifying salt with iodine is the easiest and least expensive option. In February 2010, Harold Alderman (World Bank) (Alderman 2010) estimates the payback for investments in iodizing salt to return between $12 and $30 for every dollar invested. USI is a recommendation which can be achieved in a sustainable way since salt is widely available throughout the world, salt is consumed by everyone, and its consumption is regular and constant across all socioeconomic strata. USI is achieved using simple technology, the related fortification technique is cost/effective, and quality can be monitored at production and distribution levels.

Iodine deficiency is not only a problem in developing regions; it also affects many industrialized countries. In most of them, the best strategy to control iodine deficiency is carefully-monitored salt iodization. However, whereas approximately 70% of salt consumption in these countries is from purchased processed foods, the iodization of discretionary salt only does not fit the supply adequate iodine. Globally, nearly two billion individuals have an insufficient iodine intake, and approximately 50% of continental Europe remains mildly iodine deficient (Delange 2002, S404-416). Iodine intakes in other industrialized countries, including the USA and Australia where are national targets for salt intake of 6 g salt/day, have fallen in recent years. Iodine deficiency has reappeared in Australia, as a result of declining iodine residues in milk products (iodophor use by the dairy industry). Although the general population seems iodine sufficient, it is uncertain whether iodine intakes are adequate in pregnancy, which has led to calls for iodine supplementation. In the United States, iodine has been voluntarily supplemented in table salt (70 mg/kg). Nevertheless, a large proportion of the population is unknowingly suffering from thyroid disease, which supports the usefulness of screening for early detection (Hollowell *et al.* 2002, 489-499). Available data suggest that pregnant women in the Republic of Ireland and the UK are now mildly iodine deficient, possibly as a result of reduced use of iodophor by the dairy industry, as observed in Australia.

The average salt intake of Australian adults appears to be 7–12 g/day, which is little changed from 10 years ago. Achieving adequate iodine intake is difficult, as more than 80% of intake is from salt added to food during processing. It is clear that anti-salt campaigns induce the food industry to reduce salt intake. Representative data on iodine status in children and pregnant women in Australia constitute a matter of concern. Approximately half the pregnant women participating in a recent study were able to indicate exact dietary sources of iodine, such as fish (58%) and iodized salt (51%) (Charlton 2010). However, confusion regarding other foods was evident. Only 11% of women participants reported that they had intentionally changed their diet to increase iodine intake during pregnancy. Among them, 59% indicated using supplements, of which only 35% contained iodine. Currently, no readily accessible information on iodine is available to women attending antenatal clinics in Australia. Another study was designed to evaluate Australian children status (Li *et al.* 2006, 165-169). The results confirm the existence of inadequate iodine intake in the Australian population, and experts call for the urgent implementation of mandatory iodization of all edible salt in Australia.

This situation has also been observed in Western Europe (Delange 2007, 1571-1580; discussion 81-83). The requirement for iodine by the mother during pregnancy is 250-300 μg/day, during lactation 225-350 μg/day, and during the neonatal period in infants 90 μg /day. A MEDLINE/PubMed search and compilation of all published studies since 1990 on iodine nutrition and iodine supplementation of pregnant women in Europe, as well as an Internet-based search and review on availability and legislation of iodine supplements in the European Union yielded disappointing results (Zimmermann and Delange 2004, 979-984). Although most women in Europe are iodine deficient during pregnancy, less than 50% receive supplementation with iodine. Mild-to-moderate iodine deficiency during pregnancy adversely affects thyroid function of the mother and newborn and mental development of the offspring. Public health strategies, including nutritional education and supplementation, are urgently required to improve the iodine status of pregnant women. A more exhaustive review about iodine deficiency in Europe concluded that iodine deficiency remains a public health concern (Andersson *et al.* 2007, 1-70). Salt iodization remains the recommended strategy for eliminating iodine deficiency provided that the food industry participates in.

The Technical Consultation organized by WHO reached a general consensus on several important issues (Andersson *et al.* 2007, 1606-1611). First, USI remains the key strategy to eliminate iodine deficiency disorders. Second, where USI has been effective for at least 2 years, with salt adequately iodized and consumed by more than 90% at the general population level, it can be reasonably expected that the iodine needs of women of child-bearing age, especially pregnant and lactating women, are covered by their diet, and that the iodine stored in the thyroid gland is sufficient to ensure adequate hormone synthesis and secretion. Third, where iodized salt may not provide enough iodine to meet a child's needs during complementary feeding, especially if the mother is marginally iodine sufficient, the number of foods fortified with iodine should be increased to make sure that requirements are met in the normal family food. Finally, monitoring of both iodized salt quality and iodine nutrition are important to ensure that an optimal state of iodine nutrition is reached and then sustained.

Voluntary fortification tends to be used where the risks related to nutrient deficit are not seriously taken into account within the population at large. Experience with iodine fortification of salt has shown that the strategy recommended by WHO failed to be fully implemented in certain countries due to optional provisions of the Regulations in force. Moreover, USI is a clear-cut recommendation which should not be deviated from its objectives under the fallacious argument that salt is one of the pathological factors involved in salt sensitive hypertension (Allen, de Benoist and Dary 2006, 1-341). The current push to reduce salt consumption aimed at preventing chronic diseases and the policy of salt iodization to eliminate iodine deficiency do not conflict; iodization methods can fortify salt to provide recommended iodine intakes even if per capita salt intakes are reduced to <5 g/d. Thus, it is critical that the food industry be permitted to use iodized salt in order to successfully prevent iodine deficiency in such countries.

Iodine deficiency disorders (IDD) were prevalent in China until the introduction of a governmental programme aimed at mandatory implementation of universal salt iodization. In 2010, the penetration rate of adequately iodized salt (35±15 mg/kg) exceeded 95 % and remained below 80 % in only 33 of the 2,831 counties. The last national nutrition survey which took place in 2002 showed a mean intake of 12 g/d. Since that time, a 31 % decline in salt consumption by the households was reported with a mean value of 8.4 g/d in rural and 7.7 g/d in urban areas respectively. Regular monitoring of urinary iodine excretion and complementary research are needed to ensure adequate iodine intake in spite of lower salt intake (Wu *et al.* 2012, 1728-1734). Interestingly, as regards the absence of legal incentive to implement any USI programme, mention has to be made of a quantitative survey which was performed to collect information from 21,190 households in Bangladesh (1997-1998). The results revealed that only 1.9 % of the households bought iodized salt in coastal areas because of the wide availability of coarse salt in these areas, and the lack of knowledge regarding salt iodization and IDD prevention (Rasheed *et al.* 2001, 66-72). This is the case in similar areas throughout the world where no iodine prophylaxis campaign have been undertaken because health markers were acceptable (irrespective of moderate iodine deficiency), and the health authorities did not consider these campaigns necessary provided that iodized salt was available from a legal point of view (national Regulations). Therefore, several European countries are still suffering from mild iodine deficiency in pregnancy, with potential cognitive and psychomotor deficits in neonates.

In 1998, iodized salt was introduced in Denmark. As long as iodization of table salt remained optional, less than 30% of table salt was iodized. Within 2.5 years after the introduction of compulsory iodization at an iodine concentration of 13 ppm, at least 62 % of households were using adequately iodized salt with beneficial effects in all investigated groups (Rasmussen *et al.* 2008, 166-173). The mean increase in iodine intake was close to the planned 50 μg/d. However, there are still groups with an intake below the recommended level whereas no groups with excessive iodine intake were identified. Furthermore, subjects with a low milk and salt intake are at risk of low iodine intake. Additionally, subjects with low BMI had a lower iodine intake than subjects with a high BMI. In Belgium, the use of iodized salt in bread (10-15 ppm) on a voluntary basis was endorsed by the bakers in 2009. Other sectors in the food industry have also expressed their willingness to use iodized salt for food fortification. According to a major salt supplier (esco), the sales of iodized salt to cooperative bakers increased to 41 % in 2010. Conversely, it should be observed that the penetration rate of iodized table salt amounts to only 30 %, below the WHO recommendation according to which 90 % should be reached to eliminate IDD. The situation could be improved if the legal provisions were amended to regulate salt market on a mandatory basis (Vandevijvere *et al.* 2012, 1046-1053).

In Romania, the national salt company is in a position to supply iodized salt fulfilling the national requirements. Salt iodization suffers from some imports of non-iodized salt from Ukraine. In addition, salt intended for animal feeding and/or an industrial purpose is diverted for human consumption, especially in rural areas. In various European countries the challenges remain convincing health authorities to amend the existing legal framework in order to stimulate the producers to iodize salt, the food industry to use it, and the consumer to get familiar with this type of salt instead of coarse salt.

A study was performed recently to assess global and regional iodine status by surveying the penetration rate of iodized salt in the households via measurements of urinary iodine concentration (UIC). Out of 128 countries where iodized salt was available, 37 countries showed a penetration rate meeting WHO recommendation (Zimmermann and Andersson 2012, 382-387). Iodization of cooking salt (adequate iodine level, 15-40 ppm) is an indicator which does not take into account food grade salt as a whole, as long as the food industry is not permitted or doomed to use it. ICCIDD experts explained that, as salt intake levels may vary, or as the iodine contribution of iodized salt changes within the overall diet, salt iodization programs have proved they can easily adjust the level of iodine fortification. They emphasized the necessity of systematic monitoring of iodine sufficiency either through measuring the household use of iodized salt or, better, through regular population surveys of urinary iodine excretion. There is no evidence that countries where the use of iodized salt is promoted, consume different amounts of salt than those with inadequate salt iodization, nor has salt consumption changed when a country achieved salt iodization. Irrespective of debates about salt consumption statistics, salt remains the ideal carrier for iodine and other vital nutrients because its intake is consistent and predictable. Lowering salt intake in the general population obviously increases the risk of inducing iodine deficiency.

The authors of an overview in 2013 concluded that some progress has been achieved worldwide as regards global iodine intake (Pearce, Andersson and Zimmermann 2013, 523-528). Approximately 70 % of households have currently access to iodized salt. Although 111 countries are said to have sufficient iodine intake, the problem of re-emerging iodine deficiency in several developed countries is a matter of concern. The role of the food industry could be more definite if government initiatives to lower population-wide sodium intake would not have detrimental effects on iodine status. Lack of monitoring the population at large could easily lead to neutralize WHO initiatives whose goal was to eliminate IDD. It would make sense to coordinate interventions aimed at general sodium restriction and universal salt iodization.

Germany is another example of industrialized countries where iodized salt added to processed food contributes most to iodine nutrition. Between 2004 and 2009, the 24-hr urinary iodine excretion (UIE) was repeatedly measured in 278 German children of the Dortmund area. Longitudinal regression analysis showed a plateau of UIE in 2004-2006. Afterwards, UIE significantly decreased till 2009 (Johner *et al.* 2011, 1749-1756). Such data reflect a surreptitious deterioration in the iodine status of German schoolchildren. A decreased use of iodized salt in industrially processed food may be one of the reasons for such a negative development as its market share decreased from 35 % in 2004 to 29 % recently. In both the USA and Canada, where the population is generally considered iodine-sufficient, recent data suggest again pregnant women be mildly iodine deficient. A recent study looked at the association between dietary salt restriction and iodine deficiency in women. It could be of significance because the offsprings of iodine deficient women are at risk of IDD (increased pregnancy loss, infant mortality, neonatal hypothyroidism, growth retardation etc. (Tayie and Jourdan 2010, 1095-1102). The American Thyroid Association had previously recommended that North American women receive dietary supplements containing 150 µg iodine daily during pregnancy and lactation. Neither was this recommendation endorsed nor was any initiative adopted to promote adequate iodized salt intake (Pearce 2008, 2466-2468).

Salt fluoridation

Unlike other micronutrients, fluorine is generally considered not to be essential in human nutrition. Nevertheless, fluoride is undoubtedly protective against dental decay. The prevalence of dental decay is 50 % lower in those areas of the world where water contents naturally fluorine or is fluoridated. Fluoride intake from tap water is generally low. Following ingestion, fluoride becomes incorporated into the mineral structure of the growing teeth and makes them more resistant. Besides systemic exposure, its impact on the tooth surface is also beneficial. Excessive fluoride exposure can induce enamel fluorosis. Skeletal fluorosis occurs only at fluorine intake in excess of 10 mg/day over more than ten years.

Despite observed improvements in terms of reduced prevalence of dental caries in populations worldwide, problems still persist particularly among certain sub-groups in both industrialized and developing countries as reflected by the DMFT (decayed, missing, filled, teeth) index. In many countries research and practice have demonstrated however, that dental caries can be prevented effectively through implementation of fluoride initiatives. There are several ways to get fluoride fortification implemented: water or salt or milk fluoridation play a beneficial role in dental health together with topic use of fluoridated toothpastes. The World Health Organization Oral Health Programme provides technical assistance to countries in the process of planning, implementing and evaluating salt fluoridation (Marthaler and Petersen 2005, 351-358).

Salt fluoridation is a systemic form of fluoride supplementation, leaving it to the consumer whether he wants fluoride or not. Pioneering measures in Hungary (Toth) and Switzerland (Marthaler) have paved the way to prevention programmes often combining iodide and fluoride. The introduction of salt fluoridation programmes in Latin America and in Western Europe was associated with a substantial reduction of dental decay in school children. Salt is an appropriate carrier for fluoride and fluoridated salt is used in more than 40 countries in Europe and America. Evidence supports a relevant improvement in oral health. Studies evaluating the effectiveness of salt fluoridation have concluded that fluoride delivered via salt produces decay reductions similar to that of optimally fluoridated water. In her book, the head of PAHO's Oral Health Program traces the history of salt fluoridation (Estupiñán-Day S. Promoting Oral Health: The Use of Salt Fluoridation to Prevent Dental Caries: Pan American Health Organisation 2005). She explains why salt fluoridation is better suited to practice in Latin America and the Caribbean than fluoridation of water. Experience has shown that putting fluoride in salt can reduce dental caries by as much as 84 percent at a cost of 6 cents per person per year. The practice is extremely cost-effective, saving an average of $250 per person per year in dental treatment for every $1 spent. It is also highly equitable, benefiting equally rich and poor, young and old, urban and rural dwellers, those with professional dental service and those without.

Published results of studies in many of these countries show that, for 12-year-old children, the initial level of decay reduction due to salt fluoridation is between 35% and 80%. In eight European countries there are national legal provisions which encourage the production and marketing of fluoridated salt: Austria, Czech Republic, France, Germany, Romania, Slovakia, Spain, and

Switzerland. Experience gained in Switzerland makes it evident that salt fluoridation contributes to improving oral health. On a worldwide scale, it seems particularly beneficial for developing countries due to its effectiveness and cheapness (Marthaler 2005, 785-792). Oral health education and promotion based on fluoride widely depends on socio-economic factors. Although extensive research on fluoridated salt was carried out in Hungary from 1966 to 1985, attempts to generalize its use in the population at large had little success. Most German drinking water has low fluoride concentrations. The estimated fluoride intake in German children is between 100 and 300 µg/day, and in adults, between 400 and 600 µg/day. Male subjects have higher mean intakes than females (Bergmann and Bergmann 1995, 138-143). About 70 to 90% of the salt intake of 10 to 13.5 g/day in German adults comes from processed foods. This leaves about 1 - 4 g of salt to be added as discretionary salt at the individual level and to become the source of fluoride. To increase fluoride intake it may be necessary to fortify salt at a fluoride level of around 500 µg/g or to get some popular foodstuff prepared with fluoridated salt (bread). A salt fluoride concentration of 250 µg/g does not present a risk of dental fluorosis. However, clear recommendations about systemic fluoride fortification must be issued for there are fluoride tablets, fluoride-rich mineral waters, and fluoridated salt available simultaneously.

There are a number of studies in the literature about the effectiveness of fluoride in tablets, fluoride added to frequently-consumed food like milk or salt, for the prevention of dental caries. Research results have been evaluated in different evidence-based reviews. The scope of their reviewing was to identify the most recent convincing papers on the effect in caries prevention from added fluoride in milk, salt, and fluoride tablets/drops including newer randomized controlled trials, and to synthesize the findings into practice guidelines (Espelid 2009, 149-156). The key question was whether regular use of fluoridated milk, salt, and fluoride tablets/ drops could prevent dental caries among children and adolescents. A Cochrane analysis on the caries preventive effect of fluoridated salt identified very few studies of good quality. Salt fluoridation was covered by three reviews. Two studies on fluoridated milk and seven studies dealing with fluoride tablets/drops were also analyzed. The differences between fluoride-group and control were statistically significant. Considering the protocol for the above-mentioned Cochrane review regarding salt fluoridation, it seems that the authorities will have to set fresh papers to take for granted what is taken for safe by a number of experts (Gillespie *et al.* 2009). The main objective of the review is to assess the effects of salt fluoridation in preventing caries in the population. In addition, the authors propose several specific objectives.

Commissioned to evaluate literature relating to the health effect of fluoride and fluoridation (caries prevention vs. potential health risks), another review was conducted in December 2006 by the Australian National Health and Medical Research Council (literature from 1996 onwards limited to English-language publications). Addition of fluoride to water, milk and salt and the use of topical fluoride agents were evaluated in relevant studies. As regards salt fluoridation none was identified to meet criteria for inclusion in the review. According to Australian authorities, fluoridation of drinking water remains the most effective and equitable means of achieving community-wide preventive effects of fluoride in caries. Community water-based F administration requires a centralized distribution system. Australian authorities recommend levels of 0.6 to 0.8 ppm in drinking water (Petersen, Baez and Lennon 2012, 5-10). Less blunt comments are presumably reflecting observational data in a more objective way. When evaluating caries prevention of fluoride in milk, salt and tablets, the key question should be: does regular intake prevent dental caries among children and adolescents? As regards salt fluoridation, its effectiveness remains limited to only part of the population which uses fluoridated or iodized-fluoridated salt.

There are some major hurdles that are to slow down existing or scheduled programmes of salt fluoridation, especially the absence of appropriate regulation as regards fluorine contents and the degree of food grade salt fortification, and the issue of reduced salt intake recommended for the general population. In 2011, the perspectives in the effective use of F in mass prevention were evaluated in Phan-Nga, Thailand. The participants evoked important barriers in relation to Community methods to facilitating caries prevention in Asia (Petersen, Baez and Lennon 2012, 5-10). Serious consideration continues to be given in India to the implementation of salt fluoridation whereas using salt as a carrier of iodine was a success story. Further contacts with WHO and other international organizations are expected to alleviate a number of pending problems. The authorities are contemplating to introducing salt fluoridation in Vietnam (Ho Chi Minh City excepted where drinking water is fluoridated). Cambodia instituted salt fluoridation in 2008. There is still strong need for technological and educational support. Established salt iodization paved the way to salt fluoridation in Laos with the assistance of AOI (Aide Odontologique Internationale) and WHO. The authorities are drafting a new regulation to encourage the production of iodized-fluoridated salt.

Randomized studies together with cross-sectional studies reporting on the caries preventive effect of salt fluoridation in children cohorts were pooled provided that mean DMFT scores were available with standard deviations or 90 % confidence intervals (Yeung 2011, 17-18). Two studies concerning 6 to 8 year old children showed a reduction of this index by 0.98. Eight studies involving 9 to 12 year children showed a significant reduction of 2.13. The four studies relating to 13 to 15 year old children exhibited a reduction of 4.22. The results of this meta-analysis favour salt fluoridation vs. no exposure but the very contribution of fluoridated salt as such in DMFT improvement could not be quantified.

There are good reasons to believe that fluoride in different systemic applications and topic formulas does work as caries preventive agents under supervision. There is a need for new, well-designed studies within this field, but the use of negative controls without any fluoride exposure is difficult due to ethical reasons. In particular new research is needed concerning possible caries preventive effect of fluoridated milk and salt as major carriers of fluoride. Its effectiveness is limited to only part of the population which uses fluoridated or iodized-fluoridated salt.

Salt as iron carrier

Iron-fortification programs are effective and efficient provided that the iron compound is carefully chosen and its level in the food is based on target population requirements, the amount lacking in the diet, and iron bioavailability. For monitoring, serum ferritin and transferrin receptor levels should be measured in addition to hemoglobin. Thus, recent studies of provision of iron-fortified salt to children in Morocco and in Ghana, rice to children in India, wheat flour to women in Thailand, and fish sauce in Vietnam have all proved to be efficacious. All were in non-malaria areas, and intestinal parasites were uncommon except in India, where the children were treated. Serum C-reactive protein determination was used to discriminate falsely elevated ferritin values due to infection. An efficacy study of iron-fortified salt in treated school-aged children in Ivory Coast, where the prevalence of malaria parasitemia was 55%, found no change in hemoglobin levels after 6 months, but serum ferritin increased and transferrin receptor decreased significantly, and the increase in body iron and estimated iron absorbed compared favorably with the results of a study of similar design in Morocco, where the prevalence of iron-deficiency anemia decreased from 30% to 5% after 10 months. Hence, iron-fortification programs in malaria areas may not decrease anemia prevalence but will improve iron status and, presumably, iron-dependent health outcomes. In countries such as Chile, Venezuela and Mexico, encouraging results have been obtained in reducing iron deficiency anemia in children less than 5 years of age. Iron-fortified salt has been formerly produced in Austria too.

Dual fortification of salt with iodine and iron could be a sustainable approach to eliminating iodine and iron deficiencies (Andersson *et al.* 2008, 1378-1387). Efficacy studies have shown that double-fortified salt (iodine and iron) can significantly reduce the incidence rates of iron-deficiency anemia and iodine-deficiency disorders. Double-fortified salt can be prepared by mixing microencapsulated iron compounds into conventionally iodated salt. Effective implementation of a double fortification program requires field-based analytical methods to ensure the requested levels in this type of salt. Local salt characteristics should be taken into consideration when choosing an iron carrier for dual-fortified salt (DFS) to achieve optimal iodine stability and salt color.

As regards school children living in rural Southern India, an assessment of the efficacy of DFS was carried out, using 2 proposed contrasting formulas – one fortifying with iron as micronized ground ferric pyrophosphate (MGFeP) and the other with iron as encapsulated ferrous fumarate (EFF), compared with iodized sea salt (IS) alone. After stability and acceptability testing, a double-blind, household-based intervention (Kumar and Rajagopalan 2009, 437-445) was conducted in 5-15 yr-old children (n = 458) randomly assigned into 3 groups to receive IS or DFS with iron as MGFePP or EFF, both at 2 mg/g salt. Hemoglobin, iron status, and urinary iodine were measured at baseline, 5 mo., and 10 mo. Median serum ferritin and calculated median body iron improved significantly in the 2 groups receiving iron. After 10 months, the prevalence of anemia decreased from 16.8% to 7.7% in the MGFePP group ($P < 0.05$) and from 15.1% to 5.0% in the EFF group ($P < 0.01$). The median urinary iodine concentration increased significantly in the IS and EFF groups ($P < 0.001$), but not in the MGFePP group.

Multiple micronutrient fortification

Multiple nutrient fortification having been proposed, another study was performed to test its stability during storage and cooking, and to assess its efficacy in improving the micronutrient status of schoolchildren (Kumar and Rajagopalan 2007, 505-511) and the health benefit of mass prevention. A salt fortified with multiple micronutrients was developed containing chelated ferrous sulfate and microencapsulated vitamins A, B1, B2, B6, B12, folic acid, niacin, calcium pantothenate and iodine. Its stability during 20 min of cooking, and 6 months of storage was determined. Thereafter, the efficacy of the salt was assessed in 5- to 15-year-old schoolchildren in Chennai, India. All micronutrients were stable during cooking and storage. Over the study period, there was a significant improvement ($P<0.05$) in hemoglobin, urinary iodine and serum vitamin A in the experimental group, while there was a significant drop ($P<0.05$) in the control group. The study showed that the salt fortified with multiple micronutrients appeared to be stable during storage and effective in combating multiple micronutrient deficiencies.

Optimal salt intake – debate in terms of Public Health and outlook

The above-mentioned salt debate has important implications for health policy. The most reliable evidence of the effects of a population wide reduction in salt intake comes from randomized controlled trials and systematic reviews of such trials. Recent rigorous reviews of such trials on blood pressure effects in normal subjects have shown only modest changes, ranging from 1 to 2 mm Hg for systolic blood pressure and 0.1 to 1.0 mm Hg for diastolic pressure (19-21) although authors of other meta-analyses concluded to greater effects (Strazzullo *et al.* 2009; Bibbins Domingo *et al.* 2010, 590-599; He and MacGregor 2010, 363-382). Of note, the overall modest effects of randomized controlled trials often resulted from greater reductions in sodium intake than can be achieved by population advice. The conclusions in favor of a beneficial effect may have been amplified by concomitant changes in other dietary components and by publication bias.

As pointed out above, blood pressure is only a surrogate marker of cardiovascular morbidity and mortality. What we really need to know is whether a population-wide reduction in salt intake will lead to better health and longer life expectancy. Only one randomized controlled trial has examined this issue but it has been conducted in patients with chronic heart failure, not in healthy individuals (see above). It has actually shown the opposite, namely that the group of patients allocated to salt restriction died more rapidly than those on usual salt intake. Therefore, the question remains whether there is any scientifically based, solid justification for the recommendation made by authors in the US (Bibbins Domingo *et al.* 2010, 590-599; Whelton *et al.* 2012, 2880-2889) and several other industrialized countries (Strazzullo *et al.* 2009; Carbone *et al.* 2003, 415-420) and by international health authorities such as in the WHO Expert Consultation report (WHO Expert Consultation Report, WHO editor. Reducing salt intake in population. Geneva, Switzerland: WHO; 2007) to cut back on sodium chloride intake, down to amounts of 100 mmol (6 g salt) per day or even less, irrespective of possible direct adverse effects and in addition indirect side effects through the concomitant reduction of many other nutrients contained in food ? We agree with Alderman that at present there is no compelling evidence for Public Health action to reduce salt intake since recommendations aimed at salt restriction for the general population should be based on health outcomes (Alderman and Cohen 2012, 193-201).

Detractors from the arbitrary recommendation of universal salt restriction have repeatedly called for randomized prospective trials to test the validity of the claim that this measure leads to health benefits in terms of reduced morbidity and mortality. Public Health authorities respond that such trials are not feasible because of major administrative, study design, and financial obstacles (Kotchen 1999, H1807-810). Obviously, it is much easier and cheaper to base authoritative decisions on one-sided literature interpretations. Citation analysis interestingly shows that the latter have been quoted much more frequently than rigorous reviews which reached opposite, negative conclusions on same issue, reinforcing the false impression of scientific consensus. A comprehensive and well-balanced article published in Science some years ago, entitled 'The (political) science of salt', called attention to this issue, with the debate being perpetuated by vested interests of personal agendas as well as national and international agencies (Taubes 1998, 898-907).

For those who make Public Health policy the most important question should be, prior to issuing any recommendation, whether imposing a measure such as a universal decrease in salt intake improves hard outcomes in the general population, and how the reduction of one nutrient such as salt may affect positive effects of many other nutrients and micronutrients contained in the diet. Another important question is whether it is possible to submit salt intake to Public Health dictates (McCarron *et al.* 2009, 1878-1882). Whereas a common recommendation for the treatment of hypertension is dietary salt restriction, certain zealots suggest that alternative sources of iodine should be investigated. Instead of cutting salt intake down to very low levels to achieve uncertain benefits, health professionals should inquire about dietary salt habits and other dietary behaviour that impact iodine nutrition, more particularly among women subjected to IDD.

Another, more reasonable approach might be to define different target intake levels for salt depending on individual salt sensitivity. A restriction of salt intake to amounts below 6 g per day clearly is able to reduce blood pressure in salt sensitive patients with hypertension and their need for antihypertensive drugs, and also in healthy individuals with borderline elevations of blood pressure, particularly those with overweight. However, it does not reduce blood pressure in salt-resistant hypertensive or normotensive persons. Why should the majority of the population who is unresponsive to salt restriction in terms of blood pressure be asked to radically change their dietary habits when considering the absence of convincing proof that this might lead to better health and improved life expectancy?

Other nonpharmacological interventions, such as the consumption of diets rich in fruits, vegetables and low-fat dairy products, avoidance of heavy alcohol intake, and reduction of excessive body weight by reduced energy intake and increased physical exercise are probably more effective, although not always easier to implement, in order to achieve optimal blood pressure control and improve health.

In conclusion, both excessive salt intake and excessive salt restriction are probably harmful. Both the higher and the lower limits of salt consumption need to be more clearly identified. Salt sensitivity and tolerance as well as salt intake-linked outcomes appear to be highly variable from one individual to the other. Therefore, it may be more appropriate to tailor recommendations on optimal salt consumption based on each individual's needs and tolerance rather than on simplistic extrapolations from weak, inconstant associations with blood pressure.

Addenda. Recently, the results of the Prospective Urban Rural Epidemiology (PURE) study have been reported in two articles (Menthe A et al, NEJM 2014, 371:601-611; O'Donnell M et al, NEJM 2014,371: 612-623) and commented upon (Oparil S, NEJM 2014,371: 677-679) in the New England Journal of Medicine. The study included more than 100,000 adults sampled from the general population of 17 countries. Approximately 90% of the participants had either a high sodium excretion (>6 g/ day, ie > NaCl intake >15 g/day) or moderate sodium excretion (3-6 g/day, ie NaCl intake 7.5-15 g/day). Approximately 10% excreted less than

3.00 g Na per day (ie 7.5 g NaCl per day), and only 4% had sodium excretion in the range associated with current guidelines for salt intake. The relation between sodium excretion and blood pressure was positive but nonuniform: it was strong in participants with high sodium excretion, modest in those in the moderate range, and nonsignificant in those with low sodium excretion. More important, after a mean of 3.7 years of follow-up, the composite outcome of death and cardiovascular events occurred in 3317 participants (3.3%). As compared with those who had a moderate level of sodium excretion, both those with a higher or lower level of sodium excretion had an increased risk of cardiovascular-disease outcomes including death, compatible with a J-shaped association curve. The authors concluded from the findings that a very small proportion of the worldwide population consumes a low-sodium diet and that sodium intake is not related to blood pressure in these persons, calling into question the feasibility and usefulness of reducing dietary sodium as a population-based strategy for reducing blood pressure and the risk of mortality.

References

Alderman, M.H. 2007. Presidential Address: 21st Scientific Meeting of the International Society of Hypertension: dietary sodium and cardiovascular disease: the 'J'-shaped relation. *Journal of Hypertension* May 25(5), 903-907.

Alderman, M.H. 2010a. Reducing dietary sodium: the case for caution. *Journal of the American Medical Association* February 3, 303(5), 448-449.

Alderman, M.H. 2010b. The economic cost of a poor start to life. *Journal of Developmental Origins of Health and Disease* 1(1), 19-25.

Alderman, M.H. and Cohen, H.W. 2012. Dietary sodium intake and cardiovascular mortality: controversy resolved?. *Current Hypertension Reports* June 14(3), 193-201.

Allen, L., de Benoist, B. and Dary, O. 2006. Guidelines on food fortification with micronutrients. *World Health Organization /Food and Agriculture Organization of the United Nations,* 1-341.

Alderman, M.H., Cohen, H. and Madhavan, S. 1998. Dietary sodium intake and mortality: the National Health and Nutrition Examination Survey (NHANES I). *The Lancet* March 14, 351(9105), 781-785.

Andersson, M., de Benoist, B., Darnton-Hill, I., Delange, F. 2007a. Iodine deficiency in Europe – A continuing public health problem. *World Health Organization,* 1-70.

Andersson, M., de Benoist, B., Delange, F., Zupan, J. 2007b. Prevention and control of iodine deficiency in pregnant and lactating women and in children less than 2-years-old: conclusions and recommendations of the Technical Consultation. *Public Health Nutrition* December 10(12A), 1606-1611.

Andersson, M., Thankachan, P., Muthayya, S., Goud, R.B., Kurpad, A.V., Hurrell, R.F., *et al.* 2008. Dual fortification of salt with iodine and iron: a randomized, double-blind, controlled trial of micronized ferric pyrophosphate and encapsulated ferrous fumarate in southern India. *The American Journal of Clinical Nutrition* November 88(5), 1378-1387.

Appel, L.J. and Anderson, C.A. 2010. Compelling evidence for public health action to reduce salt intake. *The New England Journal of Medicine* February 18; 362(7), 650-652.

Bergmann, K.E. and Bergmann, R.L. 1995. Salt fluoridation and general health. *Advances in Dental Research* July 9(2), 138-143.

Bernstein, A.M., Willett and W.C. 2010. Trends in 24-h urinary sodium excretion in the United States, 1957-2003: a systematic review. *The American Journal of Clinical Nutrition* November 92(5), 1172-1180.

Bibbins Domingo, K., Chertow, G.M., Coxson, P.G., Moran, A., Lightwood, J.M., Pletcher, M.J., *et al.* 2010. Projected Effect of Dietary Salt Reductions on Future Cardiovascular Disease. *The New England Journal of Medicine* 362(7), 590-599.

Borghi, L., Meschi, T., Maggiore, U., Prati, B. 2006. Dietary therapy in idiopathic nephrolithiasis. *Nutrition Reviews* July 64(7 Pt 1), 301-312.

Brenner, H., Rothenbacher, D. and Arndt, V. 2009. Epidemiology of stomach cancer. *Methods in Molecular Biology* 472, 467-477.

Brown, I.J., Tzoulaki, I., Candeias, V., Elliott, P. 2009. Salt intakes around the world: implications for public health. *International Journal of Epidemiology* June 38(3), 791-813.

Carbone, L.D., Bush, A.J., Barrow, K.D., Kang, A.H. 2003. The relationship of sodium intake to calcium and sodium excretion and bone mineral density of the hip in postmenopausal African-American and Caucasian women. *Journal of Bone and Mineral Metabolism* 21(6), 415-420.

Charlton, K.E., Gemming, L., Yeatman, H., Ma, G. 2010. Suboptimal iodine status of Australian pregnant women reflects poor knowledge and practices related to iodine nutrition. *Nutrition Journal* January 14.

Cohen, A.J. and Roe, F.J. 2000. Review of risk factors for osteoporosis with particular reference to a possible aetiological role of dietary salt. *Food and Chemical Toxicology* February-March 38(2-3), 237-253.

Cook, N.R., Cutler, J.A., Obarzanek, E., Buring, J.E., Rexrode, K.M., Kumanyika, S.K., *et al.* 2007. Long term effects of dietary sodium reduction on cardiovascular disease outcomes: observational follow-up of the trials of hypertension prevention (TOHP). *British Medical Journal* 334(7599), 885-888B.

Cook, N.R., Obarzanek, E., Cutler, J.A., Buring, J.E., Rexrode, K.M., Kumanyika, S.K., *et al.* 2009. Joint effects of sodium and potassium intake on subsequent cardiovascular disease: the Trials of Hypertension Prevention follow-up study. *Archives of Internal Medicine* January 12; 169(1), 32-40.

Delange, F. 2002. Iodine deficiency in Europe and its consequences: an update. *European Journal of Nuclear Medicine and Molecular Imaging*, August 29, Suppl. 2, S404-416.

Delange, F. 2007. Iodine requirements during pregnancy, lactation and the neonatal period and indicators of optimal iodine nutrition. *Public Health Nutrition* December 10(12A), 1571-1580; discussion 81-83.

D'Elia, L., Rossi, G., Ippolito, R., Cappuccio, F.P., Strazzullo, P. 2012. Habitual salt intake and risk of gastric cancer: a meta-analysis of prospective studies. *Clinical Nutrition* August 31(4), 489-498.

Dias-Neto, M., Pintalhao, M., Ferreira, M., Lunet, N. 2010. Salt intake and risk of gastric intestinal metaplasia: systematic review and meta-analysis. *Nutrition and Cancer* 62(2), 133-147.

Dickinson, H.O., Nicolson, D.J., Cook, J.V., Campbell, F., Beyer, F.R., Ford, G.A., *et al.* 2006a. Calcium supplementation for the management of primary hypertension in adults. *Cochrane Database of Systematic Reviews* 2, CD004639.

Dickinson, H.O., Mason, J.M., Nicolson, D.J., Campbell, F., Beyer, F.R., Cook, J.V. *et al.* 2006b. Lifestyle interventions to reduce raised blood pressure: a systematic review of randomized controlled trials. *Journal of Hypertension* 24(2), 215-233.

Espelid, I. 2009. Caries preventive effect of fluoride in milk, salt and tablets: a literature review. *European Archives of Paediatric Dentistry* September 10(3), 149-156.

Estupiñán-Day, S. 2005. Promoting Oral Health: The Use of Salt Fluoridation to Prevent Dental Caries: Pan American Health Organisation.

Fink, H.A., Akornor, J.W., Garimella, P.S., MacDonald, R., Cutting, A., Rutks, I.R., *et al.* 2009. Diet, fluid, or supplements for secondary prevention of nephrolithiasis: a systematic review and meta-analysis of randomized trials. *European Urology* July 56(1), 72-80.

Geerling, J.C. and Loewy, A.D. 2008. Central regulation of sodium appetite. *Experimental Physiology* February 93(2), 177-209.

Geleijnse, J.M., Witteman, J.C., den Breeijen, J.H., Hofman, A., de Jong, P.T., Pols, H.A., *et al.* 1996. Dietary electrolyte intake and blood pressure in older subjects: the Rotterdam Study. *Journal of Hypertension* June 14(6), 737-741.

Gillespie, G., Marinho, V., Marthaler, T.M., Holt, R., Poulsen, S., Stephen, K., *et al.* 2009. Salt fluoridation for preventing dental decay. The Cochrane Library. DOI: 10.1002/14651858.CD006846.

Graudal, N.A., Galloe, A.M. and Garred, P. 1998. Effects of sodium restriction on blood pressure, renin, aldosterone, catecholamines, cholesterols, and triglyceride: a meta-analysis. *Journal of the American Medical Association* May 6; 279(17), 1383-1391.

Graudal, N.A., Hubeck-Graudal T. and Jurgens, G. 2011. Effects of low sodium diet versus high sodium diet on blood pressure, renin, aldosterone, catecholamines, cholesterol, and triglyceride. *The Cochrane database of systematic reviews* November 9(11), CD004022.

Griffith, L.E., Guyatt, G.H., Cook, R.J., Bucher, H.C., Cook, D.J. 1999. The influence of dietary and nondietary calcium supplementation on blood pressure: an updated metaanalysis of randomized controlled trials. *American Journal of Hypertension* January 12(1 Pt 1), 84-92.

He, F.J. and MacGregor, G.A. 2006. Importance of salt in determining blood pressure in children - Meta-analysis of controlled trials. *Hypertension* 48(5), 861-869.

He, F.J. and MacGregor, G.A. 2010. Reducing population salt intake worldwide: from evidence to implementation. *Progress in Cardiovascular Diseases* 52(5), 363-382.

Heaney, R.P. 2006. Role of dietary sodium in osteoporosis. *The Journal of the American College of Nutrition* June 25(3 Suppl), 271S-276S.

Hollowell, J.G., Staehling, N.W., Flanders, W.D., Hannon, W.H., Gunter, E.W., Spencer, C.A., *et al.* 2002. Serum TSH, T(4), and thyroid antibodies in the United States population (1988 to 1994): National Health and Nutrition Examination Survey (NHANES III). *The Journal of Clinical Endocrinology and Metabolism* February 87(2), 489-499.

Hooper, L., Bartlett, C., Smith, G.D., Ebrahim, S. 2002. Systematic review of long term effects of advice to reduce dietary salt in adults. *British Medical Journal* 325(7365), 628-632.

Intersalt Cooperative Research Group. 1988. Intersalt: an international study of electrolyte excretion and blood pressure. Results for 24 hour urinary sodium and potassium excretion. *British Medical Journal* 297, 319-328.

James, W.P.T., Ralph. A. and Sanchez-Castillo. C.P. 1987. The dominance of salt in manufactured food in the sodium intake of affluent societies. *Lancet* 1, 426-429.

Johner, S.A., Gunther, A.L. and Remer, T. 2011. Current trends of 24-h urinary iodine excretion in German schoolchildren and the importance of iodised salt in processed foods. *British Journal of Nutrition* December 106(11), 1749-1756.

Kato, S., Tsukamoto, T., Mizoshita, T., Tanaka, H., Kumagai, T., Ota, H., *et al.* 2006. High salt diets dose-dependently promote gastric chemical carcinogenesis in Helicobacter pylori-infected Mongolian gerbils associated with a shift in mucin production from glandular to surface mucous cells. *International Journal of Cancer* October 1; 119(7), 1558-1566.

Kotchen, T.A. 1999. To salt, or not to salt?. *American Journal of Physiology* May 276(5 Pt 2), H1807-1810.

Kotchen, T.A., Cowley, A.W. Jr. and Frohlich. E.D. 2013. Salt in health and disease--a delicate balance. *The New England Journal of Medicine* March 28; 368(13), 1229-1237.

Kumar, M.V. and Rajagopalan, S. 2007. Multiple micronutrient fortification of salt and its effect on cognition in Chennai schoolchildren. *Asia Pacific Journal of Clinical Nutrition* 16, 505-511.

Kumar, M.V. and Rajagopalan, S. 2009. Multiple micronutrient fortification of salt. *European Journal of Clinical Nutrition* 63, 437-445.

Lambers Heerspink, H.J., Navis, G. and Ritz, E. 2012. Salt intake in kidney disease - a missed therapeutic opportunity?. *Nephrology Dialysis Transplantation* September 27(9), 3435-3442.

Law, M.R., Frost, C.D. and Wald, N.J. 1991. By how much does dietary salt reduction lower blood pressure? III – Analysis of data from trials of salt reduction. *British Medical Journal* 302, 811-815.

Li, M., Eastman, C.J., Waite, K.V., Ma, G., Zacharin, M.R., Topliss, D.J., *et al.* 2006. Are Australian children iodine deficient? Results of the Australian National Iodine Nutrition Study. *Medical Journal of Australia* February 20; 184(4), 165-169.

Lin, P.H., Ginty, F., Appel, L.J., Aickin, M., Bohannon, A., Garnero, P., *et al.* 2003. The DASH diet and sodium reduction improve markers of bone turnover and calcium metabolism in adults. *Journal of Nutrition* October 133(10), 3130-3136.

Machnik, A. *et al.* 2009. Macrophages regulate salt-dependent volume and blood pressure by a vascular endothelial growth factor-C- dependent buffering mechanism. *Nature medicine* 15(5), 545-552.

Marthaler, T.M. 2005. Increasing the public health effectiveness of fluoridated salt. *Schweiz Monatsschr Zahnmed* 115(9), 785-792.

Marthaler, T.M. and Petersen, P.E. 2005. Salt fluoridation--an alternative in automatic prevention of dental caries. *International Dental Journal* December 55(6), 351-358.

McCarron, D.A. 1985. Is calcium more important than sodium in the pathogenesis of essential hypertension?. *Hypertension* 7, 607-627.

McCarron, D.A., Geerling, J.C., Kazaks, A.G., Stern, J.S. 2009. Can dietary sodium intake be modified by public policy?. *Clinical Journal of the American Society of Nephrology* November 4(11), 1878-1882.

McCarron, D.A., Drueke, T.B. and Stricker, E.M. 2010. Science trumps politics: urinary sodium data challenge US dietary sodium guideline. *The American Journal of Clinical Nutrition* November 92(5), 1005-1006.

Midgley, J.P., Matthew, A.G., Greenwood, C.M.T., Logan, A.G. 1996. Effect of reduced dietary sodium on blood pressure. A meta-analysis of randomized controlled trials. *Journal of the American Medical Association* 275, 1590-1597.

Moinier, B.M. and Drueke. T.B. 2008. Aphrodite, sex and salt-from butterfly to man. *Nephrology Dialysis Transplantation*, July 23(7), 2154-2161.

Muntzel, M. and Drueke, T. 1992. A comprehensive review of the salt and blood pressure relationship. *American Journal of Hypertension*, April 5, 1S-42S.

Nagata, C., Takatsuka, N., Shimizu, N., Shimizu, H. 2004. Sodium intake and risk of death from stroke in Japanese men and women. *Stroke* July 35(7), 1543-1547.

O'Donnell, M.J., Yusuf, S., Mente, A., Gao, P., Mann, J.F., Teo, K., *et al.* 2011. Urinary sodium and potassium excretion and risk of cardiovascular events. *Journal of the American Medical Association* November 23; 306(20), 2229-2238.

Paterna, S., Gaspare, P., Fasullo, S., Sarullo, F.M., Di Pasquale, P. 2008. Normal-sodium diet compared with low-sodium diet in compensated congestive heart failure: is sodium an old enemy or a new friend?. *Clinical Science* (London) February 114(3), 221-230.

Pearce, E.N. 2008. Iodine in pregnancy: is salt iodization enough?. *The Journal of Clinical Endocrinology and Metabolism* July 93(7), 2466-2468.

Pearce, S.M., Andersson M. and Zimmermann, M.B. 2013. Global iodine nutrition: where do we stand in 2013?. *Thyroid* 23(5), 523-528.

Petersen, P.E., Baez, R.J. and Lennon, M.A. 2012. A Community-oriented administration of fluoride for the prevention of dental caries: a summary of the current situation in Asia. *Advances in Dental Research* 24(1), 5-10.

Ram, F.S. and Ardern, K.D. 2004. Dietary salt reduction or exclusion for allergic asthma. *Cochrane Database of Systematic Reviews* 3, CD000436.

Rasheed, S., Hanifi, M.A., Iqbal, M., Nazma, N., Bhuiya, A. 2001. Policy of universal salt iodization in Bangladesh: do coastal people benefit?. *Journal of Health, Population and Nutrition* June 19(2), 66-72.

Rasmussen, L.B., Carle, A., Jorgensen, T., Knudsen, N., Laurberg, P., Pedersen, I.B., *et al.* 2008. Iodine intake before and after mandatory iodization in Denmark: results from the Danish Investigation of Iodine Intake and Thyroid Diseases (DanThyr) study. *British Journal of Nutrition* July 100(1), 166-173.

Rogers, A.B., Taylor, N.S., Whary, M.T., Stefanich, E.D., Wang, T.C., Fox, J.G. 2005. Helicobacter pylori but not high salt induces gastric intraepithelial neoplasia in B6129 mice. *Cancer Research* December 1; 65(23), 10709-10715.

Sacks, F.M., Svetkey, L.P., Vollmer, W.M., Appel, L.J., Bray, G.A., Harsha, D., *et al.* 2001. Effects on blood pressure of reduced dietary sodium and the dietary approaches to stop hypertension (DASH) diet. *The New England Journal of Medicine* 344(1), 3-10.

Smith, W.C., Crombie, I.K., Tavendale, R.T., Gulland, S.K., Tunstall-Pedoe, H.D. 1988. Urinary electrolyte excretion, alcohol consumption, and blood pressure in the Scottish heart health study. *The British Medical Journal* July 30; 297(6644), 329-330.

Strazzullo, P. and Galletti, F. 2007. Genetics of salt-sensitive hypertension. *Current Hypertension Reports* March 9(1), 25-32.

Strazzullo, P., D'Elia, L., Kandala, N.B., Cappuccio, F.P. 2009. Salt intake, stroke, and cardiovascular disease: meta- analysis of prospective studies - art. no. b4567. *British Medical Journal* 24(339).

Strom, B.L., Yaktine, A.L. and Oria, M. (eds.) 2013 *Sodium intake in populations: Assessment of evidence.* Institute Of Medicine (IOM). Washington D.C., The National Academies Press,

Swales, J.D. 1991. Dietary salt and blood pressure: the role of meta-analyses. *Journal of Hypertension* 9 (suppl 6), S42-S46.

Taubes, G. 1998. The (political) science of salt. *Science* August 14; 281(5379), 898-907.

Tayie, F.A. and Jourdan, K. 2010. Hypertension, dietary salt restriction, and iodine deficiency among adults. *American Journal of Hypertension* October 23(10), 1095-1102.

Teucher, B. and Fairweather-Tait, S. 2003. Dietary sodium as a risk factor for osteoporosis: where is the evidence?. *Proceedings of the Nutrition Society* November 62(4), 859-866.

Teucher, B., Dainty, J.R., Spinks, C.A., Majsak-Newman, G., Berry, D.J., Hoogewerff, J.A., *et al.* 2008. Sodium and bone health: impact of moderately high and low salt intakes on calcium metabolism in postmenopausal women. *Journal of Bone and Mineral Research* Sepember 23(9), 1477-1485.

Thomas, M.C., Moran, J., Forsblom, C., Harjutsalo, V., Thorn, L., Ahola, A., *et al.* 2011. The association between dietary sodium intake, ESRD, and all-cause mortality in patients with type 1 diabetes. *Diabetes Care* April 34(4), 861-866.

Tuomilehto, J., Jousilahti, P., Rastenyte, D., Moltchanov, V., Tanskanen, A., Pietinen, P., *et al.* 2001. Urinary sodium excretion and cardiovascular mortality in Finland: a prospective study. *The Lancet* 357(9259), 848-851.

Vandevijvere, S., Mourri, A.B., Amsalkhir, S., Avni, F., Van Oyen, H., Moreno-Reyes, R. 2012. Fortification of bread with iodized salt corrected iodine deficiency in school-aged children, but not in their mothers: a national cross-sectional survey in Belgium. *Thyroid* October 22(10), 1046-1053.

Weinberger, M.H., Miller, J.Z., Luft, F.C., Grim, C.E., Fineberg, N.S. 1986. Definitions and characteristics of sodium sensitivity and blood pressure resistance. *Hypertension* June 8(6 Pt 2), II127-134.

Whelton, P.K., Appel, L.J., Sacco, R.L., Anderson, C.A., Antman, E.M., Campbell, N., *et al.* 2012. Sodium, blood pressure, and cardiovascular disease: further evidence supporting the American Heart Association sodium reduction recommendations. *Circulation* December 11; 126(24), 2880-2889.

World Health Organization. 2012. *Sodium intake for adults and children.* WHO, Geneva.

Wu, Y., Li, X., Chang, S., Liu, L., Zou, S., Hipgrave, D.B. 2012. Variable iodine intake persists in the context of universal salt iodization in China. *Journal of Nutrition* September 142(9), 1728-1734.

Yeung, C.A. 2011. Efficacy of salt fluoridation. *Evidence-Based Dentistry* 12(1), 17-18.

Zimmermann, M. and Delange, F. 2004. Iodine supplementation of pregnant women in Europe: a review and recommendations. *European Journal of Clinical Nutrition* July 58(7), 979-984.

Zimmermann, M.B. and Andersson, M. 2012. Update on iodine status worldwide. *Current Opinion in Endocrinology, Diabetes and Obesity* October 19(5), 382-387.

General Principles for the Addition of Essential Nutrients to Food CAC/GL 09-1987 (amended 1989, 1991) [database on the Internet] 1991. Available from: http://www.codexalimentarius.net/download/standards/299/CXG_009e.

Online source

Codex Alimentarius
www.codexalimentarius.net/download/standards/299/CXG_009e

Saline Aerosols: from Speleotherapy to Halotherapy

Lăcrămioara Ochiuz
Universitatea de Medicină şi Farmacie „Grigore T. Popa" Iaşi

"Science is truth, it is to know and to use natural forces in order to acquire human happiness"
Victor Babeş (bacteriologist and morpho-pathologist, 1854–1926)

Abstract
The saline aerosols have been used in the treatment of respiratory, digestive and dermatological diseases since ancient Greek medicine. Hippocrates noticed the positive effects of saline aerosols and recommended the inhalation of saline vapor for the treatment of the respiratory system diseases. Although, the therapy in salt mines and caves has been practiced from the oldest times, the speleotherapy or spelotherapy was recognized officially as complementary therapy in the middle of the 20th century in Germany. The therapeutic agent of the salt mine microclimate consists in sodium chloride aerosols together with the temperature and relative humidity of the air. Even if initially the spelotherapy was hardly practiced for the treatment of chronic inflammatory diseases of airways, over the last 30 – 40 years this holistic therapy has been increasingly developed. Nowadays, in Europe, over 50 saline mines or caves are known as places were speleotherapy is practiced successfully. In Romania, in Praid, Cacica or Ocna Mures saline mines there are genuine underground cities built of and in salt which host annually thousands of patients treated by the curative effects of saline aerosols. Halotherapy is a modern method of therapy based on an artificially-created saline microclimate. The first halochamber was used by the space agency of the former Soviet Union, in 1980. Today, the halotherapy practice is governed by the international legislation and several types of halochambers are licensed, as medical devices. The saline aerosols dispersed into the saline mine or halochamber atmosphere are inhaled through the airways or absorbed transcutaneously. Depending on the particle diameter, the inhaled aerosols act at different levels of the respiratory tract by accelerating the mucociliary clearance and by optimizing pulmonary surfactant. Halotherapy influences the respiratory system simptomatology by improving breathing. Moreover the patient feels a release of the airways as a result of cough productivity increase and pulmonary clearance, the decreasing of respiratory system infection bouts due the bacteriostatic effect and not least the decrease of sinusal edema. The saline aerosols have positive effects at tegumentary system. According to scientific literature, they balance skin microbiocenosa, increase skin elasticity, and have bacteriostatic, antiedema and antiinflammatory effects. The therapeutic potential of artificial or natural saline aerosols is researched all over the world in order to achieve new scientific supporting data for these holistic therapies.

Keywords
saline aerosols, therapeutic effect, halotherapy

Introduction
Speleotherapy and halotherapy are complementary therapeutic methods based on the curative effects of saline aerosols either inhaled by patients or absorbed through their skin. The pharmacological effects exerted by saline aerosols mainly consist in the fluidization and elimination of the bronchial secretion (mucus), the decongestion of upper and lower respiratory tracts while relieving feelings of suffocation, the reduction of the incidence of bouts of respiratory tract infections due to the bacterostatic and bactericidal action on pathogens, and finally the reduction of sinus oedema. Saline aerosols also exert positive effects on the skin system. According to data in the literature, they normalize cutaneous micro-biocenosis, increase skin elasticity and have bacteriostatic, anti-inflammatory and anti-oedematous effects.

Salt and saline aerosols through medical history
The use of saline aerosols in the treatment of respiratory, digestive and dermatological conditions has been known since the ancient Greeks. In an empirical form, halotherapy has been practised since time immemorial when man began to exploit and consume salt. The first evidence for the use of salt dates back to the pre-Christian period, to papyri of Imhotep (c. 2610 BC) that contain numerous salt-based remedies for the treatment of wounds. In the Ebers papyrus (1600 BC) salt was recommended as a laxative, anti-inflammatory and antiseptic remedy (Brandt-Rauf and Brandt-Rauf 1987, 68-70). This papyrus contains the formula for an anti-inflammatory ointment made from honey, sea salt, animal fat and seeds.

Hippocrates (the "Father of Medicine", 460-370 BC) noted the benefit of saline aerosols and recommended the inhalation of salt water vapours for the treatment of respiratory diseases. In Greek medicine, a mixture of two parts fresh cow's milk and one part water with sea salt dissolved in it drunk every morning on an empty stomach was recommended to treat spleen. For the care of wounds salt water was used, and, for healing wounds, they applied a mixture of honey and salt. Hippocrates draws attention, however, to the harmful effects of the ingestion of large quantities of salt, especially on the digestive and cardiovascular systems (Crişan and Mihăescu 2011, 36-38).

Salt was widely used and researched in Arabic medicine as well. The medieval physician Avicenna (980-1037) discovered high concentrations of iodine and iron in sea salt.

Paracelsus, doctor and alchemist (1493-1541), noted that external factors can influence bodily health and recommended salt baths to treat wounds, but also to increase skin health and to strengthen the body (Najeeba and Arakkal 2011, 128-134).

In the twentieth century it was discovered that salt water has the same properties as blood plasma, which led to intravenous (directly into the bloodstream) administration of salt solutions. Currently, a 0.9% isotonic sodium chloride solution (*Infunbile Natrii Chloridi*) is the most common preparation for infusible drugs, improperly called "normal saline" because this solution has the same osmotic pressure as blood serum, but its chemical composition is different (***Romanian Pharmacopeea 1993, 504).

Throughout history there have been many instances of people benefiting from the healing effects of saline aerosols. Thus, refugees who sheltered in caves during the European invasions of the Middle Ages survived in greater numbers compared with those who were not exposed to a cave environment. Likewise, since monks knew empirically the effects of saline aerosols on the body, they brought salt from mines for the patients in their care, to enable them to breathe salty air. We could say that these were the first steps in what we now call halotherapy.

There were, however, exaggerated claims made with regard to the curative benefits of salt and a number of theories without scientific foundation were put forward on the effectiveness of saline aerosols for the treatment of rheumatic diseases, for some Central Nervous System disorders, such as anxiety or depressive syndrome, and even for neoplastic diseases.

On the other hand there were many opinions expressed critical of the supposed miraculous effects of salt. One of the most eloquent statements was made by Paul Berg, the American biochemist and Nobel laureate for chemistry in 1980, who called salt the "white poison" or the "hell of the heart" (Chervinskaya and Zilber 1995, 221-232).

The beneficial effects are however more numerous than the negative, and it is practically impossible to imagine human existence without salt. Salt is a symbol of life and the physiology of the human body depends on the balance between salt and fluids in the body.

The mechanism of saline aerosols

Aerosols are solid or liquid microparticles, with diameters in the range 0 -10μm in suspension in a gaseous medium. Their presence in the atmosphere is caused by various phenomena such as evaporation, superficial dissolution, mechanical or chemical actions and emission. Depending on the microclimate conditions (humidity, temperature, air speed, pressure variations, UV radiation effects, etc.) Aerosolized microparticles can persist in the atmosphere for intervals varying from a few hours to days, weeks or even years. Particles with diameters larger than 10μm are deposited on the ground, due to gravity (Figure 1) (Mathay, Kolkesson and Verkman 1996, 487-503).

Saline aerosols dispersed in the atmosphere of salt mines or in special enclosures called halo chambers are inhaled in the respiratory tracts or are transcutaneously absorbed. Depending on their size, salt microparticles reach different levels of the respiratory tract (Figure 2) (Phipps *et al.* 1994, 1474-1482).

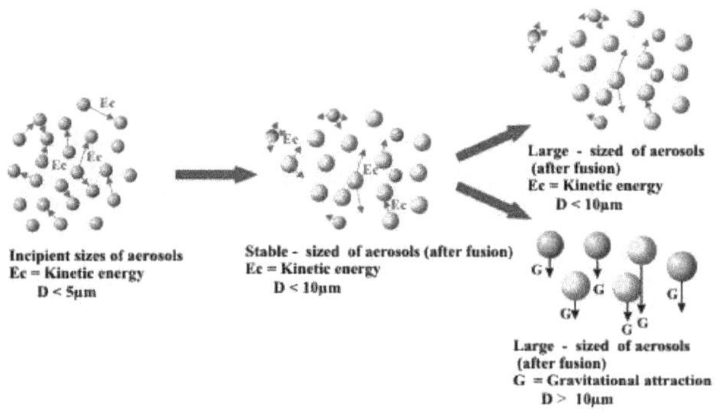

Figure 1. The kinetics of aerosols in the atmosphere.

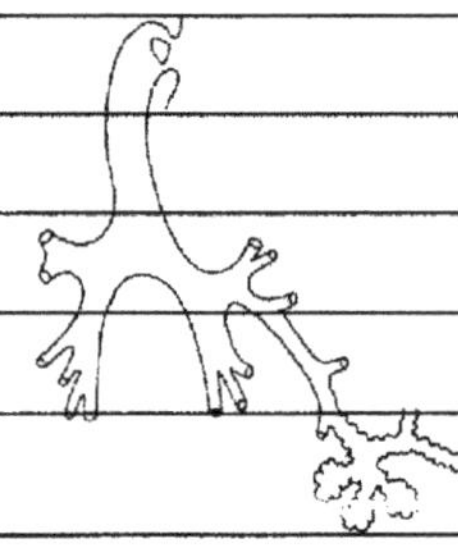

Diameter of aerosolized particles	Maximum level of penetration
>30 μm	Nasal cavity, pharynx and larynx
20-30 μm	Trachea
10-20 μm	Bronchi and bronchioles
3-5 μm	Terminal bronchioles
< 3 μm	Alveolar channels Pulmonary alveoles

Figure 2. The transit of aerosol particles along the respiratory tract depending on size.

Due to their hygroscopic properties and thanks to kinetic energy loss, large aerosols deposit in the upper respiratory tract (the nasal cavity, trachea, bronchi), while small aerosols reach the lower airways (bronchioles, lung alveoli and alveolar channels). It is important to note that due to the laminar effect, particles smaller than 3μm pass down the bronchial channels without being deposited and those smaller than 1μm in diameter will reach the pulmonary alveoli. Laminar effect is the phenomenon of increasing air velocity when moving from a larger flow section to a smaller (narrower) flow section. This flow effect drives the small aerosols and transports them to the smallest parts of the respiratory system (Phipps *et al.* 1994, 1474-1482; Najeeba and Arakkal 2011, 128-134).

Pulmonary surfactant, a phospholipoprotein complex, plays an important role in the gas exchange in the lung alveoli (Figure 3). The pulmonary surfactant molecule has an amphoteric character, it realizes the air-water interface in the alveoli and increases lung compliance (the ratio of the volume of air contained in the lung and intra-pulmonary pressure) (Daviskas *et al.* 1996, 725-732).

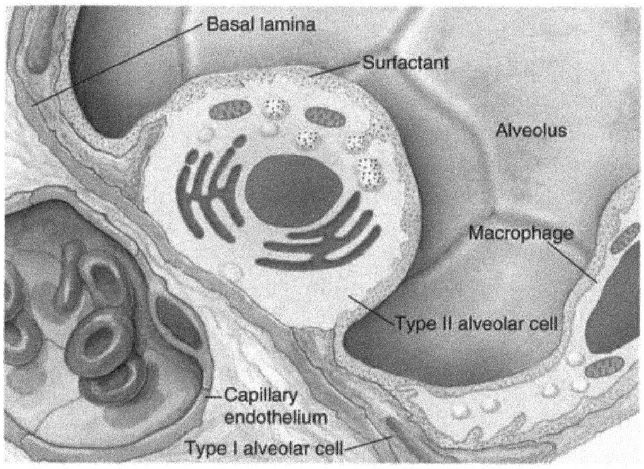

Figure 3. Pulmonary surfactant.

Pulmonary ventilation (the continuous refreshing of pulmonary air in the exchange areas where there is an intimate contact with the capillaries) is a consequence of the presence of the surfactant and of pulmonary compliance.

Na^+ ion attracts water from the interstices and increases its amount in the alveolar space. Consequently, there is a growth in the sol phase of the lung surfactant and an acceleration of mucociliary clearance. Thus, by stimulating and improving the ciliary function, the Na^+ ion increases mucociliary clearance and provides a more effective purgation of the upper and lower airways. Moreover, in smokers, who display slow to paralyzed ciliary movement, Na^+ ion manages to reactivate mucus propulsion and to cleanse the respiratory tree by cilium activation.

Last but not least, Na^+ ion is known to induce the ionization of particles around it, therefore pneumoallergenic particles that tend to deposit can be more easily eliminated by mucociliary clearance (Chervinskaya and Zilber 1995, 221-232).

From these data regarding respiratory physiology, we may identify immediate therapeutic indications for saline aerosols in chronic obstructive pulmonary disease (COPD), asthma, smoking, recurrent respiratory infections, but also in improving lung functionality in anthracosis or silicosis.

In addition, studies by American researchers have shown that a number of respiratory pathogens (*Haemophilus influenza, Steptococcus pneumonia and Staphylococcus spp.*) are highly sensitive to hypertonic salt solutions. Saline aerosols are hygroscopic, and dehydrate microbial cells by destroying their albumin structure and are subsequently removed by mucociliary clearance. These organisms adapt to the respiratory conditions of the patients with asthma and chronic obstructive bronchitis and retain an antigenic component similar to the respiratory mucosa, which produces a high risk of developing an allergic sensitization together with the inflammation. The number of these micro-organisms decreases significantly after treatment with saline aerosols (Chernova *et al.* 1996, 78-80).

Dermal exposure to a saline aerosol loaded atmosphere causes depositing followed by percutaneous absorption of saline ions through the skin capillaries and their transport to tissues and cells. Absorption through the skin of salt ions is mainly due to the osmotic process under the circumstances of bio electrovalent phenomena at molecular and atomic level.

Salt or other mineral ions are absorbed in the cells in two ways: by diffusion and by osmosis (Figure 4). Both osmosis and diffusion are vital processes by which living cells in the body "feed" on minerals, lipids, carbohydrates, proteins, water, etc. (Crişan and Mihăescu 2011, 36-38; Shani *et al.* 1985, 501-512).

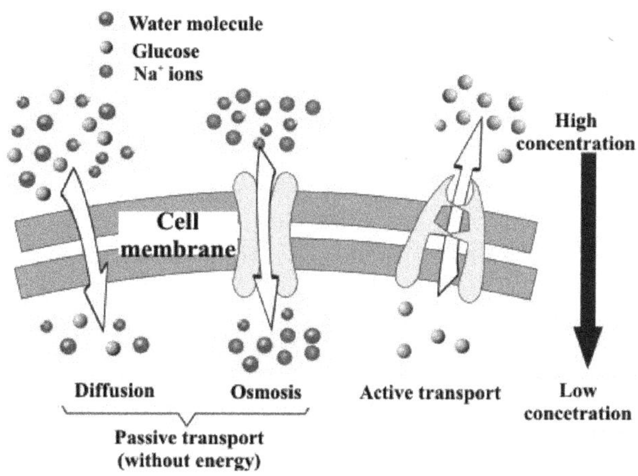

Figure 4. Graphic representation of the phenomena of diffusion, osmosis and active transport across cell membranes.

Positive influences exerted by saline aerosols in the skin system consist in normalizing skin microbiocenosis, pH adjustment, increased skin elasticity and improved barrier function, bacteriostatic, anti-inflammatory and oedematous effects.

Also, saline aerosols stimulate skin microcirculation, increase permeability and electrophysiological activity of epithelial cells resulting in stimulation of epidermal regeneration and optimization of absorption through the passive transport of drug substances in various dermato-cosmetic preparations.

Outstanding results were obtained from the application of saline aerosol therapy in the treatment of atopic dermatitis. A percentage of 65-75% of patients experienced significant improvement in the symptoms of this condition (decreased itching, drying and desquamation of scabs and reduced local xerosis).

Positive effects were reported in patients with psoriasis when after two or three sessions both inflammatory phenomena and hyperkeratosis diminished greatly (Smith *et al.* 2009, 841-856).

Emergence and development of speleotherapy

Therapy in salt mines and caves has been practiced since ancient times but speleotherapy (Greek *spelos* = cave *therapeia* = care) was officially recognized as an adjunctive therapy in mid-twentieth century Germany. Speleotherapy is based on the curative effects of sodium chloride aerosols and on the optimal conditions of the salty microclimate regarding temperature, humidity, microbiological load and the absence of allergens in it. It should be noted that every room in every mine has a distinct microclimate and the view is that the curative properties of speleotherapy are generated by a synergism between saline aerosols and microclimatic factors.

The scientific recognition and the therapeutic use of a saline environment began with a number of observations made during the nineteenth century on the absence of chronic bronchitis and asthma morbidity in miners in the salt mines in Wieliczka (Poland), the fast healing of these diseases in new employees, and the improvement or even disappearance of asthma in asthmatic refugees during the two world wars in the Kluttert Cave (Germany) used as a bomb shelter. The father of speleotherapy is the Polish doctor Felix Botchkovsky who hypothesized in 1843 that air enriched with particles of salt has a therapeutic action (Crişan and Mihăescu 2011, 36-38; Chervinskaya, and Zilber 1995, 221-232).

Modern salino-therapy was officially recognized in 1950-1960 in Germany and Poland. In Germany, the first medical reports were related to the effective therapeutic results in Kluttert Salt Cave (North Rhine Westfalia). These scientific observations generated interest in other countries, such as Hungary and Slovakia, and shortly afterwards Slovakia obtained the first favourable health outcomes observed in children with asthma and recurrent episodes of bronchitis in the Gombaek salt cave. In 1958, in the salt mine in Wieliczka (Krakow, Poland) there was established the first treatment centre for people with lung diseases (Anderson *et al.* 1997, 2465-2473).

This research enabled the evolution of salino-therapy into a system of medical treatment especially effective for the treatment of the respiratory tract. Treatment centres were opened based on the exploitation of natural underground saline cavities in Austria (Altaussee, Hallstadt, Salzbad - Salzeman), Poland (Wieliczka), Romania (Praid, Slănic Prahova), Azerbaijan (Nakhichevan), Kyrgyzstan (Chonase), Russia (Berezniky, Perm region), Ukraine (Artemovsk, in the Donetsk region and Solotvino, in the Carpathian region), Belarus (Soligorsk), at depths ranging from 100 to 400 m. The presence of a self-regeneration phenomenon of the natural factors was observed in the natural underground environment, but this phenomenon cannot be identically replicated or created in an artificial environment with a salty micro-climate with curative properties. The cumulative effect of the natural factors of the saline microclimate is also generated by the presence in the air of Ca^{2+}, K^{+}, Mg^{2+} ions with an anti-inflammatory, anti-allergic, and spasmolytic function, and by a slightly higher CO_2 concentration which stimulates deep breathing and blood circulation (Chervinskaya and Zilber 1995, 221-232).

In 1969, the Speleotherapy Commission was set up as a branch of the International Union of Speleology, with the participation of many countries in Europe and Asia. The establishment of this commission increased interest on the part of salt mining institutions in using for medical purposes mines and mining sectors closed through deposit depletion, and further to promote the development of salino-therapy or speleotherapy as a particularly effective alternative treatment method.

Currently in Europe there are about 60 known underground salt cavities and caves used for speleotherapy or are being prepared for medical use or for balneoclimatic tourism, of which 17 are in Germany, 5 in Slovakia, 4 each in Hungary and Austria, and 3 in Ukraine (Chervinskaya and Zilber 1995, 221-232; Anderson *et al.* 1997, 2465-2473).

In Romania, speleotherapy studies on climatic conditions and the evaluation of clinical outcomes began in 1970 under the supervision of specialists from the National Institute of Rehabilitation, Physical Medicine and Balneoclimatology in Bucharest.

According to publications of recent years, Romanian salt mine microclimate (Prahova Slanic, Targu Ocna) has a thermo-hygro-barrel constant (temperature 10 to 12°C, relative humidity 60-75%, atmospheric pressure depending on outside air pressure) with air velocity lower than 0.1 m/sec in the underground rooms, and about 0.3 to 0.4 m/sec near the vents, with a slight discomfort caused by cooling, slight hypotonic stress due to relatively low temperatures and slightly dehydrating due to scarcity of water vapor. In the mines in Romania, no considerable difference was reported in the concentration of air ions, especially those with negative polarity.

Small ions in air ionization is moderate, with a trend toward a predominance of positive ions and large ion concentration is relatively high, with a predominance of negative ions. Aerosol particle concentration is high, but different, with values up to 80-95% of particles below 3μm. The chemical indicators investigated, especially those that pollute underground salt air (CO_2, NO_2, SO_2, oxidizable substances), some of them monitored by staff mines, were found within permissible values or below maximal allowable concentrations. In some underground mines, microbiological determinations reveal a pure air with minimal concentrations of microorganisms in the environment. In fact, the environmental parameters of different galleries of underground salt mines in Romania vary by the geo-morphological structure of the salt layer and the methodology used to excavate salt, which influence the quality of this environment and the intensity of the curative properties and the therapeutic effect (Moscaliuc *et al.* 2004, 200-203).

In Romania, speleotherapy is practiced with good results in the Unirea mine - Slănic Prahova and Târgu Ocna salt mines. The galleries of the Praid salt mine, effectively used in the '70s for medical purposes, are now used more for tourism and "balneoclimatic tourism". In Cacica and Turda salt mines, and less in Dej, both "mining tourism" and "balneoclimatic tourism" are practiced. It is noteworthy that the National Development Programmes Agency runs a number of projects in Dej and Cacica salt mines which realize multidisciplinary studies aimed at evaluating the curative properties of salt (Simionca *et al.* 2007, 155-162; Simionca *et al.* 1989, 57-59).

Halotherapy – a modern method of treatment with saline aerosols

Halotherapy (Greek *hals*, *halos* = salt) is a treatment method based on artificially creating the conditions of the microclimate in salt mines. Although research in this area have shown that covering of walls with *halite* (rock salt, NaCl crystallized in the cubic system) or *sylvinit* (natural mixture of sylvina - KCl crystallized in the cubic system (35%) and rock salt (65%) and 5% other minerals) does not entirely replicate the curative atmosphere of salt mines, certain beneficial effects comparable to those from speleotherapy have been obtained. Halotherapy was patented in 1990 in the former Soviet Union and a year later, the Soviet Union Ministry of Health approved the methodology for halotherapy entitled "Halotherapy in therapy and prophylaxis".

In the 1980s a group of researchers in Odessa have patented the first Ukrainian therapy room with air ionized saline aerosols, called "Ionotron". Remarkable results in halotherapy devices were obtained from St. Petersburg (Russia) where the first halo chamber in the world was recorded in 1984. This was a room whose walls and floor were lined with salt brought from Solotvino salt mine (Ukraine). Later studies were aimed at modeling the microclimate of salt mine, and different types of halo chambers were patented, such as:

1."Climate chamber" for halotherapy based on sylvinit;
2."Quantum-Energetic halo chamber";

In the past 10 years, halo chambers have been patented in Germany and the USA as well. Halotherapy rooms are certified as medical devices in Russia, Eastern Europe and have begun to also spread in the United States (Crişan and Mihăescu 2011, 36-38; Chervinskaya and Zilber 1995, 221-232; Gorbenko *et al.* 1996, 24-28).

Through the nature of its organization, the halo chamber displays the following construction and treatment characteristics: due to the vertical arrangement of salt briquettes, the salt contact surface with the air is broader than in a natural salt mine; almost total lack of draught inside the halo chamber, the presence of fresh air alveoli, forced circulation of air through the salt walls with controlled porosity , large amount of saline aerosols (1-37mg/m3), optimal thermal comfort by digitally controlled constant temperature (18-24°C), monitored relative humidity between 45-75%, automatic regeneration of the therapeutic atmosphere, automatic ventilation and air filtration.

The principle of generating a saline microclimate inside halo chambers is based on relative atmospheric humidity which acts as a dispersion medium (solvent) for suspended (dispersed) liquid or solid microparticles. Obviously, in the case of the halo chambers the saline aerosols are the suspended microparticles. Following the mechanical-electrochemical process of priming the halo therapeutic atmosphere - a process that is part of the production secrets of each type of halo chamber, a microclimate is created that contains an amount of saline aerosols sufficient for triggering their ongoing self-generation cycle, respectively self-maintaining atmosphere within therapeutic values. Atmosphere self-generation is determined by diffusion phenomena, thermodynamic and electrochemical balancing naturally occurring inside the halo chamber. During these physical phenomena self-regulation of water vapor (solvent) and of dispersed material (solute - NaCl ions) pressure also takes place (Sandu *et al.* 2010, 1105-1113; Moscaliuc and Pascu 2003, 331-336).

Due to the gas solvent (water vapors that create the relative humidity) action on the salt surfaces in the walls and floor, under precise conditions of temperature, relative humidity, strictly controlled ratio of vitiated (used) and fresh (saline aerosol-free) there occurs a natural emanation of saline aerosols in the air inside, without the use of chemicals or artificial additives.

In conclusion, due to the very large contact surface between air and salt, the halo chamber atmosphere is naturally supplied with saline aerosols continuously and uniformly, through the controlled action of the solvent on the solute.

Conclusions

Both valued and charged, considered by some a "white poison" and by others the "white gold," salt remains a vital natural substance. Saline aerosols, either inhaled or transcutaneously absorbed during speleotherapy or halotherapy, certainly produce numerous therapeutic effects.

The inhalation of these aerosols, thank to their bactericidal and bacteriostatic action, cleanses the respiratory tracts of staphylococcal and streptococcal infectious nidus that trigger and maintain many diseases of the respiratory system. The stabilization of the normal body microflora in the airways causes a reduction in the inflammatory process, as increased activity of alveolar macrophages was observed and a normalization of all cells involved in providing immunity. This effect is a major benefit for patients with COPD by reducing or eliminating recurrent infectious episodes.

Also, saline aerosols positively influence mucociliary clearance by reducing mucus viscosity and by activating ciliary movement. In the lower respiratory tract, saline aerosols stimulate lung ventilation (continuous refreshing of air in the exchange pulmonary areas where there is an intimate contact with capillaries) by increasing lung surfactant and lung compliance. Consequently, saline aerosols are a genuine remedy for the treatment of respiratory diseases such as infectious and allergic asthma, asthmatic bronchitis, COPD, cystic fibrosis, chronic tracheobronchitis, rhinosinusitis, tonsillitis and chronic tonsillitis, smoking, silicosis, etc. (Anderson *et al.* 1997, 2465-2473; Jongejan *et al.* 1991, 931-937).

Saline aerosols also exert positive effects on the skin system. According to the literature, they normalize cutaneous microbiocenosis, increase skin elasticity and have a bacteriostatic, anti-edematous and anti-inflammatory effects, which recommends them for the treatment of autoimmune or unknown etiology skin disorders such as psoriasis and atopic dermatitis (Yankura *et al.* 2008, 100-101; Schlessinger 2008, 403-411).

Because it was found that during the cleavage of the sodium chloride microcrystals the transcutaneous passive transport increases in the skin, saline aerosols are expected in the future to be used as a absorption promoters for drug substances by associating halotherapy with the topical application of pharmaceutical preparations in the form of ointments, creams, lotions, etc.

The therapeutic potential of natural and artificial saline aerosols has been drawing the attention of researchers worldwide in their attempt to gather new scientific data, in support of these holistic therapies.

References

Anderson S.D., Spring J., Moore B. *et al.* 1997. The effect of inhaling a dry powder of sodium chloride on the airways of asthmatic subjects. *European Respiratory Journal* 10(11), 2465-2473.

Brandt-Rauf, P.W. and Brandt-Rauf, S.I. 1987. History of occupational medicine: relevance of Imhotep and the Edwin Smith papyrus. *British Journal of Industrial Medicine* 44/1, 68-70.

Chernova, O.P., Matiushina, S.B., Volianik, M.N. *et al.* 1996. The dynamics of the persistence characteristics of staphylococci under the action of the microclimate of a speleotherapy mine. *Zhurnal Mikrobiologii Epidemiologii I Immunobiologii* 3, 78-80.

Chervinskaya, A.V. and Zilber, N.A. 1995. Halotherapy for treatment of respiratory diseases. *Journal of Aerosol Medicine* 8/3, 221-232.

Crişan, R. and Mihăescu, T. 2011. Haloterapia – o soluţie antică şi modernă în terapia afecţiunilor respiratorii. *Medica Academica* 3/16, 36-38.

Daviskas, L., Anderson, S.D., Gonda, I. *et al.* 1996. Inhalation of hypertonic saline aerosols enhances mucociliary clearance in asthmatic and healthy subjects. *European Respiratory Journal* 9, 725-732.

Gorbenko, P.P., Adamova, I.V. and Sinitsyna T.M. 1996. Bronchial hyperactivity to inhalation of hypo- and hyperosmolar aerosols and its correction by halotherapy. *Terapevticheskii Arkhiv* 68/8, 24-28.

Jongejan, R.C., de Jongste, J.C., , R.C. *et al.* 1991. Effects of hyperosmolarity on human isolated central airway. *British Journal of Pharmacology* 102, 931-937.

Mathay, M.A., Folkesson, H.G. and Verkman, A.S. 1996. Salt and water transport across alveolar and distal airway epithelia in the adult lung. *American Journal of Physiology* 270, 487-503.

Moscaliuc, G.V. and Pascu, C. 2003. Aparatul salin – un adjuvant important în tratamentul alergiilor căilor respiratorii superioare-studiu clinic. *Revista Medico-Chirurgicală a Societăţii de Medici şi Naturalişti din Iaşi* 107/2, 331-336.

Moscaliuc, G.V., Chiruţă, R. and Pascu, C. 2004. Influenţa microclimatului realizat de aparatul Salina asupra rinopatiei alergice-studiu clinic efectuat pe un lot de 36 de pacienţi. *Revista Medico-Chirurgicală a Societăţii de Medici şi Naturalişti din Iaşi* 108/1, 200-302.

Najeeba, R. and Arakkal, F.R. 2011. Spa Therapy in dermatology. *Indian Journal of Dermatology, Venereology and Leprology* 77, 128-134.

Phipps, P.R., Gonda, I., Anderson, S.D. *et al.* 1994. Regional deposition of saline aerosols of different tonicities in normal and asthmatic subjects. *European Respiratory Journal* 7/8, 1474-1482.

Sandu, I., Chirazi, M., Sandu, G.I. *et al.* 2010. Research on NaCl saline aerosols II. New Artificial halochamber characteristics. *Environmental Engineering and Management Journal* 9/9, 1105-1113.

Schlessinger, J. 2008. Spa Dermatology: Past, present and future. *Dermatologic Clinics* 26, 403-411.

Shani, J., Barak, S., Levi, D. *et al.* 1985. Skin penetration of minerals in psoriatics and guinea pigs bathing in hypertonic salt solutions. *Pharmacological Research Communications* 17, 501-512.

Simionca, I., Chonka, I. and Pop, I. 1989. Effect of the microclimate of salt mines on T- and B- Lymphocyte function in bronchial asthma patients. *Vrachebnoe Delo* 3, 57-59.

Simionca, I., Buturuga, A., Hoteteu, M. *et al.* 2007. Particularities of speleotheraputic treatment effect in underground environment of Slanic Prahova "Unirea" salt mine on infection-inflammatory process and unspecific antiinfectious resistance factors of patients organism with non-sever bronchial asthma. *Revista de*

Recuperare, Medicină Fizică şi Balneologie, 3-4, 155-162.

Smith, A., Weymann, A., Francisko, A., Gelfand, J.M. 2009. Complementary and alternative medicine for psoriasis: A qualitative review of the clinical trial literature. *Journal of the American Academy of Dermatology* 61, 841-856.

Yankura, J.A., Marks, J.G.Jr., Anderson, B.E., Adams, D.R. 2008. Spa contact dermatitis. *Dermatitis* 19, 100-101.

*** Romanian Pharmacopeea 10th 1993. *Infundibile Natrii Chloridi Monography* 504. Bucureşti, Medicală.

The Perception of Salt Springs in the Romanian Geographic and Geologic Literature

Gheorghe Romanescu
Universitatea „Alexandru Ioan Cuza" din Iaşi

Abstract
Romania is the richest European country and one of the most important in the world when it comes to salt. This is why salt has been used from time immemorial to the present. The most important Romanian and foreign researchers in the field have approached themes related to salt deposits: genesis, repartition, exploitation methods, use, etc. The Romanian salt resources reach 400 million tonnes, and they are among the highest qualitative ones. During the Communist period, the chemical industry was based mainly on salt exploitation. In 2007, 2.6 million tonnes of salt were exploited: 1.6 million tonnes of brine–salt, used exclusively in the chemical industry; 900 000 tonnes of rock salt, used to remove snow during the winter; 100 000 tonnes of crystallized salt, for consumption. In 2007, Romania exported 30% of the production. The salt springs are intimately related to the existence of salt deposits. In Antiquity, most salt springs were considered a divine gift, reason for which churches took them over. When the Church could no longer handle the crowd ready to enjoy the divine gifts, legislation made its way. This led to most salt, mineral, and thermal springs being protected by the law. The scientific fields with a focus on salt research are dominated by geography and geology, with applicability in exploiting and using this resource. There is lapidary information on salt in the ancient and mediaeval documents, notably being Dimitrie Cantemir's work *Descriptio Moldaviae* (1716), which mentions the existences of salt marshes in Bacău County. The geographic literature studies the repartition and importance of salt within the local and regional economy, while the geologic one approaches the genesis and exploitation methods. It is important to know that the first scientific works belonged to foreign specialists, mostly German and Austrian. A special attention has been paid lately to archaeological research, revealing that the first systematic salt exploitation in the world took place in Romania, in the Moldavian sub-Carpathians. This way, Romania was the first to exploit two vital raw materials for the economic development: oil and salt (both related genetically to the same areas).

Keywords
salt springs, salt, sub-Carpathians, Romania, literature

Introduction
Salt is a mineral necessary to life. Salt plays an important role in the individual and in the economic life. This is why it has been a highly circulated product from time immemorial (Bogdan 1973; Bucur 1970; Burlacu 1973; Irimuş 2006; Mac 1972; Micalevich 1967; Paucă 1978; Stoica 1966; Visarion *et al.* 1976). It acted as currency and it represented the central piece of the barter. The Romans came to Dacia for two useful minerals: gold and salt (Ciupagea *et al.* 1970).

The obvious importance of salt made it become the property of the State. Even sea-salt is part of the State's goods for the coastal countries.

Because of the accentuated tectonics specific to the Eurasian continent, the European and Asian countries have the richest reserves of salt deposits. The Romanian territory is also part of this category, mostly the Transylvanian depression and the external sub-Carpathian area between the Moldova valley and the Motru valley.

Romania is the richest European country in salt and one of the most important on Earth concerning production and quality. This is why salt has been used continuously from time immemorial until nowadays. The most important Romanian and foreign researchers in the field have approached themes related to salt deposits: genesis, repartition, exploitation methods, use, etc. the researches in the fields have also been favoured by the fact that these deposits are most often associated with oil or gas deposits (Airinei 1959; Albu and Balteş 1983; Atudorei *et al.* 1969; Ciupagea *et al.* 1967; Contescu and Mihăilescu 1970; Grujinschi 1971; Lupu 1971; Marosi 1967; Mrazec 1907; Murgoci 1907; Neacşu and Urcan 1978; Papainopol 1982; Paraschiv 1983; Pătruţ *et al.* 1973; Pârvu and Burlacu 1972; Vasilescu *et al.* 1984).

The Romanian salt resources reach 400 million tonnes, and they are among the highest qualitative ones. Given the current consumption in Romania, the salt reserve is enough for another around 2500 years and for the entire humankind for around 400 years (Atudorei *et al.* 1971; Bogdan 1973).

In 2007, 2.6 million tonnes of salt were exploited: 1.6 million tonnes of brine–salt, used exclusively in the chemical industry; 900 000 tonnes of rock salt, used to remove snow during the winter; 100 000 tonnes of crystallized salt, for consumption. In 2007, Romania exported 30% of the production.

The many salt springs in Romania are closely connected to salt deposits (Botzan 1984, 2002; Bulgăreanu 1983; Bulgăreanu and Feurdean 1982; Bulgăreanu *et al.* 1984; 1985a,b; Burlacu 1973; Ghiţă and Năulea 1980; Ielenicz 1985; Marosi 1967; Minea 2011; Pantazică 1969, 1974; Pantazică and Schram 1981; Vasilescu *et al.* 1984). In Antiquity, most salt springs were considered a divine gift, reason for which churches took them over. When the Church could no longer handle the crowd ready to enjoy the divine gifts, legislation made its way. This led to most salt, mineral, and thermal springs being protected by the law or part of the national patrimony.

The national and international literature includes rather many special or general studies on salt deposits: Airinei 1959; Albu and Balteş 1983; Anghel 1958; Atudorei *et al.* 1969, 1971; Bâgu and Mocanu 1984; Bogdan 1973; Botezatu *et al.* 1977; Botzan 1984, 2002; Bucur 1970; Bulgăreanu and Feurdean 1982; Bulgăreanu 1983; Bulgăreanu *et al.* 1984, 1985a,b; Burlacu 1973; Ciupagea 1980; Ciupagea *et al.* 1967; Contescu and Mihăilescu 1970; Cornea *et al.* 1973; Cristea *et al.* 1977; Dicea and Dicea 1981; Dovlete *et al.* 1983; Ghiţă and Năulea 1980; Givulescu 1982; Grujinschi 1971; Ielenicz 1985; Irimuş 2006; Isopescu *et al.* 1967; Lupu 1971; Mac 1972; Marosi 1967; Mărgărit and Mărgărit 1982; Micalevich 1967; Minea, 2011; Neacşu and Urcan 1978; Neguţ 1977; Pantazică 1969, 1974; Pantazică and Schram 1981; Papaianopol 1982; Paraschiv 1983; Paucă 1968, 1978; Pătruţ *et al.* 1973; Pânzaru 1976-1977, 1982; Pârcălăbescu 1982; Pârvu and Burlacu 1972; Roman and Papaianopol 1985; Povară *et al.* 1982; Scripăţ 1984; Susnjar 1984; Vasilescu *et al.* 1984; Visarion 1967; Visarion *et al.* 1976.

This study tries to emphasize on the importance of salt in the life of the Romanian people and on the perception of the geologic and geographic literature in the sphere of knowledge.

Methodology

The scientific approach is based on consulting a rich Romanian and international specialized literature, in the fields of geology and geography. In this sense, cartographic documents presenting the repartition of salt deposits in Europe and in Romania, as well as the accompanying sources have been consulted.

Almost all salt deposits are correctly inventoried and noted in the geologic and geographic cartographic documents. Unfortunately, they also seldom note the salt springs, too. For this reason, historical, literary, and journalistic sources had to be consulted, too. One cannot identify accurately all salt springs without field trips and surveys.

Results and discussions

The special importance of salt drew the attention of the scholars in Ancient Greece early. Aristotle was the one who described in a relatively correct manner the existence of salt water in the sea, in his monumental work *Meteorologica* (350 BC). By explaining the circuit of water in the nature and the origin of rains, he concludes that precipitations also contain an important amount of salts subsequently deposited in the sea. In reality, Aristotle starts from a controversial reality even for nowadays scholars: the primordial water was it salt-water or freshwater? The author notes that sea-salt grew saltier up to a certain point of balance; at that point, the quantity of salts remained unchanged. He stresses the fact that the water evaporated from the sea is always the freshwater, just like the one evaporated from wine.
The geologic literature is divided into two currents as regards the analysis of salt" the Plutonists and the Neptunists.

The Plutonists believe that salt comes from inside the Earth and that it reaches the surface through volcanic action. However, volcanic eruptions bring a very small amount of salts and it could not account for the huge amount of salt on Earth today. Salt was formed through the sublimation of chlorine and sodium within a heat transfer (Hofmann 1888). These ideas functioned until the end of the first part of the 20th century.

The Neptunists state that salt comes from the continental deposits or from seawater. Some ideas with no scientific basis support the idea that salt comes from steppe continental deposits (Kant 1999). Other ideas circulating in the scientific world for a while stated that salt comes from the washing down of salt from plants and animals or from the salts washed down by streams and deposited within steppes. A basic condition for the emergence of salt was the existence of a lagoon or of epicontinental seas and of a steppe climate (Meruţiu 1912).

The generally accepted opinion states that salts come from the dissolution of geologic layers and from their accumulation within seawaters. The evaporation of epicontinental seas and of lagoons favours the deposition of salts as compact deposits with variable thickness. Their existence between the geologic layers is explained through the coverage due to surface erosion and to the deposition of sediment rocks in aerial or sub-aerial conditions (Cobălcescu 1883; Istrati 1889; Teisseyre 1899; Mrazec 1907; Mrazec and Teisseyre 1902; Murgoci 1907; Meruţiu 1911; Paucă 1967a,b; Pop, 1963). The emergence of salt layers at the surface is explained by the existence of a tectonics specific to salt deposits: salt has a high degree of plasticity. If there are thick geologic deposits – exerting a strong pressure and causing high temperature, the rock salt starts to move towards marginal areas. If they reach the stronger mantling rock, they rise towards the surface and they bring along sedimentary layers, which they fold. This leads to diapir folds, specific to the Transylvanian Depression or to the sub-Carpathian areas.

Both geology and geomorphology (the field of physical geography) study the genesis and evolution of salt.

The salinity of the Black Sea has varied throughout time, depending on its connection with the surrounding seas. The current value of salinity has maintained at 18‰. Near the north-western coast, where numerous rivers discharge, the salinity lowers under 10‰ (Figure 1). In front of the Danube mouths, it drops to 2–5‰, reason for which it is included in the category of freshwater up to 5–10 km (Romanescu 2003).

Sea-salt is exploited only in the areas with high salinity (usually over 30‰) and strong tide (tidal range of over

2m) (Thurman 1988; Vanney 1991). From this perspective, the Black Sea does not meet the requirements of economic profitability (average salinity of 18‰ and a tidal range of 12 cm). The sea waters at the level of the Romanian deltaic coast have salinity under 10‰, and the sea waters south from Cape Midia — under 17‰ (Romanescu 2008).

We have to specify that sea waters, at an average of 35‰ salinity, have an average amount of NaCl of 26‰, with the rest being represented by other salts (K, Mg, etc.). The entire amount of oceanic salts would be 60m thick if distributed on the entire surface of the Earth (361 mln. km^2).

The entire literature emphasizes on the origin of salts within oceanic waters: they come from the crust of the earth because of dissolution, through streams (Thurman 1988; Vanney 1991). On principle, primordial water was a freshwater. Consequently, salt follows a geologic cycle: it is dissolved from the crust of the earth, carried by rivers, accumulated in the seas and the oceans, resettled as deposits, covered by recent sediments (Figure 1). Its emergence at the surface is due to tectonics and to the lateral movements caused by plasticity. Diapir folds are specific to the Romanian territory, where the salt cores can erect the sedimentary layers to the vertical (Irimuş 2006; Mac 1972)

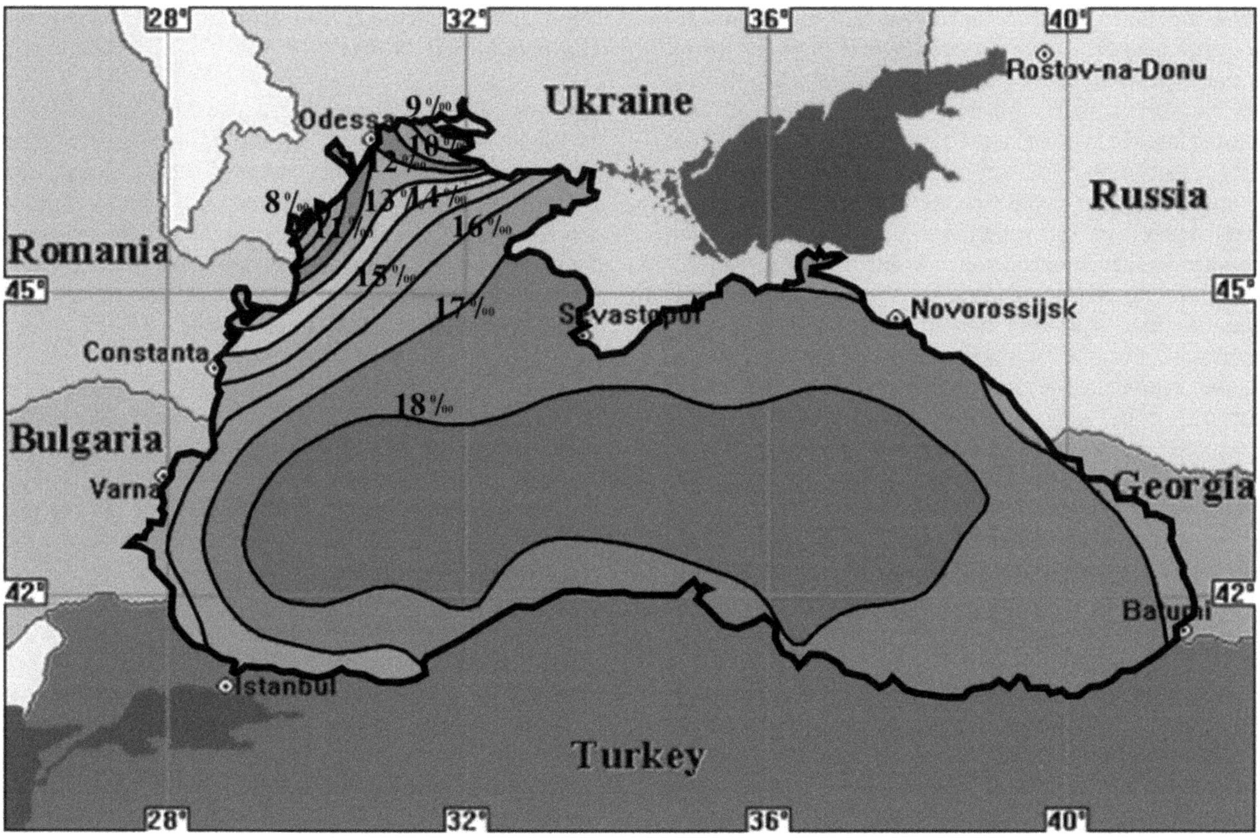

Figure 1. Salinity distribution in the surface waters of the Black Sea.

The Eocene and Miocene salt deposits on the Romanian territory are specific to the lagoon and coast facies and more rarely to the pelagic one. These testimonies clearly show the marine origin of an epicontinental sea, such as the Sarmatian one in Eastern Europe (Ciupagea *et al.* 1970).

Salt springs emerged through the washing down and dissolution of salt cores (salt rocks) or through the dissolution of salts within sedimentary deposits (loams, clays, sandstone sands, sandstones, etc.). Only in the Transylvanian depression, in the diapir zone, there are over 800 salt water springs (Ciupagea *et al.* 1970).

The first lapidary scientific information regarding the salt in Romania was noted by Dimitrie Cantemir in his monumental work *Descriptio Moldaviae* (1716): "*Aliquando etiam in his pisces integri concreti reperiuntur qui nihilo ab aliis differunt, qui in fluviis vicinis capiuntur*" ("And they [salt mines] contain sometimes whole fish planted together, which do not differ at all from the fish people catch in the nearby rivers"). Dimitrie Cantemir makes reference to the fish fossils found in the menilith schists of Târgu Ocna (Meruțiu 1912).

In the period 1900–1940, rock salt represented one of the most discussed geologic issues in Romania. The missing link is due to the fact that geologists avoided analysing the accompanying evaporates (calcium carbonates and sulphate, complex chlorides of potassium, magnesium, etc.). At that time, the focus was only on the diapir tectonics classification of the salt massifs and on determining their age (Ciupagea *et al.* 1970).

After 1940, there were detailed researches on the chemical composition of evaporates, as well as on their geographic distribution. In this period, explanations were given for the salt lakes within the old deserted mines. The lakes emerged after the collapse of the ceiling of Roman mines, exploited in chambers with a bell-shaped form. By annexing several chambers, rocks failed and an opening appeared where water accumulated meteorically. Salt dissolution led to water mineralization and to the transformation of the lake into a health resort centre (Ocna Sibiu, Ocna Mureş, Turda, Sovata, Cojocna, etc.).

Only in the 1960–1970s explanations were brought on the chemical differentiations between the salt deposits within the Transylvanian depression and those outside the Carpathian arch. The deposits within the depression are pure, while those in the sub-Carpathians are mixed with carbonates and sulphates. It is the period of the "great scientific 'convulsions' in the field of salt": what is the paleogeographic appearance of the Carpathian regions during the deposition of evaporites? What the circumstances of the separation between the various categories of evaporates were within the complex of salts dissolute in the seawaters? What is the contact route of the seawaters towards the Transylvanian depression, whose evaporation led to evaporites? What are the mounts of evaporites emerged of the main structural units and what are the causes of their repartition? What are the causes for which there is a single salt layer (400 m thick) within the Transylvanian depression, while outside the Carpathians there are 2–4 layers of sat, with different ages? (Ciupagea *et al.* 1970). The deposition of a 400 m thick layer could have occurred in 2,000 years (or in 5,000 years) (Ciupagea *et al.* 1970).

The different deposition, from one region to another, of rock salt or of potassium or magnesium salts, depends on the local physical and geographic conditions. At an average concentration of 35‰, the following categories of salts are deposited: 78% rock salt, 17.7% complex potassium and magnesium salts (chlorides, sulphates, etc.), 3.6% gypsum, 0.4% dolomites, as well as infamous amounts of bromides, iodides, etc. (Ciupagea *et al.* 1970). The different proportion of these elements makes the differentiation between the salt massifs in Romania.

The existence of isolated or large lagoons determined the emergence of salt deposits. The positive or negative tectonic movements favoured the contact between lagoons and the sea or the temporary interruption with the outside. In the main lagoon, represented by the Transylvanian depression, rock salt deposited progressively, while in the intermediate lagoons, crossed by marine currents in their way towards the central arch, other categories of salts deposited. Intermediate lagoons played the role of traps (Ciupagea *et al.* 1970). The significant thickness of the gypsum and dolomites deposits within the external region of the Curvature Carpathians is the evidence for the existence of important water current.

The alignment of the intermediate lagoons overlaps the area represented by the Curvature sub-Carpathians and the south of the Moldavian sub-Carpathians (Figure 2). These lagoons had the role of retaining seawaters (with lowered concentrations and high temperatures) in their movement towards the Transylvanian depression. In these areas, calcium and magnesium carbonates form a deposit (the least soluble and at the same time thermophilic) as dolomites. In the next stage, the calcium sulphate deposits (thermophilic). In this case, only the waters filled with sodium chloride and complex potassium and magnesium chloride manage to get to the Transylvanian depression, which appeared as a large lagoon. The waters of the Transylvanian lagoon were colder (was there was more water), thus favouring the deposition of rock salt, a relatively cryophilic mineral (Ciupagea *et al.* 1970).

The complex potassium and magnesium salts are found in small quantities only in the Miocene of northern Trotuş, where the lagoons reached high concentrations as they were not in the way of the currents directing towards the Transylvanian depression. The researches for the discovery of the potassium salts (1909) — in the Transylvanian depression — led to the discovery of pit gas, but no potassium salts were found.

The lack of potassium and magnesium salts within the Transylvanian depression is due to the existence of two contact routes with the open ocean: an entrance route in the curvature sector and an exit route of the current in the Olt corridor (Figure 3). In this case, from the great lagoon waters with high degree of potassium and magnesium concentration were eliminated.

The second cause is represented by the existence of a great water volume, which did not evaporate completely; hence, the potassium and magnesium salts were taken over by the circular current. The lack of these salts is the consequence of the fact that depression waters never reached the concentration degree necessary for the deposition of such soluble elements.

Geomorphologists draw attention on defining elements for the evolution of the landform in the east of Transylvania. This is why the landform of diapir folds is specific to the areas where salt migrated and rose at the same time with the covering geologic layers (Irimuş 2006; Mac 1972) (Figure 4). The same features are also emphasized in the case of the extra-Carpathian areas on the Slănic Valley in Buzău County (Ielenicz and Grecu 1980).

Most specialists draw attention on the impressive salt reserves on the Romanian territory and, at the same time, on its exceptional quality (99.9% NaCl), being one of the purest on Earth (Fatu, 1851; Pantazică 1974). This is why the extraction profitability is maximum, mostly considering that the most important dykes of salt are in the vicinity of the topographic surface.

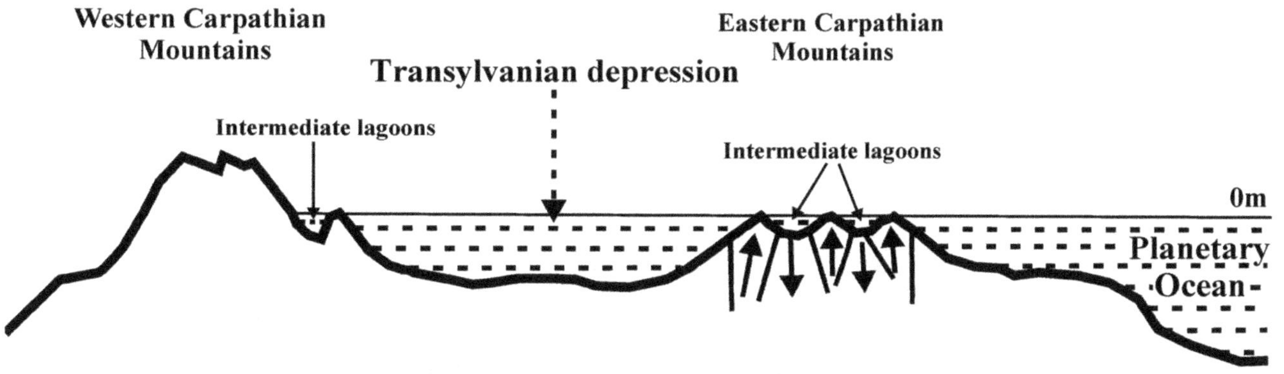

Figure 2. Localization of intermediate lagoons in the mechanism of the ocean localized on the Romanian territory in the Miocene era (adaptation after Ciupagea *et al.*, 1970).

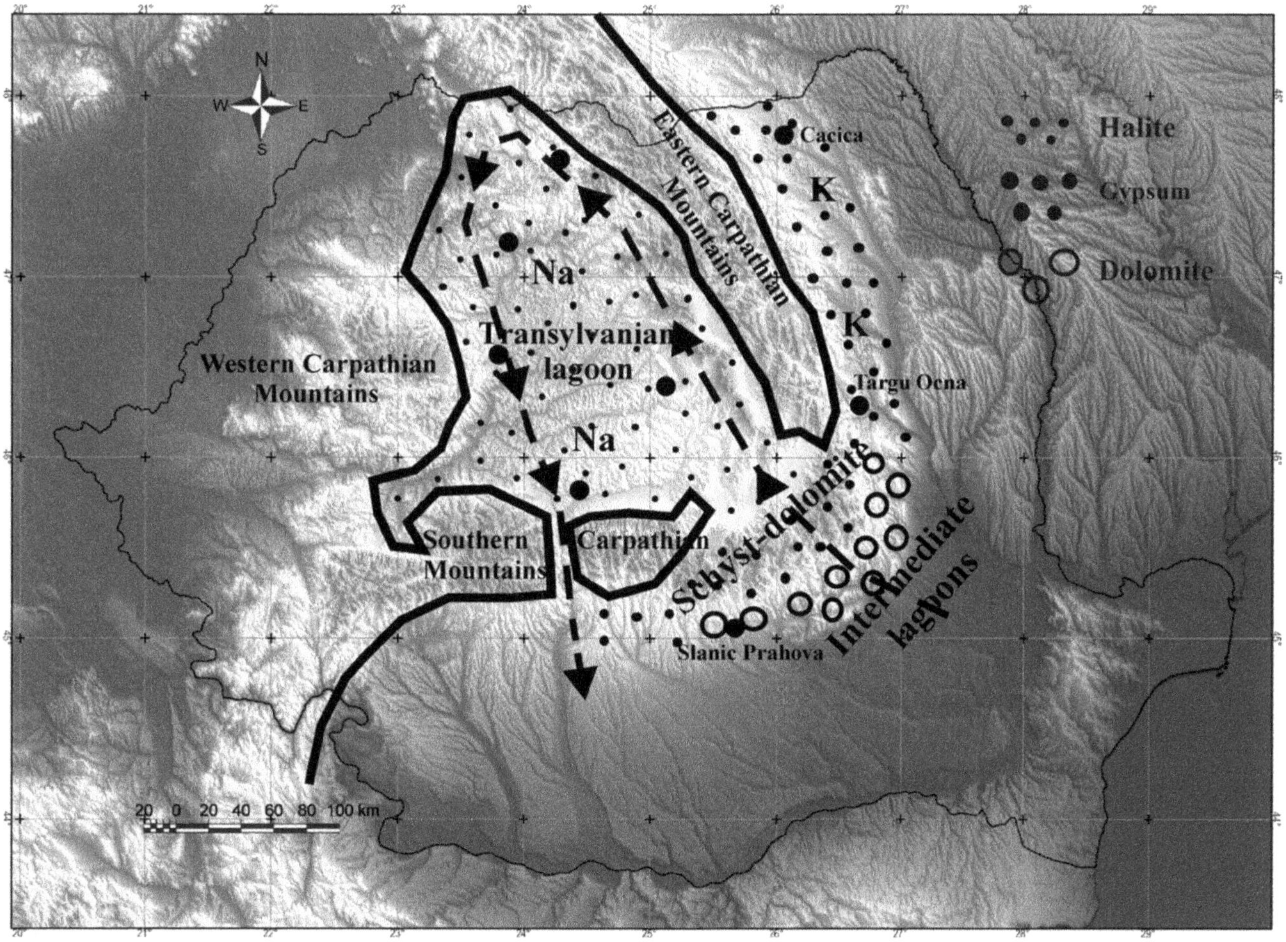

Figure 3. The Vrancea Gate and the Olt Gate as contact sector of the Planetary Ocean with the lagoon of the Transylvanian depression (adaptation after Ciupagea *et al.*, 1970).

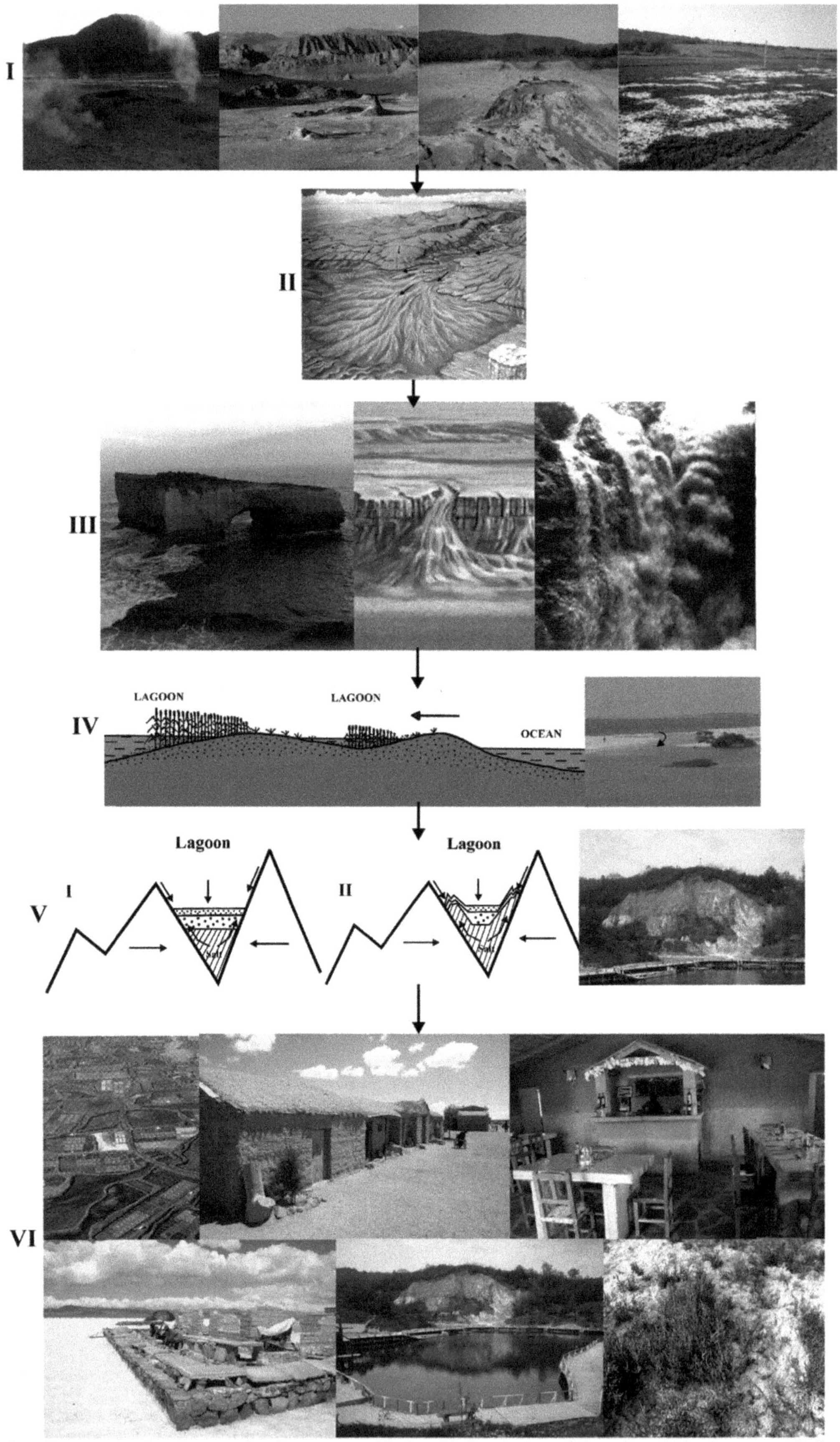

I-salts of geysers, volcanism, mud volcanoes, capillarity; II-dissolution of the salts within the crust; III-deposition of the salts within seawaters; IV-deposition of salts in intermediate lagoons; V-coverage of salts with sedimentary deposits and formation of diapir folds; VI-usage: extraction of salts within the sea; house building; manufacturing domestic devices; manufacturing handcrafts; balneo-climatology; brine-loving plants.

Figure 4. Formation and usage of salt resources.

Figure 5. Map illustrating the distribution of the salt massifs and of the main salt springs in Romania, published by Meruţiu V. in 1912.

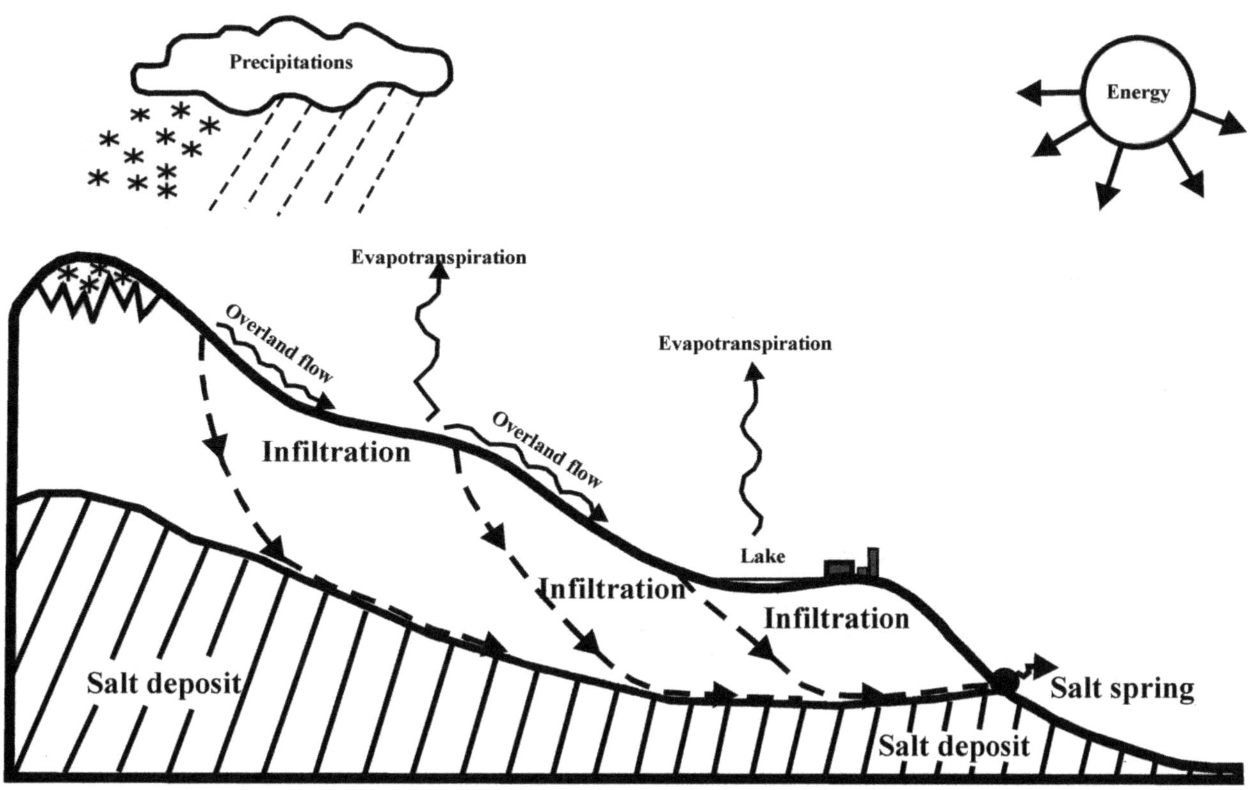

Figure 6. Formation of salt springs because of the dissolution and washing off of salts and salt deposits within the crust of the Earth.

The most complete study on the salt in Romania was elaborated by Meruţiu V. in 1912, in his reference book *Sarea în pământul românesc* (*Salt in the Romanian Land*). It is the first study which — besides genesis, evolution, and repartition — also deals with the existence of salt springs. From this perspective, it includes a complete inventory, with only very few omissions, for all the springs within the counties of Suceava, Neamţ, Bacău, Putna, Râmnicu Sărat, Buzău, Prahova, and Dâmboviţa. At the same time, the author analyses the mines within the following regions: Slănic, Târgu Ocna, and Ocnele Mari. The three salt mines were the trademark of salt exploitation in Romania at that time. This book also includes the first high-scale similarities for the Miocene salt in Romania and the salt of the same age in Switzerland, Germany, Spain, etc.

Meruţiu V. is the first and the only Romanian author who drew a map of the salt massifs and salt springs in the Moldavian, Curvature, and Getic sub-Carpathians (Figure 5).

Unfortunately, human geography makes only tangential references to salt resources and it does not bring any reference books in this field. The only studies worth mentioning are those analysing the salt exportation to the European countries, especially to Serbia and Bulgaria (mainly in the period before World War I).

The first reference books belonging to human geography were elaborated by Şandru I (1952, 1965, and 1966), where he makes references to the existence of salt springs and brines in southern Bucovina (Cacica, Pleşa-Marginea) and in the Moldavian sub-Carpathians (Târgu Ocna). He underlines the existence of traditional salt exploitations even since very old times (Neolithic, Neolithic) and he put a special emphasis on the continuity of settlements surrounding the salt water springs (Şandru 1952, 1965, and 1966).

Salt springs, just like mineral springs, have been protected and highly exploited by the local population and by the State institutions. They emerge at the surface after the dissolution of the salts within the substrate or after the direct dissolution of the compact salt deposits (Figure 6). They usually emerge at the foot of the slopes, on top of the impermeable layers (Romanescu 2005). The debits depend on the amount of precipitations and on the value of the evapotranspiration specific to the area where they form. Most salt springs in Romania have lower debits, as they form within the transition temperate-continental climate, with average precipitations between 600 and 800 mm.

From an economic perspective, the situation of salt deposits in Romania has both advantages and shortcomings: they are pure salts, used directly in the industrialization process; the salts within the Transylvanian depression do not contain potassium and magnesium salts.

Conclusions

Despite the extraordinary resources of salt deposits, Romania does not have detailed scientific studies in the field. The relative scarcity of studies on salt deposits and salt water springs is explained by the fact that salt is such an immense resource that it does not represent a special subject. Salt is somewhat like water.

The geologic and geomorphologic studies have contributed to explaining the genesis of salt deposits and their evolution in time and space. The hydrogeological studies have explained the constitution of salt springs and the way they have been used.

Within economic and human geography, there have been sporadic references to the existence of salt on the Romanian territory. There have been frequent analyses of the economic relations between Romanian and the salt-importing countries (Serbia, Bulgaria).

The most important period for the genetic explanation and for the correct determination of the chemical composition of salt deposits in Romania is represented by the years 1940-1970. It also coincides with the publication of the most important works (belonging to geology and physical geography) on salt and its use.

Acknowledgement

This work was supported by a grant of the Romanian National Authority for Scientific Research, CNCS–UEFISCDI, project number PN-II-ID-PCE-2011-3-0825, 219/5.10.2011, *The ethno-archaeology of the salt springs and salt mountains from the extra-Carpathian areas of Romania* — ethnosalro.uaic.ro. Our thanks go to the Laboratory of Geo-archaeology of the Research Platform in the Field of Archaeology — ARHEOINVEST from the "Alexandru Ioan Cuza" University of Iaşi, which provided the necessary equipment and processed the data.

References

Airinei, Ş. 1959. Recherches magnétiques pour le sel dans la région Ocnele Mari – Ocniţa. *Dări de Seamă ale Comitetului Geologic Român* XLII (1954-1955), 191-201.

Albu, E. and Balteş, N. 1983. Considérations sur l'âge du sel dans la zone des plis diapirs atténués et incipients de la Munténie et ses implications sur la genèse et la répartition des gisements d'hydrocarbures. *Carpathian Balkan Geological Association* LX, 257-264.

Anghel, V. 1958. Consideraţii asupra raionării hidrochimice a apelor curgătoare din Republica Populară Română. *Meteorologia, Hidrologia şi Gospodărirea Apelor* 1, 56-67.

Atudorei, C., Chiriţă, R. and Chiriţă, E. 1969. Criterii geologice de corelare a prospecţiunilor geofizice pentru sare în zona Ocna Mureş – Turda. *Buletinul de Geologie al Minelor* 1, 8-12.

Atudorei, C., Bocănete, E. and Miclea, P. 1971. *Cercetarea, exploatarea şi valorificarea sării.* Bucureşti, Editura Tehnică.

Bâgu, G. and Mocanu, A. 1984. *Geologia Moldovei. Stratigrafie şi consideraţii economice.* Bucureşti, Editura Tehnică.

Bogdan, A. 1973. Sarea în ţara noastră, zăcăminte, exploatare. *Terra* 25(4), 17-25.

Botezatu, R., Calotă, C. and Neguţ, A. 1977. A Model of the Slătioarele Gravity Anomaly Source. *Revue Roumaine de Géologie, Géophysique et Géographie,* Série de Géophysique, Academia S.R. Romania 21(2), 197-212.

Botzan, M. 1984. *Apele în viaţa poporului român.* Bucureşti, Editura Ceres.

Botzan, M. 2002. *Hidronimie românească sau Botezul Apelor.* Bucureşti, Editura Academiei Române.

Bucur, I. 1970. Some data concerning the geology of the Transylvanian Basin following observations in the Saliferous of the Gurghiu Valley. *Petrol şi Gaze* XXI(5), 263-265.

Bulgăreanu, V.A. and Feurdean, V. 1984. Hydroisotopic pattern of water circulation in salt lakes Băile Verzi-Slănic (Prahova county). *Studii şi Cercetări de Fizică* 34(4), 351-361.

Bulgăreanu, V.A. 1983. Limnogeology and peloidogenesis processes in the Poporului, Mihai Viteazu and Brîncoveanu lakes (Ocna Sibiului, România). *Revue Roumaine de Géologie, Géophysique et Géographie,* Série de Géophysique 27, 93-102.

Bulgăreanu, V.A., Feurdean, V., Guţu, A., Olteanu, E., Bogodiţă, A., Hannich, D. 1984. Relations between the fresh- and salt waters circulation and the geodynamics of the Ocna Şugatag karstosaline and anthroposaline alke area (Maramureş district). *Theoretical and Applied Karstology* 1, 165-171.

Bulgăreanu, V.A., Ianc, R., Ionescu, V., Kiss, S., Prodănescu, I., Ţintilă, D., Urcan, T. 1985. Roşu lake (Ocna Şugatag, Maramureş): the limnogeology of an anthroposaline, brackish-and acide water lake. *Revue Roumaine de Géologie, Géophysique et Géographie,* Série de Géologie 29, 91-95.

Bulgăreanu, V.A., Sitaru, M., Hannich, D. 1985. Physico-chemical stratification and heliothermy of (karstosaline?) lake Ursu (Sovata, Romania). *Theoretical and Applied Karstology* 2, 165-174.

Burlacu, O. 1973. *Izvoarele minerale din judeţul Buzău şi modul lor de valorificare, Geografia judeţului Buzău.* Buzău, Tipografia Buzău.

Cantemir, D. 1716. *Descriptio Moldaviae.* Berlin.

Ciupagea, D. 1980. La tectonique du sel et de la formation à gas du bassin de Transylvanie. *Revue Roumaine de Géologie, Géophysique et Géographie,* Série de Géophysique 24(1), 47-61.

Ciupagea, D., Ichim, T., Popa, M., Opran, A. 1967. L'influence du sel sur la tectonique des dépôts miocènes supérieures de la Dépression de Transylvanie. *Carpathian Balkan Geological Association, Belgrade 1967, (Géotectonique)* T.I, 67-71.

Ciupagea, D., Paucă, M. and Ichim, T. 1970. *Geologia Depresiunii Transilvaniei.* Bucureşti, Editura Academiei Române.

Cobălcescu, G. 1883. Formaţiunea saliferă sau secunda Formaţiune Mediterană. Memoriile geologice ale şcoalei militare din Iaşi.

Contescu, L. and Mihăilescu, N. 1970. Les massifs médians, importante source de clastiques pour les sillons à flysch adjacentes. *Acta Geologica Hungarica* 2, 317-324.

Cornea, I., Spânoche, S., Cristea, P., Mormu, M. 1973. Recherches sismiques souterraines et à la surface au sujet des glissements de sel gemme de Slănic-Prahova et de l'Ocna-Dej. *Studii şi Cercetări de Geologie, Geofizică, Geografie.* Seria Geofizică 11(1), 35-49.

Cristea, P., Constantinescu, P. and Spânoche, S. 1977. Recherches séismiques dans des puits pour sel. *Studii şi Cercetări de Geologie, Geofizică, Geografie.* Seria Geofizică 15, 137-149.

Dicea, O. and Dicea, M. 1981. Contributions à la connaissance de l'âge des formations de sel d'Ocniţa, a partir du nanoplancton. *Studii şi Cercetări de Geologie, Geofizică, Geografie.* Seria Geologie 26(2), 293-299.

Dovlete, C., Sonoc, S., Cuculeanu, V., Sima, O., Alexandrescu, M. 1983. Studies on the Danube and Razelm lake sediment radioactivity. *Studii şi Cercetări de Fizică* 35(1), 68-74.

Fătu, A. 1851. *Descrierea şi întrebuinţarea apei simple şi a apei minerale din Moldova.* Iaşi, Tipografia Româno-Franceză.

Ghiţă P. and Năulea, N. 1980. The anthropo-saltic lakes of Ocnele Mari depression. *Studii şi Cercetări, Conservarea Naturii, Baze Ecologice,* 137-144.

Givulescu, R. 1982. Einige Betrachtungen über das Klima des Badeniens in Zusammenhang mit der Bildung der Salzlagerstätten Rumäniens. Nymphaea, *Muzeul Ţării Crişurilor* X, 55-63.

Grujinschi, C. 1971. Observations on "the unconformity" of salt appearance at Baia Baciului (Slănic Prahova). *Buletinul Institutului de Petrol, Gaze şi Geologie* 18, 33-38.

Hofmann, K. 1888. Bericht über die im Sommer 1886 im nordwestlichen Teile des Syonlok-Dobokaer Comitates ausführten geologischen Datailfnahmen. *Jahrb. K. Ung. Geol.A.* Budapesta.

Ielenicz, M. and Grecu, F. 1980. Relieful dezvoltat pe sare şi breccias sării din bazinul hidrografic Slănic (de Buzău), *Ocrotirea Naturii* 24(1), 53-60.

Ielenicz, M. 1985. Karst et pseudokarst dans des montagnes et les collines subcarpatiques des Bassins du Buzău et du Teleajen. *Theoretical and Applied Karstology* 2, 47-53.

Irimuş, A. 2006. Hazarde şi riscuri asociate proceselor geomorfologice în aria cutelor diapire din depresiunea Transilvaniei. Cluj-Napoca, Editura Casa Cărţii de Ştiinţă.

Isopescu, A., Petrescu, D. and Stanciu, D. 1967. Sortarea prin fluorescenţă a sării geme contaminată cu incluziuni bituminoase. *Revista Minelor (C.N.I.T.)* 18(4), 170-172.

Istrati, C.I. 1889. Compoziţia chimică şi câteva observaţiuni relative la sarea gemă din cele patru mine ale României. B.S.P. Bucharest.

Kant, I. 1999. *Géographie. Bibliothèque philosophique.* Paris, Aubier.

Lupu, N.N. 1971. Le massif de sel de Cacica et sa valeur économique. *Studii şi Comunicări.* Seria Ştiintele Naturii, Muzeul Judeţean Suceava 2(1), 59-62.

Mac, I. 1972. *Subcarpaţii Transilvaniei dintre Mureş şi Olt.* Bucureşti, Editura Academiei Române.

Marosi, P. 1967. Originea mineralizării apelor freatice din câmpia interfluvială Buzău-Râmnicu Sărat. *Dări de Seamă, Comitetul Geologic, Institutul Geologic, STE* E(7).

Mărgărit, G. and Mărgărit, M. 1982. New data on the geology of the Lacu Roşu (Hăghimaş). *Studii şi Cercetări de Geologie, Geofizică, Geografie.* Seria Geologie 27, 89-97.

Meruţiu, V. 1911. Comunicare asupra masivului de sare de la Slănic-Prahova. *Dările de Seamă ale Institutului Geologic al României pe anul 1911,* Bucharest.

Meruţiu, V. 1912. Sarea în pământul românesc. *Buletinul Societăţii Regale de Geografie* XXXIII(1-2), 69-162.

Micalevich, E. 1967. Cristalul de sare – o veche bogăţie a Pământului românesc. *Natura (Geografie-Geologie)* 19, 6.

Minea, I.S. 2011. *Râurile bazinului Buzău. Consideraţii hidrografice şi hidrologice.* Buzău, Editura Alpha MDN.

Mrazec, L. 1907. Despre originea petrolului şi formarea zăcămintelor de petrol în România. Discurs de recepţiune, Academia Română, XXX.

Mrazec, L. 1937. Note préliminaire sur la salinité des marnes miocènes du Bassin Transylvain. *Buletinul Societăţii Ştiinţifice al Academiei Române.*

Mrazec, L. and Teisseyre, W. 1902. *Aperçu géologique sur les formations salifères et les gisements de sel en Roumanie.* M. P. R., Bucharest.

Murgoci, G.M. 1907. Terţiarul Olteniei cu privire la sare, petrol şi cărbuni. *Anuarul Institutului Geologic* I.

Neacşu, G. and Urcan, T. 1978. Clay Minerals in the Praid Salt Deposits. *The Second National Conference Clays (1975), Studii Tehnice şi Economice* I(14), 103-105.

Neguţ, A. 1977. The contributions to the geophysical well logs evaluation in the exploration of Slătioarele-Argeş salt dome. *Studii şi Cercetări de Geologie, Geofizică, Geografie.* Seria Geofizică 15, 167-175.

Pantazică, M. 1969. Caracterizarea hidrochimică a râurilor din nord-estul Moldovei. *Analele Ştiinţifice ale Universităţii „Alexandru Ioan Cuza" Iaşi* XV, 65-78.

Pantazică, M. 1974. *Hidrografia Câmpiei Moldovei.* Iaşi, Editura Junimea.

Pantazică, M. and Schram, M. 1981. Contributions à la caractérisation des rivières du plateau Moldave entre Siret et Prut. *Analele Ştiinţifice ale Universităţii „Alexandru Ioan Cuza" Iaşi* s.n., (s.II, b. Geologie-Geografie) XXVII, 105-109.

Papaianopol, I. 1982. On the Dosinia Beds in the Zone of the External Diapir Folds of Muntenia (the area between the Cricovul Sărat and Nişcov valleys). *Dări de Seamă ale Institutului Geologic* LXVII(4), 147-179.

Paraschiv, D. 1983. Formations salifères de la Plate-forme Moesienne (Roumanie). *Carpathian Balkan Geological Association, Congress Proceedings, XII,* LIX, 47-53.

Paucă, M. 1967a. Contribuţii la geneza zăcămintelor de săruri miocene din România. *Dări de Seamă ale Institutului Geologic* LIII(2) (1965-1966), 159-184.

Paucă, M. 1967b. Zăcămintele de evaporite din România. *Revista minelor* XVIII(1), 1-4.

Paucă, M. 1968. Beiträge yur Kenntnis der miozänen Salzlagerstätten Rumäniens. *Geol. Rundschau Bd.* 57(H2), 514-531.

Paucă, M. 1978. Die Evaporite in Romänien. Entstehung, Palaeogeographie und Tektonik. *Analele Muzeului de Ştiinte Naturale Piatra Neamţ (Geologie-Geografie)* 4, 7-47.

Pătruţ, I., Paraschiv, D. and Dicea, O. 1973. Some Considerations concerning the Formation Way of Diapyros Structures in Romania. *Petrol, Gaze* 24(9), 533-542.

Pânzaru, T.G. 1976-1977. Le complexe lacustre du massif de sel d'Ocna Sibiului. I. Lake sectors "Incinta" and "Parcul Public". *Scientific Works (A. Geogr.), Institutul de Învăţământ Superior,* Oradea, 185-204.

Pânzaru, T.G. 1982. Le complexe lacustre du massif de sel d'Ocna Sibiului (II). Lake sectors "Ocniţa-Avram Iancu" and "Ignaţiu". *Studia Universitatis Babeş-Bolyai, (Geologie-Geografie)* XXVII(1), 46-60.

Pârcălăbescu, I.D. 1982. On the chemical-mineralogical composition and building-up manner of rock salt deposit Ocnele Mari. *Mine, Petrol, Gaze, M.M., M.P.* XXXIII(8), 370-378.

Pârvu, G. and Burlacu, A. 1972. Concerning the Presence of a Diapyros Fold in the Grădişte-Româneşti Area (Gaetic Depression). *Petrol, Gaze* 23(4), 199-203.

Pop, E. 1963. *Geologische Forschungen im Becken von Petroşani.* C 163. Freiburg, Freiberger Forschungsh.

Povară, I., Cosma, R., Lascu, C., Bulgăreanu, V.A. 1982. Un cas particulier de karst dans les dépôts de sel (Slănic Prahova – Roumanie). *Travaux de l'Institute de Speologie E. Racoviţă* XXI, 87-93.

Roman, S. and Papaianopol, I. 1985. Microfloral and Macrofloral Data on the Post-Meotian Deposits in the Zone of Outer Diapir Folds in Muntenia. *Dări de Seamă ale Institutului de Geologie şi Geofizică* LXIX(3), 135-144.

Romanescu, G. 2003. *Hidrologie generală.* Iaşi, Terra Nostra.

Romanescu, G. 2008. *Oceanografie.* Iaşi, Azimuth.

Sandu, I., Poruciuc, A., Alexianu, M., Curcă, R.-G., Weller, O. 2010. Salt and Human Health: Science, Archaeology, Ancient Texts and Traditional Practices of Eastern Romania. *The Mankind Quarterly* L(3), 225-256.

Scripăţ, V. 1984. Non-metallic minerals-natural richness fully turned to account in the socialist years. *Mine, Petrol, Gaze* XXXV(8), 381-384.

Stoica, C. 1966. Le sel gemme et les sels potassiques dans la République Socialiste de Roumanie. *Revista Minelor (C.N.I.T.)* 17(1), 28-32.

Susnjar, M. 1984. Regularities in the development of salt tectonics and their importance for the ranking of salt structures in salt basins with respect to their oil and gas bearing potential. *Proceedings of the 27th International Geological Congress, Moscow* 13, 43-62.

Şandru, I. 1952. Contribuţii geografico-economice asupra exploatării slatinelor în Bucovina de Sud, *Studii şi Cercetări Ştiinţifice* 1-4(3), 407-428.

Şandru, I. 1961. Contribuţii geografico-economice asupra evoluţiei aşezărilor omeneşti în depresiunea subcarpatică Oneşti. *Scientific Annals of the "Al.I. Cuza" University in Iaşi* 7(1), 215-230.

Şandru, I. 1966. Târgu Ocna – consideraţii de geografie urbană. *Scientific Annals of the "Al.I. Cuza" University in Iaşi* 12, 207-214.

Teisseyre, W. 1899. Zur Geologie der Bacauer Karpathen. *Jahrbuch d.k. Geol.RA.* 58, 702-703.

Thurman, H.V. 1988. *Introductory oceanography.* Toronto, Merill Publishing Company A Bell & Howell Information Company Columbus.

Vanney, J.R. 1991. *La géographie de l'océan.* Paris, Institut Océanographique, Oceanis.

Vasilescu, G., Pîrvu, M. and Udicioiu, V. 1984. Hydrogeological Researches for Mineral Waters in the Govora Spa Area. *Lucrările Simpozionului de Protecţia şi Exploatarea Raţională a Apelor Minerale, Călimăneşti, 1978. Studii Tehnice şi Economice* E(14), 247-258.

Visarion, M. 1967. Réfléchissement des formes structurales à sel gemme et sels de K et Mg dans les anomalies gravimétriques des territoires de la République Socialiste de Roumanie. *Carpathian Balkan Geological Association, VIII-eme Congress Belgrade 1967 (Geophysique)* T.III, 169-177.

Visarion, M., Polonic, P. and Ali-Mehmed, E. 1976. Some Contributions to the Study of Structural Shapes of Salt within the Transylvania Depression. *Studii Tehnice şi Economice. Institutul de Geologie şi Geofizică* D(11), 29-62.

Salt Springs: Places for Salt Recrystallization and Ritual Centres for Exchange with Steppe Populations

Dan Monah

Abstract

During the last decades, the Moldavian Subcarpathians witnessed the discovery of several sites where exploitation of salt from salt springs took place. These archaeological remains unearthed near the springs reveal an intensive exploitation of salt through recrystallization during the Neolithic and the Eneolithic. The exploitation of salt in the Moldavian piedmont intensified beginning with the Cucuteni A phase, and the first briquetage appeared at the end of the phase. Cucutenian salt briquettes were conical, weighed approximately 0.75–1.5 kg, and were clearly prestige goods meant for exchange between communities. Surprisingly, important quantities of Cucuteni 'type C' ware, generally attributed to the populations from the North Pontic steppe, have also been found in the Cucutenian deposits from near the salt springs. At Solca and Cacica, Cucuteni 'C' ware comprises *ca.* 40% of the assemblage, at Lunca–*Poiana Slatinei* approximately 33%, while in contemporary regular settlements from the Subcarpathians, only rarely does Cucuteni 'C' ceramic make up more than 1% of the assemblage. Apart from the coarse Cucuteni 'type C' pottery, the salt spring sites also produced numerous fragments of high-quality, painted ceramic ware. The author attempts to explain this peculiarity through a scenario in which steppe populations organised expeditions for procuring salt briquettes. The crystallisation of the salt into briquettes was done co-jointly by the Cucutenians and their guests who travelled great distances for this occasion. In support of this hypothesis the author cites ethnographic analogies from New Guinea, where the Western Dani (Lani) linguistic group, as well as other communities, organise regular expeditions for acquiring stone and salt. The communal activity of recrystallizing salt and producing briquettes by the Cucutenians and the steppe pastoralists probably provided the occasion for complex exchanges of gifts during a ceremony similar to the *potlatch* and the *kula* systems of exchange. The exchange of gifts and also the granting of permission to cross a certain territory were validated through certain strict religious rituals. The author attempts to block out, within a reasonable margin of error, these practices and rituals.

Keywords

salt springs, Cucuteni 'C' ware, Moldavia, Pontic steppe, salt, recrystallization, potlatch, kula, exchanges, gifts

Introduction

Archaeological research conducted during the past few decades has resulted in the discovery, in the northern Moldavian Subcarpathians, of several archaeological sites located near salt springs with high salinity. These sites have been interpreted as hubs for salt recrystallization during the Eneolithic period. The archaeological remains attest to an intense human activity focused on recrystallizing salt and producing salt briquettes for long-distance transport. Of particular interest is the discovery of a great quantity of Cucuteni 'type C' ware, a kind of ceramic that is generally attributed to the pastoralist communities of the steppe lands north of the Black Sea. The ceramic assemblage attests to cohabitation and close cooperation between the Cucutenians and the steppe pastoralists with a view to producing briquetage and salt briquettes.

We have directed our attention to four such sites: Solca–*Slatina Mare*, Lunca–*Poiana Slatinei*, Ţolici–*Hălăbutoaia* and Cacica–*Salina* (Figure 1). With the exception of Cacica, all the sites have been thoroughly investigated and were the subject of extensive discussion in the archaeological literature. All four settlements produced considerable amounts of Cucuteni phase A, A-B and B material. Although Lunca–*Poiana Slatinei* and Ţolici–*Hălăbutoaia* also produced notable deposits dating from period of the Starčevo-Criş culture, the present study is only concerned with the material attributable to the Cucuteni culture.

We must highlight a common feature of the four salt springs, namely their very high salinity. Thus, an experiment conducted in 1986 during a period of drought managed to produce no less than 345 g of salt by boiling a litre of brine from Lunca–*Poiana Slatinei* (Dumitroaia 1994, 13, n. 15), 340 g from Negriteşti, and 315 g from Prohozeşti (Monah 2007, 128). The salinity of most of the Subcarpathian salt springs varies, according to the season and the precipitation regime, from 260 g to 310 g of NaCl per litre (Şandru 1952, 410). A comparison between the salinity of the Moldavian piedmont salt springs and that of the Black Sea (35 g per litre) shows the advantage of exploiting the Subcarpathian sources. Furthermore, the salt produced from these springs was a white, high-purity product.

A common feature of the sites Solca–*Slatina Mare*, Lunca–*Poiana Slatinei* and Ţolici–*Hălăbutoaia* is their location in rugged terrain where plant cultivation is not feasible. No traces of habitation structures have been found in the Cucutenian deposits, nor in the older Starčevo-Criş deposits (Monah 1991, 393; Dumitroaia *et al.*, 2008, 209). The exploitation of salt was probably only carried out during the warmer season, in a tight time window of 2–3 weeks during which the participants, namely the Cucutenians and the pastoralists, lived in light dwellings such as huts and shacks.

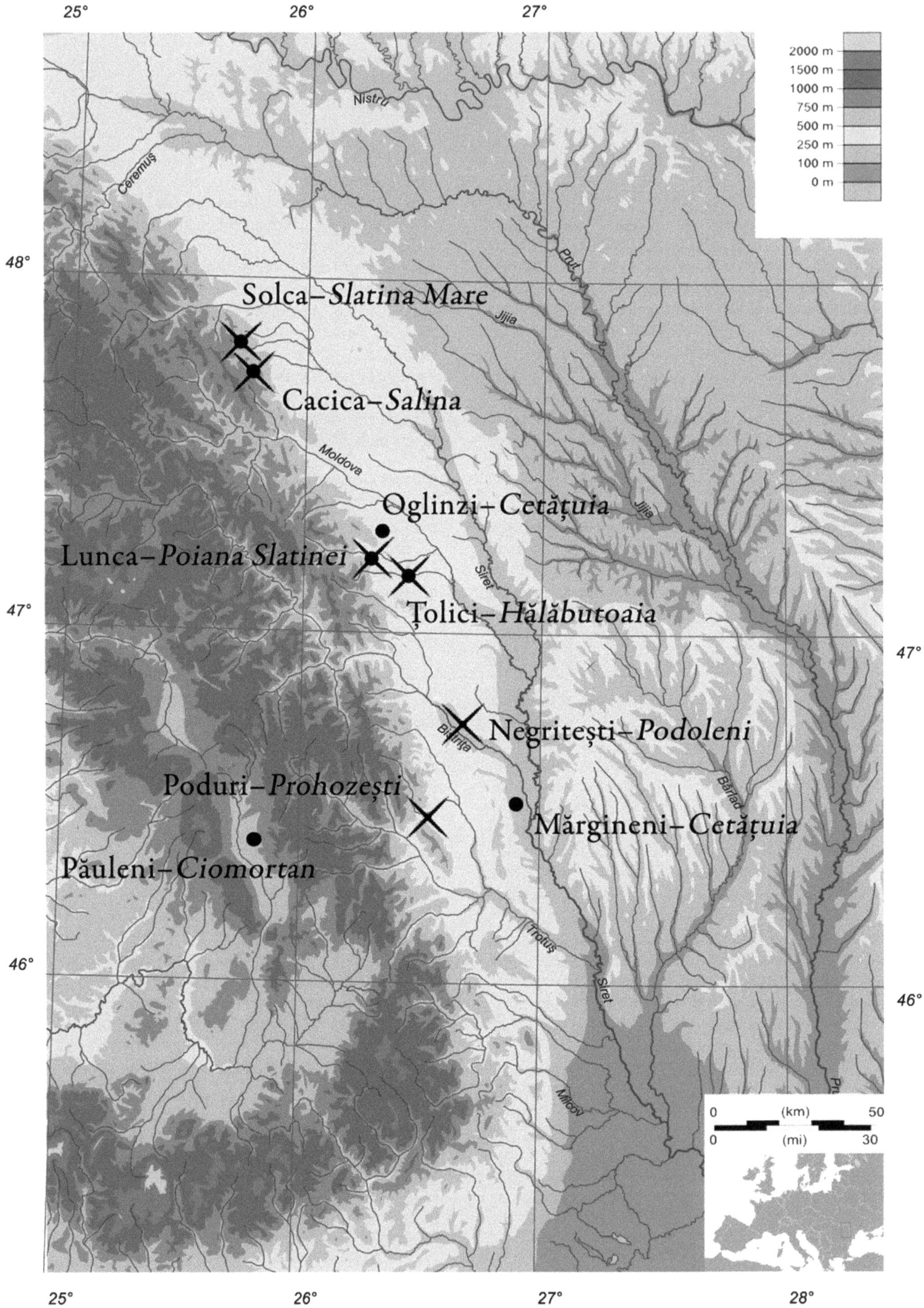

Figure 1. The Eastern Carpathians: sites (•) and salt springs (×) mentioned in the article.

Such are the solutions adopted by the ethnographic groups from Papua New Guinea during their expeditions to obtain salt or various rocks (Pétrequin *et al.*, 2000, 557). Even though the terrain around the sites is quite rough, apart from Cacica where it is wider and flatter, it permits the construction of temporary shelters and the performance of ceremonies, feasts or rituals.

An interesting observation is that the archaeological deposits from three of the sites (Solca, Lunca and Ţolici) are concentrated around modern salt springs, while in the case of the fourth (Cacica), where the salt spring was incorporated in the gallery of the salt mine, the remains are found near a building over the mine entrance (Andronic 2006, 1). Unlike the other springs, the Cucuteni deposits from Cacica are found in a flat area that is suitable for agricultural work. Nevertheless, this site too produced only vestiges that can be associated with the exploitation of salt from the spring or the resultant salty creek.

At Lunca–*Poiana Slatinei* and Ţolici–*Hălăbutoaia* the Cucuteni deposits overlie impressive layers from the Early Neolithic Starčevo-Criş culture (Dumitroaia 1987; Dumitroaia 1994, 13–15), while at Solca the Cucuteni deposits are covered by more recent but less important archaeological remains (Dumitroaia *et al.*, 2004; Munteanu 2006, 73). All the sites have been interpreted as seasonal hubs for the exploitation, recrystallization and briquetage of salt, extremely active during the Cucuteni culture. These sites produced salt briquettes destined for long-distance exchange (Monah 1991, 393; Monah 2012, 139).

State of research

As of 2010, at least 13 sites with traces of prehistoric salt exploitation were identified in the Moldavian piedmont area (Weller *et al.*, 2010, 502). As already mentioned, only four of these have been considered for the present study, namely those for which we possess a more thorough documentation. Three of the sites witnessed more-or-less extensive and systematic archaeological study, and the fourth (Cacica–*Salina Mare*) has been investigated during several fieldwork campaigns and through several control pits, during which a rather significant amount of ceramic material was collected (Weller *et al.*, 2010, 447).

Solca–*Slatina Mare*

Solca–*Slatina Mare* is found on the slope of the Cociniş hill, at an elevation of 654 m, in a glade of a mixed deciduous and coniferous forest (Şandru 1952, 413; Dumitroaia *et al.* 2004, 314; Nicola *et al.* 2007, 37, fig. 2; 6). The brine well is located on a small terrace that also serves as the entrance platform from the East into the glade. Around the salt spring, and particularly below it, there is another broad terrace suitable for raising temporary shelters and enacting ceremonies, performing rituals, etc. From the edge of the terrace, down the slope, there is a rich deposit of Cucutenian and more recent archaeological material. The ceramic remains from briquetage and Cucutenian pottery form a thick layer that slid down the slope for 80–100 m.

First mentioned by geographer Ion Şandru in 1952, the site was the subject of a limited excavation in 1968 by Mircea Ignat, the results of which were published by N. Ursulescu (1977, 310–312). In 2003, a team under the direction of Gh. Dumitroaia dug another test pit that confirmed the presence of a thick level of Cucutenian remains. The test pit, measuring 20×1.5 m, was dug 10.5 m from the salt spring. This sondage did not produce Starčevo-Criş material; the oldest archaeological remains belonged to the Precucuteni III culture. A large portion of the ceramic assemblage, approximately 98%, can be attributed to the Cucuteni culture, including a very large quantity of Cucuteni 'C' ware (Nicola *et al.*, 2007, 43). Countless fragments of briquetage were among the collected ceramic material, as well as fragments of high-quality ware, some of which were painted. Particularly relevant and exciting is that approximately 40% of the ceramic material collected from the Cucuteni deposits belonged to the Cucuteni 'type C' ceramic species, generally ascribed to the north-Pontic pastoralists (Nicola *et al.*, 2007, 43). Even though the excavations at Solca–*Slatina Mare* were confined to only 60 m^2 (Ursulescu 1977, 311; Nicola *et al.*, 2007, 39), the information yielded, taken in conjunction with what was encountered in the other sites selected for this study, permits the formulation of some interesting hypotheses.

Lunca–*Poiana Slatinei*

Poiana Slatinei is located on the Pleşului crest at an elevation of 550 m, and occupies an area of *ca.* 2 ha. The rugged terrain features a few terraces where traces of salt recrystallization activity from during the Starčevo-Criş, Linear Pottery (LBK), Precucuteni, and Cucuteni periods have been found. Three areas with archaeological deposits have been identified: area A, abutting the modern brine well, area B at a distance of 10–20m on the northern slope of the glade, and area C located to the south-east of the salt spring, and where excavations revealed a deposit *ca.* 0.65 m thick dating from the La Tène era, specifically the Canlia cultural facies (Dumitroaia 1994, 11, 58, 70). Area A elicited particular interest from the start, as it featured an ovoid mound (dimensions: 60×25 m; height: 3 m) located immediately adjacent to the modern spring. This area was the most extensively investigated, with six excavations covering a total of *ca.* 380 m^2 (Dumitroaia 1994, fig. 2; Weller, Dumitroaia 2005, 108; fig. 3). The mound revealed a complex stratigraphy comprising, in ascending order, remains from the Early Neolithic period, and from Starčevo-Criş, Precucuteni, and Cucuteni cultures. Also present in this impressive deposit were fragmentary ceramic materials attributed to the Carpi (2nd–3rd centuries AD), the Sântana de Mureş-Chernyakhov culture (3rd–4th centuries AD), and even to the 5th–7th, 10th–12th, 15th–17th and 19th centuries (Dumitroaia 1994, 11). Also collected here were samples for radiocarbon dating, and soil, pollen and coal analyses (Weller *et al.* 2007, 108; fig. 6–8). In area B, two excavations were conducted under the supervision of Gh. Dumitroaia. They covered an area of approximately 135 m^2 and produced ceramic material from the Bronze Age and the Hallstatt period, as well as a fairly sizeable Cucuteni deposit (Dumitroaia 1994, fig. 2). In the upper layers of this deposit, dated to the Cucuteni B phase, were numerous ceramic fragments ascribable to the so-called Cucuteni 'type C' ware of the steppe pastoralists (Dumitroaia 1994, 58). According to Dumitroaia, Cucuteni 'C' pottery amounted to *ca.* 33% of the entire ceramic assemblage collected from trench No. III (Dumitroaia 1994, 58). In 2000, a team lead by J. Chapman did a magnetometric survey of area B at Lunca–*Poiana Slatinei*, and in 2003 the same team dug two small trial trenches totalling 20 m^2 (Chapman 1999/2000, 17; fig. 4; Dumitroaia *et al.* 2003, 184). The present analysis only considers the remains that are certain to be Cucutenian, from areas A and B of the site.

Ţolici–*Hălăbutoaia*

The site rests on a cuesta of the partially forested Hălăbutoaia hill, on the right bank of the Slatina brook, *ca.* 2 km south-west of the village (Weller *et al.* 2007, 143; fig. 16). The most important archaeological deposit was found on the right bank of the watercourse, at *ca.* 40–80 metres north of the present-day brine well, forming an ovoid mound measuring 40×30 m (Dumitroaia *et al.*, 2008, 206). The brine has a concentration of 100 g of salt per litre (Weller *et al.*, 2007, 145). The Cucuteni remains form a layer over 2 metres thick, containing fragments of briquetage, shards from the Cucuteni B, A-B, and A phases, as well as Cucuteni 'type C' pottery fragments. The high density of ceramic remains in the Cucuteni layers renders it a brick-red colour (Dumitroaia *et al.*, 2008, 209–210). A thick Precucuteni III layer was situated beneath the Cucutenian deposits, as well as an exceptional Starčevo-Criş deposit, therefore offering "... a continuous stratigraphy of the exploitation levels for 2500 years (6000–3500 BC)" (Weller *et al.*, 2007, 145). Small agglomerations of Neolithic archaeological remains were also found on the left bank of the Slatina brook, but they are less compact. The site discovered by O. Weller, M. Alexianu and R. Brigand in 2005 (Weller *et al.*, 2005, 145) was investigated through systematic excavation by a team lead by Gh. Dumitroaia starting from 2007 (Dumitroaia *et al.*, 2008, 205). The main excavation focused on an oval mound measuring 40×30 m and containing shards, ash and charcoal. Approximately 20% of the mound was destroyed by erosion. Because of the exceptional thickness of the archaeological deposit, of over 8 m, and the difficult accessibility to the site, only *ca.* 30 m^2 of it has been excavated so far (Dumitroaia *et al.*, 2008, 206). The Cucuteni B layer, which is of particular interest for the present study, measured over 1.30 m in height at its most consistent part. The investigated area did not reveal habitation structures, but only numerous fragments of briquetage, Cucuteni fine and common pottery, as well as fragments of Cucuteni C ceramics (Dumitroaia *et al.* 2008, 209–218; fig. 5–7; 9–11). Even though the site from Hălăbutoaia is insufficiently investigated, it remains an important asset for our study.

Cacica–*Salina–Mina de sare*

The site from Cacica is located in a depression at the base of a ridge of the Obcina Mare Mountains (Andronic 2006, 61). Unlike the three sites mentioned so far, where the springs and the agglomerations of archaeological material are found in areas with rugged terrain, the Cucuteni deposits from Cacica can be found on a flat piece of land. Several areas with archaeological deposits containing ceramic fragments from briquetage, Cucutenian ceramics, including the Cucuteni type C species, and plenty of charcoal have been unearthed in the glade where the main building of the modern mine is also found. The deposits sometimes reach a thickness of 0.80 m. Several test pits and field surveys produced countless fragments from conical briquetage, common pottery and, occasionally, painted Cucutenian ware (Andronic 1989, 171–177; Andronic 2006, 61; Weller *et al.*, 2010, 445–447). The present salt mine was built after expanding and deepening a natural shaft through which salt water ran. Both inside and outside the salt mine, in the area of the former brine well, there were identified several areas with accumulations of Cucuteni 'type C' and regular pottery, fragments of briquetage, charcoal and ash (Andronic 1989, 172–175; Andronic 2006, 60–61; Weller *et al.*, 445–447). A recent field survey identified an area with important agglomerations of briquetage, Cucuteni B, and Cucuteni 'type C' ceramic fragments (Weller *et al.*, 2010, 447). Surface investigations and small-scale trial excavations were conducted on several occasions by M. Andronic, Gh. Dumitroaia, R. Munteanu, D. Garvăn, O. Weller, R. Brigand, and M. Alexianu. The collected archaeological material, subsequently deposited at the museums in Suceava and Piatra-Neamţ, provides some information of use for the present study.

The characteristics of Cucuteni ceramic assemblages

Long before we wrote these lines, as early as the '80s, following a visit of the site of Lunca–*Poiana Slatinei*, we were intrigued by the unusually large proportion of Cucuteni 'type C' pottery remains found especially in deposits from the Cucuteni B phase. In trench III at Lunca the Cucuteni 'type C' fragments amounted to 33% of the collected ceramic material (Dumitroaia 1994, 58). In the Cucuteni B layer, the percentage of Cucuteni 'type C' pottery was greater still. Considering the large number of fragments from briquetage vessels (artefacts seldom encountered in the inventories of regular Cucutenian sites), the disproportion between the Cucuteni 'type C' remains on the one side, and the common and fineware Cucuteni pottery on the other, is even more conspicuous. In regular contemporary Cucuteni B settlements of the Moldavian piedmont, the percentage of Cucuteni 'type C' pottery generally does not exceed 1%. In a previous study on the exploitation of salt in the Cucuteni-Tripolye culture, we attempted to analyse and explain this anomaly (Monah 1991, 397–398). The study in question was the first to articulate the theory of a complex system of exchange that included commodities, services and gifts (Monah 1991, 398), a theory further developed and amended here.

Returning to the issue of the Cucuteni deposits from the sites under discussion, we can observe the following: an intense exploitation of salt water resources as early as the Cucuteni A phase, the presence of briquetage, and towards the end of the Cucuteni A phase the appearance of 'type C' ceramics in the inventory. At Lunca–*Poiana Slatinei*, in area B, the Cucuteni A deposit is *ca.* 45 cm thick and contains fragments of globular vessels, bowls, and painted stems (Dumitroaia 1994, 62). Briquetage fragments are present even in this layer. Most of the briquetage fragments from Lunca come from the lower parts of vessels that ended with a short foot (Dumitroaia 1994, 62; fig. 49-50). Though the preservation conditions of the ceramics were not ideal, a number of painted shards testify to the presence in the inventory of high-quality prestige ware (Dumitroaia 1994, 62; fig. 47).

A distinctive feature of the ceramic material collected from the sites that acted as hubs for salt recrystallization activities is its extreme degree of fragmentation; this observation is valid in the case of the briquetage, of the regular Cucutenian ceramics, and of those attributed to the steppe populations (Dumitroaia 1994; 62). Nonetheless, a few typical Cucutenian vessel shapes can be discerned—fragments from cups, globular vessels, supports and bowls predominate in the Cucuteni A layers (Dumitroaia 1994, fig. 47).

In the layers corresponding to the Cucuteni A-B and B phases, the ceramic assemblage is quite similar, containing numerous fragments of briquetage, painted prestige ceramics, and an impressive amount of Cucuteni 'type C' pottery (Dumitroaia 1994, 62; fig. 48). The 'type C' pottery from Lunca is typical of the Cucuteni B phase; most distinctive are large vessels with thick walls and bowls with large mouths decorated with incisions and impressions (Dumitroaia 1994, 62; fig. 51).

The situation at Ţolici-Hălăbutoaia is similar. The layers ascribed to the Cucuteni A-B and B phases are *ca.* 1 metre thick and appear as a mass of brick-red ceramic fragments interspersed by lenses of ash. The Cucutenian ceramic assemblage contains fragments of large vessels modelled from coarse paste and decorated with barbotine and incrustation. Traces of painted decoration specific to the Cucuteni A and A-B phases are distinguishable on a small number of fragments. From among the ceramic shapes we can mention large vessels with wide mouths and three-chrome decoration, medium-size globular vessels, and several fragments that, considering their shape and painted decoration, can be attributed to the Cucuteni A-B phase (Dumitroaia *et al.* 2008, 213–214; fig. 8; 9).

Particularly intriguing is the discovery of a rather large number of fragments belonging to a ceramic category seldom encountered in Cucutenian contexts: vessels moulded from a fine and dense paste, fired evenly, and of a brick-red or grey-black colour. At Ţolici, the shapes are difficult to identify because of the fragmentary state of the remains; still, the investigators of the site managed to identify vessels with funnel-shaped mouths, rounded shoulders, and decoration limited to the upper parts of the vessels. The decoration consists in most cases of bands of incised parallel horizontal lines, alternating with rows of impressed points or triangles. Other fragments are decorated with wavy folds or impressions of twisted cord (Dumitroaia *et al.* 2008, 216; fig. 10). The only site where this ceramic category has been recorded is the Cucuteni B_1 level from Mărgineni–*Cetăţuia*. Excavations here produced a few ceramic fragments from which a conical vessel was partially reconstituted; the vessel had a wide-open mouth and was decorated with bands of deeply incised lines alternating with rows of triangular impressions (*cf.* Cucoş 1985, 70; fig. 11/2, 5). Just as at Ţolici, the paste (which is similar to the Cucutenian) does not contain shell tempering, yet the general aspect of the vessel is certainly steppe-like. Although perhaps prematurely, we venture to put the vessel from Mărgineni, and also the ceramic fragments from Ţolici, into connection with the ceramics of the Skelya culture that flourished in the area bounded by the Dnieper and the Lower Don (Rassamakin 1999, 75; fig. 3.7). We believe that the moment when the Cucutenian communities from the Moldavian Subcarpathians came into contact with Skelya elements can be placed at the end of the Cucuteni A-B phase in the case of Ţolici, and Cucuteni B_1 for Mărgineni (Cucoş 1985, fig. 11/2, 5). The peculiar ceramic fragments from Ţolici and Mărgineni seem to suggest cohabitation and mutual influence, particularly with respect to the paste and the firing, between the Cucutenians and the steppe pastoralists. The contact seems to have occurred, at least in the case of the salt recrystallization hubs, at the end of the Cucuteni A phase – the beginning of the Cucuteni A-B phase, and later during the Cucuteni B_1 phase in the case of the regular settlements. The ceramic fragments discovered at Ţolici seem to signal the early attraction and interest of the pastoralists for the crystallised salt and particularly for the salt cakes (briquettes), prestige goods held in high regard in the Pontic steppes.

At Ţolici, great quantities of ceramics specific to the Cucuteni B phase were compacted into layers that alternated with layers of ash and dirt. The typical Cucutenian ceramics consist especially of fragments from coarse, undecorated pottery. Because of the unfavourable conditions of preservation, the fine wares rarely retained their original painted decoration. Nevertheless, a few fragments can be traced to large vessels with traces of black painting (Dumitroaia *et al.* 2008, 209; fig. 4; 6). Such vessels should not be associated with the actual brine exploitation or the recrystallization activities, but rather with ritual feasting or, more likely, with storing provisions. As at Solca and Lunca, the fragments of briquetage are very numerous. A feature of the ceramic remains from Ţolici is the greater degree of fragmentation compared to Lunca and Solca (Dumitroaia *et al.* 2008, 212; fig. 7). Lastly, the amount of Cucuteni 'type C' remains is much smaller than at the other sites, only reaching *ca.* 10% of the assemblage. The ceramic shapes are typical for the type C species: vessels with wide mouths, splayed rims, often decorated with vertical grooves on the neck (Dumitroaia *et al.* 2008, 210; fig. 5).

It might be useful to highlight certain features of the Ţolici–*Hălăbutoaia* site: a salt spring with high salinity, an area unsuitable for plant cultivation, the absence of dwellings leaving archaeological traces, an impressively large deposit of ceramic material, and the presence of a large number of briquetage fragments as early as the Cucuteni A phase. Furthermore, Cucutenian fine wares have been identified among the archaeological remains, and, likewise significant, ceramic remains that can be attributed to the Skelya culture and which were discovered in Cucuteni A or A-B layers. The Cucuteni 'type C' ceramic found in the levels corresponding to the Cucuteni B phase could be attributed, with some reserve, to the Srednyi Stog II culture (Rassamakin 1999, table 3.2).

We know very little about the ceramic assemblage from Cacica–*Salina*, because of the lack of systematic excavations. The earliest works dealing with the site mention the presence of some Cucuteni B ceramic fragments. Traces of black paint were observed on some of the artefacts. Numerous fragments of briquetage and 'type C' pottery were reported, without specifying the proportions of the different ceramic categories (Andronic 1989, 174; fig. 2; 3; 4/2–3, 6; 2006; Andronic and Mareş 1996, 21). A recent (2006) survey estimated the percentage of Cucuteni 'type C' ceramics at Cacica at approximately 40%, a value equivalent to that at Solca (Munteanu and Garvăn 2008, 181; fig. 6).

Because of the difficult working conditions, the amount of information on the true situation in the areas excavated is rather limited and of questionable reliability. Nevertheless, we believe we can formulate several working hypotheses even at this stage. We should start by reviewing several pieces of information that, if confirmed, might be relevant to the issue at hand. Gh. Dumitroaia (1994, 72) mentions the discovery of fragments of briquetage in the Cucuteni A settlement from Oglinzi–*Cetăţuia* located 1 km from the Poiana Slatinei salt spring. Similar information—concerning the veracity of which we have certain doubts—is also provided for the Cucuteni A site from Păuleni–*Ciomortan* in South-Eastern Transylvania, where fragments of briquetage were discovered (Cavruc and Dumitroaia 2006, 39). If this information is accurate, then it testifies to the circulation of salt briquettes from Northern Moldavia to the west of the Eastern Carpathians mountain range, and to a spread of salt briquettes in all directions.

Finally, the discovery in the Precucuteni II level at Lunca–*Poiana Slatinei* of a fragment from a *hora* support vessel depicting an anthropomorphic caryatid (Dumitroaia 1994, 54, fig. 44/5) allows us to infer the existence, during the Precucuteni and Cucuteni period, of rituals and ceremonies that made use of items with religious associations. With respect to relations between the native Cucutenians and the incoming steppe pastoralists, the discovery at Solca–*Slatina Mare* of two briquetage fragments with fragments of shells in the body could be especially significant (Nicola *et al.*, 2007, 47; Munteanu and Garvăn 2008, fig. 8/1, 2). Such discoveries have also been made at other centres of salt recrystallization in Northern Moldavia (Munteanu and Garvăn 2011, 85; fig. 8).

Producing briquetage: where, who, how?

The answer to the first part of this question is simple and straightforward: briquetting of salt was done in the immediate vicinity of the salt springs of Northern Moldavia. The place where the briquettes were modelled and fired has not been identified, but it is safe to say that they were located not too far from the recrystallization hubs.

It is possible that the appearance of briquetage in the Cucuteni culture was due to influences coming from the Lengyel culture, but we must stress the fact that there are marked differences between the Cucuteni briquetage and those from Wielicza (Weller 2000, fig. 1). The analysis of the remains of the lower portions of the briquetage from Solca–*Slatina Mare* and Cacica (Nicola *et al.* 2007, 47; fig. 12–15; 18; Andronic 1989, 172–173; fig. 2–3) reveals a certain similarity with the Cucuteni A funnel-shaped beakers with stumpy feet. Those from Lunca–*Poiana Slatinei* resemble this type of vessel even more closely—a type that is common in the Cucuteni A phase and also present in the Cucuteni B phase. The pieces from this site have higher bases than those from Solca, and probably formed part of funnel-shaped beakers (Dumitroaia 1994, 62; fig. 49–50). Conversely, the fragments from Ţolici–*Hălăbutoaia* are shorter (Dumitroaia 2008, 212; fig. 7) and resemble more those from Solca. Nevertheless, the differences between the shapes and sizes of the briquetage of the Moldavian Subcarpathians are minimal. The reconstruction proposed by Mugur Andronic (1989, fig. 4/5) is not based on the real shape of the fragmentary lower portions of the vessels and it consequently suggests a vessel with a greater capacity than that estimated on the basis of the surviving evidence.

The paste is coarse but typically Cucutenian. The few fragments of briquetage with ground shells as temper suggest a close collaboration between the Cucutenians and the steppe pastoralists for the manufacture of these items. An interesting detail is the ratio between the briquetage fragments and the Cucuteni 'type C' ware from Solca. The first category represents 21.15% of the entire Cucutenian ceramic assemblage, while 'type C' ware reaches 30.92% of the total (Nicola *et al.* 2007, 47). The ratio is important and must be ascertained for all the sites where briquetage remains have been found. The use of briquetage for producing salt briquettes is a certainty. Recent analyses (Sandu *et al.* 2012, 152–153) revealed the presence of chlorine and a high concentration of sodium only in the internal area of reducing fire, which indicates their use for drying wet salt and obtaining salt briquettes. We believe that the brine underwent a preliminary crystallisation by boiling in 'type C' and regular Cucutenian vessels. The resulting wet salt (Rmn. *huscă*) was subsequently dried through an extended thermal process in briquetage, in order to obtain the much sought after briquettes.

The earliest briquetage remains in the Moldavian Subcarpathians date from the Cucuteni A phase (Dumitroaia 1994, 66), but the quantity increases significantly during the Cucuteni A-B and B phases, when an increased interest is revealed for salt briquettes, now serving as prestige goods (Monah 2008, 17). In the recrystallization centres for which we hold ample information, the accumulation of briquetage remains is important, attesting a comparatively intense activity which amplifies during the Cucuteni B phase. Nevertheless, if we compare the accumulations of briquetage fragments of the Moldavian piedmont with those of Provadia–*Solnitsata* (Weller 2012, 67–68) we must acknowledge that the former are quite modest. Furthermore, the briquetage from Provadia reach in some cases considerable quantities (Weller 2012, 80). Admittedly, of the 13 Moldavian sites with archaeological remains near salt springs, only four have investigated, and these only to a small extent; future investigations might lead to surprising results.

Briquetage seems to have been a creation of the Cucutenians from near the salt springs, perhaps influenced by the Lengyel culture. The technical features and forms attest to manufacture by Cucutenians, though we cannot exclude the possibility that elements from the steppes—either already integrated into the wider Cucutenian social framework around the salt springs, or arriving during provisioning expeditions—might have been involved in this process. The participation of pastoralists, alongside Cucutenians, in producing wet salt and salt briquettes would explain the unusually large quantity of Cucuteni C ceramics.

On the basis of the debris accumulated at the Subcarpathian sites, and of ethnographic analogies, we believe that this activity was practiced only during the summer and probably not on an annual basis. To produce salt briquettes from salt-spring brine, briquetage vessels were modelled and fired in another location, perhaps together with the pots ascribed to the Cucuteni 'type C' ceramic species. Wood collected from sources near the salt springs was used to obtain wet salt by boiling brine in regular or 'type C' vessels. The resulting wet salt was scraped off these vessels and then packed into salt briquettes via a thermal process. The complicated process of obtaining these compact briquettes required technological know-how and demanded the constant supervision on the part of those involved. Experiments conducted by Romanian archaeologists failed to replicate the results obtained by the Cucutenians (Bodi 2007, 94–95). Additives that facilitated the compacting of the salt must have been used; one such substance could have been cow dung, which, according to the ethnographic record, is used in certain areas of the African continent for catalysing the process of salt condensation and for facilitating the detachment of the briquettes from inside the briquetage vessels. All these activities were presumably preceded and accompanied by various ritual ceremonies based on aetiological myths, if modern ethnographic analogies are any guide (Pétrequin *et al.* 2000, 557–559; Weller *et al.* 1996, 23). In the case of the salt cakes obtained during technological processes that involved the use of fire we can conjecture a mythological weave similar to that developed around their crafts by the metal workers and alchemists of the past (Eliade 1996).

The lure of salt: provisioning expeditions?

The emergence of salt briquettes from the Moldavian Subcarpathians as prestige goods in the areas north of the Black Sea was due to an increase in demographic density, and the intensification of the relations and exchanges between the communities both from the Cucuteni-Tripolye area and from the Pontic steppe. Future research must determine whether the movement of salt briquettes was exclusively or preponderantly eastwards, or if they spread in other directions too. With the waning of the Cucuteni culture the activities of recrystallizing salt and producing salt briquettes ceased, though the Moldavian salt springs continued to be used during the Bronze Age.

In what concerns the importance of the Moldavian salt briquettes for furnishing the Cucuteni-Tripolye and steppe communities, we endorse the views put by Olivier Weller (2012, 80–81) with respect to the salt from Provadia, and assert that those from the Subcarpathians did not play a major role in human sustenance. In our opinion, the salt used for human consumption by the Cucuteni-Tripolye tribes, and perhaps by the steppe tribes, was obtained from alternative local sources, and probably from rock salt likewise quarried from Moldavian sources (Monah 2012, 129–134). The latter hypothesis must be tested in areas with rock salt outcrops.

As it is the case with the Western Dani (Lani) of New Guinea, the salt briquettes, much sought after by the Pontic steppe pastoralists, must have been used only within the prestige goods exchange network (Weller 2012, 82). The production of salt briquettes played an important role particularly with respect to relations between the Cucutenian communities from the Subcarpathians, but also from other areas, with the steppe tribes. We believe that it is not a coincidence that the appearance of Cucuteni 'type C' ware towards the end of the Cucuteni A phase, west of the Prut River, coincides with the appearance of the first briquetage.

We dismiss the possibility that the accumulations of Cucuteni 'type C' ceramics from the centres of salt recrystallization are due to steppe elements already integrated into Cucutenian communities, which in the Subcarpathians are under 1%. Considering the information at our disposal, we believe that the extremely high percentages of 'type C' ware found in areas of salt crystallization and briquetting can be explained through the attraction exerted by the salt briquettes. The briquettes were already circulating as prestige goods in the area of the Cucuteni-Tripolye culture, probably reaching the steppe pastoral communities, at first reaching those of the Skelya culture and later on those of the Srednyi Stog II culture.

Figure 2. Expedition for salt briquettes from the Pontic Steppe during the Cucuteni A/Tripolye B_1 phase.

Because the Cucuteni 'type C' ceramic from the salt springs is not identical with that of the two Pontic cultures, we might also consider the Dereivka culture (Rassamakin 1999, 90, fig. 3.24).

Since the exchange relations between the Chalcolithic communities of South-Eastern Europe and those of the Pontic-Caspian steppe developed steadily, the novel prestige commodity, imbued with magical attributes, will have quickly found its way into the complex system of exchange of the region. Other Cucuteni-Tripolye prestige goods (metal objects, shells, and even painted Tripolye ceramics) were similarly incorporated into the system.

Hypothetical reconstructions of the expeditions conducted by the steppe pastoralists for obtaining salt briquettes (Figure 2) can rely on practices attested ethnographically among certain peoples of New Guinea (Weller *et al.* 1996). We can postulate the existence of small groups of pastoralists devoted to organising such expeditions at certain intervals of time; groups that seem to have maintained close ties with the Cucutenian communities that controlled the salt springs, and also with those whose territory they had to cross. The starting point of these expeditions seems to have been located east of the Dnieper, the fact of the extremely long distance notwithstanding. Indeed, given such a starting point, the distance of these expeditions might have varied between 1000 and 1500 km. In New Guinea, expeditions that last between 8 and 15 days of strenuous march covering distances of over 250 km have been recorded (Pétrequin *et al.* 2000, 550). Marches averaging 20 km per day are not inconceivable, and therefore covering a distance of 1000 km—or less if we consider an area of transition from Tripolye to steppe tribes (Rassamakin 1999, fig. 3.42)—would require a maximum of 50 days. As already stated, such expeditions were not organised on an annual basis, but only at intervals that are difficult to establish.

Finally, we must describe how we think these complex exchange systems were organised. Generally speaking, we believe that the exchange system was organised along lines similar to those described by Marcel Mauss under the names *kula* and *potlatch*, which do not involve equivalent exchanges of goods and are not in any sense barter. Without going into too much detail, they were not only concerned with mobile prestige goods, but especially with "... courtesy formulas, feasts, rituals, military services, women, children, dances, celebrations..." (Mauss 1997, 50). Those who received these gifts were expected to reciprocate with greater profusion. At certain time intervals the cycle was repeated. This manner of organisation was specific not only to the steppe populations, but also to the settled agriculturalists.

Because most of the gifts did not leave archaeological traces, their origin and intensity are hard to ascertain. In the Cucuteni area there have been identified, besides the Cucuteni 'type C' ceramics, certain prestige artefacts such as horse-head and stylised sceptres and cruciform maces considered to have been produced by steppe communities (Rassamakin 1999, fig. 3.14). As previously stated, from the area of the Cucuteni and Tripolye cultures, perhaps also of those from the south, metal objects, prestige ceramics and even anthropomorphic idols were spread to the East (Rassamakin 1990, fig. 3.24). And we must not forget the rituals, ceremonies etc., which do not leave archaeological traces.

In our opinion, the complex systems of exchange developed in the area north of the Black Sea by settled agriculturalists and steppe pastoralists contributed to the maintenance of social stability and the emergence of hierarchies and elites. We do not claim that the abridged and hypothetical scenario outlined here answers all the questions raised by the presence in great quantity of Cucuteni 'type C' ceramics in the centres for salt recrystallization of the Moldavian foothills. Nevertheless we believe that such a scenario could constitute a foundation for future endeavours seeking to explain and interpret this phenomenon.

Acknowledgements
This work was supported by a grant from the Romanian National Authority for Scientific Research, CNCS – UEFISCDI, project number PN-II-ID-PCE-2011-3-0885. The author thanks Roxana Munteanu and Ștefan Caliniuc for help in preparing the illustration.

References

Alexianu, M., Weller, O. and Brigand, R. 2007. *Izvoarele de apă sărată din Moldova subcarpatică. Cercetări etnoarheologice*. Iași, Casa Editorială Demiurg.

Andronic, M. 1989. Cacica – un nou punct neolitic de exploatare a sării. *Studii și Cercetări de Istorie Veche și Arheologie* 40/2, 171–177.

Andronic, M. 2006. Cacica (jud. Suceava), in V. Cavruc and A. Chiri̇̆cescu (eds), *Sarea, timpul și omul*. Sfântu Gheorghe, Editura Angustia, 60–62.

Andronic, M. and Mareș, I. 1996. Cacica, jud. Suceava. *Cronica Cercetărilor Arheologice. Campania 1995*.

Bodi, G. 2007. Experiments with salt crystallization, in D. Monah, Gh. Dumitroaia, O. Weller, J. Chapman (eds.), *L'exploitation du sel à travers le temps*. Bibliotheca Memoriae Antiquitatis XVIII, 89–96. Piatra Neamț, Constantin Matasă.

Cavruc, V. and Dumitroaia, Gh. 2006. Vestigii arheologice privind exploatarea sării pe teritoriul României în perioada neo-eneolitică, in V. Cavruc and A. Chiricescu (eds.), *Sarea, timpul și omul*. Sfântu Gheorghe, Editura Angustia, 37–40.

Chapman, J., Monah, D., Dumitroaia, Gh., Amstrong, H., Milard, A, and Francis, M. 2000. The exploitation of salt in the prehistory of Moldavia, Romania. In *Archaeological Reports 1999/2000*, 23, 10–20. Universities of Durham and Newcastle-upon-Tyne.

Chapman, J. and Gaydarska, B. 2003. The provision of salt to Tripolye mega-sites. In *Tripolian Settlements-Giants. The International Symposium Materials*. Kiev, Korvin.

Chapman, J., Dumitroaia, Gh. and Monah, D. 2004. The earliest salt exploitation in the world: excavation at Lunca–Poiana Slatinei, Neamt County, Moldavia, Romania. *Archaeological Reports for Durham & Newcastle in 2002, 2003/2004*, 34–48. Universities of Durham and Newcastle-upon-Tyne.

Cucoș, Șt. 1985. Ceramica de "tip C" din aria culturii Cucuteni. *Memoria Antiquitatis* IX-XI, 63–92.

Dumitroaia, Gh. 1987. La station archéologique de Lunca–Poiana Slatinei, in M. Petrescu-Dîmbovița, N. Ursulescu, D. Monah, V. Chirica (eds.), *La Civilisation de Cucuteni en contexte européen*, Iași, 253–258. Bibliotheca Archaeologica Iassiensis, I.

Dumitroaia, Gh. 1994. Depunerile neo-eneolitice de la Lunca și Oglinzi, județul Neamț. *Memoria Antiquitatis* XIX, 7–82.

Dumitroaia, Gh., Munteanu, R., Nicola, D., Preoteasa, C., Monah, D., Chapman, J., Weller, O. 2003. Lunca, com. Vânători, jud. Neamț. Punct: Poiana Slatinei. *Cronica Cercetărilor Arheologice din România. Campania 2002*, 183–184. Cimec – Institutul de Memorie Culturală, București.

Dumitroaia, Gh., Nicola, D., Munteanu, R., Preoteasa, C., Monah, D., Boghian, D., Ignătescu, S., 2004. Solca, com. Solca, jud. Suceava. Punct: Slatina Mare. *Cronica Cercetărilor Arheologice din România. Campania 2003*, 314–315.

Dumitroaia, Gh., Munteanu, R., Weller, O., Garvăn, D., Diaconu, V., Brigand, R. 2008. Un nou punct de exploatare a apei sărate în preistorie: Țolici–Hălăbutoaia, jud. Neamț, in D. Monah, Gh. Dumitroaia and D. Garvăn (eds.), *Sarea de la prezent la trecut*, Bibliotheca Memoriae Antiquitatis XX, 203–224. Piatra Neamț, Constantin Matasă.

Eliade, M. 1996. *Făurari și alchimiști*. București, Humanitas.

Mauss, M. 1997. *Eseu despre dar*. Iași, Polirom.

Monah, D. 1991. L'exploitation du sel dans les Carpates orientales et ses rapports avec la culture de Cucuteni-Tripolye, in V. Chirica and D. Monah (eds.), *Le Paléolithique et le Néolithique de la Roumanie en contexte européen*, 387–400. Iași, Bibliotheca Archaeologica Iassiensis IV.

Monah, D. 2002. L'exploitation préhistorique du sel dans les Carpates orientales. In O. Weller (ed.) *Archéologie du sel. Techniques et sociétés dans la Préhistoire européenne. Actes du colloque international, XIVe congrès UISPP, Liège (Belgique), 2001*, 135–146. Rahden Westfalen, Verlag Marie Leidorf GmbH. Internationale Archäologie, ASTK, 3.

Monah, D. 2007. Le sel dans la préhistoire de la Roumanie. In N. Morère Molinero (ed.), *Las Salinas y sal de interior en la historia: economia, medio ambiante y sociedad*, 1, 121–164. Madrid, Dykinson.

Monah, D. 2012. L'approvisionnement en sel des tribus chalcolithiques sédentaires et des tribus des steppes du Nord de la Mer Noire. In V. Nikolov and K. Bacvarov (eds.), *Salt and Gold: The Role of Salt in Prehistoric Europe. Proceedings of the International Symposium (Humboldt-Kolleg) in Provadia, Bulgaria, 30*

September–4 October 2010, 127–142. Provadia, Veliko Tarnovo.

Munteanu, R.-E. 2006. Solca, (jud. Suceava). Punct: Slatina Mare, 73, in V. Cavruc and A. Chiricescu (eds.), *Sarea, timpul şi omul.* Sfântu Gheorghe, Angustia.

Munteanu, R. and Garvăn, D. 2008. Ceramica Cucuteni C din preajma izvoarelor sărate din Moldova. In D. Monah, Gh. Dumitroaia, D. Garvăn (eds), *Sarea de la prezent la trecut*, Bibliotheca Memoriae Antiquitatis XX, 203–224. Piatra Neamţ, Constantin Matasă.

Munteanu, R. and Garvăn, D. 2011. The Cucuteni C Pottery near the Moldavian Springs. In M. Alexianu, O. Weller and R.-G. Curcă (eds.), *Archaeology and Anthropology of Salt: A Diachronic Approach. Proceedings of the International Colloquium, 1–5 October 2008 Al. I. Cuza University (Iaşi, Romania)*, 81–87. BAR International Series 2198, Oxford, BAR Publishing.

Nicola, D., Munteanu, R., Garvăn, D., Preoteasa, C., Dumitroaia, Gh. 2007. Solca–Slatina Mare (Roumanie). Preuves archéologiques de l'exploitation du sel en préhistoire, in D. Monah, Gh. Dumitroaia, O. Weller, J. Chapman (eds.), *L'exploitation du sel à travers le temps*. Bibliotheca Memoriae Antiquitatis XVIII, 35–56. Piatra Neamţ, Constantin Matasă.

Pétrequin, P., Pétrequin, A.-M. and Weller, O. 2000. Cuire la pierre et cuire le sel en Nouvelle Guinée: des techniques actuelles de régularisation sociale. In *XXe Rencontres Internationales d'Archéologie et d'Histoire d'Antibes*. Sous la direction de P. Pétrequin, P. Fluzin, P. Benoit, Editions APDCA, Antibes, 545–564.

Rassamakin, Y. 1999. The Eneolithic of the Black Sea steppe: dynamics of cultural and economic development 4500–2300 BC, in M. Levin, Y. Rassamakin and N. Tatarintseva (eds.), *Late prehistoric exploitation of the Eurasian steppe*, 59–182. Cambridge, McDonald Institute for Archaeological Research.

Sandu, I., Weller, O., Stumbea, D., Alexianu, M. 2012. Analyses archéométriques sur les moules à sel chalcolithiques de l'est de la Roumanie, in V. Nikolov and K. Bacvarov (eds.), *Salt and Gold : The Role of Salt in Prehistoric Europe. Proceedings of the International Symposium (Humboldt-Kolleg) in Provadia, Bulgaria, 30 September–4 October 2010*, 143–154. Provadia, Veliko Tarnovo.

Şandru, I. 1952. Contribuţii geografico-economice asupra exploatării slatinelor în Bucovina de sud. *Studii şi Cercetări Ştiinţifice*, Iaşi, 1-4, III, 407–424.

Ursulescu, N. 1977. Exploatarea sării din saramură în neoliticul timpuriu, în lumina descoperirilor de la Solca (jud. Suceava). *Studii şi Cercetări de Istorie Veche şi Arheologie* 28/3, 307–317.

Weller, O. 2000. Produire du sel par le feu: techniques et enjeux socio-économiques dans le Néolithique européen, in P. Pétrequin, P. Fluzin. J. Thiriot, P. Benoit (eds.), *Arts du feu et productions artisanales*, 565-584. Antibes.

Weller, O. 2012. La production chalcolithique du sel à Provadia-Solnitsata: de la technologie céramique aux implications socio-économiques, in V. Nikolov and K. Bacarov (eds.)., *Salt and Gold: The Role of salt in Prehistoric Europe. Proceedings of the International Symopsium (Hum,bold-Kolleg) in Provadia, Bulgaria 30 September–4 October 2010*, Provadia-Veliko Tarnovo.

Weller, O., Pétrequin, P., Pétrequin, A.-M., Couturaud, A. 1996. Du sel pour les échanges sociaux. *Journal de la Société des océanistes 102/1*, 3–33.

Weller, O. and Dumitroaia, Gh. 2005. The earliest salt production in the World: an early Neolithic exploitation in Poiana Slatinei-Lunca, Romania. *Antiquity*, 79 (306). Online: www.antiquity.ac.uk./projgall/weller [accessed: 15.02.2013].

Weller, O., Dumitroaia, Gh., Monah, D., Nuninger, L. 2007a. L'exploitation des sources salées de Moldavie, un exemple de ressource structurante du territoire depuis le Néolithique, in D. Monah, Gh. Dumitroaia, O. Weller, J. Chapman (eds.), *L'expoitation du sel à travers le temps*. BMA, XVIII, Piatra Neamţ, Constantin Matasă.

Weller, O., Brigand, R. and Alexianu, M. 2007b. Cercetări sistematice asupra izvoarelor de apă sărată din Moldova. Bilanţul explorărilor din anii 2004-2007 efectuate în special în judeţul Neamţ. *Memoria Antiquitatis* XXIV, 121–190.

Weller, O., Brigand, R. and Alexianu, M. 2010. Recherches systématiques autour des sources salées de Moldavie. Bilan des prospections 2008-2010. *Memoria Antiquitatis* XXV-XXVI, 437–504.

The Role of the Salt Supply in the Location of Neolithic and Eneolithic Settlements in Moldavia (Romania)

Nicolae Ursulescu
Universitatea „Alexandru Ioan Cuza" din Iaşi

Abstract
The three principal uses of salt in the traditional life cycle (a vital ingredient of human diet, a component of fodder for domestic animals, and as the ideal preservative for products of animal origin) involve various methods for its extraction. It is natural that as part of their exploitation of the environment, communities looked to local salt resources (saline springs, running or still briny water, salty soil); additional supplies might come from sources with greater salt potential. The Neolithic/Eneolithic settlements of Moldavia reveal that their hinterland harboured a local salt source, capable of meeting at least some domestic requirements. In explaining why a certain habitat might be chosen, we must not isolate any particular factor, but instead assess the entire spectrum of requirements for daily life and the activities specific to an agrarian economy: a source of potable water; high-quality arable land and pasture; significant forest resources; finally, a reliable local salt supply (which should be one of the vital factors, and established for each settlement, as part of the reconstruction of the paleo-environment).

Keywords
local salt resources, Neolithic/Eneolithic, settlement, Moldavia

Three main uses of salt have been attested for the traditional life cycle by ethno-archaeological and experimental means: as a vital ingredient of the human diet, as a component of the fodder of domestic animals, and as the ideal preservative, especially for products of animal origin (Monah 1991; Monah 2008; Monah and Dumitroaia 2007; Alexianu, Dumitroaia and Monah 1992; Alexianu, Weller and Brigand 2007; Monah, Dumitroaia and Garvăn, eds. 2008; Cavruc and Chiricescu, eds. 2006). We must highlight the fact that all three uses are directly linked to nutrition and, by extension, to the maintenance in good order of the biological rhythm of people and of a major part of their livestock. Furthermore, salt is present in the spiritual realm, either as a material component during rituals, or as a conjured element in a series of beliefs (see O. Macari; A. Poruciuc in the present volume).

Each of these three principal uses of salt entails various methods for extracting the mineral in quantities sufficient to meet vital and economic needs. The requirements for human consumption are much smaller than those for the other two. For this purpose, both solid salt and saline water from various natural sources might be used. In the case of domestic animals the quantities of salt required are clearly greater; ideally, raw salt boulders should be used, but when these are not available, grazing in saline pastures with halophytic vegetation, or watering animals with briny water can be employed as alternatives. Finally, food preservation, apart from the other traditional methods of smoking and freezing, likewise requires large quantities of salt, especially during warmer periods of the year. This can be achieved with salt crystals, as well as with natural brine obtained from saline springs, especially in areas located at a distance convenient for its transportation. Transport would have been achieved using methods specific to each historical era, starting from human transportation in waterskins and wooden or ceramic containers (Romanian *fedeleş* = canteens[1]), to animal-drawn traction in carts or wagons, and ultimately to mechanised transport. There is, therefore, a multiplicity of ways in which people can fulfil their salt requirements, employing different methods according to the intended use. For this reason we believe that from a methodological point of view, a discussion of salt supply limited to generalities is insufficient; there must also be a differential approach, not just based on geographical conditions, but also on the intended use.

Within such a frame of reference, we consider that in order to exploit the environment as efficiently as possible, that each community would have looked primarily to local salt resources, and that if this was not enough, the community resorted, to a variable degree, to procuring salt from a distance (from sources with a greater salt potential), either from direct supply by transportation (in the case of short or medium distances), or through trade (in the case of long distances). For this reason, we believe that in investigating the environment of each archaeological site we must also consider research on the existence, in the easily-exploitable immediate hinterland of the site, of potential major or minor sources of salt, such as salt springs, running or still briny water (seas, lakes, ponds, marshes), or saline pastures with halophytic flora. We can establish a set of clues for recognising such local sources.

[1] Wooden containers flattened on one side and fitted with lugs or other means for attaching with ropes or straps in the manner of a knapsack. Such a canteen (Romanian *fedeleş*), belonging to Toader Hrib from Arbore (Suceava County), was found in the village museum and was once used for transporting brine from the Slatina Mare salt spring located in nearby Solca (Ursulescu 1977; 1996).

Thus, on numerous occasions the minor toponymy provides us with precious clues for ascertaining the presence of saline features near archaeological sites, through names such as: *La Sărături* ('at the salt lands'), *Sărata* ('the salty one'), *Sările* ('the salts'), *Valea Sării* ('salt valley'), *Sărăţel* ('little salty [spring]'), *Sărăţeni* ('folk from the salt lands'), *Sărăţica* ('little salty [water]'), *Sărişor* ('little salty [spring or place]'), *Balta Sărată* ('the salt pond'), *Lacu Sărat* ('the salt lake'), *Ocna* ('the saltworks'), *Ocniţa* ('the little saltworks'). These names derived from the basic lexicon of the Romanian language have been supplemented by a series of borrowings and calques (loan words) from Slavic, Turkic or Magyar: *Slatina* ('Salt Spring'), *Slătioara* ['little salt spring'], *Slănic* ['salt river'], *Soloneţ* ['salty place'], *Solca* ['salty river'] (with the variants: *Salca*[2], *Sălcuţa*), *Sălciua* ['brackish water'], *Sălciud* ['folk from the brackish water'], *Sălătruc* ['salty brook], *Sălăuţa* ['little salty brook'], *Bahna* ['marsh'], *Şeuşa/Şăuşa* ['salty river'], *Şieu* ['salty river'], *Valea Şoşii* ['valley of the salty river'], *Soşpatac* ['salt brook'], *Şoşcut* ['salt spring'], *Şoşdea* ['place with salt'], *Şotrile* ['places with salt'], *Şeica* ['saltland'], *Şuşiţa* ['salty water'], *Şugatag* ['salt mine'], possibly *Oituz* [Cuman 'salt route'], etc. (Iordan 1963, 125–127; Curcă 2007; Poruciuc 2011; Alexianu 2011). We can also add hydronyms that point to the presence of mineral salts in some (running or stagnant) waters, such as *Pucioasa* ('sulphurous water]'), *Ruginosu* ('rusty river'), *Puturosu* ('stinking river'), *Balta Puturoasă* ('stinking pond'), *Împuţita* ('stinky water'), *Amara* ('bitter water'), *Smârda* ('foul water'), etc. (Iordan 1963, 107–108, 120, 125). Also apt are toponyms that derive from plants that generally grow in saltland areas, such as *Pipirig* ('bulrush'), *Rogoz* ('sedge'), *Şovârna* ('[place with] bur-reed'), *Pelinu* ('wormwood'), etc. (Todor 1968, 185–191). Generally speaking, knowledge of halophytic flora is of great help in locating areas with salty soils, even more so as ethnobotanical insight has always been an integral part of the practical and intellectual horizon of traditional civilisations (Marian 2008).

In most cases, when we know the precise geographical setting and the minor toponymy/hydronymy of Neolithic and Eneolithic sites in Moldavia, there is almost invariably a local salt supply source within a given site's perimeter of economic exploitation (0–25 km: Bintliff 1999), and this is not limited to Moldavia. The best illustrations is to be found in the selection of places for settlement by the first Neolithic communities north of the Danube, both in Oltenia (the settlement of Copăcelu–*Valea Răii*, Vâlcea County, on a terrace of the Sărata: Berciu 1966, 70–72; 1994, 352), and especially in Transylvania. Thus, in the area of Sibiu (the early Starčevo settlements of Ocna Sibiului, Miercurea Sibiului, Cristian, etc.: Paul 1989, Luca and Suciu 2005; Luca, Diaconescu and Suciu 2008), of Alba (the settlement of Şeuşa: Ciută 2005), or of Cluj (the settlement of Gura Baciului: Vlassa 1972; Lazarovici and Maxim 1995), salt springs or salt lands can be found at distances of up to 5 km from the relevant sites (Lazarovici and Lazarovici 2011), which points to these first Neolithic communities on reaching what is today Romania having included among their criteria for choosing somewhere to live, the existence of saline features in the immediate vicinity.

There is a relevant passage for the understanding of the notion of "saline features" in the anonymous *Gesta Hungarorum* (Popa-Lisseanu 1934, 95–96), even though it concerns a much later period, the Early Middle Ages. A scout sent by the Magyar chieftain Tuhutum to the land of Gelou (located "beyond the forests", i.e., in the Transylvanian Plateau) reports the riches of this land: good rivers, gold and salt (including separately listed saline features, probably salt springs).

It was therefore natural that as Neolithic communities spread towards Eastern Carpathian regions they brought with them traditions relating to the choice of the best place for a settlement that satisfied as many as possible of the criteria necessary for a proper working of daily life and for the conduct of day-to-day activities.

Among the factors involved in the choice of a particular place for settlement on the part of a Neolithic community in Moldavia was presumably the salt supply, or at least the partial availability of their daily requirements. We find it hard to imagine that a community might rely on salt obtained through trade alone, which must have been a risky business involving an uncertain continuity of supply at the mercy of man-made or natural accidents. A prolonged interruption to the normal flow of an external supply would have endangered the life of the people and what is more, of the animals that constituted their main source of food. Such a scenario would have necessitated in an emergency the abandonment of the old settlement and the migration of the community to another location. Such a radical solution would have been largely avoided by a careful prospecting of the land at an early stage of selection of a place for future settlement.

Even though a definitive answer to the question of a link between the location of prehistoric settlements and the presence of local salt sources requires more research, we can nevertheless list several instances relevant to the discussion on the basis of already available data. We have not taken into consideration the Subcarpathian salt springs, which, given their particularly strong discharge and concentration, fall into a separate category, namely of intense exploitation with a very wide distribution, much greater than the economic range of a single settlement (Ursulescu 1996; Alexianu, Weller and Brigand 2007, 23–25).

Thus, from the period of the Criş culture we know of the following situations in which salt resources were possibly exploited on a local scale:

1) The settlement at Arsura (Vaslui County) located on the banks of the Ruginosu brook (in the Prut basin), a hydronym that alludes to a certain degree of water mineralisation (Ursulescu 1984, 42, no. 4).

[2] A well-known multi-level archaeological site near Oradea.

2) Four settlements in Balş, Iaşi County (two in the *Părului Valley*, one at *Bejeneasa*, and another at *Coasta Clopoţelului*: Văleanu 2003, 17–19) located near the salt spring of Arcaci (Alexianu, Weller and Brigand 2007, 25). This could have supplied, in later periods, enough salt to meet the requirements of both the Precucutenian settlement at Bejeneasa (overlapping a Criş one: Văleanu, 2003, 17), as well as the neighbouring Cucutenian settlements, including the well-known sites from Cucuteni–*Cetăţuia* and Băiceni–*Dâmbul Morii*.

3) The settlement at Leontineşti (Bacău County) located on the middle terrace of the river Tazlăul Sărat (Iacobescu 1992).

4) Two settlements at Pârcovaci, Iaşi County (Văleanu 2003, 64–65), as well as two from Strunga (Ursulescu 1984, 67, nos. 123–124; Văleanu 2003, 77–78) found near a number of mineralised water springs.

5) The settlement at Udeşti (Suceava County) close to a spot called "La Sărătură" ("at the salt land") (Ursulescu 1984, 70, no. 136).

We have restricted the list to include only Starčevo-Criş sites since, once a spectacular increase in the number of settlements occurs place during the Eneolithic (the Precucuteni and Cucuteni cultures), evidence for the existence of local salt resources grows exponentially.

To these instances, for which the presence of a local source of saline resources has been satisfactorily proved, we can add the several dozen (or even hundreds) of Starčevo-Criş settlements, and the Linear Pottery, Precucuteni and Cucuteni pottery, found near marshy expanses adjoining rivers from the Moldavian Plateau (the Jijia, Bahlui, Bârlad, etc.), where numerous ponds and fens feature halophytic vegetation that could have met, at least partially, the saline requirements of domestic animals, especially cattle. In this context, the important site of Isaiia–*Balta Popii*, Iaşi County is noteworthy. It features Neolithic occupation from the Linear Pottery and Precucuteni periods, and was continuously inhabited thereafter, and is located on part of the first non-floodable terrace of the right bank of the river Jijia (near the old offshoot into the Prut). It is situated ca. 1 km from the place known as *Balta Sărată* ('the salty pond'), which clearly attests to the existence of a possible local source for obtaining salt water (Ursulescu 2004).

Without denying in any way the importance of the major sources of salt supplies in prehistory; *viz.* saline springs and surface-exploited rock salt, we believe that on the evidence now available we should also look to any possible local alternatives, especially for those areas where the great distance from major salt deposits meant that salt was obtained only through trade, thus placing at risk the regular supply of a resource vital to both people and livestock.

These few examples and suggestions (which have the potential for augmentation if sufficient attention is paid by future researchers to the various local salt resources) suggest that in explaining why Neolithic people (indeed those of any other prehistoric period) should have chosen a particular habitat we must not isolate any particular factor, but instead assess the entire spectrum of requirements for daily life and the activities specific to a sedentary, predominantly agrarian, economy (Bolomey 1973, 624–625):

- a source of potable water;
- the availability of high-quality arable land and pasture;
- the availability of significant local forest resources;
- finally, a reliable local salt supply. This last should be categorised among the vital factors, and should be established for each settlement, as part of the reconstruction of the paleo-environment.

Other factors, such as the availability of raw materials and the proximity of major communication routes, may have contributed to the decision of a community to select a location for a settlement, but they were not mandatory, but merely ancillary factors.

In conclusion, it should be stressed that any undue emphasis placed on one of the vital factors (water, farmland, wood, or salt) by modern research can lead to a distorted view of the mindset and actions of prehistoric people. Thus, even if the presence of potable water in the immediate vicinity of a settlement has always been a critical factor, the simple presence of such water was not sufficient reason for a community to settle in a certain location. Only the cumulative presence of these elements (among which we believe that the presence of local salt resources should be included) in a particular environmental context could lead to the foundation there of a long-term settlement.

References

Alexianu, M. 2011. Sarea: reflexe onomastice în limba română. In O. Felecan (ed.), *Name and Naming, Proceedings of the International Conference on Onomastics, 1st Edition: Multiethnic Connections in Anthroponymy, Baia Mare, 19-21 September 2011*, Cluj-Napoca, 579-584.

Alexianu, M., Dumitroaia, G. and Monah, D. 1992. Exploatarea surselor de apă sărată din Moldova: o abordare etnoarheologică. *Thraco-Dacica* XIII, 159-167.

Alexianu, M., Weller, O. and Brigand, R. 2007. *Izvoarele de apă sărată din Moldova subcarpatică. Cercetări etnoarheologice*. Iaşi, Demiurg.

Berciu, D. 1966. *Zorile istoriei în Carpaţi şi la Dunăre*. Bucureşti, Editura Ştiinţifică.

Berciu D. 1994. Copăcelu. In C. Preda (coord.), *Enciclopedia arheologiei şi istoriei vechi a României*, I, 352. Bucureşti, Editura Enciclopedică.

Bintliff, J. 1999. Settlement and territory. In Graeme Barker (ed.), *Companion Encyclopedia of Archaeology*, vol.1-2, 505-526. London, New York, Routledge.

Bolomey, Al. 1973. Noi moduri de abordare a cercetării preistoriei. *Studii şi cercetări de istorie veche* 24, 4, 621-632.

Cavruc, V. and Chiricescu, A. (eds.) 2006. *Sarea, timpul şi omul*. Sfântu Gheorghe, Angustia.

Ciută, M. 2005a. *Începuturile neoliticului timpuriu în spaţiul intracarpatic transilvănean*. Bibliotheca Universitatis Apulensis XII. Alba Iulia, Aeternitas.

Ciută, M. 2005b. *Cercetări arheologice la Şeuşa – La Cărarea Morii*. Bibliotheca Brukenthal XLIII. Alba Iulia, Altip.

Curcă, R.-G. 2007. Halotoponymes et halohydronimes dans les documents médiévaux de la Moldavie (XIVe-XVIIe s.). In N. Morère Molinero (ed.), *Inland Salt Works and Salt History: Economy, Environment and Society*, II, 777-784. Madrid, Dykinson.

Iacobescu, S. 1992. O nouă aşezare Starčevo-Criş descoperită în judeţul Bacău. *Carpica* 23, 43-50.

Lazarovici, Gh. and Lazarovici, C.-M. 2011. Some salt sources in Transylvania and their connections with the archaeological sites in the area, in M. Alexianu, O. Weller, R.-G. Curcă (eds.), *Archaeology and Anthropology of Salt: a Diachronic Approach. Proceedings of the International Colloquium, 1-5 October 2008 "Al. I. Cuza" University (Iaşi, Romania)*, BAR International Series 2198, 89-110. Oxford, BAR Publishing.

Lazarovici, Gh. and Maxim, Z. 1995. *Gura Baciului*. Bibliotheca Musei Napocensis XI, Cluj-Napoca.

Luca, S.A. and Suciu, C. 2005. The beginning of the Early Neolithic in Transylvania, in V. Spinei, C.-M. Lazarovici, D. Monah (eds.), *Scripta praehistorica. Miscellanea in honorem nonagenarii magistri Mircea Petrescu-Dîmboviţa oblata*, 139-156. Iaşi, Editura Universităţii "Al. I. Cuza".

Luca, S. A., Diaconescu, D. and Suciu, C.I. 2008. Archaeological research in Miercurea Sibiului – Petriş (Sibiu County, Romania): the Starčevo-Criş level during 1997-2005 (a preliminary report). *Documenta Praehistorica* 35, 325-343.

Marian, S.F. 2008. *Botanica poporană română*, I (ed. A. Brădăţan), Suceava - Bucureşti, Editura Academiei Române.

Monah, D. 1991. L'exploitation du sel dans les Carpates Orientales et ses rapports avec la culture de Cucuteni-Tripolye, in V. Chirica and D. Monah (eds.), *Le Paléolithique et le Néolithique de la Roumanie en contexte européen*. Bibliotheca Archaeologica Iassiensis IV, 387-400.

Monah, D. 2008. Sarea în preistoria României. *Arheologia Moldovei* XXX (2007), 87-116.

Monah, D. and Dumitroaia, Gh. 2007. Recherches sur l'exploitation préhistorique du sel en Roumanie, in D. Monah et al. (eds.), *L'exploitation du sel à travers le temps*, Bibliotheca Memoriae Antiquitatis XVIII, 13-34. Piatra Neamţ, Editura Constantin Matasă.

Monah, D., Dumitroaia, Gh. and Garvăn, D. (eds.) 2008. *Sarea de la prezent la trecut*. Bibliotheca Memoriae Antiquitatis XX. Piatra Neamţ, Editura Constantin Matasă.

Paul, I. 1989. Unele probleme ale neoliticului timpuriu din zona carpato-dunăreană. *Studii şi cercetări de istorie veche şi arheologie* 40, 1, 3-25.

Popa-Lisseanu, G. 1934. *Izvoarele istoriei românilor*, I. Bucureşti.

Poruciuc, A. 2011. Etymological and historical implications of Romanian place-names referring to salt, in M. Alexianu, O. Weller and R.-G. Curcă (eds.), *Archaeology and Anthropology of Salt: a Diachronic Approach. Proceedings of the International Colloquium, 1-5 October 2008 "Al. I. Cuza" University (Iaşi, Romania)*, BAR International Series 2198, 215-218. Oxford, BAR Publishing.

Todor, I. 1968. *Mic atlas de plante din flora Republicii Socialiste România*. Bucureşti, Editura Didactică şi Pedagogică.

Ursulescu, N. 1977. Exploatarea sării din saramură în neoliticul timpuriu, în lumina descoperirilor de la Solca (jud. Suceava). *Studii şi cercetări de istorie veche şi arheologie* 28, 3, 307-317.

Ursulescu, N. 1984. *Evoluţia culturii Starčevo-Criş pe teritoriul Moldovei*. Suceava.

Ursulescu, N. 1996. L'utilisation des sources salées dans le Néolithique de la Moldavie (Roumanie). In M. Otte (ed.), *Nature et Culture*, I, 489-497. Etudes et recherches archéologiques de l'Université de Liège 68. Liège.

Ursulescu, N. 2004. Archéologie et archéozoologie dans l'habitat de Isaiia (com. de Răducăneni, dép. de Iaşi). In L. Bejenaru (ed.), *Archaeozoology and Paleozoology Summercourses*. Socrates Intensive Programme, Academic Year 2002-2003. Iaşi, Universitatea „Alexandru Ioan Cuza", 79-95.

Văleanu, M.C. 2003. *Aşezări neo-eneolitice din Moldova*. Iaşi, Helios.

Vlassa, N. 1972. Cea mai veche fază a complexului cultural Starčevo-Criş în România. *Acta Musei Napocensis* 9, 7-28.

A Pinch of Salt in the Prehistoric Eastern Carpathian Mountains (Romania)

Olivier Weller
Maison de l'Archéologie et de l'Ethnologie, UMR 8215 *Trajectoires,* Nanterre

Robin Brigand
Maison de l'Archéologie et de l'Ethnologie, UMR 8215 *Trajectoires,* Nanterre

Gheorghe Dumitroaia
Complexul Muzeal Judeţean Neamţ, Piatra Neamţ

Daniel Garvăn
Muzeul Judeţean Buzău

Roxana Munteanu
Muzeul Judeţean Buzău

Abstract
For ten years, Franco-Romanian research on salt exploitation in the Eastern Carpathians (Moldova) has allowed us to multiply approaches (archaeology, ethnology, history, geography, palaeoenvironment, chemistry...) to the study of salt resources (saline springs, rock salt) and their exploitation, beginning with the first Neolithic and still in progress. From two key areas that possess remarkable archaeological potential and which reveal a chronological framework covering the whole of the Neolithic and Chalcolithic periods (6000-3500 BC), from our knowledge of settlement patterns around salt exploitation, and from their highly informative palaeoenvironmental potential, it is possible to propose several models of human settlement and exploitation. Within these models, salt resources do not have the same function in both time and space.

Keywords
Neolithic, Cucuteni, Moldavia, salt spring, excavation, settlement pattern

With over 250 salt springs and a long tradition of archaeological research, the sub-Carpathian area of Moldavia (Eastern Romania) represents an ideal framework to perform extensive research on a lost object of archaeological research, namely salt.

By examining the type and nature of each salt resource, the diversity of archaeological evidence (briquetage, fire structure, charcoal accumulation, stone tools...), the forms of salt working, and then the types of production and territorial behaviour, we attempt to understand and characterize the first forms of salt production (6000-3500 BC), as well as to discuss their socio-economic dimensions. A number of methodological approaches are used (technology, archaeometry, palaeoenvironmental studies, geoarchaeology, ethnography, spatial analysis) in order to highlight the techniques of production, the environmental interactions, as well as the social context and the exchange system.

In the present paper, we highlight, after a brief history of research, our archaeological studies in two key areas: *Hălăbutoaia*-Ţolici and *Slatina Cozla*-Gârcina in Neamţ County.

1. Current state of research

One of the most important salt resources in the region has been the subject of archaeological and ethnographic research since the 1980s. In particular, attention has been focused on *Poiana Slatinei*-Lunca, where archaeological evidence shows exploitation from the Early Neolithic (early 6th millennium BC), intensifying through the 5th and 4th millennia BC (Dumitroaia 1987, 1994; Weller and Dumitroaia 2005). Research on this site and its landscape has been led by the Museum of Piatra Neamţ (Alexianu *et al.* 1992; Monah and Dumitroaia 2007). In 2003, after a first collaboration in 1995 as part of a PhD (O.W.), our French-Romanian team began collaborating under the aegis of several interdisciplinary programmes supported by the CNRS, the French Ministry of Foreign Affairs and the Romanian CNCSIS, and developed different approaches. In this sense, we mention the geoarchaeology and palaeoenvironmental studies in Lunca (Weller *et al.* 2008), archaeometry on briquetages (Sandu *et al.* 2012), ethnoarchaeology (Alexianu and Weller 2009; Alexianu *et al.* 2011), spatial analysis in Moldavia (Weller *et al.* 2011; Brigand *et al.* 2012; Brigand and Weller 2012).

Ten radiocarbon dates of 5900-5600 BC during the Early Neolithic have shown that the oldest salt exploitation known so far in Europe, and perhaps in the world, is that at Lunca-*Poiana Slatinei*, Neamţ County, Romania (Weller and Dumitroaia 2005; Weller *et al.* 2008). The production of ignigenous salt from salt-water springs continued even later, during the Cucuteni

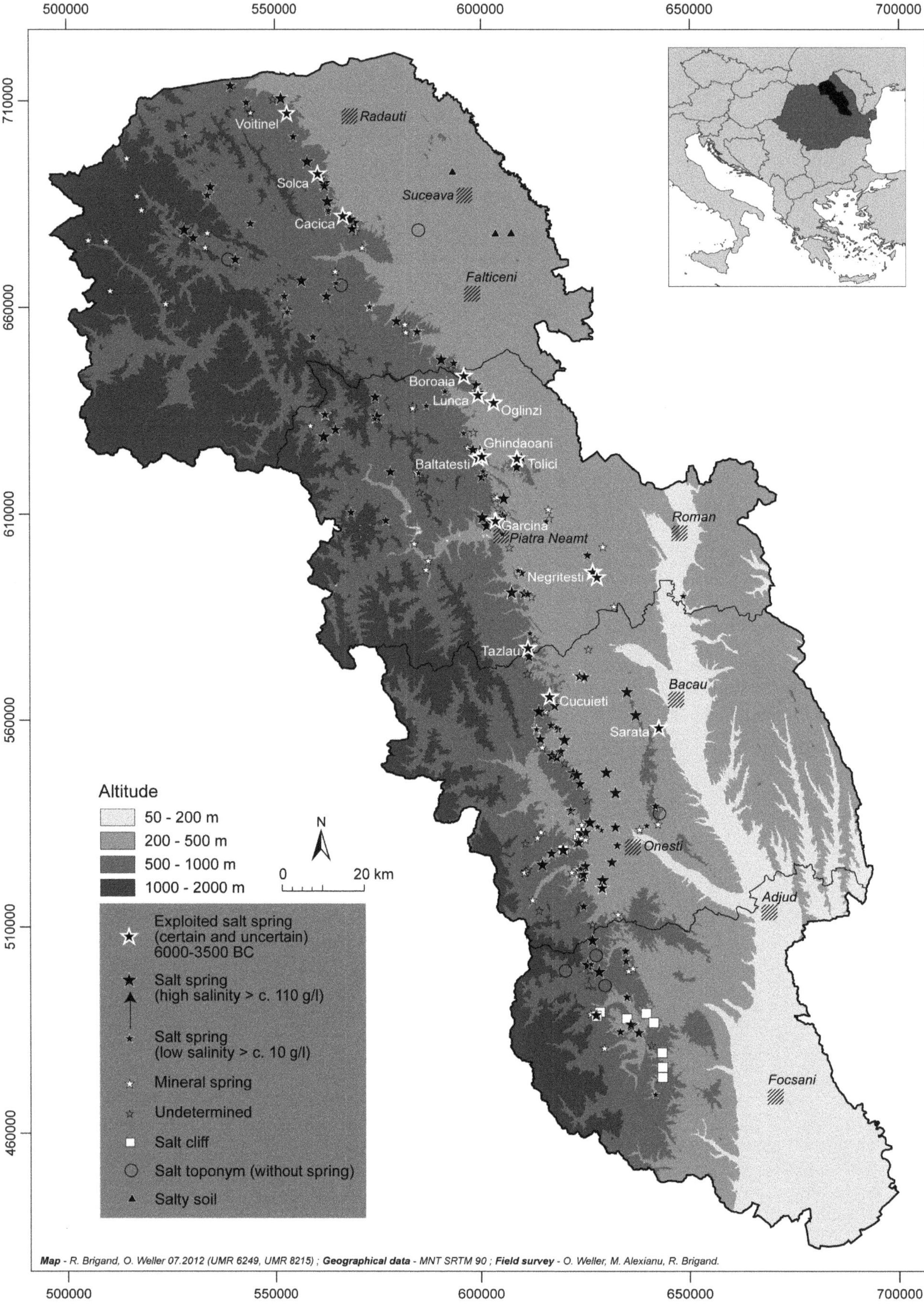

Figure 1. Salt at the feet of the Eastern Carpathians. Potential et prehistoric exploitation vestiges (6000-3500 BC).

culture, at the end of phase A, but very intensely during Cucuteni B (Dumitroaia 1994). A series of more recent discoveries brought new evidence about the use of salt-water springs in Moldavia by the Chalcolithic population as in Solca-*Slatina Mare* (Nicola *et al.* 2007) or in Cucuieţi-*Slatina Veche* (Munteanu *et al.* 2007) (see Alexianu *et al.* 2011 for more references).

At the same time, systematic field surveys since 2003 along the Eastern Carpathian mountains (4 counties) have allowed the exploration and precise description of around 280 salt resources (Weller *et al.* 2007, 2010, in press) (Figure 1).

This salt resource database includes an inventory of all mineral springs and rock salt outcrops recorded. For each spring, the following criteria were documented: access, rate of flow, pH, known uses, geographic coordinates and chemical composition of the water, the means of collection (pit, tree trunk wells, wooden square wells, stone wells, complex composite wells, cistern wells), the depth of the well, the rate of flow and the current use (none, animal husbandry, hunting, forage, food conservation, cheese industry or meat curing, crystallised salt exploitation, therapeutic use, milk curdling). In parallel, archaeological exploration has allowed the recognition of six new sites of prehistoric salt exploitation in the Subcarpathian region. The site of *Hălăbutoaia* in Ţolici (Weller *et al.* 2007, Dumitroaia *et al.* 2008) has, however, certainly the most potential from the perspectives of production technique, of palaeoenvironmental data, and of exploitation impact on the natural environment.

2. The site of *Hălăbutoaia*, Ţolici (com. Petricani, Neamţ)

The important salt spring of *Hălăbutoaia* (Figure 2) is situated in the east part of the village of Ţolici and 7 km south of the famous archaeological site of *Râpa lui Bodai* at Târpeşti (Marinescu-Bîlcu 1981), at the bottom of a small U-shaped valley, in the central part of the Slatina brook. Salt exploitation there was among the earliest in Europe.

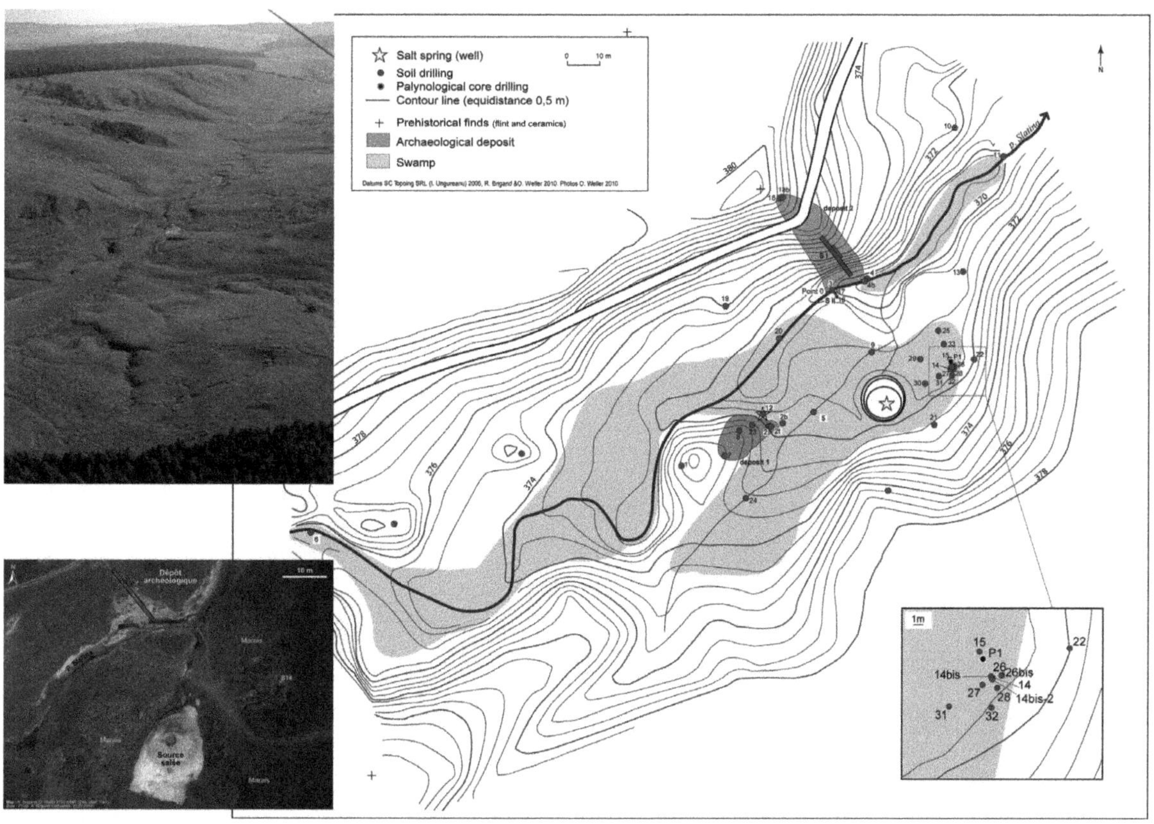

Figure 2. Plan of the archaeological area of *Hălăbutoaia* in Ţolici (salt spring, marsh, and drillings).

Discovered in 2005 (Weller *et al.* 2007), salt production is documented from the Early Neolithic to the end of Chalcolithic period (5700-3500 BC) with an important 8 m-deep stratigraphy (Figure 3). This impressive archaeological deposit is made up of multiple structures and combustion waste related to the exploitation of the salt spring. It also includes a significant amount of pottery (more than 2 tons, the study of which is still in progress). It includes salt moulds dating to the end of Cucuteni A and a remarkable amount of type C Cucuteni pottery (Munteanu and Garvăn 2012).

In 2007-2010, various core drillings were performed in the salty swamp of the spring (Danu *et al.* 2010) and numerous palaeoenvironmental studies were realized at the archaeological site. Multi-proxy approaches around the salt spring (sedimentology, pollen, non-pollen palynomorphs) and in the salt workshop (micromorphology, mineralogy, anthracology, phytoliths, chemical analysis) are currently being discussed, and they suggest a remarkable environmental impact since the Early Neolithic and highlight salt techniques and human interactions. Salt exploitation,

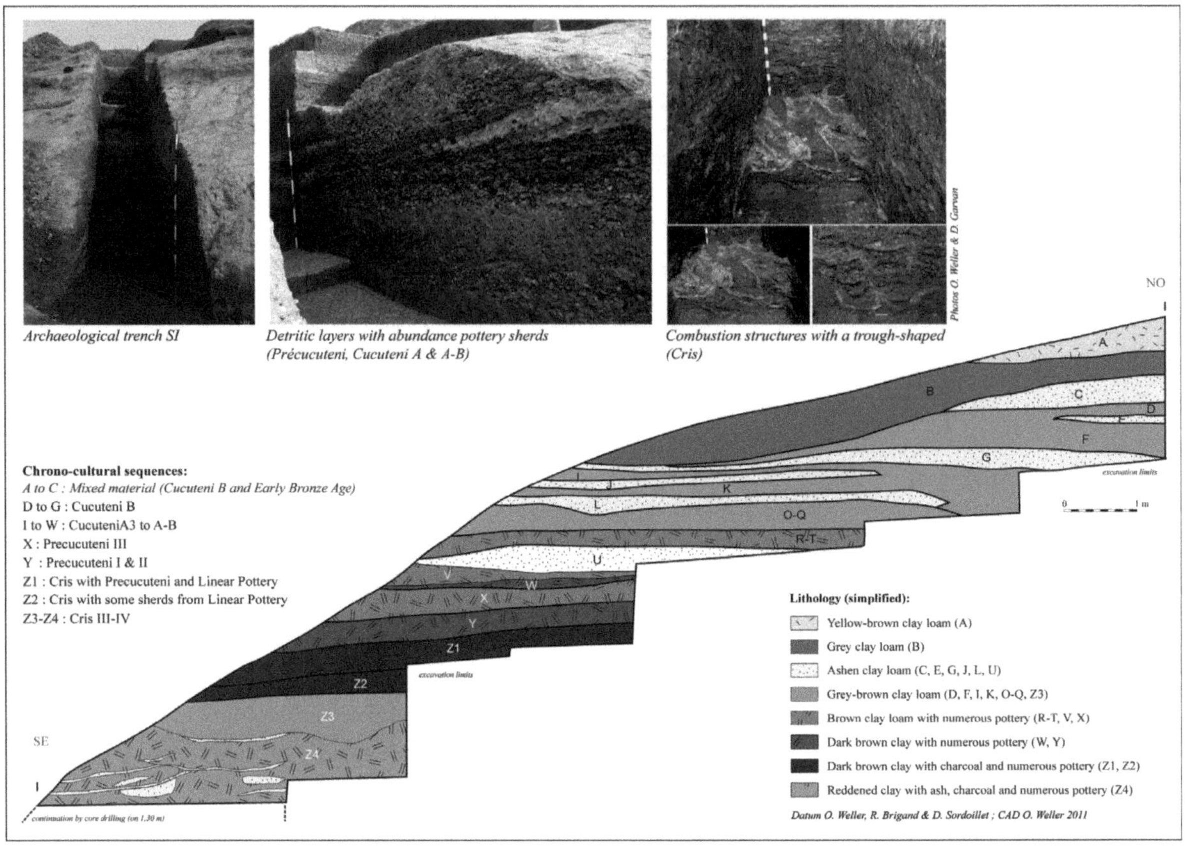

Figure 3. Archaeological deposit and synthetic cross section of the SI drilling (*Hălăbutoaia*, Ţolici).

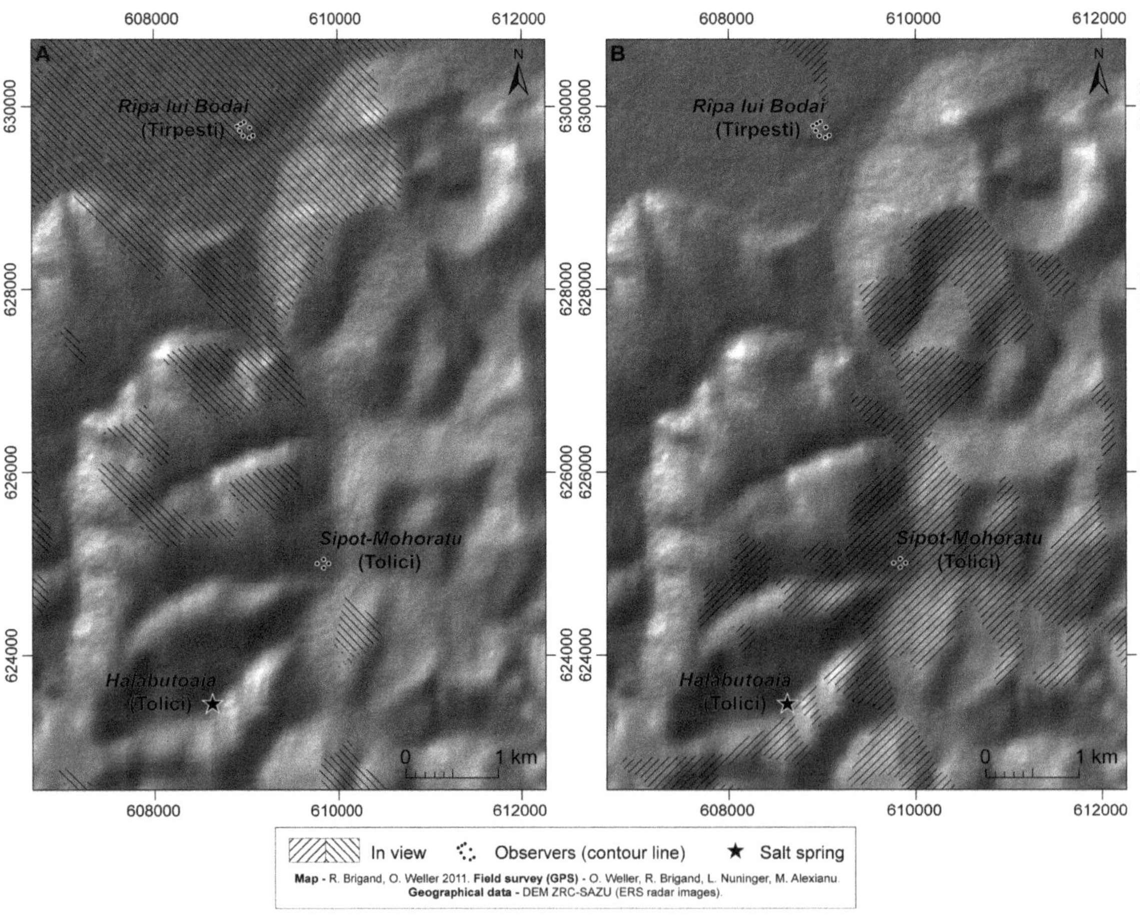

Figure 4. Territorial control and access to the salt spring: visible areas starting from the habitats of *Râpa lui Bodai*-Târpeşti and *Şipot Mohorâtu*-Ţolici.

pastoral practices since Early Neolithic and farming during Chalcolithic times are directly connected with these variations (Weller *et al.*, in prep.).

This salt exploitation is part of an exceptional archaeological environment including the famous fortified Neolithic site (*Râpa lui Bodai*-Târpeşti), less than 6 km away (Figure 4). Established at the confluence of the Ţolici Valley with the Topoliţa, this site opens towards the Moldova Valley and Neamţ Valley via the Humuleşti confluence. This ecotonal situation near the main watercourses, incorporating both axes of circulation and the most suitable land for agricultural activities, the relationship with the exploited salt spring seems to have contributed to the early occupation of *Râpa lui Bodai* during the period of Band Linear Ceramic. Nevertheless, ceramic artefacts from the Early Neolithic (Criş) observed at *Hălăbutoaia* salt springs suggest the existence of neighbouring sites, perhaps the *Deleni*-Grumazesti settlement. During the pre-Cucuteni period, a intensification of settlement can be observed in the Topoliţa valley, associated with two important fortified settlements: *Râpa lui Bodai* on one hand, and *Dealul Valea Seacă* on the other. Regarding control over salt springs, a hilltop settlement named *Şipot Mohorâtu*-Ţolici was established upstream of the confluence between the small saline valley - which is almost visible 2 km upstream - and the Ţolici valley. This hilltop settlement appears to have functioned as a relay station between the exploited salt spring and the village of *Râpa lui Bodai*-Târpeşti. During the Cucuteni culture, the settlement pattern did not evolve radically; elsewhere in Moldavia, there is a clear dissemination of habitat that now colonizes all ecosystems. Only the Topoliţa valley remained densely settled between two federating poles: *Râpa lui Bodai* and downstream, *Locul Şcolii*-Petricani. The abandonment of the site at *Şipot Mohorâtu* shows the inclusion of the source in a defined territory of salt acquisition that contrasts with the situation in the previous period, when the emergence of a hilltop site was directly related to the control and exploitation of the resource. The relative isolation of the salt valley during this period highlights the fact that, at least for Cucuteni, establishing settlements near the exploited salt springs was not as essential as controlling the access to these springs, with the settlements being positioned downstream, at the mouth of the valleys and often near confluences. The Gârcina example emphasizes that salt springs are not necessarily clearly visible from the settlements, but that accessibility and the main valley are strictly controlled.

3. The site of *Slatina Cozla*, Gârcina

In 2005, another salt exploitation site was studied at Gârcina on the slope north-north-west from the *Cozla Hill*, very close to the salt spring of *Slatina III*. This site, though significantly smaller than *Poiana Slatinei* in Lunca and *Hălăbutoaia* in Ţolici, revealed – through a small trench (4 x 1 m) and eight core drillings executed in 2011 – an archaeological deposit related to salt exploitation, ascribed exclusively to the Cucuteni culture (Figure 5).

This deposit forms a small semi-circular hill about 15 m in diameter, eroded on the west side, with 1.55 m maximum height. Besides the levels of upper colluvial deposits that include pottery sherds dating to the Early Bronze Age and to Cucuteni B, two exploitation levels have been identified: one dating to Cucuteni A2, less than 50 cm thick, without any briquetage, and the other to Cucuteni B, about 70-80 cm thick, which includes briquetage and type C pottery (Figure 6).

As in the case of other excavated salt exploitation sites (*Poiana Slatinei*-Lunca, *Hălăbutoaia*-Ţolici, Solca-*Slatina Mare* and Cucuieţi-*Slatina Veche*), the Cucuteni pottery is abundant and consists of an accumulation of sherds. At Gârcina, however, the pottery is less fragmentary and the setting is better preserved than at Ţolici, for instance. We mention the discovery of two pre-Cucuteni III sherds out of context, as well as the presence of Cucuteni A-B pottery at the base of the Cucuteni B level. Several chemical *analyses related to the pottery and to the ash layers of this deposit are in progress; their purpose is to determine the exploitation techniques and the quality of the salt.

Finally, this production site is part of a rich archaeological environment, the north area of Piatra Neamţ, well documented for its dense Cucuteni settlement pattern (Figure 7).

Despite salt exploitation demonstrated at Gârcina during Cucuteni A2, B and perhaps A-B, it is interesting to observe that there is no directly associated occupation, except perhaps at 3 and 5 km from the stream confluence of Slatina and Bistrita (*Sarata*, *Trei Coline*, *Cozla* in Piatra Neamţ). Other settlements in the area seem to be more directly associated with the salt resource: *Dealul Balaurului*-Gârcina and *Sarata*-Dobreni salt springs, Dobreni settlements, *Slatina Dealul Florii* salt springs, etc. The question is, therefore, why precisely this spring was exploited, when others seem to have been more accessible. The high salinity of the source, its location along the Cuejdiu valley and especially its central position, as well as the proximity of the sites of Piatra Neamţ seem to have justified its exploitation. Unlike the *Hălăbutoaia* salt spring, the modest operation of Gârcina and the effective control of the valleys of Cracău and Bistriţa from Cucuteni there does not appear to have been a need for more rigorous monitoring of this sector due to high density settlement in the adjacent area.

Conclusions

As illustrated by this research in progress, the study of a lost item such as salt needs manifold approaches. These approaches should focus on the different types of waste, on the fabrication techniques, on management and impact upon the natural environment, as well as on production and control methods. If salt represented a powerful economic phenomenon at certain periods, our task nowadays is to identify the attendant operations and social logistics. Far from being homogeneous, the production of salt, or rather of salt cakes, was probably an integral part of a broader process of intensification of social relations (Neolithization, transhumance, territorial appropriation, intercultural conflicts …) and exchange systems, in which control over salt resources, production and distribution of salt cakes probably became a major preoccupation.

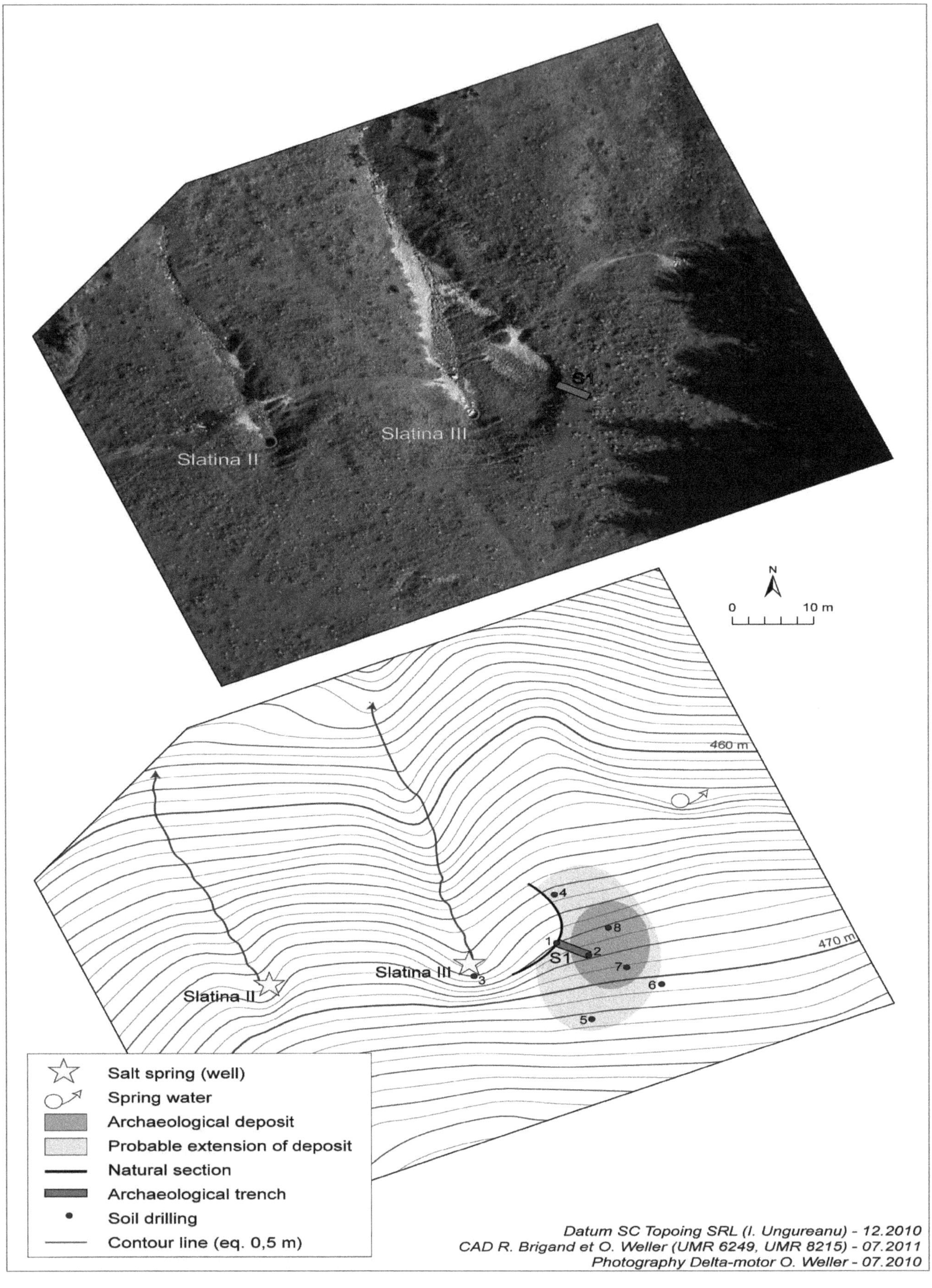

Figure 5. Plan of the archaeological area of *Slatina Cozla* II and III in Gârcina.

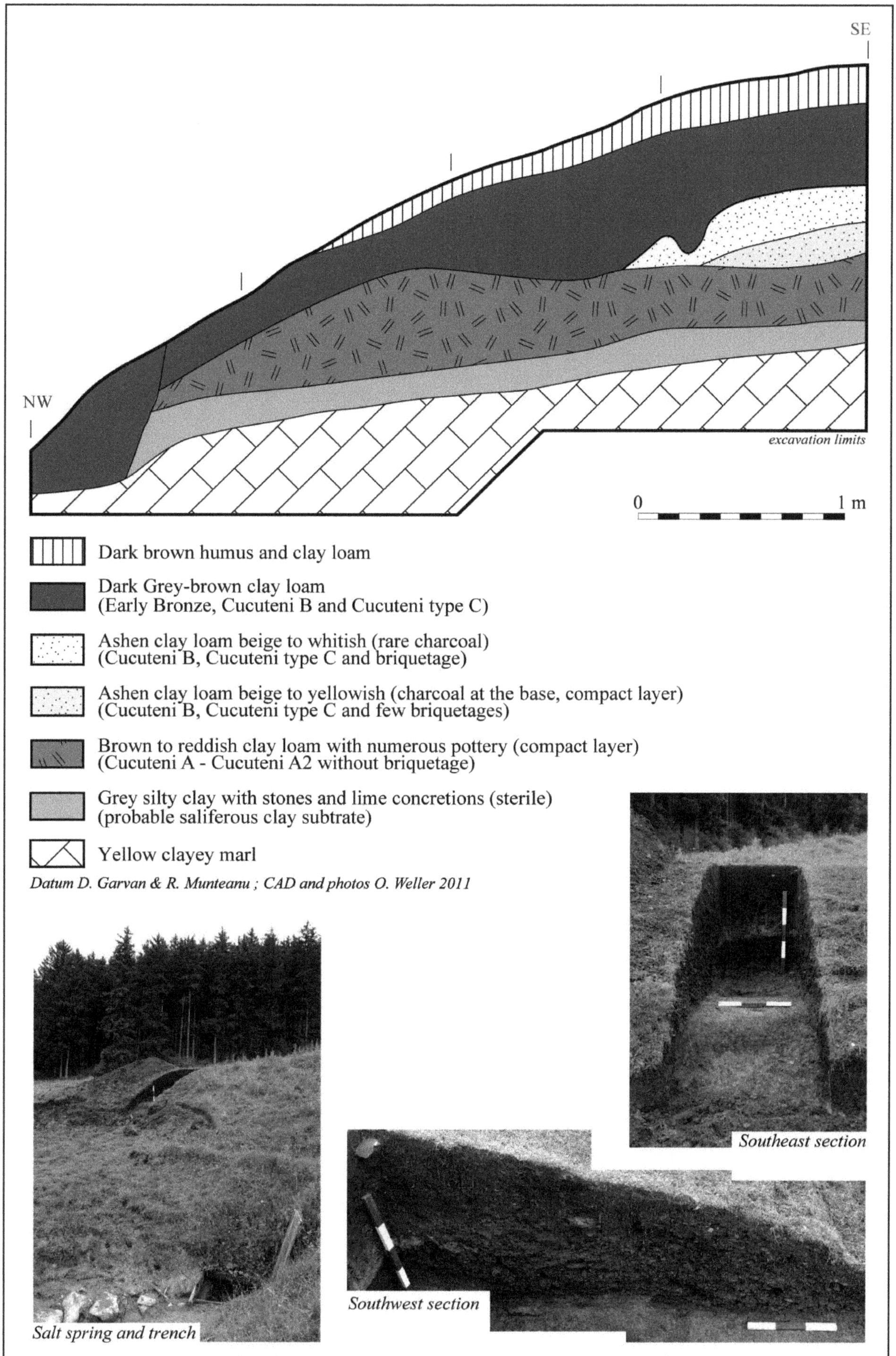

Figure 6. Cross section and SI drilling profiles (*Slatina Cozla III*, Gârcina).

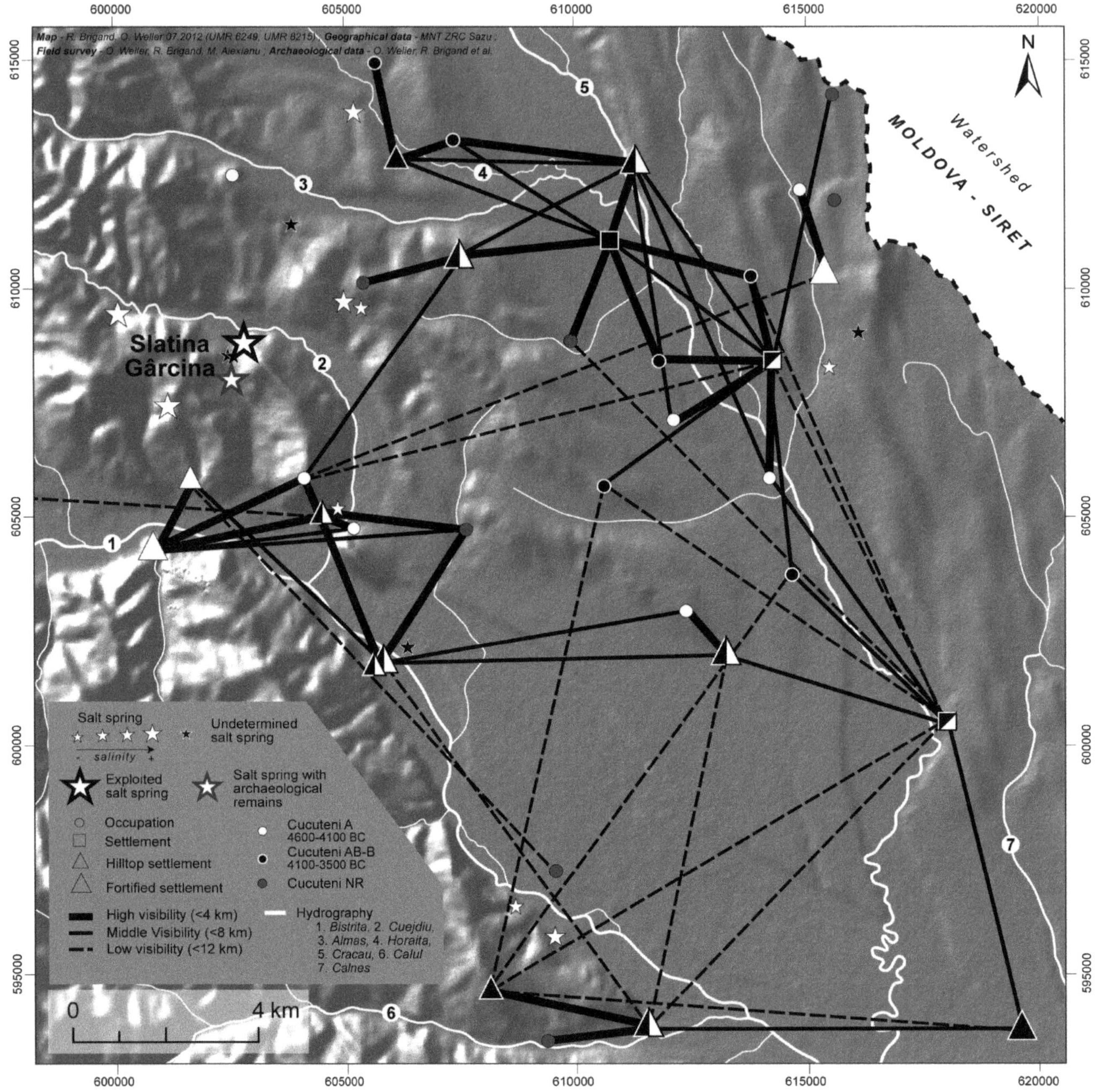

Figure 7. Salt springs and Cucuteni settlements within the Gârcina sector.

Acknowledgement

We would like to thank for the support of the University Panthéon-Sorbonne in Paris, the Commission for Foreign Excavations (French Ministry of Foreign Affairs, *Exploitations pré- et protohistoriques du sel en Roumanie orientale*, O. Weller), the National Centre of Scientific Research (CNRS, UMR 8215 *Trajectoires*) and the National Council for Scientific Research in Romania (project PN-II-ID-PCE-2011-3-0825, 219/5.10.2011, *The ethno-archaeology of the salt springs and salt mountains from the extra-Carpathian areas of Romania*).

References

Alexianu, M., Dumitroaia, Gh. and Monah, D. 1992. Exploatarea surselor de apă sărată din Moldova: o abordare etnoarheologică. *Thraco-Dacica* 13/1-2, 159-167.

Alexianu, M. and Weller, O. 2009. The *Ethnosal* project. Ethnoarchaeological investigation and the Moldavian salt springs. *Antiquity* 83 (321). Online: antiquity.ac.uk/projgall/weller321 [last accessed: 10.10.2014].

Alexianu, M., Weller, O., Brigand, R., Curcă, R.-G., Cotiugă, V., Moga, I. 2011. Salt springs in today's rural world. An ethnoarchaeological approach in Moldavia (Romania), in M. Alexianu, O. Weller and R.-G. Curca (eds.), *Archaeology and Anthropology of Salt: A Diachronic Approach*, 7–23. BAR International Series 2198. Oxford, BAR Publishing.

Brigand, R., Asăndulesei, A., Weller, O., Cotiugă, V. 2012. Contribution à l'étude du peuplement chalcolithique des bassins hydrographiques des Bahluiet et Valea Oii (dép. Iasi). *Dacia N.S.* LVI, 5-32.

Brigand, R. and Weller, O. 2012. Natural Resources and Settlements Dynamics during Later Prehistory in Central Moldavia (Romania). An Integrated GIS for Spatial Archaeological Studies, in V. Cotiuga and S. Caliniuc (eds.), *Interdisciplinary Research in Archaeology,* 1-18. BAR International Series 2433. Oxford, BAR Publishing.

Danu, M., Gauthier, E. and Weller, O. 2010. Human impact and vegetation history on salt spring exploitation (Hălăbutoaia – Ţolici, Petricani, Neamţ, Romania). *International Journal of Conservation Science* 1 (3), 167-173. Online: ijcs.uaic.ro/pub/IJCS-10-17-Danu.pdf [last accessed: 10.10.2014].

Dumitroaia, Gh. 1987. La station archéologique de Lunca-Poiana Slatinei, in M. Petrescu-Dîmbovita, N. Ursulescu, D. Monah and V. Chirica (eds.), *La civilisation de Cucuteni en contexte européen*, 253–258. Iaşi, Editura Universităţii „Alexandru Ioan Cuza".

Dumitroaia, Gh. 1994. Depunerile neo-eneolitice de la Lunca şi Oglinzi, judeţul Neamţ. *Memoria Antiquitatis* 19, 7–82.

Dumitroaia, Gh., Munteanu, R., Weller, O., Garvăn, D., Diaconu, V. and Brigand, R. 2008. Un nou punct de exploatare a apei sărate în preistorie: Ţolici-Hălăbutoaia, jud. Neamţ, in D. Monah, Gh. Dumitroaia and D. Garvăn (eds.), *Sarea. De la prezent la trecut*, 203–224. Piatra-Neamţ, Editura Constantin Matasă.

Marinescu-Bîlcu, S. 1981. *Tîrpeşti. From Prehistory to History in Eastern Romania.* Oxford, BAR Publishing.

Monah, D. and Dumitroaia, Gh. 2007. Recherches sur l'exploitation préhistorique du sel en Roumanie, in D. Monah, Gh. Dumitroaia, O. Weller and J. Chapman (eds.), *L'exploitation du sel à travers le temps*, 13-34. Biblioteca Memoria Antiquitatis, XVIII. Piatra Neamţ, Constantin Matasă.

Munteanu, R., Garvăn, D., Nicola, D., Preoteasa, C., Dumitroaia, Gh. 2007. Cucuieţi-Slatina Veche (Romania). Prehistoric Exploitation of a Salt Resource, in D. Monah, Gh. Dumitroaia, O. Weller and J. Chapman (eds.), *L'exploitation du sel à travers le temps.* Bibliotheca Memoriae Antiquitatis XVIII, 57-70. Piatra Neamţ, Constantin Matasă.

Munteanu, R. and Garvăn, D. 2011. The Cucuteni C Pottery near Moldavian Salt Springs, in M. Alexianu, O. Weller and R. Curcă (eds.), *Archaeology and Anthropology of Salt. A Diachronic Approach,* 81-87. Oxford, BAR Publishing.

Nicola, D., Munteanu, R., Garvăn, D., Preoteasa, C. and Dumitroaia, Gh. 2007. Solca – Slatina Mare (Roumanie). Preuves archéologiques de l'exploitation du sel en préhistoire, in D. Monah, Gh. Dumitroaia, O. Weller and J. Chapman (eds.), *L'exploitation du sel à travers le temps.* Bibliotheca Memoriae Antiquitatis XVIII, 35-56. Piatra Neamţ, Constantin Matasă.

Sandu, I., Weller, O., Stumbea, D. and Alexianu, M. 2012. Analyses archéométriques sur les moules à sel chalcolithiques de l'est de la Roumanie, in V. Nikolov and K. Bacvarov (eds.), *Salt and Gold: The Role of Salt in Prehistoric Europe*, 143-154. Provadia-Veliko Tarnovo, Faber.

Weller, O. and Dumitroaia, Gh. 2005. The earliest salt production in the world: an early Neolithic exploitation in Poiana Slatinei-Lunca, Romania. *Antiquity* 79 (306). Online: www.antiquity.ac.uk/ProjGall/weller [last accessed: 10.10.2014].

Weller, O., Brigand, R. and Alexianu, M. 2007. Cercetări sistematice asupra izvoarelor de apă sărată din Moldova. Bilanţul explorărilor din anii 2004-2007 efectuate în special în judeţul Neamţ. *Memoria Antiquitatis* 24, 121-190.

Weller, O., Dumitroaia, Gh., Sordoillet, D., Dufraisse, A., Gauthier, E. and Munteanu, R. 2008. Première exploitation de sel en Europe: Techniques et gestion de l'exploitation de la source salée de Poiana Slatinei à Lunca (Neamţ, Roumanie), in O. Weller, A. Dufraisse and P. Pétrequin (eds.) *Sel, eau et forêt: hier et aujourd'hui*, 205-230. Cahiers de la MSH Ledoux 12 (coll. Homme et environnement, 1). Besançon, Presses Universitaires de Franche-Comté.

Weller, O., Brigand, R. and Alexianu, M. 2010. Recherches systématiques autour des sources salées de Moldavie. Bilan des prospections 2008-2010. *Memoria Antiquitatis* 25–26, 437–504.

Weller, O., Brigand, R., Nuninger, L., Dumitroaia, Gh. 2011. Spatial analysis of prehistoric salt exploitation in Eastern Carpathians (Romania), in M. Alexianu, O. Weller and R. Curcă (eds.), *Archaeology and Anthropology of Salt. A Diachronic Approach,* 69–80. BAR International Series 2198. Oxford, BAR Publishing.

Weller, O., Brigand, R. and Alexianu, M. (in press, 2013). Recherches systématiques autour des sources salées de Moldavie. Bilan des prospections 2011-2012. *Memoria Antiquitatis.*

Salt Exploitation and Valorisation by the Human Communities of the Precucuteni – Cucuteni – Tripolye Cultural Complex

Constantin Preoteasa
Complexul Muzeal Judeţean Neamţ, Piatra-Neamţ

Abstract
Our study advances a number of considerations on the present state of the art of research on salt exploitation and valorisation by the human communities belonging to the Precucuteni-Cucuteni-Tripolye cultural complex. Within the research, we have studied a series of aspects related to the importance of this natural resource, reflected by the development of the stations situated near salt sources (seasonal settlements and exploitation points), dependent — to a certain extent — on the important permanent settlements (centres of regional polarisation). Furthermore, we have made several references regarding the cultural-chronological framing of the vestiges within the stations, related to salt exploitation and valorisation, as well as to the hypothetical need of such artefacts for the crystallisation process. At the same time, we have taken into account the presence of the artefacts specialised in salt crystallisation (such as briquetage), as well as that of the Cucuteni C pottery. Finally, we enumerated a series of experts' assessment regarding the salt intakes for certain settlements and zones belonging to this civilisation, as well as for certain so-called conventional settlements. Not least, we have proposed ourselves several assessments regarding the salt intake of the population based on our typology concerning the settlements (by surface, number of constructions and inhabitants), of the salt intake of the population within the settlements for which there are demographic assessments, as well as of the salt intake for the species of livestock (cattle, pigs, sheep and goats) within the settlements that benefit from exhaustive archaeozoological data. We mention that, in these cases, the figures are only hypothetical, and that they actually express the optimal salt intake for the population and for the livestock; they do not express the consumption per se, which may be different. Thus far, the results of the research in the field have illustrated once more the proper adjustment of the Precucuteni-Cucuteni-Tripolye human communities to the environmental conditions and the optimal valorisation of natural resources (in our case, of salt), which determined the long evolution of this civilisation, as well as their high degree of development.

Keywords
Chalcolithic, Precucuteni-Cucuteni-Tripolye, salt, seasonal settlements, exploitation points, briquetage, Cucuteni C, hypothetical assessments

1. Introduction

Halite is a chemical compound made of sodium cations (Na^+) and chloride anions (Cl^-) (Mitewa and Kolev 2012, 341, fig. 1); the first element comprises 40%, while the second around 60% (SACN 2003, 8).

Salt constitutes one of the consumption goods indispensable to man and animals, through its dietary and therapeutic role, considering it has been used since time immemorial as brine or in crystallised form, both in fresh diets and as main preservative for vegetal and animal products. At the same time, it represented one of the prestigious goods included within exchanges, through which one could obtain vegetal or animal products and various other raw materials or end products.

Considering the importance of salt for the human communities, including for those belonging to the cultural complex of Precucuteni-Cucuteni-Tripolye, the past couple of years have naturally recorded an intensification of the research (with interdisciplinary valences) on its exploitation and valorisation throughout time. The outcomes were presented during exhibitions and within publications, through prestigious research programmes, scientific reunions, and specialised works.

However, the current stage of the archaeological research in the field is not as advanced as in other fields related to the civilisation in question. Practically, thus far only one large-scale systematic investigation was carried out, on the station of Lunca-*Poiana Slatinei* (Precucuteni II and III, Cucuteni A and B). The other seasonal settlements and salt exploitation points have only benefitted from small-scale investigations and field research.

2. Sites situated in the vicinity of salt sources

For a better valorisation of natural resources, the Precucuteni-Cucuteni-Tripolye communities within the permanent settlements consolidated a series of stations throughout their millenary evolution. These stations were situated near important sources of raw materials (salt, clay, flint), which were probably directly controlled by dominant settlements, which could have represented genuine centres of regional polarisation, such as those within the *tell* of Poduri-*Dealul Ghindaru* (Precucuteni II and III, Cucuteni A and B).

In regard to the exploitation of salt, a distinction must be made between seasonal settlements and exploitation points. Furthermore, those for which there is more consistent information — situated, for the most, in the Moldavian Subcarpathians, well known for its salt sources (Stoica and Gherasie 1981, 42-60, fig. 9) — will be presented in this study.

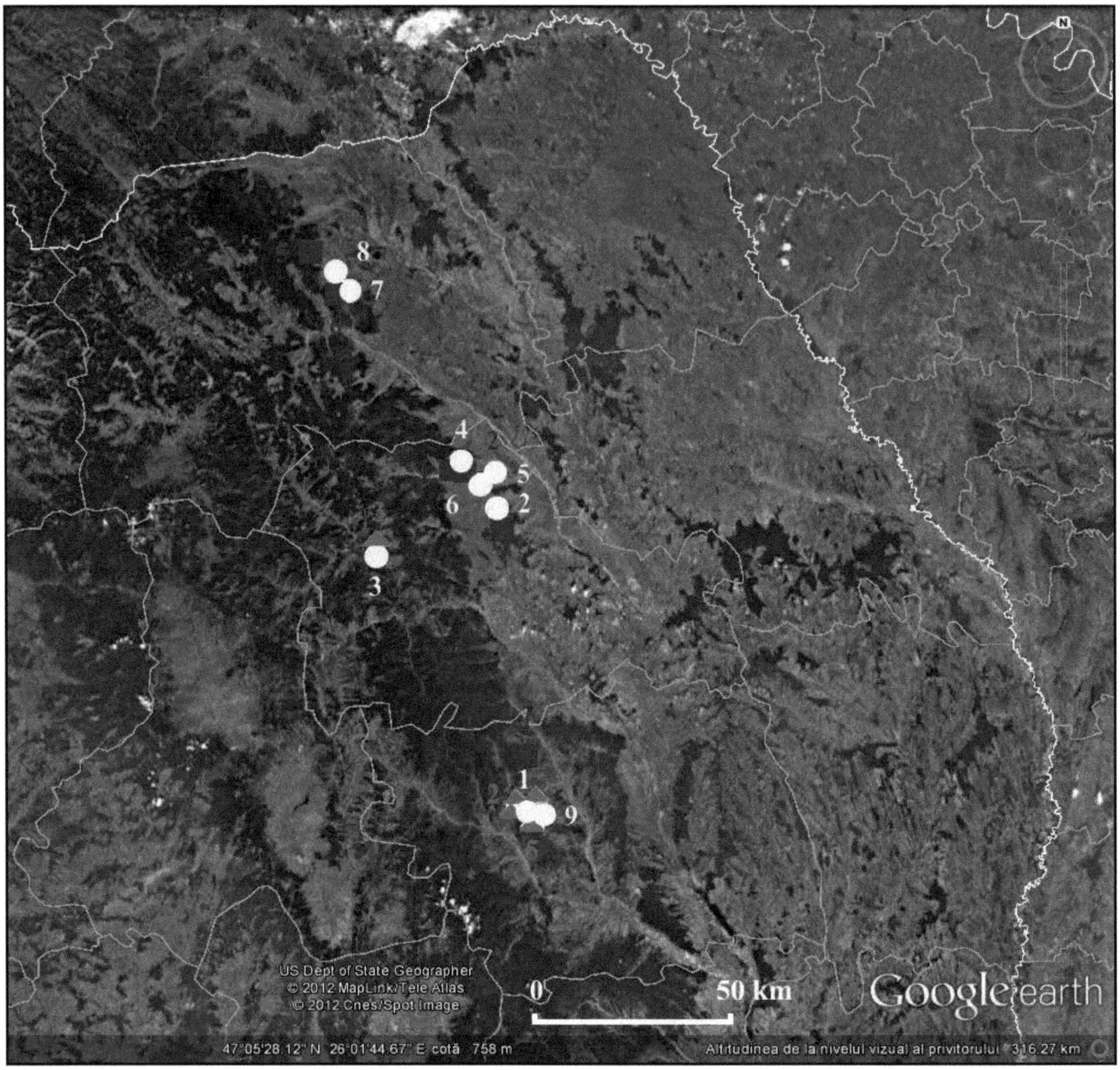

Figure 1. Sites related to salt exploitation and valorisation: permanent settlements ("yellow circle"): ***1 – Poduri-"Dealul Ghindaru" (Precucuteni II and III, Cucuteni A and B); 2 – Târpeşti-"Râpa lui Bodai" (Precucuteni II and III, Cucuteni A); 3 – Hangu-"Chiriţeni" (Cucuteni A); 4 – Oglinzi-"Cetăţuia" (Cucuteni A); 5 – Răuceşti-"Munteni" (Cucuteni A and B); 6 – Târgu Neamţ-"Dealul Pometea" (Cucuteni B); 7 – Vârvata-"Cetăţica" (Cucuteni B); 8 – Poieni-"Părţi" (Cucuteni B); 9 – Valea Şoşii-"Dealul Bujoarei" (Cucuteni B); seasonal settlements ("red triangle"): 1 – Poduri-"Pe Hăineală" (Cucuteni A); 2 – Poduri-"În Fundătură" (Cucuteni A); 3 – Hangu-"Piciorul Bocăncenilor" (Cucuteni A); 4 – Prohozeşti-"Silişte" (Cucuteni B); 5 – Valea Şoşii-Dealul Păltiniş (Cucuteni B); salt exploitation points ("blue square"): 1 – Lunca-"Poiana Slatinei" (Precucuteni II and III, Cucuteni A and B); 2 – Oglinzi-"Băi" (Precucuteni II); 3 – Ţolici-"Hălăbutoaia" (Precucuteni II, Cucuteni A and B); 4 – Cucuieţi-"Slatina Veche" (Precucuteni III, Cucuteni A); 5 – Solca-"Slatina Mare" (Precucuteni III, Cucuteni A and B); 6 – Cacica-"Salina" (Cucuteni B).***

In fact, the sites presented here are located in the western extremity of the area of this civilisation (in the Eastern Carpathians and the Moldavian sub-Carpathians), within three main areas. The first is the Poduri zone (Bacău County), where there are two stations with permanent settlements and four stations with seasonal settlements. The second is the Târgu Neamţ zone (Neamţ County), where there are four stations with permanent settlements and three exploitation points. Finally, the third is the Cacica-Solca-Soloneţ zone (Suceava County), where there are two stations with permanent settlements and two exploitation points. Besides them, it is worth mentioning the Hangu zone (Neamţ County), where there is a permanent settlement and a seasonal settlement, as well as the – apparently isolated – exploitation point from the Solonţ zone (Bacău County) (Figure 1).

2.1. Seasonal settlements

The seasonal settlements comprise those stations situated generally several hundred metres away from the salt sources, characterised by the existence of less consistent anthropogenic depositions than those within the permanent settlements, among which one may even find habitation complexes (Alexianu *et al.* 1992, 164; 2007, 291).

Among them, we mention those of Hangu-*Piciorul Bocăncenilor* (Cucuteni A) (Monah *et al.* 2008, 99; 2011, 31), Poduri – *În Fundătură* (Cucuteni A) (Monah *et al.* 2003, 70, 72) and *Pe Hăineală* (Cucuteni A) (Monah, 1991 394; Alexianu *et al.* 1992, 164; 2007, 281, 291; Monah *et al.* 2003, 70, 72) –, Prohozeşti-*Silişte* (Cucuteni B) (Monah 1991, 394; Alexianu *et al.* 1992, 164; 2007, 281, 291; Monah *et al.* 2003, 70, 71; Chapman and Monah 2007, 71-88; Monah and Dumitroaia 2007, 16, fig. 2) and Valea Şoşii-*Dealul Păltiniş* (Cucuteni B) (Monah *et al.* 2003, 70, 73; Monah and Dumitroaia 2007, 16, fig. 2) (Figure 1).

2.2. Exploitation points

The exploitation points include those stations placed in the immediate vicinity of salt sources, with anthropogenic depositions lacking habitation complexes, which occupy small (sometimes even discontinuous – *author's note*) surfaces, with a more or less important consistency in vertical plane (Alexianu *et al.* 1992, 163; 2007, 289, 290).

Among them, we mention those of Lunca-*Poiana Slatinei*[1] (Precucuteni II and III, Cucuteni A and B) (Dumitroaia 1987a, 255, 256, fig. 1/a; 1994a, 8-11, 16, 50-70, fig. 1-3, 9, 40-52; Alexianu *et al.* 1992, 163; 2007, 281, 291; Monah and Dumitroaia 2007, 18, 19, fig. 4; Monah 2008, 20, fig. 4), Oglinzi-*Băi* (Precucuteni II) (Dumitroaia 1994a, 76, 77, fig. 58), Ţolici-*Hălăbutoaia* (Precucuteni II, Cucuteni A and B) (Dumitroaia *et al.* 2008, 205, 207, 209-218, fig. 1, 2, 4-12; Monah 2008, 28, 29, fig. 7), Cucuieţi-*Slatina Veche* (Precucuteni III and Cucuteni A) (Dumitroaia 2004, 435, 436; Monah and Dumitroaia 2007, 24, fig. 11; Munteanu *et al.* 2007, 58, 59, 62-65, 69, fig. 1-3, 7-9; Monah 2008, 23, fig. 5; Garvăn 2008, 110-122, fig. 1-10), Solca-*Slatina Mare* (Precucuteni III, Cucuteni A and B) (Ursulescu 1977, 311, 313, fig. 6-11; Monah 1991, 392, 393; Dumitroaia 2004, 444-447; Monah and Dumitroaia 2007, 15, 23, fig. 10; Nicola *et al.* 2007, 36-39, 41-55, fig. 1, 2, 6-19; Monah 2008, 14-16, fig. 1) and Cacica-*Salina* (Cucuteni B) (Andronic 1989, 172-176, fig. 1-4; 2009, 19-21, 29, fig. 4; Monah and Dumitroaia 2007, 23) (Figure 1).

In a brief analysis published not long ago, despite the archaeological evidence, (that practically indicates the absence of habitation complexes within all of these stations), the author mentions them as seasonal settlements / stations (Garvăn 2008, 110-114, fig. 1).

3. Cultural-chronological framing of the vestiges

The rather limited archaeological research conducted thus far within the seasonal settlements and the salt exploitation points allow the framing of the vestiges discovered in only a few of the evolution periods of the Precucuteni and Cucuteni cultures (Table 1). In some cases, caution is required concerning any attribution, considering that the determinable pottery is scarce, fragmentary and corroded; furthermore, there are no absolute chronological dates. In other situations, as visible in the following lines, the provenance of the vestiges is contestable.

3.1. Chronology of seasonal settlements

The current stage of the research indicates that the seasonal settlements were constructed only by the Cucuteni human communities, and not by those belonging to Precucuteni, too, and that they evolved along just one phase of the culture (Cucuteni A or Cucuteni B) (Table 1).

3.1.1. Hangu-"Piciorul Bocăncenilor"

The seasonal settlement of Hangu-*Piciorul Bocăncenilor* was founded by the nearby community of the permanent settlement at Hangu-*Chiriţeni*, during Cucuteni A (Monah *et al.* 2008, 99; 2011, 31); both were situated in the mountainous area, far away from the closest Cucuteni settlements. Unfortunately, the outcomes of the investigations conducted in the two stations (Nicolăescu-Plopşor and Petrescu-Dîmboviţa 1959, 51-54, pl. IV, fig. 1, 2; Nicolăescu-Plopşor *et al.* 1959, 63-66, fig. 5, 6; Petrescu-Dîmboviţa and Teodor 2003, 83-86, 127-137, 420-429, fig. 75-85) do not allow including them within one of the Cucuteni A evolution stages. In this sense, it is worth mentioning that John Nandriş considered the settlement of Hangu-*Chiriţeni* a seasonal settlement (Nandriş 1987, 209), given that (even during the archaeological excavations), there were mentions of ... *seasonal dwellings (cabins)* (Nicolăescu-Plopşor *et al.* 1961, 42).

3.1.2. Poduri – "În Fundătură" and "Pe Hăineală"

The seasonal settlements of Poduri-*În Fundătură* (Monah *et al.* 2003, 72) and Poduri-*Pe Hăineală* (Monah *et al.* 1987, 14; Monah 1991, 394; Alexianu *et al.* 1992, 164; 2007, 281, 291; Monah *et al.* 2003, 72), situated close to each other, are also included within Cucuteni A, but the specific stage could not be determined. Taking into account the small distance between the stations and the *tell* of Poduri-*Dealul Ghindaru*, we believe that they could have been built by the human communities corresponding to the permanent settlements of the local Cucuteni A phase.

The outcomes of the archaeological research conducted in the *tell* throughout the time, corroborated with calibrated and reliable radiocarbon (^{14}C) dates, indicate the existence of four probable permanent settlements of the Cucuteni A_2 stage, which lasted about 300 years, between 4450 and 4150 BC (Monah *et al.* 2003, 38; Bem 2001, 342; 2002, 36; Preoteasa 2011, 62). Under these circumstances, we believe that the two seasonal settlements near the permanent ones (situated in the *tell*) also belong to the Cucuteni A_2 stage, and that they evolved during the aforementioned interval. However, in the absence of reliable radiocarbon (^{14}C) dates obtained on certified samples, from the seasonal settlements, we cannot determine whether they were contemporary or along which of the permanent settlements on the *tell* were founded.

[1] The salt exploitation point of Lunca is often mentioned in the literature as *Poiana Slatinii* or *Poiana Slatinei.*

Table 1. Cultural-chronological framing of the vestiges within the seasonal settlements and of salt exploitation points ("shadow"), according to the current stage of the archaeological research.

SITES	PRECUCUTENI			CUCUTENI						
				CUCUTENI A			CUCUTENI A-B		CUCUTENI B	
	PRECUCUTENI I	PRECUCUTENI II	PRECUCUTENI III	CUCUTENI A_1	CUCUTENI A_2	CUCUTENI A_3	CUCUTENI A-B_1	CUCUTENI A-B_2	CUCUTENI B_1	CUCUTENI B_2
SEASONAL SETTLEMENTS										
Hangu *Piciorul Bocăncenilor*				■	■	■				
Poduri *Pe Hăineală*					■					
Poduri *În Fundătură*					■					
Prohozești *Siliște*									■	
Valea Șoșii *Dealul Păltiniș*										■
EXPLOITATION POINTS										
Lunca *Poiana Slatinei*		■	■			■			■	
Oglinzi *Băi*		■								
Țolici *Hălăbutoaia*		■				■			■	
Cucuieți *Slatina Veche*			■			■				
Solca *Slatina Mare*			■			■				■
Cacica *Salina*									■	

3.1.3. Prohozești-"Siliște"

The seasonal settlement of Prohozești-*Siliște* was included by the specialists within Cucuteni B, with no mention on the particular stage (Popovici and Trohani 1984, 78; Monah 1991, 394; Monah *et al.* 2003, 69, 71; Chapman and Monah 2007, 74; Monah and Dumitroaia 2007, 16). It was probably built by the human communities corresponding to the permanent settlements of the Cucuteni B phase, in the *tell* of Poduri-*Dealul Ghindaru*, near which it was situated (Monah *et al.* 1980, 99; Alexianu *et al.* 1992, 164; 2007, 291; Monah *et al.* 2003, 71). The outcomes of the archaeological research conducted thus far within the *tell* illustrate the existence (in this place) of permanent settlements of the Cucuteni B_1 stage, for which – based on the only calibrated radiocarbon (^{14}C) dates – we have proposed an evolution interval of about 200 years, between 3850 and 3650 BC (Preoteasa 2011, 60, 63). Under these circumstances, all cautions considered, we ascribe the seasonal settlement of Prohozești-*Siliște* to the Cucuteni B_1 stage, without being able to determine – in the absence of an absolutely reliable chronology date resulted after analysing a single sample from it – along which of the permanent settlements in the *tell* of Poduri-*Dealul Ghindaru* it was founded.

3.1.4. Valea Șoșii-"Dealul Păltiniș"

From the very beginning, the authors of the research ascribed the seasonal settlement of Valea Șoșii-*Dealul Păltiniș* to the Cucuteni B_2 stage (Popovici and Bujor 1984, 60, 62; Monah *et al.* 2003, 73). Not far from it, there is the station of Valea Șoșii-*Dealul Bujoarei*, within which vestiges belonging to permanent settlements datable to Cucuteni A and B phases were identified (Monah *et al.* 2003, 72, 73). Under these circumstances, we could believe that the seasonal settlement on *Dealul Păltiniș* was built by the human communities of the permanent settlement on *Dealul Bujoarei*, both belonging to the Cucuteni B phase. However, this association must also be cautious, considering that – for the moment – there is no determinable pottery available to indicate that the permanent settlement belongs to the Cucuteni B_2 stage, as the seasonal one. In this case, we mention that, within the *tell* station of Poduri-*Dealul Ghindaru*– situated in the same area – three permanent settlements belonging to the Cucuteni B_1 stage were identified. The Cucuteni B_2 stage is represented only by sporadic pottery, but there are no complexes to delimit between distinct levels of habitation (Monah *et al.* 1982, 16; Monah 1987, 72; Mantu 1998, 64; Monah *et al.* 2003, 39, 41; Petrescu-Dîmbovița and Văleanu 2004, 272; Preoteasa

2011, 60). Hence, no association can be made with the seasonal settlement of Valea Șoșii-*Dealul Păltiniș*.

Thus, the current state of the research allows the dating of seasonal settlements – with a certain degree of probability, of course – only to the Cucuteni A_2, Cucuteni B_1 and Cucuteni B_2 stages. Thus far, at least, there are no seasonal settlements attested for Precucuteni I, Precucuteni II and Precucuteni III; the same goes for the Cucuteni A_1, Cucuteni A_3, Cucuteni A-B_1 and Cucuteni A-B_2 stages (Table 1).

3.2. Chronology of exploitation points

The archaeological research conducted here over time reveals a chronological reality different from that of seasonal settlements, in the sense that the exploitation points contain both Precucuteni and Cucuteni vestiges, and – for some of them – several evolution periods of the two cultures are attested (Table 1).

3.2.1. Lunca-"Poiana Slatinei"

It represents the most important and spectacular station concerning salt exploitation and valorisation; hence, it has been best researched and valorised through publications and exhibitions. The results of the research conducted here throughout time indicate that this is, in fact, the oldest known brine exploitation in the world. It dates to the beginning of the Neolithic on these lands (Starčevo-Criș culture, IIIB-IVA stage), towards 6050 BC, according to the existing calibrated radiocarbon (^{14}C) series of dates \ (Dumitroaia 1987a, 254, 257, 258, fig. 2, 3; 1994a, 11-50, fig. 4-39; Monah 1991, 393; 2008, 14, 19; Dumitroaia *et al.* 2004, 96, 97; Weller and Dumitroaia 2005, fig. 6; Monah and Dumitroaia 2007, 18, 19, fig. 5; Weller *et al.* 2007, 167; Weller *et al.* 2007, 110, fig. 8; Monah 2008, 14; Weller *et al.* 2008; 2010, 23, 33, fig. 4; Cavruc 2009, 21; Alexianu *et al.* 2011, 8).

Concerning the Precucuteni-Cucuteni-Tripolye cultural complex, we encounter here the oldest known vestiges, dating to the Precucuteni II phase, followed by those of the Precucuteni III phase. Furthermore, there are attestations for the intense use of the salt water source for the Cucuteni A and Cucuteni B phases, too (Dumitroaia 1987a, 255; 1994a, 50-70, fig. 40-52; Monah and Dumitroaia 2007, 18, 19). We underline again that, among the pottery shards discovered at Lunca-*Poiana Slatinei*, there are sporadic ones, too, without constituting a distinct deposition level; their decorations are also encountered in the Precucuteni I phase (Dumitroaia 1994a, 51-54, 56, 57, fig. 40; 41/4, 6; 43/2, 4; 44/2), but they are in fact a tradition conserved in the Precucuteni II phase. The same goes for some pottery shards painted in styles also encountered in the Cucuteni A-B_2 stage (Dumitroaia 1994a, 62, 63, fig. 48/1, 2, 4), and in the Cucuteni B_1 stage, together with the style specific to the latter. This association is a commonplace in the permanent settlements, but it allows the dating of the said vestiges in the first stage of the last evolution phase in the Cucuteni culture.

In regard to the decorated pottery of the Cucuteni A phase, it can be noticed that it is painted exclusively with three colours (white, red, black), but not with two colours, which means that the depositions are part of the Cucuteni A_3 stage, according to the classical division into periods of the Cucuteni culture proposed by Vladimir Dumitrescu, unanimously accepted (Dumitrescu 1963a, 69). At the same time, the Cucuteni B pottery, with a preponderant ε style painting with two colours (even associated with the one painted in γ style, but not with the one painted in ζ style, with three colours), imposes the inclusion of the said anthropogenic depositions in the Cucuteni B_1 stage, according to the same division into periods, elaborated by the aforementioned renowned specialist (Dumitrescu 1963b, 291).

Concerning the permanent settlements which exploited and valorised the salt source in the point of Lunca-*Poiana Slatinei*, in the current stage of the research, considering the distance and the cultural-chronological framing of the vestiges, we ascribe to the Cucuteni A_3 stage, first of all, the permanent settlements of Oglinzi-*Cetățuia*, dated during Cucuteni A, but we cannot specify the stage (Dumitroaia 1994a, 72, 75). Then, we attribute to the same stage Răucești-*Munteni*, with vestiges even since Cucuteni A_3 (Dumitroaia 1987b, 21). The Cucuteni B_1 comprises the settlements of Târgu Neamț-*Dealul Pometea*, as we have already mentioned on another occasion (Preoteasa *et al.* 2007, 324), because the painted pottery shards collected mainly during the recent field research seem to match this period[2]. For the same period, we mention Răucești-*Munteni*, with a consistent deposition dated to the last evolution phase of the culture (Dumitroaia 1987b, 21). For the Precucuteni II and Precucuteni III phases, no permanent settlements have been identified thus far in the immediate vicinity of the exploitation point of Lunca-*Poiana Slatinei*; some distance away, the settlements of Târpești-*Râpa lui Bodai* – dated as mentioned above – are placed (Marinescu-Bîlcu 1974, 18, 168, 169; 1981, 24, 82).

3.2.2. Oglinzi-"Băi"

The archaeological research conducted in this point allows the framing of the anthropogenic depositions only in the Precucuteni II phase (Dumitroaia 1994a, 76, 77, fig. 58/a).

This time, too, the relation with a permanent settlement regards the one of Târpești-*Râpa lui Bodai* (Marinescu-Bîlcu 1974, 18, 168, 169; 1981, 24, 82).

3.2.3. Ţolici-"Hălăbutoaia"

It represents another salt exploitation point where several anthropogenic levels belonging to the cultural complex of Precucuteni-Cucuteni-Tripolye have been identified. Among them, the earliest belong to the Precucuteni II phase, followed by those of the Cucuteni A and Cucuteni B phases (Dumitroaia *et al.* 2008, 206,

[2] This was kindly transmitted to us by Vasile Diaconu (Museum of History and Ethnography Târgu Neamț), the author of the research, to whom we express our gratitude on this occasion.

207, 209-218, fig. 2, 4, 6, 8, 9, 12; Monah 2008, 28, 29, fig. 7).

In this case, too, sporadic pottery shards with a similar decoration to those of the Precucuteni I phase were identified, but they do not form a distinct deposition level and they are, in fact, traditions preserved in the Precucuteni II phase (Dumitroaia *et al.* 2008, 218). Considering the two-colour painted decoration (in ε style), the vestiges of the Cucuteni B phase can be included with certainty in the Cucuteni B_1 stage, according to the division into periods proposed by Vladimir Dumitrescu (Dumitrescu 1963b, 291).

At the same time, among other finds in the station, it is worth mentioning pottery shards decorated ... *with two colours or with three colours, with thin ribbons of black, red and white, which can be ascribed to both the* [Cucuteni] *A phase and the* [Cucuteni] *A-B phase* (Dumitroaia *et al.* 2008, 214, 215, fig. 9). It is also worth noting pottery shards from ... *vessels with very flattened body, a cylindrical neck and a grooving at its base* (Dumitroaia *et al.* 2008, 214) or a pottery shard on which ... *two parallel rows made of "x" letters are painted in black* (Dumitroaia *et al.* 2008, 214). The last are considered ... *by shape and decoration ... specific to the Cucuteni A-B phase* (Dumitroaia *et al.* 2008, 214). Hence, ... *given the presence of the aforementioned fabrics in the same level*, the fact that the proportion ... *of the fabrics of the* [Cucuteni] *A-B phase in relation to those in the* [Cucuteni] *A phase* is not known, as well as ... *other reasons* (unspecified – *author's note*), the research authors place the said level ... *somewhere at the end of the* [Cucuteni] *A and the beginning of the Cucuteni A-B phase* (Dumitroaia *et al.* 2008, 214). Furthermore, it is noted that ... *the closest analogies for the pottery of the* [Cucuteni] *A phase* (of Ţolici-*Hălăbutoaia* – *author's note*) *are those with Târpeşti-Râpa lui Bodai* (Dumitroaia *et al.* 2008, 214).

In regard to the aforementioned opinions, expressed by the research authors of the salt exploitation point of Ţolici-*Hălăbutoaia*, we believe that it is worth making several observations. Obviously, it is impossible to ascribe the same artefacts – irrespective of their nature – to both the Cucuteni A and the Cucuteni A-B phase. The pottery shards of the Cucuteni A phase in this place are painted with three colours, but not with two colours, too, reason for which they have to be included, together with their level, in the Cucuteni A_3 stage, according to the division into periods by Vladimir Dumitrescu (Dumitrescu 1963a, 69). For the research authors to date the vestiges, it is unnatural to associate the pottery shards with three-colour painting specific to the Cucuteni A phase (Cucuteni A_3 stage) to those with γ style painting (which appeared during Cucuteni A-B_2 and continued to the Cucuteni B_1 stage, a commonplace actually for the permanent settlements). It is equally wrong to associate them with shards of vessel shapes which were also found for Cucuteni A-B (Cucuteni A-B_2 stage), because between Cucuteni A_3 and Cucuteni A-B_2, there is Cucuteni A-B_1 (characterised by pottery painted in the α and β style, without any evidence at Ţolici-*Hălăbutoaia*). According to the calibrated (with 1 σ) and reliable radiocarbon (^{14}C) dates available for the Cucuteni culture, the Cucuteni A_3 stage would have had an approximate 300-year evolution and it can be placed in the interval 4350-4050 BC (Dumitrescu 1974, 31, 38; Monah 1987, 76, 79, fig. 1; László 1997, 263; Mantu 1998, 247, 255, Tab. 7, 8; Bem 2001, 350, 356, Tab. 2, 7; 2002, 70, 76, 77, Tab. 1, 2; Ursulescu and Ignătescu 2003, 107; Suciu and Luca 2007; Lazarovici 2010, 73, 81, 82, fig. 4, Tab. 1). Furthermore, the Cucuteni A-B_2 stage ends towards 3800 BC (Mantu 1998, 250, 256, Tab. 7, 8; Bem 2002, 47, 71, 78, Tab. 1, 2; 2007, 118; Suciu and Luca 2007; Lazarovici 2010, 73, 82, fig. 5, Tab. 1). Hence, the end of the Cucuteni A phase can be placed around 4050 BC, while that of the Cucuteni A-B phase towards 3800 BC, which means that the entire Cucuteni A-B phase would have evolved for around 250 years, in the interval 4050-3800 BC. Under these circumstances, the Cucuteni A-B_2 stage may not have lasted for more than half of this 250-year period, while the time interval elapsed since the end of the Cucuteni A_3 stage until the beginning of the Cucuteni A-B_2 stage exceeded 100 years, corresponding to the evolution of the Cucuteni A-B_1 stage. Therefore, within this station, as well as within any other that included such Cucuteni vestiges (from the Cucuteni A_3 and from the Cucuteni A-B_2 stage, respectively), it is impossible to date the anthropogenic level – regardless of the reasons invoked – to the end of the Cucuteni A phase and the beginning of the Cucuteni A-B phase. In this case, the pottery shards – whose vessel shape and painting (γ style) remind of the Cucuteni A-B phase (Cucuteni A-B_2 stage) – had to be associated to those painted in ε style, specific to the Cucuteni B phase (to the Cucuteni B_1 stage), thus dated to the first stage of the last evolution phase in the Cucuteni culture. Considering the chronological realities of Ţolici-*Hălăbutoaia*, the analogy with the station of Târpeşti-*Râpa lui Bodai* is not justified, because the Cucuteni A vestiges within the salt exploitation point (with three-colour painting) date to the Cucuteni A_3 stage, while those within the permanent settlements referred to (with two-colour and with three-colour painting) certainly belong to the Cucuteni A_2 stage and probably to the Cucuteni A_1 stage, too (Marinescu-Bîlcu 1981, 82; Popovici 2003, 314-320). Hence, the anthropogenic deposition of the Cucuteni A phase (Cucuteni A_3 stage) of Ţolici-*Hălăbutoaia* is more recent than the anthropogenic depositions of the Cucuteni A phase (Cucuteni A_2 stage and probably Cucuteni A_1 stage) at Târpeşti-*Râpa lui Bodai*.

Regarding the permanent settlements that exploited salt in the point of Ţolici-*Hălăbutoaia*, for the Precucuteni II phase, we associate them first of all to the nearby one with similar vestiges, of Târpeşti-*Râpa lui Bodai* (Marinescu-Bîlcu 1974, 18, 168, 169; 1981, 24, 82). For the Cucuteni A_3 and Cucuteni B_1 stages, the association is made with the contemporary ones of Răuceşti-*Munteni* (Dumitroaia 1987b, 21), situated to a greater distance. At the same time, for the Cucuteni B_1 stage, we also mention the possible relation with the permanent settlement of Târgu Neamţ-*Dealul Pometea*.

3.2.4. Cucuieți-"Slatina Veche"

The archaeological research conducted here has led, among others, to the discovery of certain vestiges dated phase Precucuteni III and Cucuteni A phase, respectively (Monah and Dumitroaia 2007, 24; Munteanu *et al.* 2007, 62-65, 69, fig. 7, 8; Monah 2008, 23; Garvăn 2008, 110-122, fig. 1-10).

In the last case, though the painted pottery is exclusively with three colours, not with two colours, too, the anthropogenic deposition to which it belonged was included in the Cucuteni A_2 stage by the research authors (Munteanu *et al.* 2007, 63, 64, 69, fig. 8), not in the Cucuteni A_3 stage, according to the classical division into periods of the Cucuteni culture, elaborated by Vladimir Dumitrescu (Dumitrescu 1963a, 69) and subsequently accepted by most specialists. Furthermore, we mention that the assumptions regarding the presence within this salt exploitation point of certain artefacts belonging to the Cucuteni B phase (Munteanu *et al.* 2007, 63) are not confirmed.

Unfortunately, in the current stage of the research, it is virtually impossible to identify the permanent settlements that exploited salt in the Precucuteni III phase and in the Cucuteni A_3 stage in the point of Cucuieți-*Slatina Veche* because, for the moment, in the area no stations with vestiges from the aforementioned periods have been discovered.

3.2.5. Solca-"Slatina Mare"

It represents the first archaeologically investigated salt exploitation point on the current Romanian territory. The investigation began after the place was mentioned, around the half of the past century, by the geographer Ion Șandru, who identified here an important anthropogenic deposition with vestiges initially determined as belonging to the Neo-Eneolithic cultures of Criș, Precucuteni and Cucuteni (phase B, in association with Cucuteni C pottery). It was also determined to have belonged to the Hallstatt, probably La Tène, as well as to the Romanian ancient and migration periods (Șandru 1952, 414; Ursulescu 1977, 310; Monah and Dumitroaia 2007, 14; Nicola *et al.* 2007, 35). From the very beginning, the discoverer realised that the area was not suitable for agriculture, and the presence of the vestiges was related to the exploitation of brine (Șandru 1952, 417; Ursulescu 1977, 310, 311; Monah and Dumitroaia 2007, 14; Monah 2008, 15). In this case, we are talking mostly about natural brine, though the human communities of the aforementioned periods ... *also had here all the conditions to industrialise brine, such as the necessary wood and vessels* (Șandru 1952, 417; Ursulescu 1977, 310, 311).

Afterwards, in 1968, the station was researched by Mircea Ignat and Nicolae Ursulescu (Ursulescu 1977, 311, n. 26; Monah and Dumitroaia 2007, 14; Nicola *et al.* 2007, 36; Monah 2008, 15, 16). The outcomes of the research were published a decade later (Ursulescu 1977, 307-317). On this occasion, mentions were made on the discovery of archaeological fabrics belonging to the Neo-Eneolithic, and to the cultures Criș, Precucuteni and Cucuteni (phase B, in association with Cucuteni C), to Hallstatt and to the medieval period (17th century) (Ursulescu 1977, 311; Monah 1991, 392, 393; Monah and Dumitroaia 2007, 15; Nicola *et al.* 2007, 36).

The most recent of the investigations of Solca-*Slatina Mare* have eventually highlighted the existence, in this salt exploitation point, of anthropogenic depositions dating to the Precucuteni III phase, to the Cucuteni A and Cucuteni B phases, to the Bronze Age and to the medieval period (14th century and 17th-18th centuries, respectively). (Nicola *et al.* 2007, 39-55, fig. 4, 5, 8-20) Initially, there was a mention that no vestiges older than those of the Cucuteni B phase had been discovered on the said occasion; there were also mentions of late medieval vestiges (17th-18th centuries) (Dumitroaia 2004, 445, 447).

As far as we are concerned, we will focus on the outcomes of the archaeological investigations conducted at Solca-*Slatina Mare* in 2003 – to which we have directly participated, as a member of the research collective. The outcomes were published in 2007 within the volume reuniting the acts of the international colloquium *Pre- and protohistorical archaeology of salt*, organised by the History and Archaeology Museum of Piatra-Neamț in the period 21-24 October 2004 (Nicola *et al.* 2007, 35-56).

Thus, the oldest vestiges belonging to the cultural complex of Precucuteni-Cucuteni-Tripolye found in this salt exploitation are those of the Precucuteni III phase (Nicola *et al.* 2007, 43, 44, fig. 8).

Concerning the determinable pottery of the Cucuteni A phase – scarce, fragmentary and corroded –, it was painted in three colours (red, white and black), reason for which it dates Cucuteni A_3, according to the aforementioned division into periods elaborated by Vladimir Dumitrescu (Dumitrescu 1963a, 69). Hence, it does not belong to Cucuteni A_2, as stated in the study illustrating the outcomes of the archaeological research of 2003 (Nicola *et al.* 2007, 43, 44, fig. 9). The painted pottery shards of the Cucuteni B are mainly decorated in the ε style (Nicola *et al.* 2007, 45, fig. 10/1-6, 8). The exception is represented by two artefacts probably belonging to bitronconical vessels with fine fabric (burnt in an oxidising atmosphere, evenly, thus tile-red), with decorative motifs painted in three colours, in ζ style, disposed on registers and metopes. One of the pottery shards in question probably comes from a medium-sized bitronconical vessel, with thin walls, decorated with parallel brown vertical lines and asymmetrical red horizontal lines, on a white-yellowish background (Nicola *et al.* 2007, 45, fig. 1/7). The second shard probably belonged to a large-sized bitronconical vessel, with thicker walls and flaring rim, decorated on the inside with a brown scalloped ribbon, and on the outside covered in white and bordered (in the neck area) by a brown circular ribbon; on the body, there are traces of red painting. Such an association of the painted pottery decoration (between the ε and ζ painting style) determines us to include the anthropogenic deposition in the final cultural phase of the Cucuteni B_2 stage, according to the division into periods of Vladimir Dumitrescu (Dumitrescu 1963b, 291).

In the current stage of the research, it is practically impossible to identify the permanent settlements which exploited the point of Solca-*Slatina Mare* in the Precucuteni III phase and in the Cucuteni A_3 stage, respectively, considering that no such stations have been discovered in the area. For the Cucuteni B_2 stage, we could be cautiously tempted to believe that the exploitation of salt in this point may have been ensured by the human communities within the nearby permanent settlement, of Poieni-*Părţi*. However, the discoverer only ascribed it to the Cucuteni culture, without specifying any phase or stage, considering the degree of fragmentation and corrosion of the archaeological vestiges discovered here (Andronic 1997, 71, no. 3.6.127). We ascribe this permanent settlement to the last phase of the Cucuteni culture, based on the fact that, in that area (when it was possible to state the evolution period), mentions were made only of Cucuteni B vestiges.

3.2.6. Cacica-"Salina"

The archaeological research conducted over time within this salt exploitation point led to the discovery of vestiges datable exclusively Cucuteni B (Andronic 1989, 171-176, fig. 1-4; 1997, 65, 66, no. 3.5.114; 2009, 19-21, 29, fig. 4; Monah and Dumitroaia 2007, 23; Monah 2008, 21).

Taking into account the few determinable shards published thus far, painted in ε style (Andronic 2009, 29, fig. 4/6, 8), we ascribe the anthropogenic deposition here to the Cucuteni B_1 stage, in agreement with the culture division into periods proposed by Vladimir Dumitrescu (Dumitrescu 1963b, 291).

In regard to the permanent settlement that exploited salt in the point of Cacica-*Salina*, we considered, first, the closest and the most important one in the area, meaning that of Vârvata-*Cetăţica*. However, it has been ascribed to Cucuteni B, without any details on the stage, considering the high degree of fragmentation and corrosion of its artefacts (Andronic 1997, 59, no. 3.4.94; Niculică *et al.* 1999, 42, 43, 47, 48, 58, no. 14, fig. II/1; III/8; XIII/1).

Hence, the current stage of the research permit to date the salt exploitation points – with a certain degree of probability, of course – only Precucuteni II and Precucuteni III phases, Cucuteni A_3, Cucuteni B_1 and Cucuteni B_2 stages, respectively. Thus far, at least, no salt exploitation points belonging to Precucuteni I phase and to Cucuteni A_1, Cucuteni A_2, Cucuteni A-B_1, and Cucuteni A-B_2 stages have been discovered (Table 1).

We also have to mention the alleged existence within the exploitation points of vestiges belonging to the Cucuteni A_2 stage; this categorisation was done by the authors of the finds based on the decoration painted with three colours, present at some shards, but none with two colours was discovered. The piece of information was taken over by other specialists, too. However, as we have already pinpointed, according to the culture division into periods Cucuteni elaborated by Vladimir Dumitrescu, this stage is characterised by the association between the pottery painted with two colours and with three colours (Dumitrescu 1963a, 69). The pottery painted exclusively with two colours is characteristic to the Cucuteni A_1 stage, while the pottery painted exclusively with three colours is characteristic to the Cucuteni A_3 stage (Dumitrescu 1963a, 69).

For an accurate as possible cultural-chronological framing of the anthropogenic depositions within an evolution phase / stage of the Cucuteni culture, more extended and minute (interdisciplinary) research within the stations would be required. The purpose of such endeavour would be to correlate the relative chronology and the absolute chronology dates. Thus, associations could be made between the pottery decorations and the potential reliable radiocarbon (^{14}C) dates, calibrated if possible. Only one of the two types of chronology (relative or absolute) may be insufficient sometimes for an exact dating of the vestiges.

In regard to the relative chronology, it is well-known that a field research or a survey could pinpoint only one category of painted pottery decoration (with two colours or with three colours), not necessarily the co-existence within the anthropogenic depositions of both, too. Thus far, there is no way of knowing for sure whether for the Cucuteni A phase, the two-colour pottery has not been discovered yet alongside the one with three colours or, on the contrary, the two-colour pottery – more fragile and corrosion-sensitive, implicitly – was not preserved on the shards. This could have happened because of the precarious soil conditions, which only allowed the conservation of the decoration with three colours, more enduring. Furthermore, a characteristic of the anthropogenic depositions within the seasonal settlements and mostly of exploitation points is the high amount of common use Precucuteni and Cucuteni pottery. On the contrary, the fine, decorated pottery is very scarce and it belongs to a very small number of vessels.

In regard to the absolute chronology, based on the existing calibrated (with 1 σ) and reliable radiocarbon (^{14}C) dates, the Cucuteni A_2 stage had an approximate 350-year evolution, between 4500 and 4150 BC (Monah 1978, 37, 40; 1987, 72, 73, 75, 76, 78, fig. 1; László 1997, 262, 263; 2009, 208-215, Tab. 1; Mantu 1998, 246, 247, 255, Tab. 7, 8; Bem 2001, 342, 349, 350, 356, Tab. 2, 7; 2002, 35, 36, 69, 70, 74, 75, 76, fig. 3, Tab. 1, 2; Monah *et al.* 2003, 63; Petrescu-Dîmboviţa and Văleanu 2004, 271, 273; Suciu and Luca 2007; Lazarovici 2010, 72, 80, 81, Tab. 1, fig. 2, 3; Preoteasa 2011, 61, 62, Tab. 1). Moreover, the Cucuteni A_3 stage had an approximate 300-year evolution, between 4350 and 4050 BC (Dumitrescu 1974, 31, 38; Monah 1987, 76, 79, fig. 1; László 1997, 263; Mantu 1998, 247, 255, Tab. 7, 8; Bem

2001, 350, 356, Tab. 2, 7; 2002, 70, 76, 77, Tab. 1, 2; Ursulescu and Ignătescu 2003, 107; Suciu and Luca 2007; Lazarovici 2010, 73, 81, 82, fig. 4, Tab. 1). Hence, in this case, there is a common 200-year time interval for the two periods – 4350-4150 BC – from the beginning of the Cucuteni A_3 stage to the end of the Cucuteni A_2 stage. Under these circumstances, a date within the common time interval could be problematic concerning the temporal framing of the respective anthropogenic deposition within one period or the other of the Cucuteni culture.

In regard to the cultural-chronological provenance of the vestiges within the seasonal settlements and of salt exploitation points, we pinpoint that our idea is not to suggest that they could be included only within the evolution periods of the Precucuteni-Cucuteni-Tripolye cultural complex highlighted by the outcomes of the archaeological research conducted thus far. On the contrary, we are persuaded that salt was exploited and valorised by the human communities belonging to this civilisation throughout all its evolution periods. In our opinion, these periods concern a time interval of approximately 1,500 years, between 5000 and 3500 BC, according to the calibrated (with 1 σ) and reliable radiocarbon (^{14}C) dates (Dumitrescu 1964, 28, 30-34, 39, Tab. 1; Monah 1987, 74-79, fig. 1, 2; László 1997, 262-265; Mantu 1998, 246-253, 255, 256, Tab. 7, 8; Bem 2001, 341-344, 349-352, 356, 357, Tab. 2, 7, 8; 2002, 69-82, Tab. 1, 2; Suciu and Luca 2007; Lazarovici 2010, 72-74, 80-88, Tab. 1, 2, fig. 1-6). The realities illustrated in the present study reflect only the current stage of the archaeological research. It could be significantly improved in the future, through large-scale complex interdisciplinary research programmes, within the stations related to salt exploitation and valorisation.

Also related to the cultural-chronological aspect, we believe that a permanent settlement could have exploited not just one, but even several salt sources concomitantly. We also believe that a salt-water spring could have been exploited along the same evolution period (phase or stage) of the Precucuteni-Cucuteni-Tripolye cultural complex not only by one, but also by several permanent settlements, though probably not at the same time. In this sense, we take into account that many of the salt springs valorised then have been exploitable to the present day, that the evolution periods of this civilisation generally lasted several hundred years – as shown by the existing absolute chronology dates. We also consider the fact that a permanent settlement has an average 50-year lifespan (Monah 1978, 36-38; Markevič 1981, 10; Kruc 1989, 127; Preoteasa 2011, 63).

4. The briquetages

Within some of the sites related to the exploitation of salt on the current territory of Romania, which include Cucuteni vestiges, a series of vessels have been discovered over time, which are not part of the usual vessels belonging to this civilisation; they are called briquetages. The special interest manifested by the specialists for this type of artefacts has often led to fierce controversies, especially concerning their cultural-chronological framing.

4.1. Definition

From the beginning, we mention that we will use the term *briquetage* (a French term) in a restricted sense, in order to designate those rather small tapered vessels with pedestal, which constitute artefacts specialised for the salt crystallisation process (Ursulescu 1977, 313; Monah 2008, 16; Cavruc 2009, 24, n. 1; Cavruc and Dumitroaia 2006, 38; Monah and Dumitroaia 2007, 15; Nicola *et al.* 2007, 37). They are generally very coarse vessels; they have uneven and secondary burns, with yellowish-reddish to greyish-blackish hues. The composition of the fabric probably made them more enduring thermally and over time. The reconstitution of such a container illustrated that it had a volume of around 1.5 l (Andronic 1989, 173, 175, 176, fig. 4/1; 2009, 21, 27, fig. 2/1; Cavruc and Dumitroaia 2006, 25).

In the Occidental specialised literature, the term *briquetage* has a broader meaning, as it represents the totality of earthenware artefacts (vessels, columns, tubs, pots, bowls, holders, fasteners, etc) and of shards from fire facilities (hearths, furnaces) used for the crystallisation of salt (Ursulescu 1977, 307-309, fig. 2; Cavruc 2009, 24, n. 1).

4.2. Functionality

Whereas initially the foreign specialists interpreted the briquetages as possible lamps or cult items, they were subsequently reconsidered as vessels or parts of complex devices related to the salt crystallisation process (Ursulescu 1977, 309). The briquetages in question (the tapered vessels with pedestal) were used in order to obtain salt bars, which constituted genuine prestigious goods destined to more important distance trades (Monah 1991, 390; 2008, 17, 33; Monah and Dumitroaia 2007, 24). The human communities situated near salt sources used directly the brine or the *huscă*.

The important amount of briquetage shards (belonging mostly to the lower side) within the exploitation points, as well as the absence of such whole items, are due to the fact that the vessel had to be broken in order to take the salt bar out (Cavruc and Dumitroaia 2006, 24; Cavruc 2009, 24). The crystallisation process probably comprised two phases. During the first phase, the brine was boiled in common use vessels (the rests of which are abundant within the exploitation points) until obtaining a semi-fluid substance called *huscă*, which was subsequently put in vessels and gradually turned into salt bars, at high temperature (Cavruc and Dumitroaia 2006, 38; Cavruc 2009, 24). The hypothesis according to which briquetages were used to boil directly the brine to obtain salt bars, and that they were refilled as the evaporation process unfolded (Andronic 1989, 172, 176) is yet to be confirmed, even experimentally (Bodi 2007, 93-95, fig. 5-7).

In relation to the last archaeological aspect, we do not find plausible the hypothesis of the salt crystallisation by pouring brine on wood and then setting it on fire, as proven by the experiment of Cucuteni (Bodi 2007, 91-93, fig. 1-4). In this case, the amount of crystallised salt obtained from a significant amount of brine was really small and full of impurities (salt mixed with cinder and charcoal), to which one could add great efforts and a rather long duration of the entire process. It was significantly more effective to boil a smaller amount of brine directly in the vessels; the amount of *huscă* obtained was obviously higher and purer, not to mention the minimal effort and duration, as shown in the experiment of Cucuieți (Munteanu *et al.* 2007, 69, fig. 15).

4.3. Context of the finds

The current stage of the archaeological research in the field shows that the stations where the briquetages were discovered are situated in the areas with salt springs, but not also in those with salt outcrops, which proves that they were used to turn brine into crystallised salt (Cavruc 2009, 26).

The most numerous such items (always fragmentary) have been identified within the brine exploitation points. Thus far, at least, no such artefacts have been discovered within the seasonal settlements near salt springs. However, only a few such items have been identified within the permanent settlements, which probably controlled the salt sources.

4.3.1. Permanent settlements

Only a few briquetage shards (some of which reconstituted) have been identified within the permanent settlements that controlled the salt sources, and the communities that used them partially created the anthropogenic depositions within the exploitation points. Such is the case for the permanent settlements of Răucești-*Munteni* (Cucuteni A and B) (Dumitroaia 1994a, 66; Cavruc and Chiricescu 2006, 202, no. 47, 48), which probably exploited the sources of Lunca-*Poiana Slatinei* and Țolici-*Hălăbutoaia*, or Negritești-*Movila Flocoasă* (Cucuteni A and B), which exploited the salt source of Negritești-*Slatina*, situated 2-3 km away, northwards (Dumitroaia 1994a, 66; 1994b, 517, 518, no. 11; Weller *et al.* 2007, 156, 157).

The presence of briquetages within the permanent settlements can provide clues regarding the Cucuteni human communities which valorised the salt water sources in the vicinity, marked by the anthropogenic depositions within the exploitation points where such items were discovered. These vessels, like the ones common for the Cucuteni repertory (some of which decorated), discovered only in shards within the exploitation points, were definitely created within the settlements, from where they were brought near salt sources in order to exploit and valorise it. Our opinion is based on the outcomes of the archaeological research conducted thus far, which failed to find vestiges (clay extraction holes or ware burning holes, clay flakes, temper, small clay rings, rotating modelling devices, rests of furnaces, pigments for painting, specific tools necessary to make pottery) able to show that the pottery within the exploitation points was made in that spot. Under these circumstances, we may pinpoint that the stations with anthropogenic depositions but no habitation complexes near salt sources represented specialised points for the salt crystallisation process exclusively.

The alleged briquetage find in one of the permanent settlements within the *tell* of Poduri-*Dealul Ghindaru* (Precucuteni II and III, Cucuteni A and B) (Dumitroaia 1994a, 66) is yet to be confirmed. Concerning the finds of Păuleni-Ciomortan - *Dâmbul Cetății* (Cucuteni A) (Cavruc and Dumitroaia 2006, 39; Cavruc and Chiricescu 2006, 203, no. 51, 52), Olteni-*Cariera de nisip* (Cucuteni A) (Cavruc and Chiricescu 2006, 202, no. 49, 50), and Oglinzi-*Cetățuia* (Cucuteni A and B) (Dumitroaia 1994a, 66, 72), we believe that the items are, in fact, not briquetage.

4.3.2. Exploitation points

Within the exploitation points, the first briquetages were discovered at Solca-*Slatina Mare* (Precucuteni III, Cucuteni A and B) and ascribed by Nicolae Ursulescu to the Criș culture (Ursulescu 1977, 313; 1996, 492; Andronic; 1989, 172; 2009, 20; Monah 1991, 393; 2008, 16; Dumitroaia 2004, 444; Nicola *et al.* 2007, 36, 37). Actually, this opinion has been reiterated recently by the renowned specialist (Ursulescu 2011, 1). However, as we have already mentioned in this study, the most recent archaeological research conducted within this salt exploitation point failed to confirm the previous outcomes, considering that no anthropogenic depositions belonging to Starčevo-Criș were discovered. As for the Precucuteni-Cucuteni-Tripolye cultural complex, only vestiges of the Precucuteni III phase, of the Cucuteni A phase (Cucuteni A_3 stage – *author's note*) and of the Cucuteni B phase (Cucuteni B_2 stage – *author's note*) (Nicola *et al.* 2007, 39-55, fig. 8-19) were found. Following subsequent archaeological excavations, several similar items were discovered within the salt exploitation points of Lunca-*Poiana Slatinei* (Precucuteni II and III, Cucuteni A and B) (Dumitroaia 1987a, 255; 1994a, 62-66, fig. 49, 50; Monah and Dumitroaia 2007, 19, 20; Monah 2008, 20), Țolici-*Hălăbutoaia* (Precucuteni II, Cucuteni A and B) (Dumitroaia *et al.* 2008, 212, 216, 217, fig. 7, 11; Monah 2008, 29) and Cacica-*Salina* (Cucuteni B) (Andronic 1989, 172-176, fig. 1-3; 4/1; 2009, 19-21, 27, 29, fig. 2/1-6, 19, 20, 22, 23, 25, 26; 4/2, 3, 7, 10-12; Monah and Dumitroaia 2007, 23). The same cannot be said of Oglinzi-*Băi* (Precucuteni II) (Monah and Dumitroaia 2007, 21) or Cucuieți-*Slatina Veche* (Precucuteni III and Cucuteni A) (Monah and Dumitroaia 2007, 24; Munteanu *et al.* 2007, 63; Monah 2008, 23; Garvăn 2008, 110-122, fig. 1-10).

At the same time, it is worth highlighting the discovery of briquetage shards tempered with crushed shells, specific to the so-called Cucuteni C pottery – within the salt exploitation point of Solca-*Slatina Mare* (Precucuteni III, Cucuteni A and B). They could have been the result of certain experiments elaborated by the Cucuteni potter who made them (Nicola *et al.* 2007, 47;

Munteanu and Garvăn 2008, 195, 196, fig. 8; 2011, 84, 86, fig. 6).
Actually, the briquetages, just as the artefacts specialised in the salt crystallisation process, are probably the outcome of long experiments of the Cucuteni people.

4.4. Cultural-chronological framing

As we have already mentioned, the cultural-chronological framing of briquetages has been the source of fierce disputes among the specialists; furthermore, the opinions expressed over time have been anything but unitary.

4.4.1. Solca-"Slatina Mare"

The first such artefacts were discovered at Solca-*Slatina Mare* (Precucuteni III, Cucuteni A and B) and ascribed by Nicolae Ursulescu to the Criş culture (Starčevo-Criş), considering them ... *the oldest briquetage pottery in the entire Europe* and ... *not more recent than the second half of the* 5th *millennium BC* (Ursulescu 1977, 313; 1996, 492; Andronic 1989, 172; 2009, 20; Monah 1991, 393; 2008, 16; Dumitroaia 2004, 444; Nicola *et al.* 2007, 36, 37; Cotoi 2009, 8). The researcher pinpointed the same idea recently (Ursulescu 2011, 1). The new research conducted within this salt exploitation point allowed a reconsideration, and the vestiges were included in Cucuteni B (Dumitroaia 2004, 445; Monah and Dumitroaia 2007, 24; Nicola *et al.* 47, 51).

4.4.2. Lunca-"Poiana Slatinei"

Following the first investigations at Lunca-*Poiana Slatinei* (Precucuteni II and III, Cucuteni A and B), the briquetages identified were initially ascribed to Cucuteni A (Dumitroaia 1987, 255). Subsequently, as the results became more consistent, such items discovered not only here, but near salt sources, too (concerning the permanent settlements – Răuceşti-*Munteni* (Cucuteni A and B) or Negriteşti-*Movila Flocoasă* (Cucuteni A and B), and the exploitation points – Solca-*Slatina Mare* (Precucuteni III, Cucuteni A and B) or Cacica-*Salina* (Cucuteni B) were ascribed to the Cucuteni culture in general, without additional mentions of a specific phase or stage in its evolution (Dumitroaia 1994a, 66; 2004, 444; Monah and Dumitroaia 2007, 19, 21; Monah 2008, 17, 20; Cotoi 2009, 8). On a relatively recent date, these artefacts of Lunca-*Poiana Slatinei* (Precucuteni II and III, Cucuteni A and B) have been included again in Cucuteni A (Cavruc and Chiricescu 2006, 200, 201, no. 42-45).

4.4.3. Cacica-"Salina"

The archaeological research conducted at Cacica-*Salina* (Cucuteni B) confirmed the presence of briquetages during Cucuteni B (Andronic 1989, 172-176; 1996; 65, no. 3.5.114 2009, 19-21; Monah and Dumitroaia 2007, 23), more precisely in the Cucuteni B_1 stage (Cavruc and Chiricescu 2006, 197, 198, no. 27-31).

4.4.4. Ţolici-"Hălăbutoaia"

Such artefacts were also identified at Ţolici-*Hălăbutoaia* (Precucuteni II, Cucuteni A and B), and they were ascribed to all three stages of the culture, practically (A, A-B and B) (Dumitroaia *et al.* 2008, 212, 216). The cultural-chronological aspects of the anthropogenic depositions belonging to the cultural complex of Precucuteni-Cucuteni-Tripolye of this salt exploitation point have already been discussed, reason for which we will not detail the discussion any further.

4.4.5. Răuceşti-"Munteni"

Such items within the permanent settlement of Răuceşti-*Munteni* (Cucuteni A and B) were initially included in the Cucuteni culture generally, without mentioning an evolution phase (Dumitroaia 1994a, 66); subsequently, they were ascribed to Cucuteni A (Cavruc and Chiricescu 2006, 201, 202, no. 46-48).

4.4.6. Negriteşti-"Movila Flocoasă"

The briquetages at Negriteşti-*Movila Flocoasă* (Cucuteni A and B) were also included in the Cucuteni culture, but no other details were provided on the evolution phase or stage (Dumitroaia 1994a, 66).

Referring to briquetage in general, Dan Monah and Gheorghe Dumitroaia stated, rather recently, that it was an invention of the human communities that lived in Cucuteni B (Monah and Dumitroaia 2007, 22). Later, Dan Monah mentioned that they had appeared at the end of the Cucuteni A phase (Monah 2008, 17; 2012, 139) and that they were intensely used in Cucuteni B (Monah 2008, 17). Other specialists admit the existence of this type of items throughout all the three Cucuteni phases (A, A-B and B) (Garvăn 2008, 112; Sandu *et al.* 2012, 143). Another perspective pinpointed in the specialised literature is that the briquetage should be associated to the presence of the so-called Cucuteni C pottery during Cucuteni B (Cotoi 2009, 10).

As it can be seen, according to the current stage of the research conducted in the field, the briquetages were discovered only in certain permanent settlements situated near salt springs – Răuceşti-*Munteni* (Cucuteni A and B) or Negriteşti-*Movila Flocoasă* (Cucuteni A and B). Furthermore, the discoveries were made mostly in some salt exploitation points – Lunca-*Poiana Slatinei* (Precucuteni II and III, Cucuteni A and B), Ţolici-*Hălăbutoaia* (Precucuteni II, Cucuteni A and B), Solca-*Slatina Mare* (Precucuteni III, Cucuteni A and B), or Cacica-*Salina* (Cucuteni B), but not within the seasonal settlements with Cucuteni vestiges. Moreover, these artefacts specialised in salt crystallisation were found only within the permanent settlements and salt exploitation points that comprise anthropogenic depositions belonging to Cucuteni B phase. It is worth noting that the salt exploitation point of Cucuieţi-*Slatina Veche* comprises vestiges of the Cucuteni A phase (Cucuteni A_3 stage), but none since Cucuteni B, the briquetages are absent. Such items were discovered in the salt exploitation point of Cacica-*Salina*, that point also comprises vestiges of the Cucuteni B phase (Cucuteni B_1 stage). Under these circumstances, at least for now, from a cultural-chronological perspective, we can include the briquetages exclusively in the Cucuteni B phase, not in Cucuteni A and Cucuteni A-B phases (Table 2).

Table 2. Cultural-chronological framing of the vestiges within permanent settlements and salt exploitation points ("shadow"), including those with briquetages ("B"), according to the current stage of the archaeological research.

SITES	PRECUCUTENI			CUCUTENI						
				CUCUTENI A			CUCUTENI A-B		CUCUTENI B	
	PRECUCUTENI I	PRECUCUTENI II	PRECUCUTENI III	CUCUTENI A_1	CUCUTENI A_2	CUCUTENI A_3	CUCUTENI A-B_1	CUCUTENI A-B_2	CUCUTENI B_1	CUCUTENI B_2
PERMANENT SETTLEMENTS										
Răucești *Munteni*						■			B	
Negritești *Movila Flocoasă*						■			B	
EXPLOITATION POINTS										
Lunca *Poiana Slatinei*		■	■			■			B	
Oglinzi *Băi*		■								
Țolici *Hălăbutoaia*		■				■			B	
Cucuieți *Slatina Veche*			■			■				
Solca *Slatina Mare*			■			■				B
Cacica *Salina*									B	

5. Cucuteni C pottery

This category of pottery is distinct from the classical Cucuteni repertory. It is generally a rare find in the permanent and seasonal settlements, but a commonplace in the salt exploitation points.

5.1. Definition

In the specialised literature of the Cucuteni C pottery, two categories of vessels are included (by the composition of the fabric): the vessels tempered with crushed shells and the vessels tempered with sand, crushed stones or glass. However, from a typological-formal and decorative perspective, the first and the second categories are rather similar. In both cases, the vessels are made of very coarse fabric, and the burning manner (including the secondary one) ranges their hues from yellowish-reddish to greyish-blackish. In the few situations (with an exclusive provenance from settlements) where a reconstitution was possible, the conclusions were the following: the shape is tapered or spherical, while the decoration striated, incised, imprinted, or honeycombed. Unfortunately, the Cucuteni C pottery from salt exploitation points is fragmentary and corroded, reason for which no reconstitution has been possible.

5.2. Provenance

The obvious differences in regard to the fabric composition, to burning, to shape and to decorations between the Cucuteni C pottery and the classical Cucuteni pottery made the specialists ascribe to the former an origin distinct from the Cucuteni cultural setting.

Concerning the Cucuteni-Tripolye area, the specialists' opinions over time have regarded the northern origin of the Cucuteni C pottery (Schmidt 1932, 42-45; Nițu *et al.* 1971, 67; Marinescu-Bîlcu 1981, 88, n. 23; Crîșmaru 1977, 61, 62; Dodd-Oprițescu 1980, 554; 1982, 79). Other experts considered a southern origin generally related to the Srednij Stog culture (Dumitrescu 1963b, 286-289; Danilenko 1974, 71; Dragomir 1982, 426; Cucoș 1985, 63; Rassamakin 1999, 102; Marinescu-Bîlcu and Bolomey 2000, 108-110; Sorochin 2002, 180-184; Tsvek and Rassamakin 2005, 175; Alaiba 2006, 194; Palaguta 2007, 67-72). Actually, the last perspective, which we agree with, seems to be currently accepted by most researchers.

However, some of them admit that at a certain point, determined sometime at the beginning of the Cucuteni B phase (thus in the Cucuteni B_1 stage – *author's note*), a "cucutenisation" process of the Cucuteni C pottery began. Hence, this type of pottery would become part of the Cucuteni pottery, by giving up on the tempering with crushed shells and by taking over the Cucuteni shapes and decorations (Nestor and Zaharia 1968, 27; Dodd-Oprițescu 1980, 549, 550; Dragomir 1982, 427; Cotoi 2007, 155).

Others do not agree with the "Cucutenisation" idea, but they consider the Cucuteni C pottery ... *one of the categories of the Cucuteni pottery*, through they admit that ... *it may represent an initial "technological import"* ..., but subsequently ... *the Cucuteni people adopted and integrated it in their own repertory* (Munteanu and Garvăn 2008, 195; 2011, 85).

In regard to the "cucutenisation" process, we agree that it may have occurred for the vessels not tempered with crushed shells, but only concerning the composition of the fabric. However, the shapes are relatively similar to the vessels tempered with crushed shells, which preserved much of their originality. Concerning the ornamentation, only the application of the protomas on the vessel walls may bring to mind to a certain extent the classical Cucuteni decorations, with crushed shells. The potters of the Cucuteni environment probably elaborated the vessels without shells, sometimes also decorated with zoomorphic protomas, per se, but we tend to consider the ones tempered with crushed shells as "imports" from the native cultural environment (Srednij Stog).

It is also worth pointing out that the specialised literature has already discussed the idea of a possible exchange that would have included various goods from the part of the Cucuteni population and women from the part of steppe shepherds. This could justify the presence of the Cucuteni C pottery in the Cucuteni cultural setting (Monah 1991, 397, 398; 2008, 21). On the other hand, we consider improbable the hypothesis that important groups of steppe shepherds would have crossed almost the entire Cucuteni area in order to exploit and valorise its significant western salt resources, even alongside the local Cucuteni communities, within some sort of temporary ... *condominium* ... on the salt sources (Cotoi 2007, 159). In our opinion, there is no way of justifying in this manner the significant presence of the Cucuteni C pottery within the salt exploitation points.

5.3. Functionality

The functionality of the Cucuteni C pottery is probably the crucial factor in order to determine its significant presence within the salt exploitation points, unlike within the (permanent or seasonal) settlements, where it is only scarce.

The composition of the fabric and mostly the presence crushed shells as temper seems to have made the vessels more resistant to physical and thermal shocks. Moreover, the porosity of the fabric favoured the water evaporation process, hence the crystallisation of salt, and of turning the *brine* into *huscă* (Rowlett and Show 2005, 165; Munteanu and Garvăn 2008, 194; 2011, 85; Cavruc 2009, 25).

Also in regard to the functional aspect of the Cucuteni C pottery, we mention that within the exploitation points it is destined exclusively to salt crystallisation (hence to common use). However, in the settlements per se, at least partially, it can also be included in the category of prestigious pottery (not due to the artistic qualities, much inferior to the classical Cucuteni pottery, but probably precisely due to its scarcity). This is illustrated by the ritual depositions of Drăguşeni (Cucuteni A), where a vessel deposited in a hole constituted the foundation rite of a construction (Marinescu-Bîlcu and Bolomey 2000, 38, fig. 113/1; 151/1; Frînculeasa 2006, 36; Lazarovici 2009, 226). It is also illustrated by the depositions at Traian (Cucuteni A-B) where, in similar contexts, shards of the same pottery species were discovered (Dumitrescu 1960, 100, 101; Frînculeasa 2006, 36; Lazarovici 2009, 227, 228).

5.4. Context of the finds

The archaeological research conducted thus far in the Cucuteni area illustrates that the Cucuteni C pottery is present in all the three categories of stations related to the valorisation of salt (permanent settlements, seasonal settlements, and exploitation points). In the following lines, we will only analyse those mentioned in the present study (Figure 1).

Concerning both the nomination and the percentage, this pottery species is much more present within the salt exploitation points, where it reaches 10-40% of the entire existing pottery (Dumitroaia 1994a, 58; Nicola *et al.* 2007, 47; Dumitroaia *et al.* 2008, 210; Munteanu and Garvăn 2008, 180, 181; 2011, 81, 83; Monah 2012, 139). At the same time, within the settlements (generally the permanent ones, but the same probably goes for the seasonal ones), its percentage is estimated to only 1% (Monah 2012, 139).

5.4.1. Permanent settlements

In the current stage of the research, the Cucuteni C pottery is documented only for some of the permanent settlements related to the exploitation and valorisation of salt, which we have considered for the present study (Figure 1).

Among those assumed to have constructed the seasonal settlements, only the one of Poduri-*Dealul Ghindaru* provided such finds (Cucoş 1999, 120, 122, 280, fig. 59/3, 7; Monah *et al.* 2003, 230, no. 331. Concerning the settlements which controlled the salt exploitation points, there is evidence in this sense at Răuceşti-*Munteni* (Cucoş 1999, 120, 280, fig. 59/2) and Târgu Neamţ-*Dealul Pometea* (Preoteasa *et al.* 2007, 303, 304, 309, 310, fig. 11/13-22; 12).

5.4.2. Seasonal settlements

For the time being, the only seasonal settlement where shards of this category were discovered is the one of Prohozeşti-*Silişte* (Popovici and Trohani 1984, 76; Chapman and Monah 2007, 82).

5.4.3. Exploitation points

A significant amount of Cucuteni C pottery – unfortunately entirely in shards – was been discovered thus far within the salt exploitation points of Solca-*Slatina Mare* (Şandru 1952, 414; Ursulescu 1977, 310, 311; Cucoş 1985, 69; Monah 1991, 393; 2008, 25; Dumitroaia 2004, 445; Cotoi 2007, 157, n. 24; Monah and Dumitroaia 2007, 14, 15; Nicola *et al.* 2007, 35-37, 39, 41, 43, 45, 47, 50-52, 54, 55, fig. 16, 17, 19; Munteanu and Garvăn 2008, 180, 184, 187, 188, 190, fig. 2-4; 5/1, 2; 2011, 81-84, fig. 2-4; 5/1, 2), Cacica-*Salina* (Andronic 1989, 174; 1997, 65, no. 3.5.114; 2009, 21; Monah and Dumitroaia 2007, 23; Monah

2008, 21; Munteanu and Garvăn 2008, 181, 183, 185, 190, 191, fig. 5/3-7; 6; 2011, 81, 82, 84, 85, fig. 5/3-7; 6) and Lunca-*Poiana Slatinei* (Dumitroaia 1987a, 255; 1994a, 58, 62, 67, fig. 51; Monah and Dumitroaia 2007, 19, 22; Monah 2008, 21, 25; Munteanu and Garvăn 2008, 181-183, fig. 1; 2011, 81, 82, fig. 1). Less significant amounts of pottery have been discovered at Ţolici-*Hălăbutoaia* (Dumitroaia *et al.* 2008, 210, 221, fig. 5; Monah 2008, 25, 29; Munteanu and Garvăn 2008, 181, 185, 192, fig. 7; 2011, 81, 82, 85, fig. 7).

5.5. Cultural-chronological framing

Within the permanent settlements of the Cucuteni area, the Cucuteni C pottery emerged as early as the end of the Cucuteni A phase (Cucuteni A_3 stage) and it was discovered in the other evolution periods, too (Alaiba 2006, 188). Nonetheless, within the stations related to salt valorisation (permanent settlements, seasonal settlements, and exploitation points), this pottery species is present only during Cucuteni B (Cucuteni B_1 and Cucuteni B_2 stages), as pinpointed including in the data presented here.

Hence, in the current stage of the research, corroborated with the existing calibrated (with 1 σ) and reliable radiocarbon (^{14}C) dates, we underline that both the Cucuteni C pottery and the briquetage were used in the salt crystallisation process during Cucuteni B (Cucuteni B_1 and Cucuteni B_2 stage). This means they were used approximately 300 years, between 3800-3500 BC (Dumitrescu 1974, 32, 38; Monah 1987, 76, 79, fig. 1; László 1997, 263, 264; Mantu 1998, 249-251, 256, Tab. 7, 8; Bem 2001, 343, 344, 351, 356, 357, Tab. 2, 7, 8; 2002, 48, 71, 72, 79-81, Tab. 1, 2; Petrescu-Dîmboviţa and Văleanu 2004, 274; Lazarovici 2010, 74, 82, 83, fig. 6, Tab. 1; Preoteasa 2011, 61, 63, Tab. 1) (Table 3).

6. Case studies

Within this study, we will try to assess the amount of artefacts necessary and the efforts made by certain human communities at a certain point in time for salt crystallisation, with the hope of being able to draw a realistic – thus reliable – picture of this aspect. However, the figures are only hypothetical, of course. In this sense, in our opinion, the most relevant evolution period of the culture – considering the consistency of the anthropogenic depositions – would be the Cucuteni B_1 stage, which includes the briquetages and the Cucuteni C pottery.

Our attempt is based on the existing statistical data for the salt exploitation points of Solca-*Slatina Mare* (Precucuteni III, Cucuteni A and B) (Nicola *et al.* 2007, 47) and Lunca-*Poiana Slatinei*, respectively (Precucuteni II and III, Cucuteni A and B) (Dumitroaia 1994a 58, 62). Such statistical data – very important, actually – may be corroborated with the probable exploitation duration of a salt-water source throughout an evolution phase of the Cucuteni culture (a period that could be better reflected by a series of calibrated and reliable radiocarbon [^{14}C] dates). The data could also be corroborated with the average lifespan of permanent settlements, accepted as approximately 50 years (Monah 1978, 36-38; Markevič 1981, 10; Kruc 1989, 127; Preoteasa 2011, 63), which could have controlled the respective salt source.

6.1. Solca-"Slatina Mare"

The new archaeological research conducted in 2003 at Solca-*Slatina Mare* (Precucuteni III, Cucuteni A and B) illustrated that the briquetages represent about 20%, while the Cucuteni C pottery approximately 40% of the entire Eneolithic artefacts (Nicola *et al.* 2007, 47), case in which the classical pottery would constitute around 40%. However, it is worth pinpointing that the amount of fabrics of the Precucuteni III and Cucuteni A phases is insignificant, unlike the one of the Cucuteni B phase (very well represented).

6.2. Lunca-"Poiana Slatinei"

In the B area of Lunca-*Poiana Slatinei* (Precucuteni II and III, Cucuteni A and B), the classical Cucuteni pottery (from the first and from the last culture phase, respectively) comprises around 50%, the briquetages – 16% (represented by over 420 shards probably belonging to as much items), while the Cucuteni C pottery, around 33% (Dumitroaia 1994a 58, 62) (Table 4). All of these artefacts have resulted from a field research on about 50 m^2, on a 1 m depth, between 0.35-1.45 m, through section III (S. III) (Dumitroaia 1994a, 62, fig. 2). This would suppose that the volume of the investigated Cucuteni anthropogenic deposition was approximately 50 m^3.

6.3. Hypothetical assessments

Within the hypothetical assessments proposed, we take into account the permanent settlement of Târgu Neamţ-*Dealul Pometea*, assumed to have controlled the salt exploitation point of Lunca-*Poiana Slatinei* in the Cucuteni B_1 stage (Table 5).

We believe that the permanent settlement of Târgu Neamţ-*Dealul Pometea* lasted about 50 years, during which it valorised the salt-water source of Lunca-*Poiana Slatinei*. Practically, its human community created the anthropogenic deposition within the exploitation point following the repeated salt crystallisation processes occurred during the Cucuteni B_1 stage. Out of the entire Cucuteni fabric within the exploitation point of Lunca-*Poiana Slatinei*, the classical pottery of the Cucuteni A phase (Cucuteni A_3 stage) would have represented 16%, the classical pottery of the Cucuteni B phase (Cucuteni B_1 stage) represented 34%, while the briquetages summed up 16%, and the Cucuteni C pottery comprised 34%. Naturally, the briquetages and the Cucuteni C pottery are associated to the pottery of the Cucuteni B phase (Cucuteni B_1 stage).

Table 3. Cultural-chronological framing of the vestiges within certain permanent settlements, seasonal settlements, and salt exploitation points ("shadow"), including those within which Cucuteni C pottery ("C") was discovered, according to the current stage of the archaeological research.

SITES	PRECUCUTENI			CUCUTENI						
				CUCUTENI A			CUCUTENI A-B		CUCUTENI B	
	PRECUCUTENI I	PRECUCUTENI II	PRECUCUTENI III	CUCUTENI A_1	CUCUTENI A_2	CUCUTENI A_3	CUCUTENI $A\text{-}B_1$	CUCUTENI $A\text{-}B_2$	CUCUTENI B_1	CUCUTENI B_2
PERMANENT SETTLEMENTS										
Târpeşti *Râpa lui Bodai*		■	■	■	■					
Poduri *Dealul Ghindaru*		■	■		■				C	
Răuceşti *Munteni*					■				C	
Hangu *Chiriţeni*				■	■	■				
Oglinzi *Cetăţuia*				■	■	■				
Târgu Neamţ *Dealul Pometea*									C	
Vârvata *Cetăţica*									■	■
Poieni *Părţi*									■	■
Valea Şoşii *Dealul Bujoarei*									■	■
SEASONAL SETTLEMENTS										
Poduri *În Fundătură*					■					
Poduri *Pe Hăineală*					■					
Hangu *Picior Bocănceni*				■	■	■				
Prohozeşti *Silişte*									C	
Valea Şoşii *Dealul Păltiniş*										■
EXPLOITATION POINTS										
Lunca *Poiana Slatinei*		■	■			■			C	
Oglinzi *Băi*		■								
Ţolici *Hălăbutoaia*		■				■			C	
Cucuieţi *Slatina Veche*			■			■				
Solca *Slatina Mare*			■			■				C
Cacica *Salina*									C	

Table 4. Statistical data on pottery, reflected by the outcomes of the archaeological research conducted within the salt exploitation point of Lunca-"Poiana Slatinei", B area, section III; the anthropogenic deposition belongs to the Cucuteni culture (after Dumitroaia 1994, modified).

CUCUTENI							
CUCUTENI A_3		CUCUTENI B_1					
CLASSICAL POTTERY		CLASSICAL POTTERY		BRIQUETAGES		CUCUTENI C	
NUMBER	PERCENTAGE	NUMBER	PERCENTAGE	NUMBER	PERCENTAGE	NUMBER	PERCENTAGE

400	16	850	34	400	16	850	34
Total artefacts (no.)				2500			
Total percentages (%)				100			
Total volume section (m^3)				50			

Table 5. Hypothetical quantitative assessments on the necessary of artefacts of certain Cucuteni human settlements within permanent settlements for salt crystallisation in an exploitation point.

YEARS	CUCUTENI							
	CUCUTENI A_3		CUCUTENI B_1					
	CLASSICAL POTTERY		CLASSICAL POTTERY		BRIQUETAGES		CUCUTENI C	
	NO.	%	NO.	%	NO.	%	NO.	%
50	1,500	15	3,500	35	1,500	15	3,500	35
1	30	0.3	70	0.7	30	0.3	70	0.7
50	**Total artefacts (no.)**				10,000			
	Total percentages (%)				100			
	Total area volume (m^3)				200			
1	**Total artefacts (no.)**				200			
	Total percentages (%)				2			
	Total area volume (m^3)				4			

Within the surface investigated, with the total volume of circa 50 m^3, the briquetages counted 400 items, representing circa 16%, which means that the classical pottery of the Cucuteni A phase (Cucuteni A_3 stage) could have also been represented by circa 400 vessels, representing also 16%. In this case, the classical pottery of the Cucuteni B phase (Cucuteni B_1 stage) may have counted around 850 vessels, representing 34%, while the Cucuteni C pottery would have been represented by circa 850 vessels, which also means 34%. Hence, the supposed total of the Cucuteni artefacts within the excavated surface would have been circa 2,500 items.

Considering that the subsequent (limited) archaeological research in the B area of Lunca-*Poiana Slatinei* – with a 3,000 m^2 total estimated surface (Dumitroaia 1994a, fig. 2) – failed to provide equally spectacular outcomes, we assume that the investigated section (S. III) could represent approximately 25% of the portion with a consistent Cucuteni anthropogenic deposition.

Under these circumstances, we believe that the total volume of the consistent Cucuteni anthropogenic deposition in the B area within this salt exploitation point would have been circa 200 m^3. Hence, rafter rounding the proportions of the various categories of artefacts (16% to 15% and 34% to 35%), the classical Cucuteni vessels of the Cucuteni A phase (Cucuteni A_3 stage), representing 15%, would have meant around 1,500. The briquetages – also representing 15% – would have also counted 1,500 items; the classical Cucuteni vessels of the Cucuteni B phase (Cucuteni B_1 stage), representing 35%, would have comprised circa 3,500 items; the Cucuteni C vessels – also representing 35% – would have also counted 3,500. Hence, 10,000 artefacts overall – representing, naturally, 100% – from the Cucuteni period.

This means that, during the Cucuteni B_1 stage, the anthropogenic deposition would include a potential number of around 8,500 artefacts, among which: 3,500 classical Cucuteni vessels, 3,500 Cucuteni C vessels, and 1,500 briquetages. All of them would have accumulated within the salt exploitation point along around 50 years (the accepted average lifespan of a permanent Cucuteni settlement).

Hence, for salt crystallisation in the exploitation point of Lunca-*Poiana Slatinei* during the Cucuteni B_1 stage, the human community within the permanent settlement of Târgu Neamţ-*Dealul Pometea* would have needed annually around 170 artefacts. That means 2% of the estimated total for the entire period, of which: 70 classical Cucuteni vessels, 70 Cucuteni C vessels, and 30 briquetages.

Of the total of the artefacts belonging to the Cucuteni culture, these values would mean circa 0.3% for briquetages, 0.7% for the classical Cucuteni pottery of the Cucuteni B phase (Cucuteni B_1 stage), and 0.7% for the Cucuteni C pottery.

Assuming that the Cucuteni anthropogenic deposition within this exploitation point had a constant constitution rate – considering the salt crystallisation process conducted by the human communities within a permanent settlement –, its total annual volume would represent around 4 m^3.

After analyzing these – though hypothetical – values, which could represent the necessary of artefacts used by the Cucuteni human communities in the salt crystallisation process, the initial impression made by the seemingly significant anthropogenic depositions within the exploitation points changes. In regard to them, it must not be forgotten that they concerning rather limited surfaces (generally several hundred or even thousand square metres). Furthermore, sometimes they are discontinuous; reason for which they had a more flourishing development on the vertical, in the areas situated the closest to the salt source. The volume of their anthropogenic depositions also comprises several hundred or thousand cubic metres, but they are the result of rather extended periods (several hundred or even thousand calendar years), which means that the annual accumulations are rather modest.

In our opinion, the impressive aspect is the technological level reached by the human communities of the Cucuteni B phase; they crystallised salt by using specialised vessels (the briquetages). Though they may seem coarsely made artefacts, we believe that specialised people made them, after creating the composition of the fabric and the optimal shape for salt bars, probably after extensive experiments. Taking into account the advanced knowledge involved in the rather complex salt crystallisation, finalised with the production of the precious bars, it may have also been elaborated by genuine specialists in the field.

7. Quantitative assessments

Considering the importance of salt in the diet of humans and animals, throughout time, specialists have attempted to elaborate a series of – hypothetical, naturally – quantitative assessments in this sense. The figures are characterised by important differences.

The amount of salt destined to consumption is influenced by a series of factors, such as age, weight, gender, body health state, physical effort, or environmental conditions; in the case of animals, it is also different depending on the species or secondary products provided.

The salt consumption may be direct (10-15%) and indirect (85-90%). The latter is represented by vegetal and animal products that contain naturally a certain amount of salt (10-15%) and by foods with added salt (75%) (WHO 2007, 31).

It is well known that both the excess and the deficit of salt may cause health problems, some of which serious (Mitewa and Kolev 2012, 341-344, fig. 2-5), sometimes leading to death, if – for instance – the salt retention in the body reaches 0.5-1 g/ kilogram of body weight (EFSA 2005, 8).

7.1. The salt intakes for humans

In regard to daily human salt intakes, researchers have recommended an amount of 1-2 g (Monah 2012, 127), 2-5 g (Monah 1991, 387), 1-20 g (Chapman and Gaydarska 2003, 204), or even 25-35 g (Cavruc and Chiricescu 2007, 26).

According to the recommendations of the World Health Organisation (WHO) and of the Food and Agriculture Organisation (FAO) of the Unites Nations (UN), the average daily consumption of salt for humans should be under 5 g (WHO 2007, 3, 23, 33, 36, 45, 49; 2010, 6, 11, 13). In the same sense, the European Food Safety Authority (EFSA) recommends up to 6 g of per day (EFSA 2009, 3, 5, 7, 10, a. XI, Tab. 1). Currently, the average values recommended in the various countries of the world are different; the general consumption trend is to exceed the normal values, mainly because of the foods with added salt.

Based on the above-stated figures, in certain cases, the specialists attempted to assess the salt intakes for settlements or areas within the Precucuteni-Cucuteni-Tripolye cultural complex for which there are – hypothetical – estimates regarding the population (Chapman and Gaydarska 2003, 205, 208, Tab. 6). Furthermore, there are assessments for the so-called conventional settlements, for which the population and potential number of livestock (by species) has been approximated (Chapman and Gaydarska 2003, 204, 207, Tab. 5; Monah *et al.* 2008, 98).

We have made such assessments ourselves, for a series of stations belonging to the cultural complex of Precucuteni-Cucuteni-Tripolye for which there are demographic estimates considered constant for the entire lifespan of those settlements. However, we underline that they are hypothetical and that they express the optimal salt intake for those human communities, not the actual consumption (that could have been different). Some of the data obtained in this sense are rounded.

7.1.1. Assessments by specialists

They concern five Cucuteni-Tripolye settlements, as well as two zones within the Tripolye area on the current territory of Ukraine (Chapman and Gaydarska 2003, 205, 208, Tab. 6) (Table 6).

In their study, the two authors mentioned, for the Uman (Ukraine) region, the surface (530 km^2) and density of the population (5 inhabitants per km^2). Based on these data, we have calculated ourselves the number of inhabitants; for the region of the Dniester - Southern Bug interfluves, the authors studied three settlements: one giant and the other two smaller (Chapman and Gaydarska 2003, 205, 208, Tab. 6).

In regard to the cultural-chronological framing of the three Triploye settlements included in the Tripolye B phase, we believe that they should be better dated (regarding the stages). In this sense, we mention that the settlements of Pesčanaja and Jatranovka belong to the Tripolye BII stage, while the giant-settlement of Majdaneckoe actually belongs to the Tripolye CI stage.

In order to complete the data of the table, we also mention that the station of Târpeşti (Cucuteni A) occupied around 1 ha, while the one of Hăbăşeşti (Cucuteni A) about 1.5 ha.

7.1.2. Our own assessments

Admitting that the optimal salt intake for humans is 1-6 g per day (as recommended by the aforementioned prestigious international bodies), we can estimate the intake for the population – daily, monthly, annual, and total (for 50 years) – for the settlements belonging to the cultural complex of Precucuteni-Cucuteni-Tripolye for which there are demographic assessments (Table 7).

Hence, we believe that the optimal salt intake for a person is 1-6 g daily, 30-180 g monthly, 360-2,160 g annually, and 18,000-108,000 g for 50 years.

As it can be noticed, for the giant settlement of Majdaneckoe (Tripolye CI), several demographic assessments have been made throughout time, which we have considered in the present approach.

At the same time, we have assessed hypothetically the salt intakes for the settlements of the Precucuteni-Cucuteni-Tripolye cultural complex – daily, monthly, annual, and total (for 50 years) –, considering, in this sense, the typology we have proposed by surface, number of constructions, and number of inhabitants (Preoteasa 2009, 111) (Table 8).

Table 6. Evaluation of the annual salt intakes (expressed in kg) for certain settlements and zones within the Precucuteni-Cucuteni-Tripolye cultural complex for which there are demographic estimates (after Chapman and Gaydarska 2003 – modified).

SETTLEMENT ZONE	PERIOD	SURFACE (ha / km^2)	POPULATION	MINIMUM	AVERAGE	MAXIMUM	ELITE
Târpeşti	Cucuteni A		165	569	1,404	2,240	1,758
Hăbăşeşti	Cucuteni A		500	2,262	4,255	6,787	5,327
Pesčanaja	Tripolye B	20	> 600	> 2,715	> 5,160	> 8,145	> 6,393
Jatranovka	Tripolye B	50	1,500	7,335	13,860	22,005	16,750
Majdaneckoe	Tripolye B	180	8,000	36,200	68,080	108,600	85,240
Uman (Ukraine)	Tripolye	530	2,650	11,991	22,551	35,973	28,235
Bug - Dniester (3 settlements)	Tripolye			41,630	78,292	124,890	98,026

Table 7. Hypothetical assessments regarding the salt intakes (expressed in kg) of the population within settlements belonging to the cultural complex of Precucuteni-Cucuteni-Tripolye for which there are demographic estimates, as follows: (0) – our own hypothetical demographic assessment; (1) – according to Videiko 2003; (2) – according to Monah and Cucoş 1985; (3) – according to Monah *et al.* 2008; (4) – according to Sorokin 1997; (5) – according to Kruc 1989; (6) – according to Markevič 1981; (7) – according to Bibikov 1965; (8) – according to Gaydarska 2003; (9) – according to Chapman and Gaydarska 2003; (10) and (11) – according to Šmaglij 2001.

SETTLEMENT	PERIOD	POPULATION	HYPOTHETICAL SALT INTAKE			
			1 DAY	1 MONTH	1 YEAR	50 YEARS
Mogyl'na III	Tripolye A	500-800 (1)	0.5/1-3/5	15/24-90/144	180/288-1,080/1,728	9,000/14,400-54,000/86,400
Târpeşti	Precucuteni III	75-80 (2)	0.1-0.5	2.5-13.5/14.5	27/29-162/173	1,350/1,440-8,100/8,640
Târpeşti	Cucuteni A	160-165 (2)	0.2-1	5-29/30	58/61-346/356	2,880/3,060-17,280/17,820
Hangu	Cucuteni A	150-200 (3)	0.2-1	4.5/6-27/36	54/72-324/432	2,700/3,600-16,200/21,600
Hăbăşeşti	Cucuteni A	440-500 (2)	0.5-3	13/15-79/90	158/180-950/1,080	7,920/9,000-47,520/54,000
Truşeşti	Cucuteni A	930 (2)	1-5.5	28-167	335-2,009	16,740-100,440
Druţa I	Cucuteni A	100-120 (4)	0.1-0.5/0.75	3/3.5-18/21.5	36/43-216/259	1,800/2,160-10,800/12,960
Pesčanaja	Tripolye BII	620 (5)	0.5-3.5	19-112	223-1,339	11,160-66,960
Peregonovka	Tripolye BII	920 (5)	1-5.5	28-166	331-1,987	16,560-99,360
Jatranovka	Tripolye BII	1,235 (5)	1.5-7.5	37-222	445-2,668	22,230-133,380
Jampol'	Tripolye BII	1,540 (5)	1.5-9	46-277	554-3,326	27,720-166,300
Kolodistoe	Tripolye BII	1,540 (5)	1.5-9	46-277	554-3,326	27,720-166,300
Krivje Kolena	Tripolye BII	1,540 (5)	1.5-9	46-277	554-3,326	27,720-166,300
Glybočok	Tripolye BII	3,090 (5)	3-18.5	93-556	1112-6674	55,620-333,720
Veselij Kut	Tripolye BII	4,630 (5)	4.5-28	139-833	1,667-10,001	83,340-500,040
Nebelevka	Tripolye BII	6,790 (5)	7-41	204-1,222	2,444-14,666	122,220-733,320
Truşeşti	Cucuteni B	50 (0)	0.1-0.3	0.2-1	2-11	90-540

Petreni	Cucuteni B	4,000 (6)	4-24	120-720	1,440-8,640	72,000-432,000
Mošurov	Tripolye CI	210 (5)	0.2-1.3	6.5-38	76-454	3,780-22,680
Kočerincy *Pankovka*	Tripolye CI	370 (5)	0.5-2.5	11-67	133-799	6,660-39,960
Popudnja	Tripolye CI	370 (5)	0.5-2.5	11-67	133-799	6,660-39,960
Tal'noe	Tripolye CI	460 (5)	0.5-3	14-83	166-994	8,280-49,680
Kolomijščina	Tripolye CI	500 (7)	0.5-3	15-90	180-1,080	9,000-54,000
Mihailovka	Tripolye CI	2,160 (5)	2-13	65-389	778-4,666	38,880-233,280
Vladimirovka	Tripolye CI	2,160 (5)	2-13	65-389	778-4,666	38,880-233,280
Kosenovka	Tripolye CI	3,090 (5)	3-18.5	93-556	1,112-6,674	55,620-333,720
Šuškovka	Tripolye CI	3,090 (5)	3-18.5	93-556	1,112-6,674	55,620-333,720
Gordašovka	Tripolye CI	4,010 (5)	4-24	120-722	1,444-8,662	72,180-433,080
Dobrovody	Tripolye CI	7,720 (5)	8-46	232-1,390	2,779-16,675	138,960-833,760
Tomaševka	Tripolye CI	7,720 (5)	8-46	232-1,390	2,779-16,675	138,960-833,760
Tal'janki	Tripolye CI	14,175 (5)	14-85	425-2,552	5,103-30,618	255,150-1,530,900
Majdaneckoe	Tripolye CI	6,000-9,000 (8)	6/9-36/54	180/270-1,080/1,620	2,160/3,240-12,960/19,440	108,000/162,000-648,000/972,000
		8,000-8,250 (9)	8/8,5-48/50	240/248-1,440/1,485	2,880/2,970-17,280/17,820	144,000/148,500-864,000/891,000
		10,000-12,000 (10)	10/12-60/72	300/360-1,800/2,160	3,600/4,320-21,600/25,920	180,000/216,000-1,080,000/1,296,000
		20,000-24,000 (11)	20/24-120/144	600/720-3,600/4,320	7,200/8,640-43,200/51,840	360,000/432,000-2,160,000/2,592,000
Kočerincy *Šul'govka*	Tripolye CII	770 (5)	1-4.5	23-139	277-1,663	13,860-83,160
Sverdlikovo	Tripolye CII	770 (5)	1-4.5	23-139	277-1,663	13,860-83,160

Table 8. Hypothetical assessments regarding the salt intakes for the settlements belonging to the cultural complex of Precucuteni-Cucuteni-Tripolye, according to our own typology.

SETTLEMENT (type)	SURFACE (ha)	CONSTRUCTIONS (no.)	POPULATION (no.)	1 DAY (kg)	1 MONTH (kg)	1 YEAR (kg)	50 YEARS (kg)
very small	< 1	< 50	< 500	0.5/3	15/90	180/1,080	9,000/54,000
small	1-5	50-100	500-1,000	0.5/1-3/6	15/30-90/180	180/360-1,080/2,160	9,000/18,000-54,000/108,000
medium	5-10	100-250	1,000-2,500	1/2.5-6/15	30/75-180/450	360/900-2,160/5,400	18,000/45,000-108,000/270,000
large	10-50	250-500	2,500-5,000	2.5/5-15/30	75/150-450/900	900/1,800-5,400/10,800	45,000/90,000-270,000-540,000
very large	50-100	500-1,000	5,000-10,000	5/10-30/60	150/300-900/1,800	1,800/3,600-10,800/21,600	90,000/180,000-540,000/1,080,000
giant	> 100	> 1000	> 10,000	10/60	300/1,800	3,600/21,600	180,000/1,080,000

7.2. The salt intakes for the livestock

In regard to the daily salt intake for the livestock, some archaeologists believe it is reduced, thus ensured by the daily diet, in the case of pigs; 10 g for sheep and goats; 28-100 g for cattle; 7-50 g for horses (Chapman and Gaydarska 2003, 205, Tab. 4). Others pinpoint that the necessary salt intake is 5-10 g for pigs, 7-15 g for sheep and goats, and 10-30 g for cattle (Cavruc and Chiricescu 2007, 24). It is also admitted that a 500 kg cow would have needed 25 g salt daily for maintenance, to which 2 g are added for each litre of milk (Monah 1991, 388).

According to the recommendations of the specialists in animal production science, the daily salt intake of the livestock would be 5-15 g for sheep and goats, 5-40 g for pigs, 20-40 g for horses, and 25-50 g for cattle (Colă 2010, 158).

In certain cases, based on the aforementioned figures, researchers have created conventional settlements for which they estimated including the salt intakes for the species of livestock in question (Chapman and Gaydarska 2003, 204, 207, Tab. 5; Monah *et al.* 2008, 98). However, considering that they failed to consider, in their studies, the archaeozoological data, we believe that the figures they concluded are practically irrelevant.

We have made such assessments ourselves for stations belonging to the cultural complex of Precucuteni-Cucuteni-Tripolye, researched exhaustively and for which there are estimates on the minimal number of the livestock, considered constant for the entire lifespan of those settlements. In this case, too, we underline that they are merely hypothetical and that they express the optimal salt intake of the livestock, not the real consumption (that could have been different). Some of the data obtained in this sense are rounded.

7.2.1. Assessments by specialists

A conventional settlement analysed was hypothetically made of people, cattle and sheep, and goats, for which the daily, monthly, and annual salt intakes was assessed for each category, as well as for the entire module (Chapman and Gaydarska 2003, 204, 207, Tab. 5) (Table 9).

Another such example concerns the settlement of Hangu-*Chiriţeni* (Cucuteni A); the module also comprises people, cattle and sheep, and goats, and it was inspired by the aforementioned module (Monah *et al.* 2008, 98) (Table 10).

7.2.2. Our own assessments

In regard to the salt intakes of the livestock, based on the values accepted by animal production science specialists, we can make such hypothetical assessments – daily, monthly, annual, and for their entire lifespan (for 50 years) – only for three of the Cucuteni settlements studied totally, which also provide exhaustive archaeozoological information. We refer here to the settlements of Târpeşti (Precucuteni III and Cucuteni A) and Truşeşti (Cucuteni A) (Table 11).

Hence, we consider the optimal salt intake for the livestock as follows: bovine – 25-50 g daily, 750-1,500 g monthly, 9,000-18,000 g annually, and 450,000-900,000 g for 50 years; pigs – 5-40 g daily, 150-1,200 g monthly, 1,800-14,400 g annually, and 90,000-720,000 g for 50 years; sheep and goats – 5-15 g daily, 150-450 g monthly, 1,800-5,400 g annually, and 90,000-270,000 g for 50 years.

As noticeable in the aforementioned fauna-related data, the number of presumed items for livestock for the settlements of Târpeşti (Precucuteni III and Cucuteni A), and mostly for that of Truşeşti (Cucuteni A) is very small, all the more as, in the second case, it is practically the largest Cucuteni settlement studied exhaustively. The rapport between the number of inhabitants and that of the livestock is insufficient to have satisfied (over time) the animal-related primary and secondary needs of these human communities. As consequence, the salt intakes of the livestock cannot represent realistic values, either. An explanation in this sense could be the selective sampling of the fauna remains and / or their partial archaeozoological determination.

Table 9. Salt intake estimate (expressed in kg) for a conventional settlement (according to Chapman and Gaydarska 2003).

MODULE	NUMBER	LOW	AVERAGE	LARGE	ELITE
DAILY					
PEOPLE	100	0.5	1	1.5	0.7
CATTLE	30	0.3	0.45	0.75	0.75
SHEEP AND GOATS	150	0.45	0.9	1.5	1.5
TOTAL		1.25	2.35	3.75	2.95
MONTHLY					
PEOPLE	100	15	30	45	21
CATTLE	30	9	13.5	22.5	22.5
SHEEP AND GOATS	150	13.5	27	45	45
TOTAL		37.5	70.5	112.5	88.5
ANNUAL					
PEOPLE	100	180	360	540	252
CATTLE	30	108	162	270	270
SHEEP AND GOATS	150	162	324	540	540
TOTAL		450	846	1,350	1,062

Table 10. Salt intake estimate (expressed in kg) for a conventional settlement (Hangu-"Chiriţeni" – Cucuteni A) (according to Monah *et al.* 2008).

MODULE	NUMBER	MINIMUM	MAXIMUM
DAILY			
PEOPLE	150	0.75	

CATTLE	45	0.45	
SHEEP AND GOATS	225	0.675	
TOTAL		1.875	
ANNUAL			
PEOPLE	150	273.75	810
CATTLE	45	410.62	1,215
SHEEP AND GOATS	225		
TOTAL		684.47	2,025

Table 11. Hypothetical assessments on the salt intakes of the livestock (expressed in kg) within the settlements belonging to the cultural complex of Precucuteni-Cucuteni-Tripolye that benefit from exhaustive archaeozoological data, as follows: (1) – according to Marinescu-Bîlcu 1981; (2) – according to Petrescu-Dîmbovița *et al.* 1999.

SETTLEMENT	PERIOD	SPECIES	NUMBER	HYPOTHETICAL SALT INTAKE			
				1 DAY	1 MONTH	1 YEAR	50 YEARS
Târpești [(1)]	Precucuteni III	Cattle	150	3.75-7.5	113-225	1,350-2,700	67,500-135,000
		Pigs	60	0.25-2.5	9-72	108-864	5,400-43,200
		Sheep and goats	50	0.25-0.75	7.5-22.5	90-270	4,500-13,500
		Total	260	4-11	129-320	1,548-3,834	77,400-191,700
Târpești [(1)]	Cucuteni A	Cattle	52	1.25-2.5	39-78	468-936	23,400-46,800
		Pigs	24	0.1-1	3.5-29	43-346	2150-17,300
		Sheep and goats	11	0.05-0.2	1.75-5	20-60	1000-3000
		Total	87	1.5-3.75	44-112	511-1,342	26,550-67,100
Trușești [(2)]	Cucuteni A	Cattle	13	0.25-0.75	10-20	117-234	5,850-11,700
		Pigs	10-11	0.05-0.5	1.5/1.75-12/13	18/20-144/158	900/1,000-7,200-7,900
		Sheep and goats	7	0.025-0.1	1-3.25	13-38	650-1,900
		Total	30-31	0.5-1.25	12/13-35/36	148/150-416/430	7,400/7,500-20,800/21,500

8. Conclusions

The presence, especially, in the Moldavian Subcarpathians, of an essential raw material for people and for the livestock, such as salt, has always constituted one of the favourable factors for the anthropogenic character of a settlement.

The archaeological research conducted thus far illustrates that prehistory (especially the Neo-Eneolithic) was one of the periods when salt was intensely exploited and valorised.

In this sense, the human communities belonging to the cultural complex of Precucuteni-Cucuteni-Tripolye – for a better valorisation of this natural resource – played a crucial role, as they constructed a series of seasonal settlements and exploitation points in their immediate vicinity. These were probably under the direct control of the important permanent settlements, which represented genuine centres of regional polarisation. In the same period, it is worth noting the presence of specialised artefacts for the salt crystallisation process (the briquetages), as well as of a pottery species (Cucuteni C) which – through the special fabric composition (shell-tempered) – probably favoured the transformation of *salt water* into *huscă*.

As pinpointed on a previous occasion (Ellis 1984, 205), the presence of rich salt resources in the very constitution area of this civilisation provided to the human communities stability and a superior life quality, compared to other cultures.

Its dietary and therapeutic role has influenced directly an important demographic increase, thus allowing the existence of a high natural growth rate, as well as the increase of the life expectancy of the population. More precisely, the better health state meant women were fertile for a longer period, leading to an increase in the birth rate; at the same time, it led to a decrease in the morbidity and mortality rates, including infant mortality.

Furthermore, the inclusion of salt in the exchange relations contributed to the creation of intense contacts with contemporary civilisations and it provided supplies for the human communities belonging to the cultural complex of Precucuteni-Cucuteni-Tripolye. We refer here to consumption goods (grains, livestock), to various other raw materials (flint, obsidian, other categories of volcanic rocks, copper, gold), or prestigious artefacts (vessels, tools and guns, cultic objects and ornaments).

An example in this sense may be the *tell* of Poduri-*Dealul Ghindaru*, which comprises several Precucuteni and Cucuteni levels, which constitute a settlement per se, due precisely to the presence of rich salt-water sources in the area. The *tell* was placed strategically, and the human communities controlled for over a millennium the entire micro-depression of Tazlăul Sărat and the *salt water* sources in the vicinity, implicitly. They also controlled one of the communication routes connecting the East-Carpathian and the Intra-Carpathian one, where salt – metal or even salt – volcanic rocks trades took place, as proven by the discoveries made thus far (Monah *et al.* 1982, 20; Monah 1991, 394; Cavruc and Dumitroaia 2006, 39). Moreover, within this station, numerous "imports" were discovered, which illustrate the complex trading system of these human communities. Another evidence in this sense is the presence within the successive settlements in the *tell* of an impressive amount of macrovegetal remains (especially carbonised grain caryopsis) (Monah and Monah 1995, 311-317; 1996, 49-62; 2003, 252-256; 2005a, 335-356; 2005b, 247-276; 2008, 55-89; Cârciumaru 1996, 101-105), corroborated with a very large number of grinders. As the Poduri zone was not even by far the ideal area for grain cultivation within this civilisation, the (justified) hypothesis was issued that most of the amount of grains would have resulted from the trades with other human communities, which involved the presence of salt (Monah *et al.* 1987, 258, 260. We believe that, in the case of the settlements in the *tell*, even trades such as salt – livestock (especially cattle) may have occurred, which may explain the high percentage of fauna remains.

Practically, the outcomes of the archaeological research conducted thus far regarding the exploitation of salt by the human communities belonging to the cultural complex of Precucuteni-Cucuteni-Tripolye illustrate, once again, their good adjustment to the environmental conditions and the efficient valorisation of natural resources (in our case, salt). This allowed, on one side, a long-term evolution of the civilisation, and a high degree of development, on the other.

Abbreviations

EFSA– *European Food Safety Authority*, Parma, Italy.
FAO–*Food and Agriculture Organisation*, Roma, Italy
IPCTE–*Institute for the Research and Valorisation of the Transylvanian Cultural Patrimony in a European Context*, Sibiu, Romania.
UN–*United Nations*, New York, USA.
SACN–*Scientific Advisory Committee on Nutrition*, Norwich, United Kingdom.
WHO–*World Health Organisation*, Geneva, Switzerland.

References

Alaiba, R. 2006. *Ceramica de tip Cucuteni C*, in N. Ursulescu and C.-M. Lazarovici (coord.), *Cucuteni 120 – valori universale. Lucrările simpozionului național, Iași, 30 septembrie 2004*, 181-208. Bibliotheca Archaeologica Iassiensis, XVII. Iași, Editura Pim.

Alexianu, M., Dumitroaia, Gh. and Monah, D. 1992. Exploatarea surselor de apă sărată din Moldova: o abordare etnoarheologică. *Thraco-Dacica* XIII, 1-2, 159-167.

Alexianu, M., Dumitroaia, Gh. and Monah, D. 2007. *The Exploitation of the Salt-Water Sources in Moldavia: an Ethnoarchaeological Approach*, in D. Monah, Gh. Dumitroaia, O. Weller and J. Chapman (eds.), *L'exploitation du sel à travers le temps*, 279-298. Bibliotheca Memoriae Antiquitatis, XVIII. Piatra-Neamț, Editura Constantin Matasă.

Alexianu, M., Weller, O., Brigand, R., Curcă, R.-G., Cotiugă, V. and Moga, I. 2011. Salt Springs in Today's Rural World. An Ethnoarchaeological Approach in Moldavia (Romania), in M. Alexianu, O. Weller, R.-G. Curcă (eds.), *Archaeology and Anthropology of Salt: A Diachronic Approach. Proceedings of the International Colloquium, 1-5 October 2008, Al. I. Cuza University (Iași, Romania)*, 7-23. British Archaeological Reports, International Series, 2198. Oxford, BAR Publishing.

Andronic, M. 1989. Cacica – un nou punct neolitic de exploatare a sării. *Studii și Cercetări de Istorie Veche și Arheologie* 40, 2, 171-177.

Andronic, M. 1997. Evoluția habitatului uman în Bazinul Hidrografic Soloneț din paleolitic până la sfârșitul secolului al XVIII-lea. *Suceava*, XXII-XXIII (1995-1996), Supliment VI. Iași, Editura Dosoftei.

Andronic, M. 2009. L'exploitation du sel à l'est des Carpates Orientaux dans la préhistoire et dans la protohistoire. *Studii și Cercetări de Istorie Veche și Arheologie* 59-60 (2008-2009), 17-32.

Bem, C. 2001. Elemente de cronologie radiocarbon. Ariile culturale Boian-Gumelnița-Cernavoda I și Precucuteni-Cucuteni/ Tripolie. *Cercetări Arheologice* XI, I, 337-359.

Bem, C. 2002. Noi propuneri pentru o schiță cronologică a eneoliticului românesc. *Pontica* XXXIII-XXXIV, 25-121.

Bem, C. 2007. *Traian Dealul Fântânilor. Fenomenul Cucuteni A-B*. Târgoviște, Editura Cetatea de Scaun.

Bibikov, S.N. 1965. Hozja'stvenno-ekonomičeski' komplex razvitogo Tripol'ja (Opyt izučenija pervobytno' tekhniki). *Sovetskaja Arkheologija* 1, 48-62.

Bodi, G. 2007. Experiments with Salt Crystallization, in D. Monah, Gh. Dumitroaia, O. Weller and J. Chapman (eds.), *L'exploitation du sel à travers le temps*, 89-97. Bibliotheca Memoriae Antiquitatis, XVIII. Piatra-Neamț, Editura Constantin Matasă.

Cavruc, V. 2009. Stadiul actual al cercetărilor privind exploatarea preistorică a sării în spaţiul carpato-dunărean, in A. Zanoci, T. Arnăut and M. Băţ (eds.), *Studia Archeologiae et Historiae Antiquae. Doctissimo viro Scientiarum Archeologiae et Historiae Ion Niculiţă, anno septuagesimo aetatis suae, dedicatur*, 21-36. Chişinău, Casa Editorial-Poligrafică "Bonus Offices".

Cavruc, V. and Chiricescu, A. 2006. Fişele exponatelor, in V. Cavruc and A. Chiricescu (eds.), *Sarea, timpul şi omul. Catalog de expoziţie*, 189-247. Sfântu Gheorghe, Editura Angustia.

Cavruc, V. and Chiricescu, A. 2007. *Sarea, timpul şi omul*. Power-point presentation of the exhibition shown at the Museum of Eastern Carpathians in 2007. Sfântu Gheorghe, Editura Angustia. *http://www.cimec.ro/Arheologie/sarea/index.html*

Cavruc, V. and Dumitroaia, Gh. 2006. Vestigii arheologice privind exploatarea sării pe teritoriul României în perioada neo-eneolitică, in V. Cavruc and A. Chiricescu (eds.), *Sarea, timpul şi omul. Catalog de expoziţie*, 37-40. Sfântu Gheorghe, Editura Angustia.

Cârciumaru, M. 1996. *Paleoetnobotanica. Studii în preistoria şi protoistoria României (Istoria agriculturii din România)*. Iaşi, Editura Glasul Bucovinei – Helios.

Chapman, J. and Gaydarska, B. 2003. The Provision of Salt to Tripolye Mega-Sites, in O.G. Korvin-Piotrovs'kyj, V.O. Kruc and S.M. Ryžov (eds.), *Trypil's'ki poselennja-giganty. Materialy mižnarodnoj konferencij / Tripolian Settlements-Giants. The International Symposium Materials*, 203-211. Kiev, Korvin Press.

Chapman, J. and Monah, D. 2007. A Seasonal Cucuteni Occupation at Silişte-Prohozeşti, Romania, in D. Monah, Gh. Dumitroaia, O. Weller and J. Chapman (eds.), *L'exploitation du sel à travers le temps*, 71-88. Bibliotheca Memoriae Antiquitatis, XVIII. Piatra-Neamţ, Editura Constantin Matasă.

Colă, M. 2010. *Zootehnie I + II. Manual universitar pentru învăţământul la distanţă*. Craiova.

Cotoi, O. 2007. Observaţii privitoare la tehnologia şi materiile prime utilizate la prepararea pastei ceramicii Cucuteni C. *Annales Universitatis Apulensis*, Seria Historica, 11, I, 153-160.

Cotoi, O. 2009. Considerations on the Origin of the Briquetage Vessels. *Analele Universităţii "Dunărea de Jos" Galaţi*, Istorie, 19, VIII, 7-16.

Crîşmaru, A. 1977. *Drăguşeni. Contribuţii la o monografie arheologică*. Bacău.

Cucoş, Şt. 1985. Ceramica de "tip C" din aria culturii Cucuteni. *Memoria Antiquitatis* IX-XI (1977-1979), 63-92.

Cucoş, Şt. 1999. *Faza Cucuteni B în zona subcarpatică a Moldovei*. Bibliotheca Memoriae Antiquitatis VI. Piatra-Neamţ, Editura Constantin Matasă.

Danilenko, V.N. 1974. *Eneolit Ukrainy*. Kiev.

Dodd-Opriţescu, A. 1980. Consideraţii asupra ceramicii Cucuteni C. *Studii şi Cercetări de Istorie Veche şi Arheologie* 31, 4, 547-557.

Dodd-Opriţescu, A. 1982. La céramique Cucuteni C. Son origine. Sa signification historico-culturelle. *Thracia Praehistorica. Supplementum Pulpudeva* 3, 70-79.

Dragomir, I.T. 1982. Elemente stepice "Cucuteni C", descoperite la Bereşti (jud. Galaţi). *Studii şi Cercetări de Istorie Veche şi Arheologie* 33, 4, 422-429.

Dumitrescu, H. 1960. Şantierul arheologic Traian. *Materiale şi Cercetări Arheologice* VII, 91-106.

Dumitrescu, Vl. 1963a. Originea şi evoluţia culturii Cucuteni-Tripolie (I). *Studii şi Cercetări de Istorie Veche* XIV, 1, 51-78.

Dumitrescu, Vl. 1963b. Originea şi evoluţia culturii Cucuteni-Tripolie (II). *Studii şi Cercetări de Istorie Veche* XIV, 2, 285-308.

Dumitrescu, Vl. 1974. Cronologia absolută a eneoliticului românesc în lumina datelor C14. *Apulum* XII, 23-39.

Dumitroaia, Gh. 1987a. La station archéologique de la Lunca-Poiana Slatinii, in M. Petrescu-Dîmboviţa, N. Ursulescu, D. Monah, V. Chirica (eds.), *La civilisation de Cucuteni en contexte européen. Session scientifique dédiée au centenaire des premières découvertes de Cucuteni (Iaşi - Piatra Neamţ, 24-28 septembre 1984)*, 253-258. Bibliotheca Archaeologica Iassiensis, I. Iaşi.

Dumitroaia, Gh. 1987b. Plastica antropomorfă din aşezarea de la Răuceşti-"Munteni", jud. Neamţ. *Memoria Antiquitatis* XV-XVII (1983-1985), 21-42.

Dumitroaia, Gh. 1994a. Depunerile neo-eneolitice de la Lunca şi Oglinzi, judeţul Neamţ. *Memoria Antiquitatis* XIX, 7-82.

Dumitroaia, Gh. 1994b. Săpături şi cercetări arheologice de suprafaţă în judeţul Neamţ (1992-1993). *Memoria Antiquitatis* XIX, 509-521.

Dumitroaia, Gh. 2004. Săpături arheologice sistematice şi periegheze din judeţele Neamţ, Bacău şi Suceava (2002-2003). *Memoria Antiquitatis* XXIII, 431-449.

Dumitroaia, Gh., Munteanu, R., Nicola, D., Preoteasa, C., Weller, O. 2004. Lunca-"Poiana Slatinei", le plus ancien point d'exploitation du sel de l'Europe. In *Rezumatele comunicărilor colocviilor internaţionale "Cucuteni. 120 de ani de cercetări. Timpul Bilanţului" şi "Arheologia pre- şi protoistorică a sării". Muzeul de*

Istorie şi Arheologie Piatra-Neamţ, 21-24 octombrie 2004, 95-97. Piatra-Neamţ, Editura Constantin Matasă.

Dumitroaia, Gh., Munteanu, R., Weller, O., Garvăn, D., Diaconu, V., Brigand, R. 2008. Un nou punct de exploatare a apei sărate în preistorie: Ţolici-Hălăbutoaia, jud. Neamţ, in D. Monah, Gh. Dumitroaia and D. Garvăn (eds.), *Sarea, de la prezent la trecut*, 203-224. Bibliotheca Memoriae Antiquitatis, XX. Piatra-Neamţ, Editura Constantin Matasă.

Ellis, L. 1984. *The Cucuteni-Tripolye Culture. A Study in Technology and the Origins of Complex Society*. British Archaeological Reports, International Series, 217. Oxford, BAR Publishing.

European Food Safety Authority 2005. Opinion of the Scientific Panel on Dietetic Products, Nutrition and Allergies on a Request from the Commission Related to the Tolerable Upper Intake Level of Sodium. *The EFSA Journal* 209, 1-26.

European Food Safety Authority 2009. Scientific Opinion of the Panel on Dietetic Products, Nutrition and Allergies on a Request from European Commission on the Review of Labelling Reference Intake Values for Elected Nutritional Elements. *The EFSA Journal* 1008, 1-14.

Frînculeasa, A. 2006. Complexe de cult descoperite în aşezările Precucuteni şi Cucuteni (cu privire specială asupra gropilor de cult). *Carpica* XXXV, 23-38.

Garvăn, D. 2008. Locuiri sezoniere precucuteniene. Cucuieţi-Slatina Veche, judeţul Bacău. *Carpica* XXXVII, 110-122.

Gaydarska, B. 2003. Application of GIS in Settlement Archaeology: an Integrated Approach to Prehistoric Subsistence Strategies, in O.G. Korvin-Piotrovs'kyj, V.O. Kruc and S.M. Ryžov (eds.), *Trypil's'ki poselennja-giganty. Materialy mižnarodnoj konferencij / Tripolian Settlements-Giants. The International Symposium Materials*, 212-215. Kiev, Korvin Press.

Kruc, V.A. 1989. K istorii naselenija Tripol'sko' kultury v meždureč'e jujnogo Buga i Dnepra. *Pervobitnaja Arkheologija*, 117-132.

László, A., 1997. *Datarea prin radiocarbon în arheologie, Biblioteca Muzeului Naţional*, II, Bucureşti.

László, A., 2009. Radiocarbon Dating of Malnaş Băi Settlement and some Problems of Development and Chronology of Ariuşd-Cucuteni Culture. In G. Bodi (ed.), *In medias res praehistoriae. Miscellanea in honorem annos LXV peragentis Professoris Dan Monah oblata*, 205-222. Iaşi, Editura "Alexandru Ioan Cuza".

Lazarovici, C.-M. 2009. Ritualuri de fundare, ritualuri de abandonare din cultura Cucuteni. In G. Bodi (ed.), *In medias res praehistoriae. Miscellanea in honorem annos LXV peragentis Professoris Dan Monah oblata*, 223-238. Iaşi, Editura "Alexandru Ioan Cuza".

Lazarovici, C.-M. 2010. New Data Regarding the Chronology of the Precucuteni, Cucuteni and Horodiştea-Erbiceni Cultures, in J. Šuteková, P. Pavúk, P. Kálabková, B. Kovár (eds.), *Panta Rhei. Studies on the Chronology and Cultural Development of South-Eastern and Central Europe in Earlier Prehistory Presented to Juraj Pavúk on the Occasion of his 75th Birthday*. 71-94. Studia Archaeologica et Mediaevalia, XI. Bratislava.

Mantu, C.-M. 1998. *Cultura Cucuteni. Evoluţie, cronologie, legături*. Bibliotheca Memoriae Antiquitatis, V. Piatra-Neamţ, Editura Nona.

Marinescu-Bîlcu, S. 1974. *Cultura Precucuteni pe teritoriul României*. Bucharest, Academy Publishing House.

Marinescu-Bîlcu, S. 1981. *Tîrpeşti. From Prehistory to History in Eastern Romania*. British Archaeological Reports, International Series, 107. Oxford, BAR Publishing.

Marinescu-Bîlcu, S. and Bolomey, Al. 2000. *Drăguşeni. A Cucutenian Community*. Bucureşti – Tübingen, Editura Enciclopedică – Wasmuth Verlag.

Markevič, V.I. 1981. *Pozdnetripol'skie plemena severnoj Moldavii*. Chişinău, Editura Ştiinţa.

Mitewa, M. and Kolev, H. 2012. Sodium Chloride: food and poison, in V. Nikolov and K. Bacvarov (eds.), *Salz und Gold: die Rolle des Salzes im prähistorischen Europa. Akten der internationaler Fachtagung (Humboldt-Kolleg) in Provadia, Bulgarien. 30 September – 4 October 2010 / Salt and Gold: The Role of Salt in Prehistoric Europe. Proceedings of the International Symposium (Humboldt-Kolleg) in Provadia, Bulgaria. 30 September – 4 October 2010*, 341-344. Provadia – Veliko Târnovo, Verlag Faber.

Monah, D. 1978. Datarea prin C_{14} a etapei Cucuteni A_2. *Studii şi Cercetări de Istorie Veche şi Arheologie*, 29, 1, 33-42.

Monah, D. 1987. La datation par C_{14} du complexe culturel Cucuteni-Tripolye, in M. Petrescu-Dîmboviţa, N. Ursulescu, D. Monah, V. Chirica (eds.), *La civilisation de Cucuteni en contexte européen. Session scientifique dédiée au centenaire des premières découvertes de Cucuteni (Iaşi - Piatra Neamţ, 24-28 septembre 1984)*, 67-79. Bibliotheca Archaeologica Iassiensis, I, Iaşi.

Monah, D. 1991. L'exploitation du sel dans les Carpates Orientales et ses rapports avec la culture Cucuteni-Tripolye, in V. Chirica and D. Monah (eds.), *Le paléolithique et le néolithique de la Roumanie en contexte européen*, 387-400. Bibliotheca Archaeologica Iassiensis, IV. Iaşi.

Monah, D. 2008. Arheologia preistorică a sării în România. Scurt istoric, in D. Monah, Gh. Dumitroaia and D. Garvăn (eds.), *Sarea, de la prezent la trecut*, 13-39. Bibliotheca Memoriae Antiquitatis, XX. Piatra-Neamţ, Editura Constantin Matasă.

Monah, D., Antonescu, S. and Bujor, A. 1980. Raport preliminar asupra cercetărilor arheologice din comuna Poduri, jud. Bacău. *Materiale şi Cercetări Arheologice*, XIV, 86-99.

Monah, D. and Cucoş, Şt. 1985. *Aşezările culturii Cucuteni din România*. Iaşi, Editura Junimea.

Monah, D., Cucoş, Şt., Popovici, D., Antonescu, S. 1982. Săpăturile arheologice din tell-ul cucutenian Dealul Ghindaru, com. Poduri, jud. Bacău. *Cercetări Arheologice* V, 9-22.

Monah, D. and Dumitroaia, Gh. 2007. Recherches sur l'exploitation préhistorique du sel en Roumanie, in D. Monah, Gh. Dumitroaia, O. Weller and J. Chapman (eds.), *L'exploitation du sel à travers le temps*, 13-34. Bibliotheca Memoriae Antiquitatis, XVIII. Piatra-Neamţ, Editura Constantin Matasă.

Monah, D., Dumitroaia, Gh., Monah, F., Preoteasa, C., Munteanu, R., Nicola, D. 2003. *Poduri-Dealul Ghindaru. O Troie în Subcarpaţii Moldovei*. Bibliotheca Memoriae Antiquitatis, XIII. Piatra-Neamţ, Editura Constantin Matasă.

Monah, D., Dumitroaia, Gh. and Nicola, D. 2008. Noi investigaţii etnoarheologice asupra izvoarelor sărate de pe Valea Muntelui, in D. Monah, Gh. Dumitroaia and D. Garvăn (eds.), *Sarea, de la prezent la trecut*, 81-105. Bibliotheca Memoriae Antiquitatis, XX. Piatra-Neamţ, Editura Constantin Matasă.

Monah, D., Dumitroaia, Gh. and Nicola, D. 2011. New Ethnoarchaeological Investigations upon the Salt Springs in Valea Muntelui, Romania, in M. Alexianu, O. Weller, R.-G. Curcă (eds.), *Archaeology and Anthropology of Salt: A Diachronic Approach. Proceedings of the International Colloquium, 1-5 October 2008, Al. I. Cuza University (Iaşi, Romania)*, 25-33. British Archaeological Reports, International Series, 2198. Oxford, BAR Publishing.

Monah, F., Băra, I. and Monah, D. 1987. Observaţii asupra compoziţiei depozitelor de cereale din aşezarea Precucuteni III de la Poduri-Dealul Ghindaru. *Memoria Antiquitatis* XV-XVII (1983-1985), 249-261.

Monah, F. and Monah, D. 1995. Macroresturi vegetale descoperite în nivelurile Cucuteni A_2 şi B_1 de la Poduri-Dealul Ghindaru. *Arheologia Moldovei* XVIII, 311-317.

Monah, F. and Monah, D. 1996. Macrorestes vgtaux dcouvertes dans les niveaux Cucuteni A_2 et B_1 de Poduri-"Dealul Ghindaru", in Gh. Dumitroaia and D. Monah (eds.), *Cucuteni aujourd'hui. 110 ans depuis la découverte en 1884 du site eponyme*, 49-62. Bibliotheca Memoriae Antiquitatis, II. Piatra-Neamţ, Editura Constantin Matasă.

Monah, F. and Monah, D. 2003. Nouvelles dterminations archobotaniques concernant le tell Poduri-Dealul Ghindaru, in O.G. Korvin-Piotrovs'kyj, V.O. Kruc and S.M. Ryžov (eds.), *Trypil's'ki poselennja-giganty. Materialy mižnarodnoj konferencij /Tripolian Settlements-Giants. The International Symposium Materials*, 252-256. Kiev, Korvin Press.

Monah, F. and Monah, D. 2005a. Determinări arheobotanice pentru tell-ul calcolitic Poduri-Dealul Ghindaru. *Arheologia Moldovei* XXVIII, 335-356.

Monah, F. and Monah, D. 2005b. Le rpertoire des dterminations archobotaniques du tell chalcolithique de Poduri-Dealul Ghindaru, in Gh. Dumitroaia, J. Chapman, O. Weller, C. Preoteasa, R. Munteanu, D. Nicola and D. Monah (eds.), *Cucuteni. 120 ans de recherches. Le temps du bilan / Cucuteni. 120 Years of Research. Time to sum up*, 247-276. Bibliotheca Memoriae Antiquitatis, XVI. Piatra-Neamţ, Editura Constantin Matasă.

Monah, F. and Monah, D. 2008. *Cercetări arheobotanice în tell-ul calcolitic Poduri-Dealul Ghindaru*. Bibliotheca Memoriae Antiquitatis, XIX. Piatra-Neamţ, Editura Constantin Matasă.

Munteanu, R., Garvăn, D., Nicola, D., Preoteasa, C., Dumitroaia, Gh. 2007. Cucuieţi-Slatina Veche (Romania). Prehistoric Exploitation of a Salt Resource, in D. Monah, Gh. Dumitroaia, O. Weller and J. Chapman (eds.), *L'exploitation du sel à travers le temps*, 57-70. Bibliotheca Memoriae Antiquitatis, XVIII. Piatra-Neamţ, Editura Constantin Matasă.

Munteanu, R. and Garvăn, D. 2008. Ceramica Cucuteni C din preajma izvoarelor de apă sărată din Moldova, in D. Monah, Gh. Dumitroaia and D. Garvăn (eds.), *Sarea, de la prezent la trecut*, 179-202. Bibliotheca Memoriae Antiquitatis, XX. Piatra-Neamţ, Editura Constantin Matasă.

Munteanu, R. and Garvăn, D. 2011. The Cucuteni C Pottery near the Moldavian Salt Springs, in M. Alexianu, O. Weller and R.-G. Curcă (eds.), *Archaeology and Anthropology of Salt: A Diachronic Approach. Proceedings of the International Colloquium, 1-5 October 2008, Al. I. Cuza University (Iaşi, Romania)*, 81-87. British Archaeological Reports, International Series, 2198. Oxford, BAR Publishing.

Nandriş, J. 1987. Romanian Ethnoarchaeology and the Emergence and Development of Cucuteni in the European Context. In M. Petrescu-Dîmboviţa, N. Ursulescu, D. Monah, V. Chirica (eds.), *La civilisation de Cucuteni en contexte européen. Session scientifique dédiée au centenaire des premières découvertes de Cucuteni (Iaşi - Piatra Neamţ, 24-28 septembre 1984)*, 201-222. Bibliotheca Archaeologica Iassiensis, I, Iaşi.

Nestor, I. and Zaharia, E. 1968. Sur la priode de transition du nolithique l'age du bronze dans l'aire des civilisations de Cucuteni et de Gumelnița. *Dacia*, Nouvelle Srie XII, 17-43.

Nicola, D., Munteanu, R., Garvăn, D., Preoteasa, C., Dumitroaia, Gh. 2007. Solca-Slatina Mare (Roumanie). Préues archélogiques de l'exploitation du sel en prhistoire, in D. Monah, Gh. Dumitroaia, O. Weller and J. Chapman (eds.), *L'exploitation du sel à travers le temps*, 35-56. Bibliotheca Memoriae Antiquitatis, XVIII. Piatra-Neamț, Editura Constantin Matasă.

Nicolăescu-Plopșor, C.S. and Petrescu-Dîmbovița, M. 1959. Principalele rezultate ale cercetărilor arheologice de la Bicaz (r. Piatra-Neamț, reg. Bacău). *Materiale și Cercetări Arheologice* V, 45-60.

Nicolăescu-Plopșor, C.S., Petrescu-Dîmbovița, M., Constantinescu, N., Teodoru, D.G., Zaharia, Em., Florescu, A., Florescu, M., Diaconu, Gh. 1959. Șantierul arheologic Bicaz (reg. Bacău). *Materiale și Cercetări Arheologice* VI, 57-84.

Nicolăescu-Plopșor, C.S., Păunescu, Al., Mogoșanu, Fl., Bitiri, M., Bolomey-Paul, Al., Teodoru, D., Teodoru, S., Florescu, M., Berlescu, N. 1961. Șantierul arheologic Bicaz. *Materiale și Cercetări Arheologice* VII, 37-48.

Niculică, B., Ignătescu, S. and Boghian, D. 1999. Recunoașteri arheologice în Podișul Sucevei. *Codrul Cosminului*, Serie Nouă 5 (15), 39-59.

Nițu, A., Buzdugan, C. and Eminovici, C. 1971. Descoperirile arheologice de la Gura Văii (Municipiul Gheorghe Gheorghiu-Dej). *Carpica* IV, 31-80.

Palaguta, I. 2007. *Tripolye Culture During the Beginning of the Middle Period (BI). The Relative Chronology and Local Grouping of Sites.* British Archaeological Reports, International Series, 1666. Oxford, BAR Publishing.

Petrescu-Dîmbovița, M., Florescu, M. and Florescu, A.C. 1999. *Trușești. Monografie arheologică.* București–Iași, Complexul Muzeal Național "Moldova".

Petrescu-Dîmbovița, M. and Teodor, D.Gh. 2003. Eneoliticul, in M. Petrescu-Dîmbovița and V. Spinei (coord.), *Cercetări arheologice și istorice din zona lacului de acumulare Bicaz*, 83-86, 127-137, 420-429, fig. 75-85. Bibliotheca Memoriae Antiquitatis, XII. Piatra-Neamț, Editura Constantin Matasă.

Petrescu-Dîmbovița, M. and Văleanu M.-C. 2004. *Cucuteni-Cetățuie. Monografie arheologică. Săpăturile din anii 1961-1966.* Bibliotheca Memoriae Antiquitatis, XIV. Piatra-Neamț, Editura Constantin Matasă.

Popovici, D. 2003. Area Organization, Arrangement and Use in the Cucuteni, Phase A Culture (I). *Cercetări Arheologice* XII, 303-322.

Popovici, D. and Bujor, A. 1984. Șantierul arheologic Poduri-Valea Șoșii "Dealul Păltiniș", jud. Bacău. *Cercetări Arheologice* VII, 45-64.

Popovici, D. and Trohani, G. 1984. Șantierul arheologic Poduri-Prohozești "Siliște", jud. Bacău. *Cercetări Arheologice* VII, 65-79.

Preoteasa, C. 2009. Considérations d'ordre démographique et social concernant le complexe culturel Precucuteni-Cucuteni-Tripolye. *Annales d'Université "Valahia" Târgoviște*, Section d'Archéologie et d'Histoire XI, 2, 105-118.

Preoteasa, C. 2011. Nouveaux repères chronologiques concernant l'habitation chalcolithique du tell de Poduri-Dealul Ghindaru (dép. de Bacău, Roumanie). *Studii de Preistorie* 8, 59-68.

Preoteasa, C., Munteanu, R., Nicola, D., Dumitroaia, Gh. 2007. Cercetările arheologice de la Târgu Neamț-Dealul Pometea. *Memoria Antiquitatis* XXIV, 293-327.

Rassamakin, Y. 1999. The Eneolithic of the Black Sea Steppe: the Dynamics of Cultural and Economic Development 4500-2300 BC, in M. Levine, Y. Rassamakin, A. Kislenko and N. Tatarintseva (eds.), *Late prehistoric exploitation of the Eurasian steppe*, 59-181. Cambridge, Cambridge University Press.

Rowlett, R.M. and Shaw, M. 2005. Shell tempered pottery in the Cucuteni-Tripolye areal culture, in Gh. Dumitroaia, J. Chapman, O. Weller, C. Preoteasa, R. Munteanu, D. Nicola and D. Monah (eds.), *Cucuteni. 120 ans de recherches. Le temps du bilan / Cucuteni. 120 Years of Research. Time to sum up*, 157-171. Bibliotheca Memoriae Antiquitatis, XVI. Piatra-Neamț, Editura Constantin Matasă.

Sandu, I., Weller, O., Stumbea, D., Alexianu, M. 2012. Analyses archéométriques sur les moules à sel chalcolithique de l'est de la Roumanie, in V. Nikolov and K. Bacvarov (eds.), *Salz und Gold: die Rolle des Salzes im prähistorischen Europa. Akten der internationaler Fachtagung (Humboldt-Kolleg) in Provadia, Bulgarien. 30 September – 4 October 2010 / Salt and Gold: The Role of Salt in Prehistoric Europe. Proceedings of the International Symposium (Humboldt-Kolleg) in Provadia, Bulgaria. 30 September – 4 October 2010*, 143-154. Provadia – Veliko Târnovo, Verlag Faber.

Schmidt, H. 1932. *Cucuteni in der Oberen Moldau, Rumänien. Die Befestigte Siedlung mit Bernalter Keramik von der Steinkupferzeit bis in die Vollentwickelte Bronzezeit.* Berlin-Leipzig, Verlag Walter de Gruyter & Co.

Scientific Advisory Commitee on Nutrition 2003. *Salt and Health.* London, The Stationery Office.

Sorochin, V. 1997. *Consideraţii referitoare la aşezările fazei Cucuteni A - Tripolie B_1 din Ucraina şi Republica Moldova. Memoria Antiquitatis* XXI, 7-83.

Sorochin, V. 2002. *Aspectul regional cucutenian Drăguşeni-Jura.* Bibliotheca Memoriae Antiquitatis XI. Piatra-Neamţ, Editura Constantin Matasă.

Stoica, C. and Gherasie, I. 1981. *Sarea şi sărurile de potasiu şi magneziu din România.* Bucureşti, Editura Tehnică.

Suciu, C.I. and Luca, S.A. 2007. *IPCTE Radiocarbon Database. Romania and Nearby Areas from Neolithic to Eneolithic.* Universitatea "Lucian Blaga", Institutul Pentru Cercetarea şi Valorificarea Patrimoniului Cultural Transilvănean în Context European. Sibiu. *http://www.arheologie.ulbsibiu.ro/radiocarbon/download.htm*

Šmaglij, N.M. 2001. *Veliki Tripil'ski poselenija i problema rannih form urbanizacij / The Large Tripillya Culture Settlements and Problem of Early Urbanization.* Kiev.

Şandru, I. 1952. Contribuţii geografico-economice asupra exploatării slatinelor în Bucovina de Sud. *Studii şi Cercetări Ştiinţifice* III, 1-4, 407-428.

Tsvek, E.V. and Rassamakin, I.Iu. 2005. The Interactions between the Eastern Tripolye Culture and the Pontic Steppe Area: Some Aspects of the Problem, in Gh. Dumitroaia, J. Chapman, O. Weller, C. Preoteasa, R. Munteanu, D. Nicola and D. Monah (eds.), *Cucuteni. 120 ans de recherches. Le temps du bilan / Cucuteni. 120 Years of Research. Time to sum up*, 173-192. Bibliotheca Memoriae Antiquitatis, XVI. Piatra-Neamţ, Editura Constantin Matasă.

Ursulescu, N. 1977. Exploatarea sării din saramură în neoliticul timpuriu, în lumina descoperirilor de la Solca (jud. Suceava). *Studii şi Cercetări de Istorie Veche şi Arheologie* 28, 3, 307-317.

Ursulescu, N. 1996. L'utilisation des sources salées dans le Nolithique de la Moldavie (Roumanie). In M. Otte (ed.), *Nature et Culture*, I, 489-497. *Colloque de Liège (13-17 décembre 1993). Etudes et Recherches Archéologiques de l'Université de Liège*, 68. Liège.

Ursulescu, N. 2011. Welcoming speech, in M. Alexianu, O. Weller and R.-G. Curcă (eds.), *Archaeology and Anthropology of Salt: A Diachronic Approach. Proceedings of the International Colloquium, 1-5 October 2008, Al. I. Cuza University (Iaşi, Romania)*, 1-3. British Archaeological Reports, International Series, 2198. Oxford, BAR Publishing.

Ursulescu, N. and Ignătescu, S. 2003. *Preuteşti-Haltă. O aşezare cucuteniană pe Valea Şomuzului Mare.* Iaşi, Casa Editorială "Demiurg".

Videiko, M.Ju. 2003. Processes of Urbanization in Old Europe and Trypillya Culture Proto-Cities, in O.G. Korvin-Piotrovs'kyj, V.O. Kruc and S.M. Ryžov (eds.), *Trypil's'ki poselennja-giganty. Materialy mižnarodnoj konferencij / Tripolian Settlements-Giants. The International Symposium Materials*, 256-261. Kiev, Korvin Press.

Weller, O. and Dumitroaia, Gh. 2005. The Earliest Salt Production in the World: an Early Neolithic Exploitation in Poiana Slatinei-Lunca, Romania. *Antiquity*, 79, 306. *http://www.antiquity.ac.uk/projgall/weller/index.html*

Weller, O., Brigand, R. and Alexianu, M. 2007a. Cercetări sistematice asupra izvoarelor de apă sărată din Moldova. Bilanţul explorărilor din anii 2004-2007, efectuate în special în judeţul Neamţ. *Memoria Antiquitatis*, XXIV, 121-190.

Weller, O., Dumitroaia, Gh., Monah, D. and Nuninger, L. 2007b. L'exploitation des sources salées de Moldavie: un exemple de ressource structurante du territoire depuis le Néolithique, in D. Monah, Gh. Dumitroaia, O. Weller and J. Chapman (eds.), *L'exploitation du sel à travers le temps*, 99-113. Bibliotheca Memoriae Antiquitatis, XVIII. Piatra-Neamţ, Editura Constantin Matasă.

Weller, O., Dumitroaia, Gh., Sordoillet, D., Dufraisse, A., Gauthier, E., Munteanu, R. 2008. Première exploitation de sel en Europe: Techniques et gestion de l'exploitation de la source salée de Poiana Slatinei à Lunca (Neamţ, Roumanie), in O. Weller, A. Dufraisse and P. Pétrequin (eds.) *Sel, eau et forêt: hier et aujourd'hui*, 205-230. Cahiers de la MSH Ledoux 12 (coll. Homme et environnement, 1). Besançon, Presses Universitaires de Franche-Comté.

Weller, O., Dumitroaia, Gh., Sordoillet, D., Dufraisse, A., Gauthier, E., Munteanu, R. 2010. Lunca-Poiana Slatinei (jud. Neamţ): cel mai vechi sit de exploatare a sării din preistoria europeană. *Arheologia Moldovei* XXXII (2009), 21-39.

World Health Organization 2007. Reducing Salt Intake in Populations. *Report of a WHO Forum and Technical Meeting, 5-7 October 2006, Paris, France.* Geneva, WHO Press.

World Health Organization 2010. Creating an Enabling Environment for Population-Based Salt Reduction Strategies. *Report of a Joint Technical Meeting held by WHO and the Food Standards Agency, United Kingdom, July 2010.* Geneva, WHO Press.

Vessels for Transportation or Preservation of Salt Water (Brine) in the Starčevo-Criş Culture? – an Ethno-Archaeological Study

Gheorghe Lazarovici
Universitatea „Lucian Blaga" Sibiu

Cornelia-Magda Lazarovici
Institutul de Arheologie Iaşi

Abstract
Different archaeologists (ourselves included) underlined that one of the many reasons for the neolithization process was the requirements for salt. While many stages of Romanian neolithization process were related to salt, we have been interested to explore topics related to transportation of this important resource. We have analyzed several pot types, especially from Early (Dudeştii Vechi, Sővenyháza, Csátalia, Ókécke, Kotocpart, Kopancs, Mađgiari, Vršnik, Stence) and Developed (Vinča, Cioka) Neolithic periods - vessels whose distinctive shape sahows their use as a sort of knapsack. These types of pots permit long-distance transportation of brine as well as other liquids[1]. At Mađgiari (Macedonia), many such pots have been discovered in a sanctuary, proving again the relation of salt to religious life. Besides transportation on the back, we have presented other possible way of transportation (on the head, as in Gumelniţa or other contemporary civilizations).

Keywords
salt exploitation, salt transport, Neolithic, special pots used for transport (on the back or on the head)

We and other authors have underlined several times the reasons for migrations from the South Balkans to the north because of the need for salt, without excluding other important factors. The fact that some stages of the neolithisation process in Romania are related to salt sources, which may even be one of the essential factors, have led us to the possibilities of trade and transport, not just those related to the use of salt. Well-known drying processes and modeling of the drying of salt blocks (briquetage[2]) have already been reported in the Romanian archaeological literature (***Sarea ...2006, 208, catalogue 50, 71; Buzea, Andrea (Chiricescu) Deák 2008, 69, pl. VII.). The first use of briquetage in Romania has been dated later - to the Cucuteni B phase. There are several hypotheses on previous drying processes, with some colleagues supposing the use of other types of vessels less frequent in the Neolithic in general, but present in the sites of the salt source areas (Schuster *et al,*. 2010 from area Govora – Ocnele Mari). Almost all Early Neolithic migrations from the South have their main settlements located in areas rich in salt. Migration I is attested in the areas of Râmnicu Vâlcea, Sibiu, Cluj-Napoca and Târgu Mureş, whereas migrations III and IV have sites in the Râmnicu Vâlcea, Focşani and Piatra Neamţ areas (Lazarovici Gh. 1996; Lazarovici Gh., Maxim 1995, the fourth migration, 41-43, 49, 140, 156, 165, 169, 184, 195). There are seasonal and secondary habitations that can hardly be placed exactly in current chronological stages, where material culture is related to functionality rather than fashion; here there are few materials - a fact compounded by the small size of excavations. Many small settlements are at the head of valleys, in the vicinity of springs with what now appears as favorable grazing areas; however, the Early Neolithic environment is unknown, notably the relationship between forest and pasture. However, in the pit houses with dark filling and depths of over 60 cm, the dark, humus-rich soil shows that the vegetation had not exhausted local soils. In these sites, there are few monochrome ceramics but the fragmentary state of the ceramics due to poor combustion does not allow more accurate reconstructions of the shapes and classifications. Hence, there is a need for data and statistical analysis for more accurate time-framing and to know the dynamics of different types of ceramic (Figure 1).

These statistic data allow us to make some observations, especially for the colleagues tempted by the radiocarbon data to ascribe an early date to these Neolithic habitations.

Archaeological data are often vague, sometimes time-framing is incorrect, whether because based on old chronological systems or owing to selective publication of materials which, take into account what authors found important for that site or assignment to a certain phase.

Sometimes the publication includes more radiocarbon data than material. The chronological table here is based on our database and includes over 30.000 records from South Europe as far North as the Balkans.

In the first column are specified stages of evolution - sometimes clear and sometimes vague because of incorrect or insufficient publication. We made some

[1] Salt is dissolved by water and, because of the salt water transport, it is almost impossible to demonstrate the presence of salt in some containers. In certain types of vessels, transport of liquids other than brine was possible.
[2] The meaning of 'briquetage' is not quite correct, because the operation involves moulding and pressing, which is not the case in the Romanian Neolithic - but rather evaporation.

remarks on the manner of publication: old dating, global, not on complexes; old materials without explanations or details, secondary habitations; fewer materials, without closed complexes. We have noted in several studies and syntheses the chronological and cultural links with Southern areas, often related through migrations and diffusions.

Ia/phase	No. fragments	Monochrome	Painted	Unpainted	Specification technology ornaments
unspecified **	416	267	149	267	
SC IA *	69	28 *	41		
SC IB, IC.....	1286	810	434		
SC IB	182	134	41		
SC IBIC	550	430	122		
SC IB IIIB	1286		434		
SC IC***	8333		433	7910	2021
SC IC	372		63	372	22
SC IC IIA	7496		337	5500	1494
SC I II	185	90	35		
SC IIIA	278		31	26	158
SC III...	6625		565		189

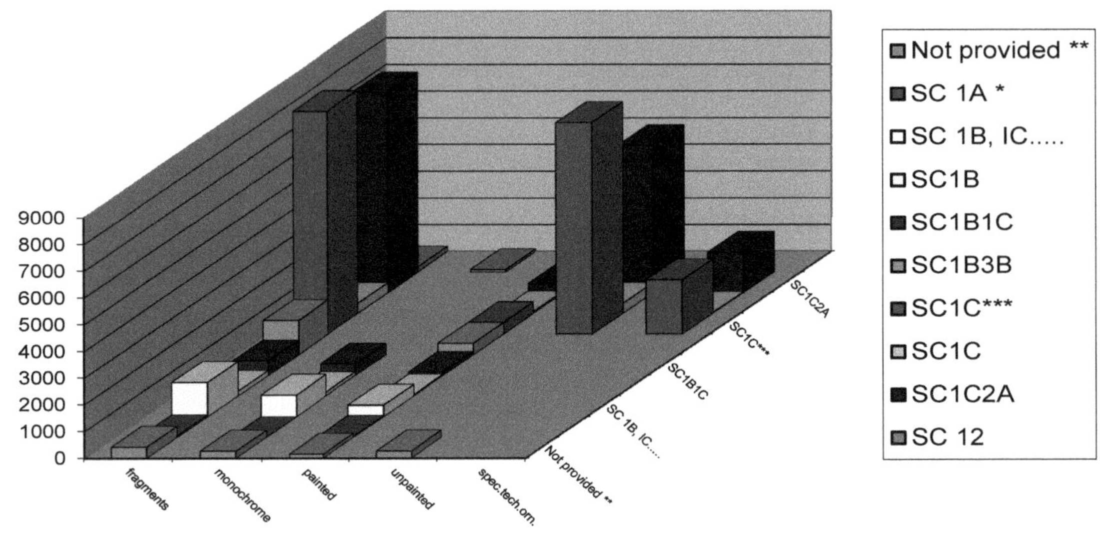

Figure 1. Starčevo-Criş culture, chronological table with frequency.

Migration IA is less secure, with insufficient material in closed complexes, but stratigraphic observations show that they are as early as the B1 materials at Gura Baciului (Studies summarizing neolithization: Lazarovici Gh. 1996 (Forli); 2000 (Bulgaria); 2006; Lazarovici C.-M., Lazarovici Gh. 2006, chapter I and bibliography; and others). Recent discoveries made by our Sibiu colleagues prove that, starting with the first migration and continuing with the others, in the Sibiu – Miercurea Sibiului area there are several early Neolithic sites, all in areas with salt sources: 4 sites at Miercurea Sibiului, Cristian (that seems to be the earliest) and Ocna Sibiului. These data are supplemented by findings from the Cluj-Napoca area (Gura Baciului, Fundătura and Vultureni). These settlements show migrations that occur through the Olt Valley with routes to the South at Copăcele – Valea Răii, via Cârcea to the Balkans, but also to the Banat and the Pannonian Plain on large rivers (Donja Branjevina, Szarvas 23, and others), as well as to Central Serbia along the Morava Valley (Grivac etc.). Publication of these data will change opinions and current statistics. The maintenance of some habitations from Starčevo-Criş (SC) stage IA to IIIB SC, continuing into Vinča A, shows not only ways of entry, but possible "Neolithic roads" of salt on shepherds' roads.

Methods of transport and exchange of salt

Ethnographic studies, archaeological and historical researches have demonstrated many means, methods, techniques and procedures for the exploitation and transportation of salt.

Briquetage

This methodology is proven in regions with large numbers of such briquetages found in the area of some sources and ancient salt exploitation, allowing large transport distances, but also serves as "monetary" exchange or element of prestige (At Lunca - Oglinzi,

Solca, Cacica and others: Ursulescu 1977; 1995; Andronic 1989; *** Sarea... 2006; *** Sarea... 2008; Munteanu 2006a; 2006b; 2008; Munteanu, Garvăn 2008, 196 etc.; *** Archaeology and Anthropology of Salt 2011). Briquetage is proven in Moldova starting with Cucuteni B phase, viz., the Copper Age, during its most developed period (Monah D. 1991; 2002 etc.; Dumitroaia 1987; 1994; Dumitroaia Gh. *et al.*, 2002; 2003; 2004a; 2004b; 2008 etc. Monah D., Dumitroaia G., Nicola D. 2011 and bibliography; Monah D., Dumitroaia Gh., Nicola D. 2011 etc.) - the time of other crafts and prestige goods, in particular the production of copper and gold.

Some pots with a conical shape are reminiscent of salt briquetage, as in the case of the mould from Răuceşti - Munteni (*** Sarea... 2006, records made by R. Munteanu, 73, 202 and others). We cannot guarantee that such vessels were used in briquetage, but the presence of holes in some vessels allows their suspension and the preservation of valuable goods (liquid or powder), such as salt or oils. At Lunca–Poiana Slatinii, such a vessel with holes on the mouth (*** Sarea ... 2006, 202, Munteanu, catalogue 196) has been ascribed to the early Neolithic. Such sort of pots are present at other Early Neolithic sites related to salt, viz., Şeuşa in house L1, a site located in a salt-rich area, as is suggested by the Hungarian name (so salt; şos viz salted water) and other sites such as Valea Şeuşei in the Tureni-Miceşti area (Ciută 2000, Figure 25/1; toponyms found by us at Bădeni and Deleni; for toponymy and linguistic data, see Poruciuc 2008; 2011) - (In sites with pottery painted with white and dots: Vaksevo, Čohađiev Şt. 2001, pl. 47/ 2, 4; Lepenski Vir: Perišić, Nikolić 2004, pl. II/6; Cuina Turcului I, II, Păunescu 1979, Figure 20/5; Lazarovici Gh. 1979, pl. V/B5; Aiman: Staljo 1986; Lazarovici Gh. PB, Figure II.18a.3; or others: Pesterica, Kitanovski 1976, Figure VII/10; Obre I, III: Benac 1973, Figure 30/10)

A characteristic of these briquetage is a base with wide legs, allowing the salt to dry out naturally in the sun, and a conical interior that allows removal of content in the shape of a cone (Figure 4e: suitable for transport, especially for grating over dishes)[3].

At Hallstatt, briquetages are represented by short, wide vessels with a flat bottom (Figure 3b), shapes common for several periods; if they are in areas with evidence of salt exploitation and processing, they may be considered as briquetage.

Some studies identify as briquetage those vessels with conical interior, found in the area with salt sources at Govora - *Dealul Săpunarului* (Schuster *et al.*, 2010, pots from the Govora – *Dealul Săpunarului* area) (Figure 3a) and dated to the Starčevo-Criş culture.

Such pots with a conical interior, used for salt or with a different function have been discovered at Cluj-Napoca - Stăvilar in stage SC IVA (Figure 2a) - a site located on the salt road to the Pannonian Plain, some 5-7 km from the salt-rich area at Someşeni – Apahida (Lazarovici Gh., Lazarovici C.-M. 2011). A pot of similar shape was found at Ostrovu Golu (Figure 2b), two others at Traian - Dealul Viei (Precucuteni I) (Dumitrescu H. 1976, 47-54, Figure 2, being related with Bug-Dnestr culture and regarded as imports) and another in the Cucuteni culture, at Mitoc - Pârâul lui Istrati (Figure 2c) (Cucuteni-Trypillia 2008, 93, 208). The sites of Ostrovu Golu, Lepenski Vir and Cuina Turcului had no nearby sources of salt, but needed salt to smoke and preserve fish. But proving such cases is difficult.

Transport of brine from springs

Ethno-archeological analogies show the transport of salt as salty water/brine, as salt blocks and re-crystallized salt. Such transport involved animals, either pulling carts or as pack-animals, but salt was also carried by people.

There are ample documents on salt transport from the Roman period until our times (Marc 2006, 155). Some current examples show the controlled use of salt sources.

There are some inconveniences caused by salt water control made by the local community, local or zonal authority (e.g., in Filea de Jos, Cluj district)[4].

Sometimes there is regulation of the days when salt can be exploited by a community (Marc 2006, 153-155).

Individual collection of the salt water from pits (funnel-shaped, similar to mines, but smaller, with a base of about 3 m and depth of 1,80 m) from outside the area recently exploited but with rich salt sources (e.g., at Uioara de Sus, near salt sources at Ocna Mureş) (Lazarovici Gh., Cristea 1979, 431-446).

Valeriu Cavruc believes that many holes from salt springs areas at Figa were produced by old exploitation prior to the Bronze Age.

We proposed to analyze some types of vessels, especially from the Early Neolithic (Dudeştii Vechi, Sővenyháza, Csátalia, Ókécske, Kotocpart, Mađiari, Vršnik, Stence etc.) or from the Developed or Late Neolithic (Vinča, Turdaş, Cioka), whose asymmetrical form, their narrow neck and especially the position and form of their handles, suggest a use as a knapsack, in this case for brine transport.

---

[3] From my childhood, I remember my grandmother (in the village of Milcoveni, Caraş-Severin district), scratching from the salt or sugar block held on a shelf, as in the tale „*Soacra cu trei nurori*"/Mother in law with three daughters (Gh. Lazarovici).

[4] Our informer in Filea de Jos, Chirilă, told us to be careful not to pour brine on driveways.

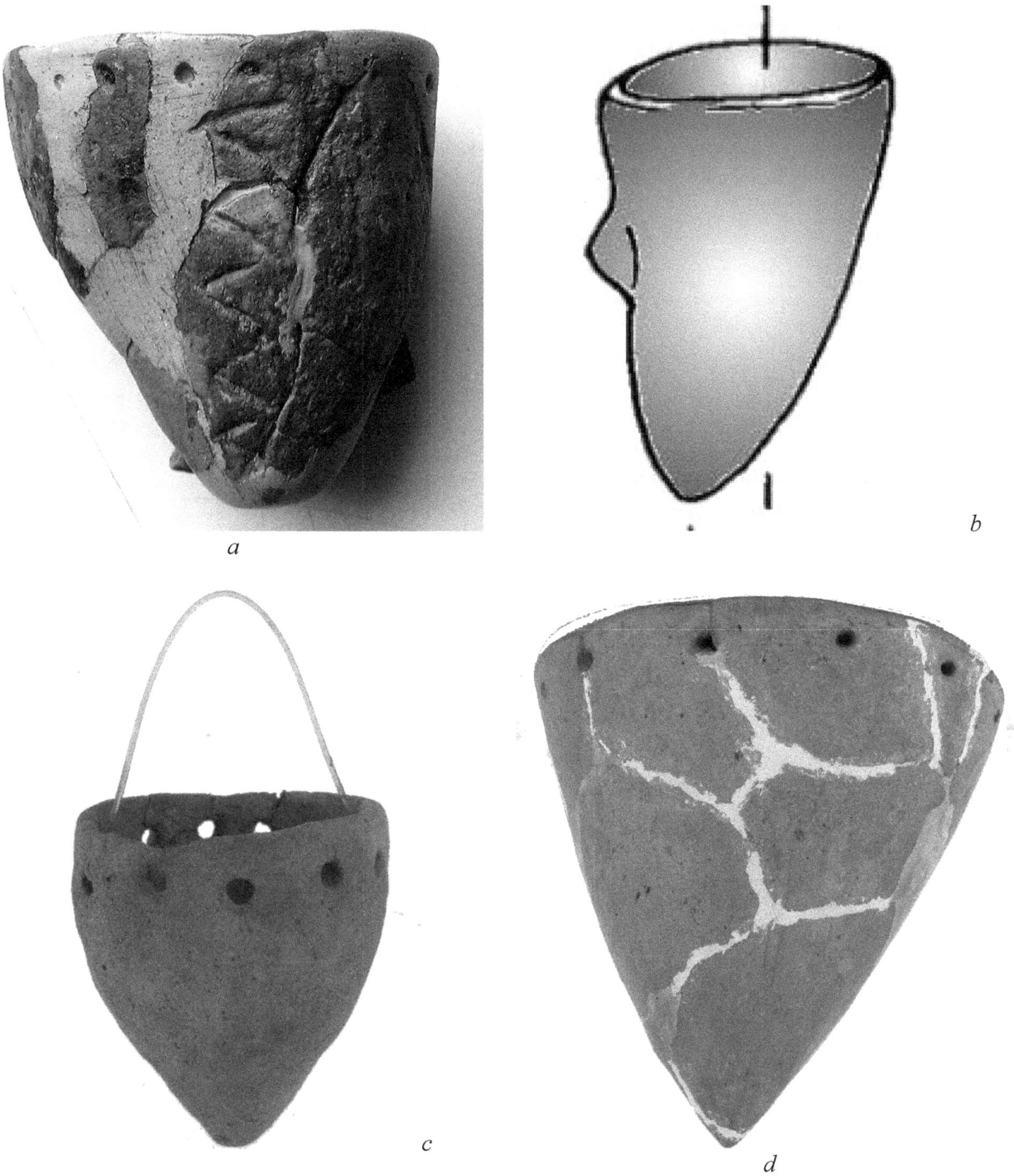

Figure 2. Starčevo-Criş culture, IVA phase, pointed-bottomed vessels: a) Cluj- Stăvilar; b) Ostrovu Golu; c) Pericei, Zau culture; d) Cucuteni B, Mitoc-Pârâul lui Istrati site, after Cucuteni-Trypillia 2008, catalogue no. R71.

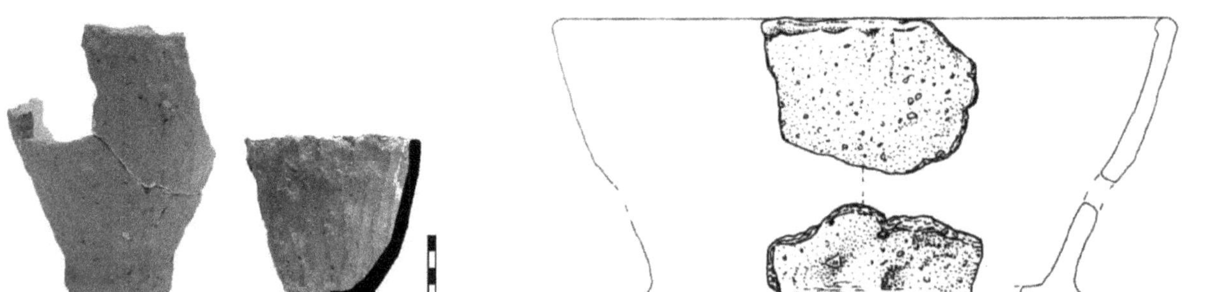

Figure 3. a. Copăcele, pots for briquetage, after Ch. Schuster *et al.*, 2010; b. Hallstatt, pot for briquetage, Kingdom of Salt 2009, p. 86.

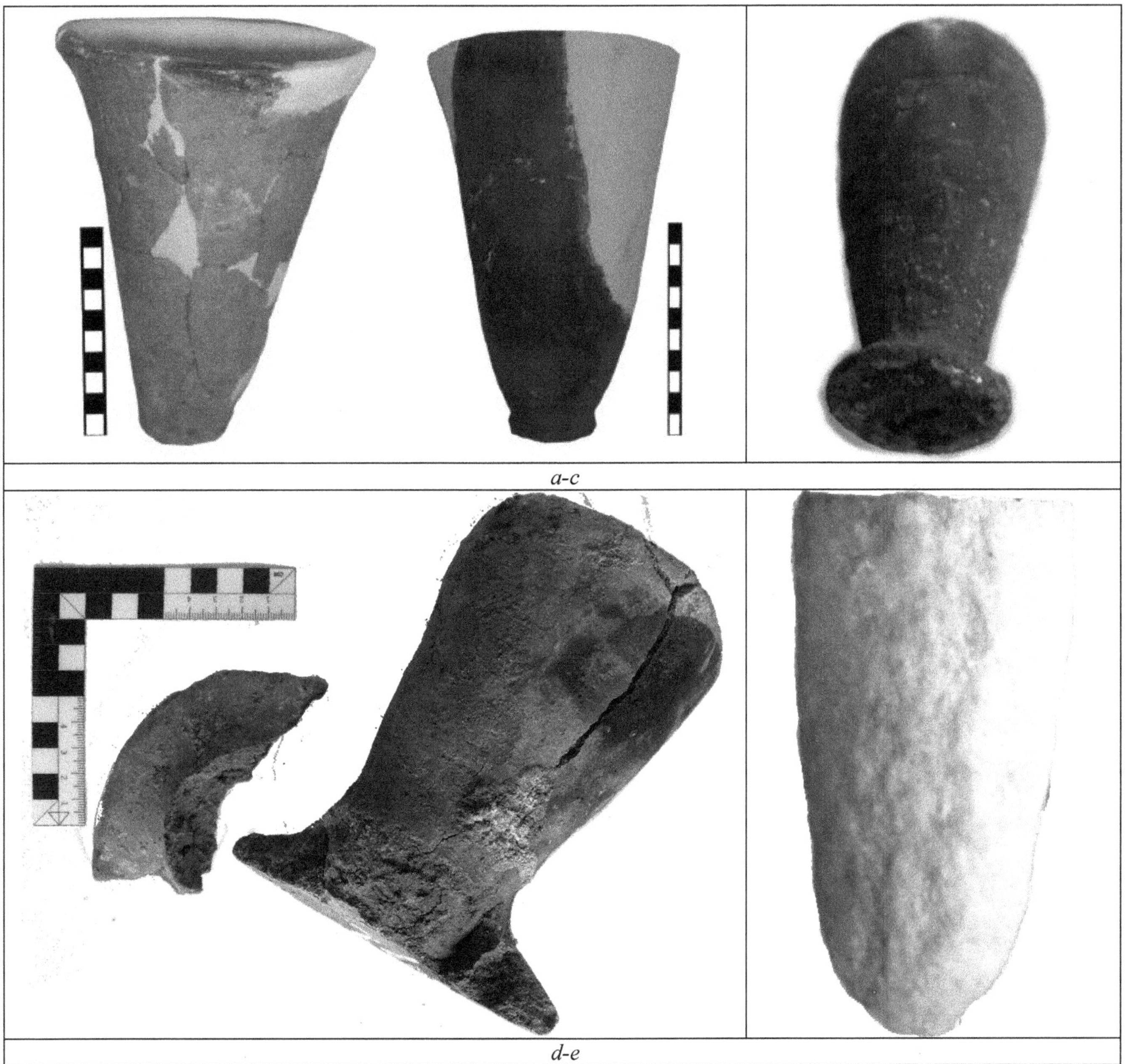

Figure 4. Briquetage: a-b) Govora Museum; d) archaeological experiments, Figa; c, d) Răucești – Munteni; e) salt cone.

Figure 5. Hallstatt, knapsack used for salt transportation from the mine into exterior, after Kingdom of Salt 2009, 61.

Using back-packs is documented from the Bronze Age, when the back of a menhir at Ciceu – Corabia (Figure 7) (Baia de Criş, 3 examples (remembered by Téglas 1884; 1985; Floca 1966: apud Glodariu *** ***Enciclopedia*** 1994, 152 and bibliography; Wollmann 1996, pl. VII, XXVII.3)) shows a longish, narrow knapsack. The mentioned menhir was located at the mouth of a valley and near some ridges marking maybe routes leading to the Gutâi Mountains, an area where there are copper sources as well as a centre of salt sources (e.g., Dej, Figa). Such pieces are also known in the Apuseni Mountains at Baia de Criş (Figure 6a-c), dated to the Iron Age or to the beginning of the Bronze Age (Alexandrescu 1963, 148), a date to which we subscribe for the piece discovered at Ciceu-Corabia. These menhirs, as well as one discovered in Spain and ascribed to the Bronze Age, have a knapsack for the transport of ore (Figure 6d).

In Deva museum, from the Zsófia Torma collection from Turdaş there is an unpublished figurine bearing on her backside a vessel. We must also not forget the Early Iron Age leather knapsack found in the salt mines at Hallstatt, used to transport salt from the mine to the surface (Figure 5) (*** Kingdom of Salt 2009, 60-61).

We mentioned above a number of vessels used to transport brine but we do not exclude the possibility that water transport was also involved. The Neolithic sites mentioned above were located on smaller or larger rivers. On small streams that apparently have no watercourse or a very low flow, peasants were digging pits 1-2 m from the edge of the stream for water filtration before covering the pits which were occasionally used[5]. Equally, on the floodplains of large rivers were located springs in the bank or on the mouth of old-channels, which operated at times of high flood (e.g., the Timiş River, near the Parţa excavations).

Reasons for Neolithic migrations to the North were determined by several factors: global warming and the desertification process that took place in the South, the formation of an optimal climate in the Danube regions, and rich sources of salt good for animals and the preservation of meat, as well as favorable places for agriculture and livestock.

These data allow us to support such migrations, but do not help in defining exactly the place where it started. The time-scale of these migrations was defined by S. Dimitrijević. For Hungary (Ykers *et al.,* 2009 apud http://arheologie.ulbsibiu.ro/radiocarbon/2007%2010%20ian%20data.htm and bibliography) and the south Romanian Plain, there are new radiocarbon data, showing a possible migration in stage SC IIIA, but until now there is very little published material and that not from closed complexes, so is very difficult to support this idea through statistical analysis.

After an initial cooling time at circa 6400-5900 B.C., an optimal climate for shepherds can be defined cca. 6100 - 5900 CAL B.C. - the time of the first radiocarbon dates from Gura Baciului and Ocna Sibului - allowing the use of lowland water in the spring and autumn and mountain water during the summer (Figure 8).

For farmers and shepherds, the SC IC/IIA phase (after 5800 B. C.) is the time of a milder climate, when there is an extension of the neolithisation process in the Early Neolithic.

We note that sites are grouped by major rivers forming easy communication routes, not forgetting that mountain peaks are also favored by shepherds in the summer for the same purpose.

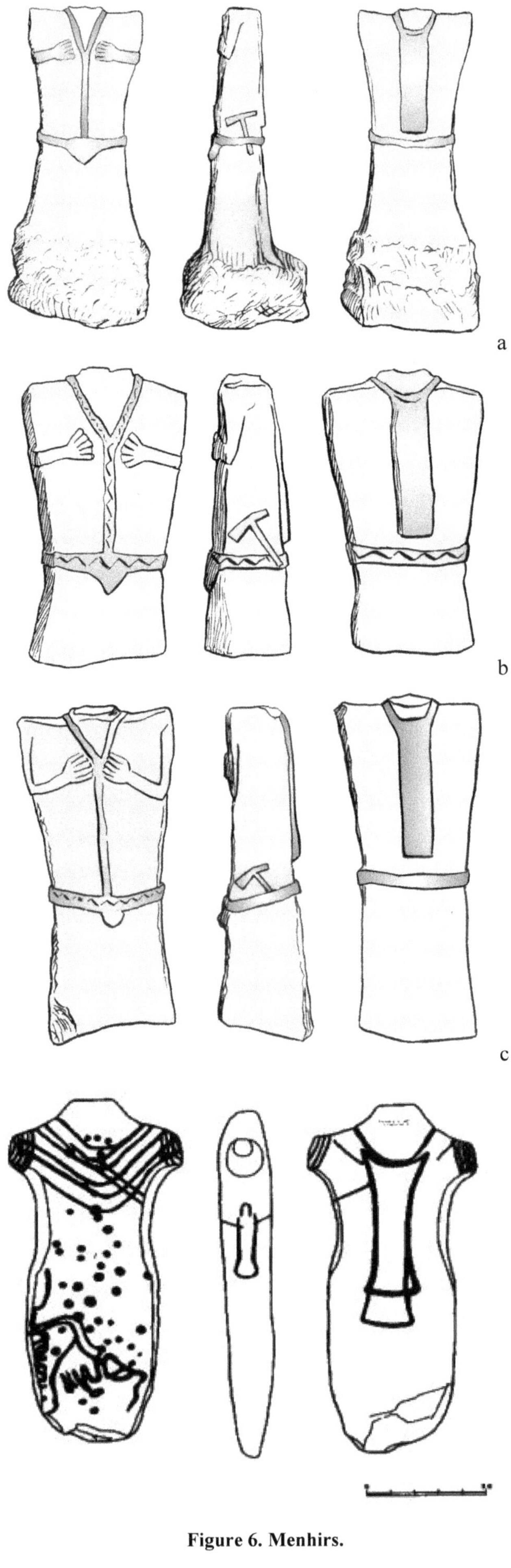

Figure 6. Menhirs.

[5] We note that in our childhood in the village Milcoveni, on the streams *Săliştiuţă* and *Crivaia*, the pit dug by my grandfather was named *ştiubei* (Romanian).

Figure 7. Ciceu – Corabia, menhir representing a miner with a knapsack on his back tied by a belt-harness. Photos and processing made by Gh. Lazarovici.

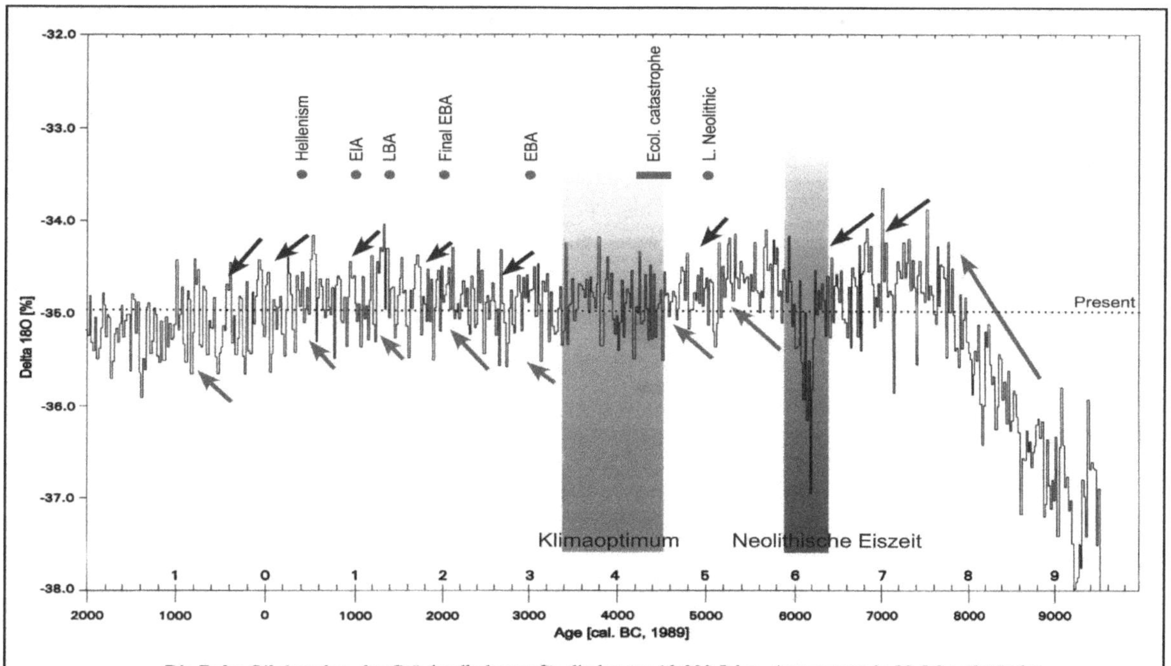

Die Delta O[18]-Angaben der Grönlandbohrung für die letzten 10 000 Jahre. Auswertung in 20-Jahr-Abständen (nach Dansgaard u.a. 1993). – Graphik: I. Vajsov, AIM-BAW.

Figure 8. Climate Change in the ratio of C14 and Bulgarian chronology (after H. Todorova and I. Vajsov 1993).

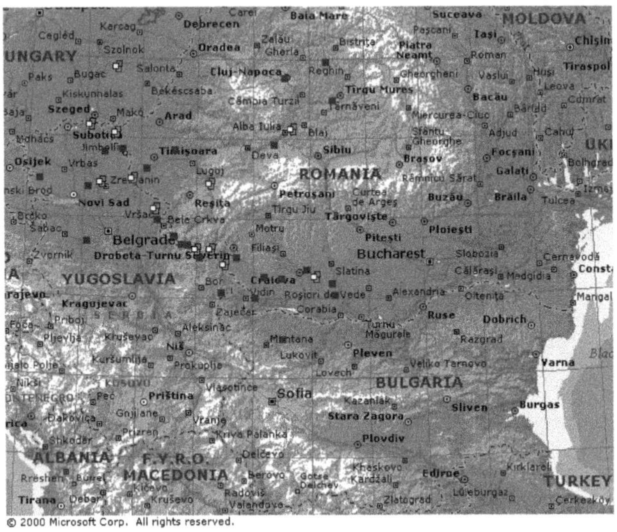

Figure 9. Extension of the neolithisation process in SC I-IIA in South Central Europe (3 small squares show more points concentrated in an area).

A new migration followed during stage SC IIB-IIIA. To these periods correspond archaeological materials from Endrőd 3/119 (pit 9, trench 19), most of the complexes here being dated to phases SC IIIA and IIIAB (Makkay, Starnini 2008, Figure 206-300; Lazarovici Gh., Maxim 1995: SC IIIA: 51, 58, 73, 82, 93- 96, 113); this being a local ceramic evolution, the length of time these vessels were used in phase IIIB is difficult to establish but in some complexes appear Szatmár imports (Makkay, Starnini 2008, in pit G9 from Section XIX, Figure 281/8). Such horizons were not systematically excavated in Romania, except at Dudeştii Vechi: radiocarbon data are around 5800 CAL B.C. at Dudeştii Vechi, Foeni-Gaz, Şeuşa, etc. (Lazarovici C.-M., Lazarovici Gh. 2006, 83 pl. II.23; Lazarovici C.- M. 2006 see and bibliography; Suciu, Luca on: arheologie .ulbsibiu. ro/radiocarbon /2007 %2010%20ian%20data.htm). The pottery at Gilvăz I and II continued to be in use into phase SC IIIB. Some shapes at Endrőd parallel vessel forms in Bulgaria. But such details cannot be made without an analytical study of databases.

Figure 10. Dudeştii Vechi (Beşenova Veche), pot for salt transport on the back, after I. Kutzián 1944.

The site of Dudeştii Vechi (Figure 10) is located some 100m from a stream water course.

If the vessels we are talking about (with four vertical holes arranged two by two, on the top and bottom of the vessel) were used for carrying water, we should expect to find more such containers in Körös houses.

These holes allow the introduction of thick cords made of fibers or wool, making them convenient to be transported over long distances. The curved form and flat back of the pot allows a woman of low height, as is exemplified by the Mediterranean type of women in the Starčevo-Criş culture, to position the vessel perfectly on the curvature of the spine. Macedonian artifacts from Mađgiari also had a hole that allowed binding at the neck, which reduces movement while walking and the loss of contents. Such vessels are numerous in the Hungarian sites of the Early Neolithic at Kotocpart – Vata (Figure 11a), Sővényháza –Baja (Figure 11b), Csàtalja (Figure 11c), Szentes and Ókécske (Figure 12a) (Kutzián 1944, pl. XVIII/2,

XIII/15a-b.). All have four vertically oriented handles, an asymmetrical form and a narrow neck; one of them has four short legs to ensure greater stability (Figure 11) (Kutziàn 1944, pl. I, XI, XXV).

The same kind of vessels with large vertically-perforated lugs is present in the Developed Neolithic, as in several Vinča sites: Vinča - *Belo Brdo*, Cioca (Figure 13) and Crnokalačka Bara (Kutziàn 1944, pl. LVII). Of course this type of vessel was not given much attention[6], because amphorae were numerous. We must specify that only asymmetrical amphorae with four vertical handles were used for long-distance transport. There are some amphorae with three handles, all with narrow necks, which are suited for transport with a yoke.

Such pots have been discovered at the Vinča sites of Zorlenţu Mare and Liubcova (Lazarovici Gh. 1979, pl. XVI F33; XVII A34 and others). Of course, the location of water sources at Zorlenţu Mare, some 50-100 m lower than the site, indicated the use of such amphorae for water transport using a yoke. In an Early Neolithic sanctuary in Macedonia (the Anzabegovo culture) at Mađgiari were found dozens of large vessels (Angeleski 2012, type Ak1, Ak2, pl. 2, 224, pl. 47/I, 48a.), most of them reconstructed having an asymmetrical shape and five handles (Figure 124/a-c) (Sanev 1988, 9-10; Kitanovski *et al.,*1990, 107-112, Figure 1-7; Lazarovici Gh. *et al.*, 2001, vol. I.1, 238).

All these vessels have four thick vertical handles for the insertion of cords and a handle on the neck on the opposite side to stabilize carrying over longer distances (Figure 14).

The presence of such pots in the Early Neolithic sanctuary at Mađgiari (Kitanovski *et al.*, 1990, 107-112, Figure 1-7; Sanev 1988, 9-10; Lazarovici Gh *et al.,* 2001, 270; Angeleski 2012, pl. 47/I, 48/a, type Ak1, Ak2, pl. 2, 224), along with other evidence, shows the importance that community members gave to the content of these vessels and underlines the religious role of salt. On the other hand, the natural question is what was deposited in so many vessels and which product was brought from a long distance. Our answer is salt. This was necessary, given the lack of salt in the Balkans (Monah 2008a, 125; Chapman, Gyadarska 2003). In fact, different archaeologists have often written about ethno-cultural links between the Danube Region (especially but not only Oltenia-Transylvania) and Macedonia – Greece, related with several stages of the neolithisation processes (Garašanin D. 1959; Rodden 1962, 285, Makkay 1965, 8; 1966, 58; Vlassa 1972; 1972a; 1980, 692, 694-695; Nica 1976; 1977; Garašanin M. 1979, 83-84, 90-92, Figure 9/5-6, pl. XIII/1, 4, 132-133; Lazarovici Gh. 1986, 3, 11; 1996, 22; 1998; 2000; 2006 etc.; Dumitrescu Vl. 1987-1988; Paul 1995, pl. XXVIII/1-4) where grazing and salt have played an important role.

If we start from the present to the past, without going too much into details, we note that salt roads remained the same for ancient Southern communities wishing to reach the Danube areas. Whether salt was transported by rafts or on ships, taken with carts, or carried by humans on the back or on the head - salt roads passed along major traffic routes - the same routes used for neolithisation processes in the Balkans.

An ethno-historic study written by Dorel Marc indicates salt transport on rafts from Transylvania to the north or south Pannonian Plain, from Transylvania to Oltenia until the Danube, as well as for Moldova, on Bistriţa and Siret rivers and from there further south. For Transylvania, some studies mention rafting and salt block transport on rafts, with animals or on the back. Especially during prehistoric times, human transport was a natural thing (Figure 15).

Medieval documents from Moldova (Vitcu 1987; Ciobanu 2011; Alexianu *et al.,* 1992; 1993; 2007; 2008; Monah D. 2008, 13; Istrate 2008, vol. I, 97, 103, 245, 246, 310, 362 etc.; vol. II, 105, 142, 271, 461, 573 etc.; vol. III, 35; etc.) talk about salt exploitation and salt transport to the Balkans, to the East (Ukraine), to the North (Poland), and to the South (Greece and Ottoman Empire). For Transylvania, there are documents, prints and various studies covering the AD 17th - 19th centuries (Figure 5). Earlier in time, we note the first early medieval documents governing salt export and salt customs (Marc 2006), as well as documents referring to the Roman period (Wollmann 1996, 240ff.) when the exploitation and transportation of salt were well organized, involving officials, customs and localities (Wollmann 1996, 240ff. and old bibliography; Wollmann, Ciugudean 2005; Cavruc *et al.,* 2006b; etc.). We cannot neglect the salt mining of the Pre-Feudal period (Goths, Gepids, Huns, Avars and Slavs): so, the treasures of Apahida or the discoveries of cemeteries and gold artifacts (e.g., Cluj-Napoca, Fundătura, Turda, Cluj-Napoca – *Polus* etc.) (***Polus 2008, 38-49) are located along the terrestrial salt roads (e.g., for the Mureş, Arieş and Someş, see Figure 15) which control its export to the Pannonian Plain. The first states of the Early Medieval period also showed a strong interest in salt, as the Moravian kingdom. Later, Hungarian kings built fortresses on the salt roads where the fortifications at Dăbâca, Cluj-Napoca - Mânăştur, Moldoveneşti and others controlled salt or established customs points for salt (e.g., Apahida, Cojocna, Dej and Turda). Is it possible that the Neolithic communities of Iclod or Fundătura controlled the Neolithic salt roads to the Pannonian Plain or central Transylvania (Pascu *et al.*, 1968; Rusu 1977; Wollmann 1996, 240ff.; Chiricescu 2006; see our study with comparative analyzes on the intensity of living in different historical periods in areas of salt, Lazarovici Gh., Lazarovici C.-M. 2011 and bibliography)? Later still, the Hungarian kings regulated salt trade, salt customs, as well as who were the beneficiaries of such customs (monasteries, royalty).

[6] Chapman 1981 considers that handeled vessels are characteristic of the late Vinča phase, as the ones at Cioca.

Figure 11. Pots for transporting on the back: a) Kotocpart - Vata; b) Sővényhàza -Baja; c) Csàtalja, after I. Kutzián 1944.

a

b-c

Figure 12 a) Ókécske, b-c) Vinča, asymmetrical amphorae with handles, after I. Kutzián 1944.

Figure 13. Cioca, Vinča culture, asymmetric amphora with vertical handles, after I. Kutzián 1944.

Figure 14. Ceramic knapsacks for transporting salt or other liquids over large distances: a-c) Mađgiari, d) Stenče, Early Neolithic, after S. Angeleski 2012.

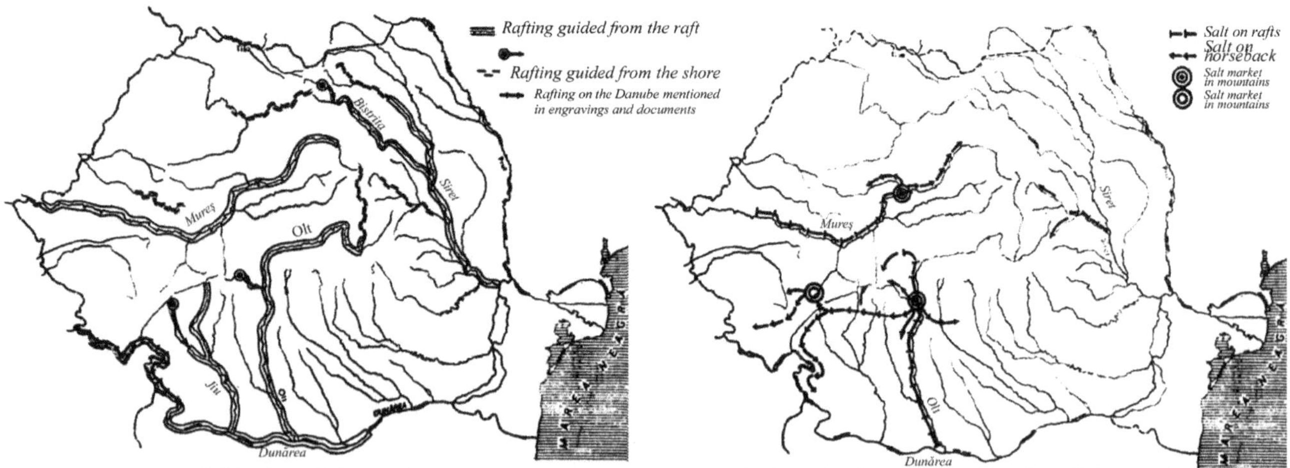

Figure 15. Rafting and salt trading centers in Romanian Middle Ages (after D. Marc 2006, 157).

For example, Catholic knights (Teutonic Knights) enjoyed certain rights from the Hungarian royalty in exchange for guarding roads or passes into southern Transylvanian areas via the Mureş and the Olt (cf. others (Cavruc, Chiricescu 2006; Chiricescu 2006 and bibliography; Marc 2006, 153-154 and bibliography)).

Salt is found in other parts of Europe, but the quantity and quality of salt in our regions have caused interest in this area.

Continuing the journey backward in time to see the interest of salt extraction, salt-related localities, baths and salt-rich areas, we have ample evidence, functions and customs for the major rivers used in the Roman period. The Roman interest in salt and their intensity of exploitation has been studied by different colleagues (Wollmann 1996, 240ff. and old bibliography; Cavruc *et al.,* 2006; 2006a; 2006b). Dacian salt exploitation is documented by the discoveries at Valea Florilor (Wollmann 1996, 243; Lazarovici Gh., Lazarovici C.-M. 2011, 2- 4, 10, 22; Cavruc 2012, 3-12, especially 5) Cluj district, as well as in the Maramureş belonging to the free Dacians, from the same period of time or earlier (Wollmann 1996, 243; Kacso 2006; Dăncuş 2006).

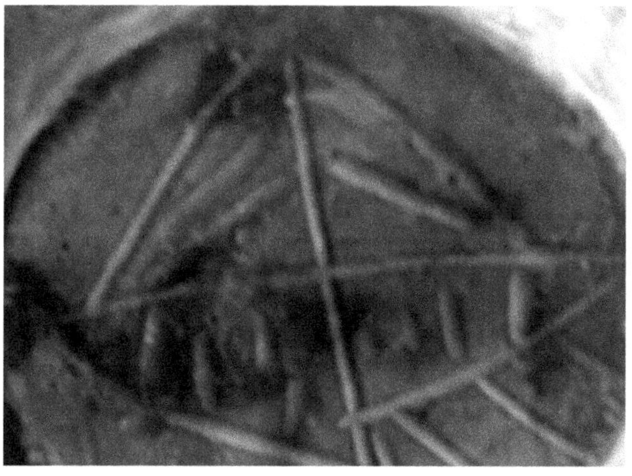

Figure 16. Sailboat from Cluj-Napoca, Middle Neolithic, photo Gh. Lazarovici.

a

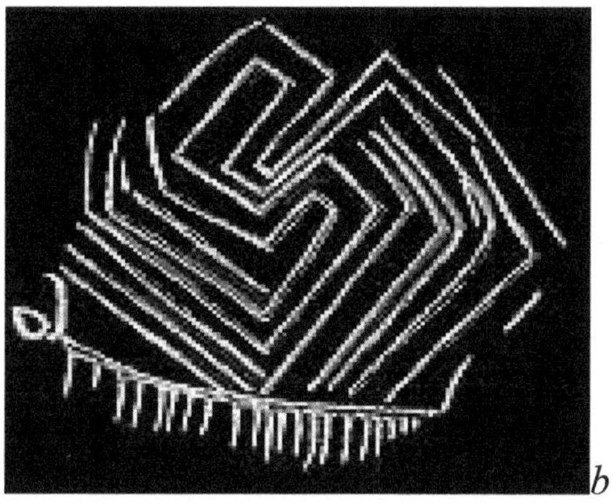

b

Figure 17. Other boats: a) Poduri, monoxile, after D. Monah et. al. 2003, Figure 83, 155; b) Korbovo, Vinča C, after J. Šarić 2003, Figure 2.

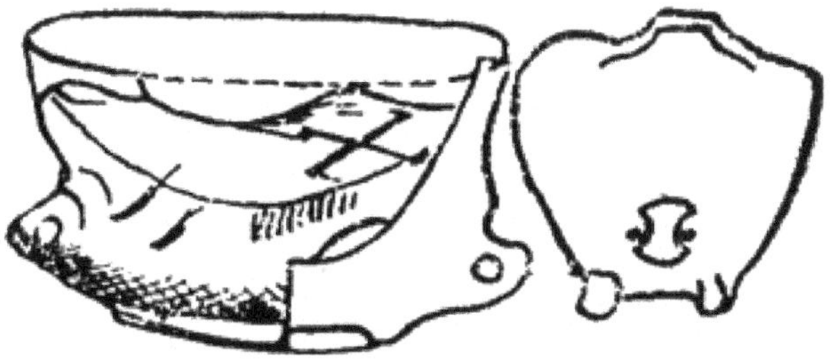

Figure 18. Boat from Zau culture, Iclod group, after Gh. Lazarovici, Z. Kalmar 1987, Figure 10/21.

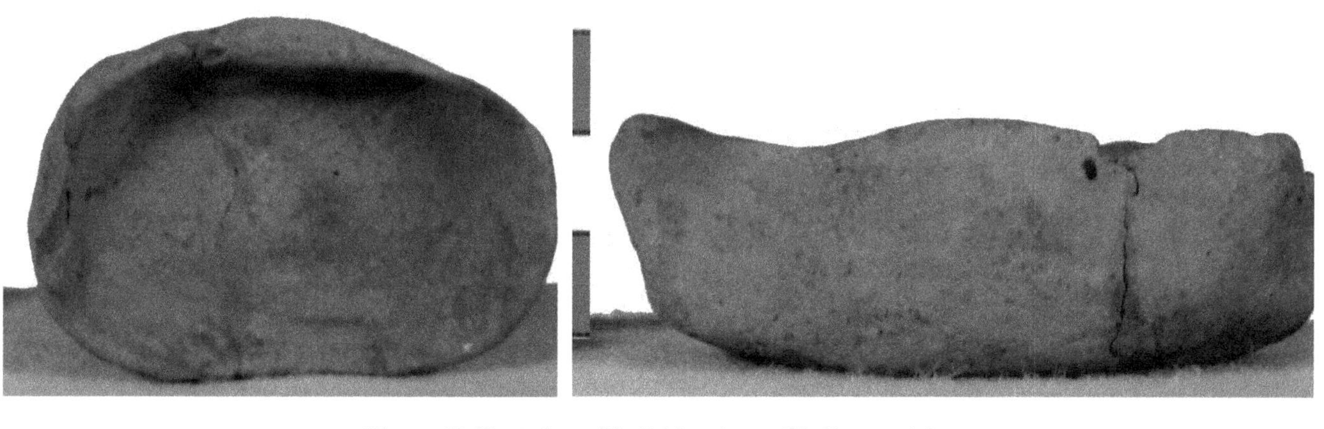

Figure 19. Parţa boat, Neolithic, photos Gh. Lazarovici.

Recently, complex investigations at Figa, with chemical, dating and dendrocronological analysis has brought new information and important data on operating systems and salt mining in the Late Bronze Age (Cavruc, Chiricescu 2006, 7 ff.; Cavruc 2006, 33ff.; Cavruc *et al.,* 2006, 41ff.; 2006a, 50ff.; Cavruc 2012, 6), with some wells dated to the Early Bronze Age (At Figa it is a well where have been discovered ceramic fragments decorated with a small sweep, ascribed to Early Bronze Age: Cavruc 2012) and real mines in the Late Bronze Age; all these studies show many new dimensions of recent salt research. We should not disregard the fact that some of the largest bronze hoards in Europe, from Uioara, Şpălnaca and other large hoards (Wollmann 1996, 241, no. 7 apud A. Popa, ing. Chiriţă Radu), were deposited in salt-rich areas, all of which with discoveries from the Neolithic until Early Medieval times (Wollmann 1996, 242-244; Petrescu-Dîmboviţa 1977, 336-337).

There is also knowledge from early times of the well-known curative properties of salt (halotherapy), as well as the religious importance of salt. Already in the Bible are mentioned customs and practices related to salt (Monah 2008a, 122, Leviticus 2.13 and bibliography).

Given the special role played by salt in human and animal life, we believe that the vessels discussed in this article were used for the transport of brine. The presence of these vessels in the Mađgiari sanctuary shows the great importance given to food, and the therapeutic qualities of salt and brine (Alexianu *et al.,* 2007, 103, 112-123).

The askos vessel shape, with its potential for handles for carrying, constitutes evidence for long-distance transport of liquids, including brine. All these show us that we should not be surprised at the presence of pots for salt transport in the Neolithic. Transportation of salt blocks is hard to prove for the Neolithic, but we must not forget that we have from this period evidence not only for pedestrian transport, but also for the navigation of large rivers by boats or monoxiles (Figure 14-19).

An example is offered by an artifact discovered at Cluj-Napoca, depicting the earliest boat (Figure 16) and published by N. Vlassa (Vlassa 1970 (= Vlassa 1976), 170), who proposed that other vessels depicted boats, including the "fish-pans" from Turdaş, and examples from the Cycladic area, Minoan Crete and Mesopotamia (Vlassa 1970 (= Vlassa 1976, 170, no. 28; and others)). To these examples, we can add other boats, such as the one discovered at Parţa, Neolithic Korbovo (Sarić J. http://www. ai.sanu. ac.yu/~ jsaric /boat/), Iclod (Zau culture), in the Precucuteni culture or in the Copper Age (Poduri, Isaiia, Mărgineni etc.) (Monah 2005 and bibliography; Höckmann 1993; Radunčeva 2003, Figure 155-2157; Zorlenţu Mare, Vinča culture, unpublished). These artifacts led to the conclusion that such boats or monoxiles allowed the movement of salt blocks. Evidence for salt blocks and their exploitation requires further research, for instance studies of the tools used for cutting the blocks and their use-wear traces (Vlassa 1976, 170ff.).

Recently, complex investigations at Figa, with chemical, dating and dendrocronological analysis has brought new information and important data on operating systems and salt mining in the Late Bronze Age (Cavruc, Chiricescu 2006, 7 ff.; Cavruc 2006, 33ff.; Cavruc *et al.,* 2006, 41ff.; 2006a, 50ff.; Cavruc 2012, 6), with some wells dated to the Early Bronze Age (At Figa it is a well where have been discovered ceramic fragments decorated with a small sweep, ascribed to Early Bronze Age: Cavruc 2012) and real mines in the Late Bronze Age; all these studies show many new dimensions of recent salt research. We should not disregard the fact that some of the largest bronze hoards in Europe, from Uioara, Şpălnaca and other large hoards (Wollmann 1996, 241, no. 7 apud A. Popa, ing. Chiriţă Radu), were deposited in salt-rich areas, all of which with discoveries from the Neolithic until Early Medieval times (Wollmann 1996, 242-244; Petrescu – Dîmboviţa M. 1977, 336-337).

There is also knowledge from early times of the well-known curative properties of salt (halotherapy), as well as the religious importance of salt. Already in the Bible are mentioned customs and practices related to salt (Monah 2008a, 122, Leviticus 2.13 and bibliography).

Given the special role played by salt in human and animal life, we believe that the vessels discussed in this article were used for the transport of brine. The presence of these vessels in the Mađgiari sanctuary shows the great importance given to food, and the therapeutic qualities of salt and brine (Alexianu *et al.,* 2007, 103, 112-123).

The askos vessel shape, with its potential for handles for carrying, constitutes evidence for long-distance transport of liquids, including brine. All these show us that we should not be surprised at the presence of pots for salt transport in the Neolithic. Transportation of salt blocks is hard to prove for the Neolithic, but we must not forget that we have from this period evidence not only for pedestrian transport, but also for the navigation of large rivers by boats or monoxiles (Figure 14-19).

An example is offered by an artifact discovered at Cluj-Napoca, depicting the earliest boat (Figure 16) and published by N. Vlassa (Vlassa 1970 (= Vlassa 1976), 170), who proposed that other vessels depicted boats, including the "fish-pans" from Turdaş, and examples from the Cycladic area, Minoan Crete and Mesopotamia (Vlassa 1970 (= Vlassa 1976, 170, no. 28; and others)). To these examples, we can add other boats, such as the one discovered at Parţa, Neolithic Korbovo (Sarić J. http://www. ai.sanu .ac.yu/~ jsaric /boat/), Iclod (Zau culture), in the Precucuteni culture or in the Copper Age (Poduri, Isaiia, Mărgineni etc.) (Monah 2005 and bibliography; Höckmann 1993; Radunčeva 2003, Figure 155-2157; Zorlenţu Mare, Vinča culture, unpublished). These artifacts led to the conclusion that such boats or monoxiles allowed the movement of salt blocks. Evidence for salt blocks and their exploitation requires further research, for instance studies of the tools used for cutting the blocks and their use-wear traces (Vlassa 1976, 170ff.).

Figure 20. XX century, baskets with different products, carried by women and men on the head, after R. A. Roman 2008.

In the group of very large Walzenbeil-type axes from a Cluj-Napoca collection, one axe weighs over 2 kg; given the improbability of its use as a „scepter", it is possible that the axe could have been used in the separation of salt blocks. A search for large, heavy tools would be a profitable line of enquiry in salt-rich areas.

There is good evidence for the use of saltwater sources from the early Neolithic, with some radiocarbon data showing the possibility of an even earlier phase of exploitation by Mesolithic communities (We are thinking of the radiocarbon data from Lunca - *Poiana Slatinii*: Dumitroaia 1987; 1994; 2000; Weller, Dumitroaia 2005; Weller O. *et al.*, 2007b, 110-111; Cavruc 2012, 3-12, especially 2, where radiocarbon data are earlier then the earliest pottery discovered here; our opinions, Traore 1993; Lazarovici Gh., Maxim 1995; 1996; Lazarovici C.-M., Lazarovici Gh. 2006).

Transport on the head (Figure 20-22)
Transport on the head is the last problem we wish to discuss here, without entering into a great debate. There are numerous examples in ethno-archaeology of this practice, whose functionality is extremely important and should not be ignored, even if not directly related to salt but because it raises issues of ethno-religion.

From the present to the past, the surest way to avoid mistakes in the reconstruction of an artifact's function is remain silent for lack of direct evidence.

Long distance transport on the back or on the head of liquids (milk) or other products (clothes from washing, baskets with vegetables, fruit etc.) was extensively used until recently by both genders[7]. Ethnographic museums have many documents and pictures, so an ethno-historical overview should not be ignored. In the images from our time, we have peddlers carrying baskets of food on the head, women who bring washing in wooden pails on their heads from the river.

The Romanian term for a wooden pail - '*cofă*' - is quite old and is present in several Balkan languages, as is shown in a DEX variant (Apud DEX online, *cófă (cófe)*: http://dexonline.ro/definitie/cof%C4%83).

The oldest proofs for carrying different things on the head are related with Çatal Höyük PPN, where some characteristics of human vertebrae (their welding or their detachment) indicate such sort of transport custom (Figure 22a) (Başak Boz 2007, p. 240 -241).

In the Gumelniţa culture, several anthropomorphic pots show individuals carrying vessels on the head (Figure 21b, d) (Voinea 2005, 63-72, pl. 109 and bibliography). Such a method can be used to transport liquids (water, brine, oil etc.) over long distances. The principal requirement for the trip is the stability of the vessel, with a cloth coil used for stability. The Romanian term used for such a container equates to 'barrel, (*putină*), again a word with forms and equivalent meanings in many Indo-European languages as DEX examples (Apud DEX online, *pútină (pútini)*: http://dexonline.ro/definitie/putin%C4%83).

[7] In the 50' in Reşiţa, a common practice called "subscription to milk" referred to carrying dairy products, either from Ţărova (4 km) or Târnova (7 km), on the head on baskets or on *cantă de lapte*; the same method was used to transport sour cream or *urdă*).

Figure 21. Pots or statuettes with pots on the head: a) El-Ghassul, Ghassulian culture (after Al. H. Joffe *et al.* 2001); b-d) Gumelniţa culture, Gumelniţa, Glina, photos Gh. Lazarovici

Figure 22a. Çatal Höyük, person carrying things on the head (after Başak Boz 2007, p. 241).

Figure 22b. Bronze Age period, woman carrying a pot on her head, reconstruction (after J. Lolk 2009, Figure 11).

An example of another type of vessel, oval with side handles and a small narrow bottom shaped on a head of a figurine is from the Ghassulian culture, Israel (Figure 21a). This pot shape is quite typical for this culture, with anthropomorphic and zoomorphic pots carrying such a vessel interpreted as used for the preparation of cheese. This shape appears at Beer-Seba, Azor and Teleilat El-Ghassul (Müller-Karpe 1968, 433, Kat. 69, Taf. 108, Kat. 67, 69, 71, Taf. 106/18, 107/20, 108/17-18. http://apd.farli.org/the-southern-levant/The-Chalcolithic-Period/ghassulian-pottery; Joffe *et al.,* 2001) dating to the transition from Neolithic to Copper Age (between 4700-4300 CAL B.C.). If we interpret the attributes of that figurine (breasts, sex, belly, navel) as shown in the image (Figure 21a), this can evokes a variant of the Great Mother, a goddess symbol of sexuality and motherhood (Joffe *et al.,* 2001, 18). Radiocarbon data show that this civilization is contemporary with Gumelniţa A2 (Lazarovici C.-M., Lazarovici Gh. 2007, 82). From the Iron Age, we have a reconstruction of a human carrying a pot on the head (Figure 22b), with other related findings (Figure 21c).

Salt and salt roads in Neolithic (Figure 15)
All discoveries of the early Neolithic in Transylvania, from the time of the second migration and the use of polychrome painted pottery are located in areas of concentrated salt sources, areas already known exploited from the first migration of the Early Neolithic (Miercurea Sibiului, Ocna Sibiului, Şeuşa, Gura Baciului) (Lazarovici Gh., Maxim 1995; Lazarovici Gh. 2001; 2005; 2006; Lazarovici Gh., Lazarovici C.- M. 2006; Ciută 2001; 2002; 2005) - the period that led to neolithisation of Moldova, as exemplified at Lunca - *Poiana Slatinii* (Dumitroaia 1986; 1987 etc.; Monah D. 1991; Alexianu M. *et al.,* 1993); or from the time of the migration that occurred during the Balkan-Anatolian Chalcolithic (Ocna Sibiului, Miercurea Sibiului, the Râmnicu Vâlcea area near Copăcele (Valea Răii). At Lunca, radiocarbon data indicates that habitation here is considered even earlier than the first migration (Weller, Dumitroaia 2005 WEB); however, pottery related to this early phase either does not exist or has not yet been published. During the second migration in Transylvania, areas with salt attracted a rapid spread of small communities, such as those from Ţaga - *Valea Mileului*, Vultureni, Fundătura, Moruţ and Suceag. This process started from the *II migration* at Gura Baciului, settling suitable places for shepherds (Lazarovici Gh. 2006). Salt circulated large distances, being known even later in the Cucuteni culture, at the eastern margins of its distribution (Chapman, Gaydarska 2003).

For Hallstatt period there are several proofs indicating transportation on the head.

At Hallstatt site, through the anthropological analysis, Anton Kern' collaborators (the leader of today investigation) have noticed some spinal deformities caused by osteo-arthritis or neck vertebrae deformation which is more evident especially for children[8].

They also offer some ethnographic analogies regarding transportation on the head of salt or other products. Some Hallstatt situla as the one at Certosa (Figure 23) offers other examples regarding the transportation on the head of different containers; other examples are illustrated by ethnographic and ethno-historical data with which we can make analogies. On of such example is Hallstatt area in Austria: here until XX century salt was transported on the head with a shoulder support (Figure 24), for balancing and fixation to the body the system was provided with a belt; this system remember us pots from Anzabegovo (Macedonia Neolithic) or the one from Starčevo-Criş (Figure 12-14). Other analogies are related to ethno-historical data in Portugal (Figure 26a-b) or to ethnographic ones in Nepal (Pany-Kucera *et al.* 2010, Figure 10-17) (Figure 25). In Portugal at the beginning of the XXth century when unloading ships, people to have free hands to maintain balance (or to carry other things) used transportation of the luggage on the head; this system is very much similar to the one used by Oltenia women or those from other Romanian provinces, carrying on the head baskets, stretchers or vessels (Figure 20).

The problem of salt in Europe and Romania has been, and continues to be the stimulus for systematic expeditions (Weller, Dumitroaia 2005; *** Sarea... 2008; Cavruc, Harding 2008; *** *Archaeology and Anthropology of Salt 2011*), symposia (Piatra Neamţ 21-24 october 2004, summaries, 85-112; *** *Archaeology and Anthropology of Salt 2011*), monographs and studies (Expeditions made by C.-M. Lazarovici, Gh. Lazarovici, G. Trnka during 2006-2010; Nikolov *et al.*, 2008 etc. ; *** Provadia 2005-2007; Petrova 2011; etc.).

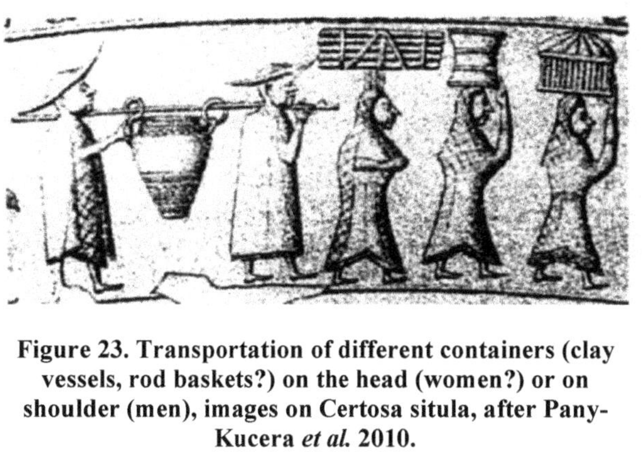

Figure 23. Transportation of different containers (clay vessels, rod baskets?) on the head (women?) or on shoulder (men), images on Certosa situla, after Pany-Kucera *et al.* 2010.

[8] Pany-Kucera *et al.* 2010, Figure 3-7. We would like to thank again our colleague Marius Alexianu that kindly offers us more information related with transportation on the head or backside of salt and different products, in relation with diverse geographical and cultural areas.

Figure 24. Hallstatt, salt transport on the head with a shoulder support (XIX century), after Pany-Kucera *et al.* 2010.

Figure 26a-b. Portugal, beginning of XXth century, transport of some luggage, baskets or bales, after Pany-Kucera *et al.* 2010.

Figure 25. Nepal, children transporting wood on their back and having a supporting band on the head, after Pany-Kucera *et al.* 2010.

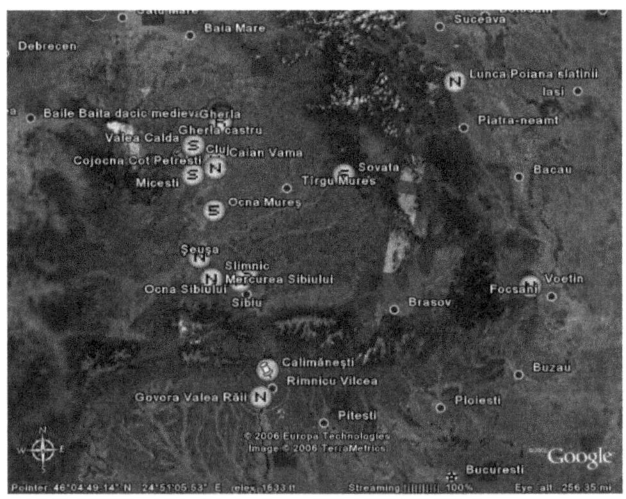

Figure 27. Early Neolithic settlements located in salt source areas we visited; legend: S –salt sources; N – Neolithic sites.

Today, there are very rich documentary sources operating in Romania on salt from the Neolithic until modern times, there are several projects completed or underway especially for Moldova (Weller, Dumitroaia 2005; *** Sarea... 2008) and Transylvania (*** Sarea... 2006; Cavruc, Harding 2008).

During the fourth migration in Moldova, all the salt sources in Romania were integrated into a single network, since the Polychrome communities were very much interested in searching for salt sources. From the map below (Figure 27), one can see the location of salt sources (symbol S) and Polychrome Neolithic sites from their areas (symbol N).

Salt

Salt continued to play an important role in Neolithic times. In previous periods and now are present pots with vertically perforated handles. There are cases in the archaeological literature of pots with banded handles (Milojčić 1949, 288, Figure 2/12; Berciu 1961, 30). Such pots are present in the later phases at Perieni, Trestiana (Popuşoi 2005, Figure 148/7), Dudeştii Vechi (Figure 10), Idjoš (Figure 28) and other sites - all dated to the SC III-IV phase. Of course, pots with three or four handles, with smaller holes could be used in transport with the yoke.

We have often spoken of Neolithic communities' interest in using salt for both humans and animals. Of course, sheep prefer to lick the salt block, but they also lick earth mixed with salt (Lazarovici Gh., Lazarovici C.- M. 2011, Figure 9).

From our recent ethno-archeological research, we found that recent communities prefer to use brine springs for food preservation instead of rock salt; with people carrying brine from tens of kilometers away, such activity is very intense in December, especially around the St. Ignat holy day, when pigs are butchered for the preparation of meat for Christmas.

We believe that, during the Neolithic, brine was also transported, as exemplified by the large amphorae at Lepenski Vir, especially in pots having wide handles, their asymmetric shapes suited to be carried on the back and their wide vertical holes suitable for attaching a harness, as seen in the example from Dudeştii Vechi and other pots with wide handles (Kutzián 1944, pl. XXV/6a, XXVI/2, XXX/).

Such types of vessels have also been found in the Early Neolithic sanctuary at Mađiari, Skoplje (Angeleski 2012, type Ak1, Ak2, 224, 47/I), as well as in other Hungarian sites where salt was necessary, as is proven by several discoveries from Kőrős culture, published by Ida Kutzián: Sővényháza, Csátalja - Baja, Ókécske - Kecskemet, Devavanya, Szentes – *Nagytőke*, Kotokpart – *Vata* and Kopancs – *Zsoldos* (Kutzián 1944 s. v.). The presence of such large numbers of these vessels in the Hungarian Plain can be linked with salt trade in the Early Neolithic.

Ida Kutzián presented analogies for these findings in the south, at Vinča and even farther south until Molfetta, where there are identical materials with those from the Danube region in period SC III-IV (Kutzián 1944, pl. LVIII/1-2, LXXIII-LXXIV).

The fourth migration integrated the search for salt sources in Romania, with Polychrome communities specifically looking for salt sources.

The presence of vessels that we believe were used for the transport of liquids and particularly brine in sanctuaries shows the cultic importance of this mineral. Many scholars have written about the curative qualities and rituals or about its use in treatment, the birth of children and others who have passed over time from prehistory until today, from Biblical information up to our time (Monah 2008, 117-131).

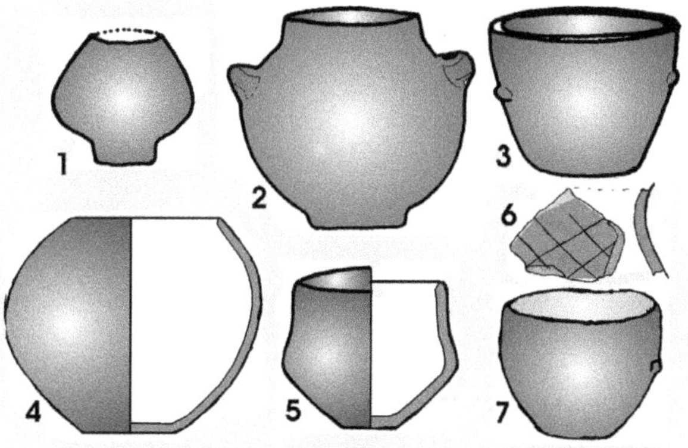

Figure 28. Idjoš (processing made by Gh. Lazarovici, Preistoria Banatului II).

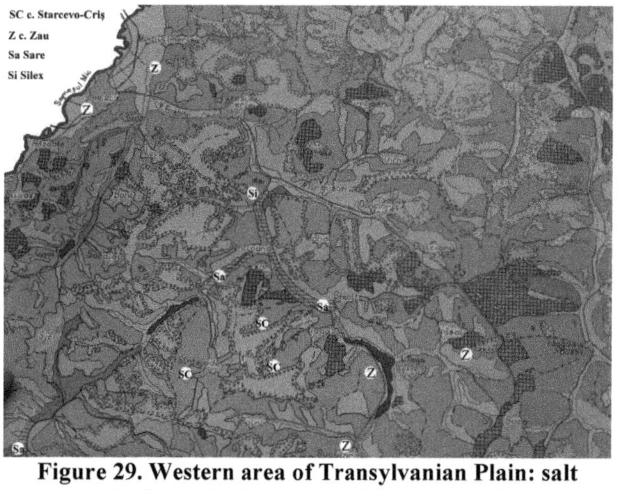

Figure 29. Western area of Transylvanian Plain: salt sources (yellow with hachure and Sa marks); Neolithic sites (see the legend; Sa = salt, Si = flint).

Figure 30. Neolithic axes from Hallstatt site (after Kingdom of Salt 2009, p. 44).

We also want to remember some sort of axes types (Figure 30) discovered nearby salt exploitation at Hallstatt (*Kingdom of Salt 2009*, p. 44). Such sorts of artefacts are also present in many areas related with salt exploitation in Romanian territory, so we should reflect on it.

We also believe that in the future are necessary more investigation regarding toponimes around Ţaga: *Sós Dombom* (Şoşdàmbu) Ghiolţ hamlet, surrounding of the salt lake at Sântioana, Sântejude (creek with two wells, now a salt lake: Romanian, *Sărătura* = Hungarian *Là Széréture*, maybe Şugău, Là Teusare, Hungarian 1828; Hungarian 1796-1797 *Teusallye*, *Kitre Téusaré* from Romanian *Către tău [cu] sare* etc.) (Mârza 2009, p. 63; Laşonţi 2009, p. 487, 492).

References

***Cucuteni-Trypillia. *Cucuteni-Trypillia: A Great civilization of Old Europe, Palazzo della Cancelleria, Rome-Vatican 16 September-31 October 2008*, Bucureşti, 2008.

***Enciclopedia: *Enciclopedia arheologiei şi istoriei vechi a României*, vol. L-M, Bucureşti, 1994.

***Provadia 2005-2007: Nikolov V. (ed.) *Praistoričeski solodobiveh Centîr Provadija – Solonicata. Raskopki 2005-2007*, Sofia, 2008.

Alexandrescu, A.D. 1963. În legătură cu statuile - menhir de la Baia de Criş. *Studii şi Cercetări de Istorie Veche* 14, 1, 145-150.

Alexianu, M., Dumitroaia, Gh. and Monah, D. 1992. Exploatarea surselor de apă sărată din Moldova: o abordare etnoarheologica. *Thraco-Dacica* XIII/1-2, 159-167.

Alexianu, M., Dumitroaia, Gh. and Monah, D. 1993. L'exploitation des sources salées dans l'est de la Roumanie: un démarche ethnoarchéologique In J. Pavuk (ed.), *Actes du XIIe Congrès International des Sciences Préhistoriques et Protohistoriques, Bratislava 1-7 septembre 1991*, 88-90. Bratislava.

Alexianu, M., Weller, O., Brigand, R., Curcă, R.-G., Cotiugă, V., Moga, I. 2011. Salt Springs in Today's Rural World. An Etnoarchaeological Approach in Moldavia (Romania), in M. Alexianu, O. Weller, R.-G. Curcă (eds.), *Archaeology and Anthropology of Salt: a Diachronic Approach Proceedings of the International Colloquium, 1–5 October 2008 "Al. I. Cuza" University (Iaşi, Romania)*, 7-23. Oxford, BAR Publishing.

Alexianu, M., Weller, O. and Brigand, R. 2007. *Izvoarele de apă sărată din Moldova subcarpatică. Cercetări etnoarheologice.* Iaşi, Casa Editorială Demiurg.

Alexianu, M., Weller, O. and Brigand, R. 2008. Usages et enjeux autour des sources salées de Moldavie précarpatique, Roumanie, in O. Weller, D. Dufraisse, R. Pétrequin (eds.), *Sel, eau et forêt d'hier et aujourd'hui. Actes du colloque international de la Saline royale d'Arc-et-Senans 3-5 Octobre 2006, Cahiers de la MSH Ledoux 12*, 49-72. Besançon, Presses Universitaires de Franche-Comté, (coll. Homme et environnement).

Alicu, D. (ed.) 2008. *Polus. Istorie pierdută –istorie regăsită.* Cluj-Napoca, Muzeul Naţional de Istorie a Transilvaniei.

Andronic, M. 1989. Cacica - un nou punct neolitic de exploatare a sării. *Studii şi Cercetări de Istorie Veche şi Arheologie* 40/2, 171-177.

Angeleski, S. 2012. *The Early and Middle Neolithic in Macedonia*, BAR International Series 2332. Oxford, BAR Publishing.

Başak, B. 2007. Aus dem Leben im Neolithikum. Einblicke in die damalige Lebenswaise nach Befunden von Sktlettüberresten. In *Vor 12.000 Jahern in Anatolien. Die ältesten Monumente der Menchheit*, 238-245. Stuttgart, Badisches Landesmuseum, Karlsruhe, Konrad Theiss Verlag.

Benac, Al. 1973. *Obre*, Wissenschaftliche Mitteilungen des Bosnisch-Herzegowinisch Landesmuseum, Band III, Heft A, Archäologie. Sarajevo.

Berciu, D. 1961. *Contribuţii la problemele neoliticului din România în lumina noilor cercetări.* Bucureşti, Editura Academiei Republicii Populare Române.

Bodi, G. 2007. Experiments with Salt Crystallization, in D. Monah, Gh. Dumitroaia, O. Weller, J. Chapman (eds.), *L'exploitation du sel à travers le temps*, Bibliotheca Memoriae Antiquitatis XVIII, 89-98. Piatra Neamţ, Constantin Matasă.

Buzea, D.L. 2011. Arheologie experimentală. Construirea unei instalaţii de foc (vatră) după modelul celor descoperite la Păuleni Ciuc - Ciomortan "Dâmbul Cetăţii", jud. Harghita. *Angustia* 12, 1-12.

Buzea, D. and (Chiricescu) Deák A. 2008. Cercetările etnoarheologice de la Olteni, jud. Covasna, in D. Monah, Gh. Dumitroaia and D. Garvăn (eds.), *Sarea de la prezent la trecut*, 41-80. Bibliotheca Memoriae Antiquitatis XX, Piatra Neamţ, Contantin Matasă.

Carusi, C. 2011. Hypotheses, Considerations – and unknown Factors – regarding the Demand for Salt in Ancient Greece, in M. Alexianu, O. Weller, R.-G. Curcă (eds.), *Archaeology and Anthropology of Salt: A Diachronic Approach, Proceedings of the International Colloquium, 1-5 October 2008, Al. I. Cuza University (Iaşi, Romania)*, BAR IS 2198, 149-154. Oxford, BAR Publishing.

Cavruc, V. 2006. Sarea în societăţile arhaice. Consideraţii generale, in V. Cavruc and A. Chiricescu (eds.), *Sarea, Timpul şi Omul. Catalog de expoziţie*, 33-36. Sfântu Gheorghe, Angustia.

Cavruc, V. 2012. The Present Stage of the Researches regarding Prehistoric Salt Production in the Carpatho-Danubian Region. *Angustia*, Arheologie 12, 79-90.

Cavruc, V. and Chiricescu, A. (eds.) 2006. *Sarea, timpul şi omul. Catalog de expoziţie.* Sfântu Gheorghe, Angustia.

Cavruc, V. and Harding, A.F. 2008. Noi cercetări arheologice privind exploatarea sării în nord-estul Transilvaniei. Raport preliminar, in D. Monah, Gh. Dumitroaia and D. Garvăn (eds.), *Sarea de la prezent la trecut,* 149-178. Bibliotheca Memoriae Antiquitatis XX. Piatra Neamţ, Constantin Matasă.

Cavruc, V. and Harding, A.F. 2011. New Archaeological Researches concerning Saltworking in Transylvania. Preliminary Report, in M. Alexianu, O. Weller and R.-G. Curcă (eds.), *Archaeology and Anthropology of Salt: A Diachronic Approach, Proceedings of the International Colloquium, 1-5 October 2008, Al. I. Cuza University (Iaşi, Romania),* 111-122. BAR International Series 2198. Oxford, BAR Publishing.

Cavruc, V., Ciugudean, H. and Harding, A.F. 2006. Vestigiile arheologice privind exploatarea sării pe teritoriul României în epoca bronzului, in V. Cavruc and A. Chiricescu (eds.), *Sarea, Timpul şi Omul. Catalog de expoziţie,* 41-49. Sfântu Gheorghe, Angustia.

Cavruc, V., Munteanu, R.E. and Ciugudean, H. 2006a. Vestigii arheologice privind exploatarea sării pe teritoriul României în epoca timpurie a fierului, in V. Cavruc, A. Chiricescu (eds.), *Sarea, Timpul şi Omul. Catalog de expoziţie,* 50-52. Sfântu Gheorghe, Angustia.

Cavruc, V., Moga, V. and Stănescu, Al. 2006b. Exploatarea sării în Dacia Romană, in V. Cavruc and A. Chiricescu (eds.), *Sarea, Timpul şi Omul. Catalog de expoziţie,* 53-55. Sfântu Gheorghe, Angustia.

Ceauşu Şt.M. 1982. Contribuţii la istoricul exploatării sării la sfârşitul secolului al XVIII-lea şi în prima jumătate a secolului al XIX-lea în Bucovina. *Anuarul Muzeului Judeţean Suceava* IX, 377-392.

Ceja Acosta J.A. 2011. El Salado-Ixtahuehue and Benito Juárez-Soconusco: an Ethno-Archaeological Study of Salt Pre-Industries of Southeast Veracruz, Mexico, in M. Alexianu, O. Weller, R.-G. Curcă (eds.), *Archaeology and Anthropology of Salt: A Diachronic Approach, Proceedings of the International Colloquium, 1-5 October 2008, Al. I. Cuza University (Iaşi, Romania),* 37-48. BAR International Series 2198. Oxford, BAR Publishing.

Chapman, J. 1981. *The Vinča Culture of South-East Europe*. BAR International Series 117, I. BAR Publishing.

Chapman, J. and Gayadarska B. 2003. The Provision of Salt to Tripolye Mega-Sites. In *Tripolian settlements-giants. The international symposium materials, Kiev,* 203-211.

Chapman, J., Monah, D., Dumitroaia, Gh., Armstrong, H., Millard, A, Francis, M. 2000. The Exploitation of Salt in the Prehistory of Moldavia, Romania. *Archaeological Reports 1999/2000,* 23, 10-20. University of Durham and University of Newcastle upon Tyne.

Chiricescu, A. 2006. Exploatarea tradiţională a sării în Transilvania, in V. Cavruc, A. Chiricescu (eds.), *Sarea, Timpul şi Omul. Catalog de expoziţie,* 158-163. Sfântu Gheorghe, Angustia.

Ciobanu, D. 2011. Traditional Methods of Salt Mining în Buzău County, Romania in the 21st Century, in M. Alexianu, O. Weller, R.-G. Curcă (eds.), *Archaeology and Anthropology of Salt: A Diachronic Approach, Proceedings of the International Colloquium, 1-5 October 2008, Al. I. Cuza University (Iaşi, Romania),* 35-36. BAR International Series 2198. Oxford, BAR Publishing.

Ciută, M.M. 2000. Contribuţii la cunoaşterea celui mai vechi orizont al neoliticului timpuriu din România, cultura Precriş. Descoperirile arheologice de la Şeuşa-La cărarea Morii. *Apulum* XXXVII/ 1, 51– 102.

Ciută, M.M. 2001. Contribuţii la cunoaşterea celui mai vechi orizont al neoliticului timpuriu din România: cultura Precriş. *Apulum* XXXVIII/1, 9-26.

Ciută, M.M. 2002. *Cronica Cercetărilor Arheologice din România. Campania 2001.* Bucureşti, Cimec.

Ciută, M.M. 2005. *Începuturile neoliticului timpuriu în spaţiul intracarpatic transilvănean.* Alba Iulia, Aeternitas.

Čohađiev, Şt. 2001. *Vaksevo. Praistoričeski Selišča,* Veliko Târnovo.

Curcă, R.-G. 2008. Halotoponime şi halohidronime în documentele medievale din Moldova (sec. XIV-XVII), in D. Monah, Gh. Dumitroaia, D. Garvăn (eds.), *Sarea de la prezent la trecut,* 107-116. Bibliotheca Memoriae Antiquitatis XX. Piatra Neamţ, Constantin Matasă.

Dăncuş, M. 2006. Sarea – fragmentarium etnologicum, in V. Cavruc, A. Chiricescu (eds.), *Sarea, Timpul şi Omul. Catalog de expoziţie,* 147-148. Sfântu Gheorghe, Angustia.

Dumitrescu, H. 1976. Două vase de o formă deosebită descoperite în aşezarea de la Traian-Dealul Viei. *Memoria Antiquitatis* IV-V (1972-1973), 47-54.

Dumitrescu, Vl. 1987-1988. Plastica neolitică din aşezarea de la Rast (jud. Dolj). *Acta Musei Napocensis* XXIV-XXV, 29-68.

Dumitroaia, Gh. 1986. Săpăturile arheologice din judeţul Neamţ (1984-1985). *Memoria Antiquitatis* XII-XIV (1980-1982), 317-321.

Dumitroaia, Gh. 1987. La station archeologique de Lunca-Poiana Slatinii, in M. Petrescu-Dîmboviţa, N. Ursulescu, D. Monah, V. Chirica (eds.), *La civilisation de Cucuteni en contexte européen. Session scientifique dédié au centenaire des premières découvertes de Cucuteni (Iaşi – Piatra Neamţ 24-28 septembre 1984)*, 235-258. Iaşi, Bibliotheca Archaeologica Iassiensis I.

Dumitroaia, Gh. 1994. Depunerile neo-eneolitice de la Lunca şi Oglinzi, judeţul Neamţ. *Memoria Antiquitatis* XIX, 7-82.

Dumitroaia, Gh. 2000. *Comunităţi preistorice din nord-estul României. De la cultura Cucuteni până la bronzul mijiociu.* Bibliotheca Memoriae Antiquitatis VII. Piatra-Neamt.

Dumitroaia, Gh. 2004. Săpături arheologice sistematice şi periegheze din judeţele Neamt, Bacău şi Suceava (2002-2003). *Memoria Antiquitatis* XXIII, 431-449.

Dumitroaia, Gh., Munteanu, R., Nicola, D., Monah, D., Chapman, J., Weller, O. 2002. Lunca, com. Vânători-Neamţ, jud, Neamţ. Punct: Poiana Slatinii. *Cronica Cercetărilor Arheologice din România. Campania 2001*, 183-184. Bucureşti, Cimec.

Dumitroaia, Gh., Munteanu, R., Nicola, D., Preoteasa, C., Monah, D. 2003. Cucuieţi, com. Solonţ, jud. Bacău. Punct: Slatina Veche. *Cronica Cercetărilor Arheologice din România. Campania 2002*, 110-111. Bucureşti, Cimec.

Dumitroaia, Gh., Munteanu, R., Nicola, D., Preoteasa, C., Monah, D., Chapman, J., Weller, O. 2003a. Lunca, com. Vânători-Neamţ, jud. Neamt. Punct: Poiana Slatinii. *Cronica Cercetărilor Arheologice din România. Campania 2002*, index 112. Bucureşti, Cimec.

Dumitroaia, Gh., Nicola, D., Munteanu, R., Preoteasa, C., Monah, D., Boghian, D., Ignătescu, S. 2004a. Solca, com. Solca, jud. Suceava. Punct: Slatina Mare. *Cronica Cercetărilor Arheologice din România. Campania 2003*, 314-315. Bucureşti, Cimec.

Dumitroaia, Gh., Munteanu, R., Nicola, D., Preoteasa, C., Monah, D. 2004b. Cucuieţi, com. Solonţ, jud. Bacău. Punct: Slatina Veche. *Cronica Cercetărilor Arheologice din România. Campania 2003*, 110-111. Bucureşti, Cimec.

Dumitroaia, Gh., Munteanu, R., Weller, O., Garvăn, D., Diaconu, V., Brigand, R. 2008. Un nou punct de exploatare a sării în preistorie: Ţolici-Hălăbutoaia, jud. Neamţ, in D. Monah, Gh. Dumitroaia, D. Garvăn (eds.), *Sarea de la prezent la trecut,* 203-224. Bibliotheca Memoriae Antiquitatis XX, Piatra Neamţ, Constantin Matasă.

Ellis, L. 1984. The Cucuteni - Tripolye Culture. A Study in Technology and the Origins of Complex Society. BAR International Series 217. Oxford, BAR Publishing.

Garašanin, D. 1959. Nosa-Biserba Obala. *Arheološki Preglad* 1, 9-12.

Garašanin, M. 1979. Centralnobalkanoška Zona. In Benac A. (ed.), *Praistorija Jugoslovenskih Zemalja,* 79-212. Sarajevo, ANU BiH-Centar za Balkanološka ispitivanja.

Gaydarska, B. and Chapman, J. 2007. Salt Research in Bulgaria, in D. Monah, Gh. Dumitroaia, O. Weller, J. Chapman (eds.), *L'exploitation du sel à travers le temps,* 147-160. Bibliotheca Memoriae Antiquitatis XVIII. Piatra Neamţ, Constantin Matasă.

Guijarro, J.J. 2011. *The Beginning of the Salt Exploitation in Spain: Thinking about the Salt Exploitation in the Iberian Peninsula during Prehistoric Times,* in M. Alexianu, O. Weller, R.-G. Curcă (eds.), *Archaeology and Anthropology of Salt: A Diachronic Approach, Proceedings of the International Colloqvium, 1-5 October 2008, Al. I. Cuza University (Iaşi, Romania),* 123-133. BAR International Series 2198. Oxford, BAR Publishing.

Harding, A.F. and Cavruc, V. 2006. Băile Figa (or. Beclean, jud. Bistriţa Năsăud), in V. Cavruc, A. Chiricescu (eds.), *Sarea, Timpul şi Omul. Catalog de expoziţie,* 56-59. Sfântu Gheorghe, Angustia.

Hewitt, W.P., Winter, M.C. and Peterson, D.A. 1987. Salt Production at Hierve El Agua, Oaxaca. *American Antiquity* 52, 4, 799-816.

Höckmann O. 1993. Urzeitschiffahrt –Archäologische Zeugnisse zur ältesten Schiffhart in Europa. *Jahrbuch RGZM* 40, 2, 656-658.

Israeli Y. and Tadmor M. 1986. *Treasures of the Holy Land, Ancient Art from Israel Museum.* New York, Metropolitan Museum of Art.

Istrate C. 2008. *Condica lui Mavrocordat*, vol. I-III. Iaşi, Editura Universităţii „Alexandru Ioan Cuza".

Joffe, Al.H., Dessel, J.P. and Hallote, R.S. 2001. The Gilat woman. Female iconography, Chalcolithic cult and the End of Southern Levantine Prehistory. *New Eastern Archaeology* 64/1-2, 9-23.

Kacso, C. 2006. Date cu privire la exploatările timpurii de sare din Maramureş, in V. Cavruc, A. Chiricescu (eds.), *Sarea, Timpul şi Omul. Catalog de expoziţie,* 97-121. Sfântu Gheorghe, Angustia.

Kern, A., Kovarik, K., Rausch, A.V., Reschreiter, H. (eds.) 2009. *Kingdom of Salt. 7000 years of Hallstatt,* Natural History Museum, Veröffentlichen der Prähistorischen Abteilung 3. Vienna.

Kitanovski, B. 1976. Ranoneolitskata naselba pešterica kai Prilep. *Materijali* XIV, Prilep.

Kitanovski, B., Simoska, D. and Todorović, J. 1978. Novi Arheološki Istraživanja na naselbata Ćuka vo Topolčani kai Prilep. *Acta Archeologica Macedoniae* 4, 9-32.

Kitanovski, B., Simoska, D. and Todorović, J. 1990. Der Kultplatz auf der Fundstätte Verbjanska Čuka bei Prilep, in D. Srejović, N. Tasić (eds.), *Vinča and its World. International Symposium. The Danubian Region from 6000 to 3000 B.C.*, Volume LI, Book 14, 107-112. Serbian Akademy of Science and Arts, Centre for Archaeological Research, Faculty of Philosophy. Belgrade.

Kutzián, I. 1944. *A Körös Kultura*. Budapest.

Lazarovici, C.-M. 2006. Absolute Chronology of the Late Vinča Culture in Romania and its Role in the Development of the Early Cooper Age in Romania, in N. Tasić, C. Grazdanov (eds.), *Homage to Milutin Garašanin*, 277-294. Serbian Academy of Science and Arts, Macedonian Academy of Science and Arts. Belgrade, SASA, Special Editions.

Lazarovici, C.-M. and Lazarovici, Gh. 2006. *Arhitectura Neoliticului şi Epocii Cuprului în România. I. Neoliticul,* in V. Spinei and V. Mihailescu-Bîrliba (eds.). Iaşi, Editura Trinitas.

Lazarovici, C.-M. and Lazarovici, Gh. 2007. *Arhitectura Neoliticului şi Epocii Cuprului în România. II. Epoca Cuprului,* in V. Spinei and V. Mihailescu-Bîrliba (eds.). Iaşi, Editura Trinitas.

Lazarovici, Gh. 1979. *Neoliticul Banatului*. Bibliotheca Musei Napocensis III, Cluj-Napoca.

Lazarovici, Gh. 1986. Neoliticul târziu din nord-vestul României. Sălajul, Clujul, Bihorul. *Acta Musei Porolisensis* X, 15-46.

Lazarovici, Gh. 1996. The Process of the Neolithisation and the Development of the first Neolithic Civilisations in the Balkans, in R. Grifoni Cremonesi, J. Guilaine, J. L'Helgouach (eds.), *The Neolithic in the Near East and Europe, Colloquium XVII, 9, XIII ICPPS, Forli, Italia 4-14 September 1996,* 21-38. Abaco.

Lazarovici, Gh. 1998. About the neolithisation process of the second migration of the Early Neolithic. In Fl. Draşovean (ed.), I*nternational Symposium on the Problems of the Neolithic in the Middle Danube Region, June 1997,* 7-37. Museum Banaticum Temesiensis, Bibliotheca Historica et Archaeologica Banatica XIV. Timişoara, Eurobit.

Lazarovici, Gh. 2000. Links between the Early Neolithic from Romania and Bulgaria. *Karanovo III, Beiträge zum Neolithikum in Südosteuropa*. 273-286. Wien, Phoibos Verlag.

Lazarovici, Gh. 2001. Despre locul descoperirilor de la Şeuşa în neoliticul timpuriu din sud-estul Europei. *Studia Archaeologica et Historica Nicolao Gudea Dicata Omagiu Profesorului Nicolae Gudea la 60 de ani,* Zalău, 37-47.

Lazarovici, Gh. 2005. Culturile Precriş I, Precriş II şi Postcriş I, Postcriş II. *Acta Terrae Septemcastrensis* IV, Sibiu, 23-78.

Lazarovici, Gh. 2006. The Anzabegovo – Gura Baciului Axis and the First Stage of the neolithisation Process in the Southern – Central Europe, in N. Tasić, C. Grazdanov (eds.), *Hommage to Milutin Garašanin,* 111-158. Serbian Academy of Science and Arts, Macedonian Academy of Science and Arts. Belgrade, SASA, Special Editions.

Lazarovici, Gh. 2009. The Zau Culture, in Fl. Draşovean, D. Leopold Ciobotaru and M. Maddison (eds.) *Ten years after: The Neolithic of the Balkans, as uncovered by the last decade of Research. Proceedings of the Conference held at the Museum of Banat on November 9th – 10th*, Timişoara, 179-217.

Lazarovici, Gh., PB: Preistoria Banatului, II, manuscript.

Lazarovici, Gh. and Cristea N. 1979. Contribuţii arheologice la istoria străveche a comunei Uioara de Jos, Ciunga, jud. Alba. *Acta Musei Napocensis* XVI, 431-446.

Lazarovici, Gh. and Kalmar, Z. 1987. Şantierul arheologic Iclod. Campania 1985. *Apulum* XXIV, 9-39.

Lazarovici, Gh. and Lazarovici, C.-M. 2011. Some Salt Sources in Transylvania and their Connections with the Archaeological Sites in the Area, in M. Alexianu, O. Weller and R.-G. Curcă (eds.), *Archaeology and Anthropology of Salt: A Diachronic Approach, Proceedings of the International Colloquium, 1-5 October 2008, Al. I. Cuza University (Iaşi, Romania),* 89-110. BAR International Series 2198. Oxford, BAR Publishing.

Lazarovici, Gh. and Maxim, Z. 1995. *Gura Baciului. Monografie arheologică,* Bibliotheca Musei Napocensis XI, Cluj-Napoca.

Lazarovici, Gh., Tarcea, L., Piciu, T., Ghergari, L., Mârza, I., El Suzi, G., Bindea, D., Kesler, E., Cantemir, P., Sângerean, C., Radu, S., Mogoş, L., Hosu, Al. 1995. Anexele 1-8. In *Gura Baciului. Monografie arheologică,* 209-304. Bibliotheca Musei Napocensis XI, Cluj-Napoca.

Lazarovici, Gh., Draşovean, Fl., Maxim, Z. 2001. *Parţa. Monografie arheologică,* Vol. 1.1, Vol. 1.2. Bibliotheca Historica et Archaeologica Banatica XII. Timişoara, Editura Waldpress.

Lolk, J. 2009. Bronzealderens - En Kilde tip forståelse of Kommunication og social interaktion i bronzealderen. *Kuml*, 79-84.

Makkay, J. 1965. Die Wichtigsten Fragen der Körös-Starčevo-Periode. *Acta Prehistorica et Archaeologica* VIII, Szeged.

Makkay, J. 1966. Die bemalte Keramik der Körös Gruppe. Mora Ferencz Museum Évkonyve.

Makkay, J. and Trogmayer, O. 1966. Die bemalte Keramik der Körös-Gruppe. *Móra Ferenc Múzeum Évkönyve* 1964–65, 47-58.

Makkay, J. and Starnini, E. 2008. *The Excavations of Early Neolithic sites of the Körös culture in the Körös Valley, Hungary: the final Report. Vol. II: The pottery asamblages and Vol. III: The small finds*, Budapest.

Malpica Cuello, A. 2011. Salt Production in Mediterranean Andalusia in the Transition from Late Antiquity to the Early Middle Ages. in M. Alexianu, O. Weller and R.-G. Curcă (eds.), *Archaeology and Anthropology of Salt: A Diachronic Approach, Proceedings of the International Colloquium, 1-5 October 2008, Al. I. Cuza University (Iaşi, Romania)*, 171-177. BAR International Series 2198. Oxford, BAR Publishing.

Malpica Cuello, A., Morère Molinero, N., Fábregas Garcia, A., Jiménez Guijarro, J. 2011. Land Organisation and Salt Production in Region of the Salado River (Sigüenza, Province of Guadalajara, Spain): Ancient and Medieval Times. Results of the First Campaign 2008, in M. Alexianu, O. Weller and R.-G. Curcă (eds.), *Archaeology and Anthropology of Salt: A Diachronic Approach, Proceedings of the International Colloquium, 1-5 October 2008, Al. I. Cuza University (Iaşi, Romania)*, 179-186. BAR International Series 2198. Oxford, BAR Publishing.

Mantu, C.-M. 1998. *Cultura Cucuteni. Evoluţie, cronologie, legături.* Bibliotheca Memoriae Antiquitatis V. Piatra-Neamţ, Nona.

Marc, D. 2006. Sisteme de transport şi comercializare tradiţională a sării, in V. Cavruc and A. Chiricescu (eds.), *Sarea, timpul şi omul. Catalog de expoziţie*, 152-157. Sfântu Gheorghe, Angustia.

Marta, L. and Levente, S.P. (eds.), 2007. *Catalogul colecţiei de arheologie.* Muzeul Judeţean Satu Mare, Satu Mare.

Martell Contreras, M.L. 2011. The Saltmakers of Soconusco and Benito Juárez: An Interpretation of Ethnoarchaeological Data from the Perspective of Gender and Identity, in M. Alexianu, O. Weller and R.-G. Curcă (eds.), *Archaeology and Anthropology of Salt: A Diachronic Approach, Proceedings of the International Colloquium, 1-5 October 2008, Al. I. Cuza University (Iaşi, Romania)*, 49-58. BAR International Series 2198. Oxford, BAR Publishing.

Maxim, A.I. 1971. Un depozit de unelte dacice pentru exploatarea sării. *Acta Musei Napocensis* VIII, 457-463.

Milojčić,Vl. 1949. *Chronologie der jungeren Steinzeit Mittel-und Südosteuropas*, Berlin.

Mircea, M. 2003. O enigmă la gura fluviului Borysthenes - cristalizarea naturală a sării în grămezi uriaşe de bulgări (Hdt., IV, 53, 10). In E. Sava (ed.), *Interferenţe cultural-cronologice în spaţiul nord-pontic*, 236-251. Chişinău, Academy of Sciences of Moldova.

Monah, D. 1991. L'exploitation du sel dans les Carpathes Orientales et ses rapports avec la culture Cucuteni-Tripolye, in V. Chirica and D. Monah (eds.), *Le Paléolithique et le Néolithique de la Roumanie en contexte européen*, 387-400. Bibliotheca Archaeologica Iasiensis IV, Iaşi.

Monah, D. 2002. Exploitation du sel dans les Carpates Orientales. In O. Weller (ed.), *Archéologie du sel: techniques et sociétés, Internationale Archäologie, ASTK 3. Colloque 12.2, XVe Congrès UISPP, Liège, Septembre 2002*, Rahden-Westfalie, 135-146.

Monah, D. 2005. Machete de ambarcaţiuni din culturile Precucuteni şi Cucuteni, in V. Spinei, C.-M. Lazarovici and D. Monah (eds.), *Studia praehistorica. Miscellanea in honorem nonagenari magistri Mircea Petrescu-Dîmboviţa oblata*, 271-294. Honoraria 1. Iaşi, Trinitas.

Monah, D. 2007. Proiectul „Exploatarea sării preistorice în Romania şi Anatolia". *Arheologia Moldovei* XXIX, 353-358.

Monah, D. 2008. Arheologia preistorică a sării în Romania. Scurt istoric, in D. Monah, Gh. Dumitroaia and D. Garvăn (eds.), *Sarea de la prezent la trecut*, 13-40. Bibliotheca Memoriae Antiquitatis XX. Piatra Neamţ, Constantin Matasă.

Monah, D. 2008a. O confirmare etnografică a versetului Iezechiel 16/4. Sărarea noilor născuţi, in D. Monah, Gh. Dumitroaia and D. Garvăn (eds.), *Sarea de la prezent la trecut*, 117-132. Bibliotheca Memoriae Antiquitatis XX. Piatra Neamţ, Constantin Matasă.

Monah, D. and Dumitroaia, Gh. 2007. Recherches sur l'exploitation préhistorique du sel en Roumanie, in D. Monah, Gh. Dumitroaia, O. Weller, J. Chapman (eds.), *L'exploitation du sel à travers le temps,* 13-34. Bibliotheca Memoriae Antiquitatis XVIII. Piatra Neamţ, Constantin Matasă.

Monah, D., Antonescu, S. and Bujor, A. 1980. Raport preliminar asupra cercetărilor arheologice din comuna Poduri, jud. Bacău. *Materiale şi Cercetări Arheologice*, Tulcea, 86-99.

Monah, D., Dumitroaia, Gh., Monah, F., Preoteasa, C., Munteanu, R., Nicola, D. 2003. *Poduri-Dealul Ghindaru. O Troie în Subcarpaţii Moldovei*, Bibliotheca Memoriae Antiquitatis XIII. Piatra-Neamţ, Constantin Matasă.

Monah, D., Dumitroaia, Gh. and Garvăn, D. (eds.) 2008. *Sarea de la prezent la trecut,* Bibliotheca Memoriae Antiquitatis XX. Piatra Neamţ, Editura Constantin Matasă.

Monah, D., Popovici, D.N. and Dumitroaia, Gh. 2004. Poduri-Dealul Ghindaru: un tell chalcolithique dans l'est de la Roumanie. In *Le Néolithique au Proche Orient et en Europe /L'âge du cuivre au Proche Orient et en Europe. Actes du XIVeme Congrès UISPP, Université de Liège, Belgique, 2-8 Septembre 2001*, 349-357. BAR International Series 1303. Oxford, BAR Publishing.

Monah, D., Dumitroaia, Gh. and Nicola, D. 2008. Noi investigaţii etnoarheologice asupra izvoarelor sărate de pe Valea Muntelui, in D. Monah, Gh. Dumitroaia and D. Garvăn (eds.), *Sarea de la prezent la trecut*, 81-106. Bibliotheca Memoriae Antiquitatis XX. Piatra Neamţ, Constantin Matasă.

Monah, D., Dumitroaia, Gh. and Nicola, D. 2011. New Ethnoarchaeological Investigations upon the Salt Springs in Valea Muntelui, Romania, in M. Alexianu, O. Weller and R.-G. Curcă (eds.), *Archaeology and Anthropology of Salt: A Diachronic Approach, Proceedings of the International Colloquium, 1-5 October 2008, Al. I. Cuza University (Iaşi, Romania)*, 25-34. BAR International Series 2198. Oxford, BAR Publishing.

Munteanu, R. 2006a. Première exploitation de sel en Europe. Techniques et gestion de l'exploitation de la source salée de Poiana Slatinei à Lunca (Neamt, Roumanie), in O. Weller, A. Dufraisse and P. Pétrequin (eds.), *Sel, eau et forêt. Hier et aujourd'hui. Actes du colloque international de la Saline royale d'Arc-et-Senans, 3-5 Octobre 2006.* Cahiers de la MSH Ledoux, Collection Homme et environnement. Besançon, Presses Universitaires de Franche-Comté.

Munteanu, R. 2006b. Cucuieţi, com. Solonţ, jud. Bacău. Punct: Slatina Veche, in V. Cavruc and A. Chiricescu (eds.), *Sarea, timpul şi omul. Catalog de expoziţie,* 62-64. Sfântu Gheorghe, Angustia.

Munteanu, R. and Garvăn, D. 2008. Ceramica Cucuteni C din preajma izvoarelor sărate din Moldova, in D. Monah, Gh. Dumitroaia and D. Garvăn (eds.), *Sarea de la prezent la trecut*, 179-202. Bibliotheca Memoriae Antiquitatis XX. Piatra Neamţ, Constantin Matasă.

Munteanu, R. and Garvăn D. 2011. The Cucuteni C Pottery near the Moldavian Salt Springs, in M. Alexianu, O. Weller, R.-G. Curcă (eds.), *Archaeology and Anthropology of Salt: A Diachronic Approach, Proceedings of the International Colloquium, 1-5 October 2008, Al. I. Cuza University (Iaşi, Romania),* 81-88. BAR International Series 2198. Oxford, BAR Publishing.

Munteanu, R., Garvăn, D., Nicola, D., Preoteasa, C., Dumitroaia, Gh. 2007. Cucuieţi-Slatina Veche (Romania). Prehistoric Exploitation of a Salt Resource, in D. Monah, Gh. Dumitroaia, O. Weller, J. Chapman (eds.), *L'exploitation du sel à travers le temps,* 57-70. Bibliotheca Memoriae Antiquitatis XVIII. Piatra Neamţ, Constantin Matasă.

Müller-Karpe, H. 1968. *Handbuch der Vorgeschichte, I,* München.

Nandriş, J. 1987. Romanian Ethnoarchaeology and the emergence and development of Cucuteni in the European context, in M. Petrescu-Dîmboviţa, N. Ursulescu, D. Monah, V. Chirica (eds.), *La civilisation de Cucuteni en contexte européen. Session scientifique dédiée au centenaire des premières découvertes de Cucuteni (Iaşi – Piatra Neamţ 24-28 septembre 1984),* 201-222. Bibliotheca Archaeologica Iassiensis I, Iaşi.

Nica, M. 1976. Cârcea, cea mai veche aşezare neolitică de la sud de Carpaţi. *Studii şi Cercetări de Istorie Veche* 27, 4, 435-463.

Nica, M. 1977. Nouvelles données sur le néolithique ancien d'Olténie. *Dacia* N. S. XXI, 13-53.

Nicola, D., Munteanu, R., Garvăn, D., Preoteasa, C., Dumitroaia, Gh. 2007. Solca – Slatina Mare (Roumanie). Preuves archéologiques de l'exploatation du sel en prehistoire, in D. Monah, Gh. Dumitroaia, O. Weller, J. Chapman (eds.), *L'exploitation du sel à travers le temps,* 35-56. Bibliotheca Memoriae Antiquitatis XVIII. Piatra Neamţ, Constantin Matasă.

Nikolov, V. 2011. Provadia-Solnitsata (NE Bulgaria): A Salt-Producing Center of the 6th and 5th Millennia BC, in M. Alexianu, O. Weller and R.-G. Curcă (eds.), *Archaeology and Anthropology of Salt: A Diachronic Approach, Proceedings of the International Colloquium, 1-5 October 2008, Al. I. Cuza University (Iaşi, Romania),* 59-64. BAR International Series 2198. Oxford, BAR Publishing.

Nikolov, V. *et al.* (eds.), 2008. Praistoričeski solodobiveh Centîr Provadija – Solonicata. *Raskopki 2005-2007*, Sofia.

Paraschiv, M. 2011. Salt in the Greek and Latin Aphoristic Phrase, in M. Alexianu, O. Weller and R.-G. Curcă (eds.), *Archaeology and Anthropology of Salt: A Diachronic Approach, Proceeding of the International Colloqvium, 1-5 October 2008, Al. I. Cuza University (Iaşi, Romania),* 219-224. BAR International Series 2198. Oxford, BAR Publishing.

Pascu, Şt., Rusu, M., Edroiu, N., Iambor, P., Gyulai, P., Wollmann, V., Matei, Şt. 1968. Cetatea Dăbâca. *Acta Musei Napocensis* V, 153-199.

Paul, I. 1995. *Vorgeschichtliche Untersuchungen in Siebenburgen*. Alba Iulia, Universitatea Alba Iulia.

Păunescu, Al. 1979. Cercetările arheologice de la Cuina Turcului-Dubova (jud. Mehedinţi). *Tibiscus* V, 11-56.

Perić, S. and Nikolić, D. 2004. Stratigraphic, Cultural and Chronological Characteristics of the Pottery from Lepenski Vir 1965 excavation. In S. Perić (ed.), *The Central Pomoravlje in Neolithisation of South-East Europe. The Neolithic in the Middle Morava Valley 1*, 157-217. Belgrade, Archeological Institute.

Petrescu-Dîmboviţa, M. 1977. *Depozitele de bronzuri din România*. Bucureşti, Editura Academiei Române.

Petrova, V. 2011. Tell Provadia-Solnitsata (Bulgaria): Data on Chalcolithic Salt Extraction, in M. Alexianu, O. Weller and R.-G. Curcă (eds.), *Archaeology and Anthropology of Salt: A Diachronic Approach, Proceedings of the International Colloquium, 1-5 October 2008, Al. I. Cuza University (Iaşi, Romania),* 65-68. BAR International Series 2198. Oxford, BAR Publishing.

Popuşoi, E. 2005. *Trestiana. Monografie arhaeologică.* Bârlad, Sfera.

Poruciuc, A. 2008. Denumirile sării şi implicaţiile lor în plan (indo)european, in D. Monah, Gh. Dumitroaia and D. Garvăn (eds.), *Sarea de la prezent la trecut,* 133-148. Bibliotheca Memoriae Antiquitatis XX. Piatra Neamţ, Constanin Matasă.

Poruciuc, A. 2011. Etymological and Historical Implications of Romanian Place-Names Referring to Salt, in M. Alexianu, O. Weller and R.-G. Curcă (eds.), *Archaeology and Anthropology of Salt: A Diachronic Approach, Proceedings of the International Colloquium, 1-5 October 2008, Al. I. Cuza University (Iaşi, Romania),* 215-218. BAR International Series 2198. Oxford, BAR Publishing.

Radunčeva, A. 2003. Kîsnoeneolitnoto obšcestvo v Bîlgarskite zemi. *Raskopki i proučvanija* XXXII, Sofija.

Rodden, R. 1962. Excavation at Early Neolithic Site at Nea Nicomedea, (1961 season). *Proceedings of Prehistoric Society* XXVIII, 267-288.

Roman, R.A. 2008. *Romanian dishes, wines and customs.* Bucureşti, Paideia.

Roman, C., Tincu, S. and Bucur, E. 2012. Consideraţii etno-arheologice privind exploatarea surselor saline din Ţara Făgăraşului (I). *Angustia* 12, 91-100.

Rusu, M. 1977. Transilvania şi Banatul în secolele VI-

IX. *Banatica* IV, 169-213.

Saile, T. 2010. Early Salt/Making in Central Europe: Pattern of salt Production and Trade in the Neolithic. *Analecta Archaeologica Ressovienska* 3, 208, 97-117.

Sanev, V. 1988. Neolitska Svetilište od Tumba vo Madjari. Srpska, *Macedonica Acta Archaeologica* 9, 9-10.

Šarić, J. 2003. A log boat representation on a Neolithic vessel from Korbovo (East Serbia). *Archäologisches Korrespondezblatt* 33, 187-192.

Schuster, C., Tuţulescu, I. and Dumitrescu, I. 2010. Câteva gânduri cu privire la exploatarea sării în nord-estul Olteniei. Din preistorie şi până în epoca modernă. *Angustia* 14, 261-270.

Staljo, B. 1986. Le site préhistorique Ajmana á Mala Vrbica. *Dzerdapske Svečke* III, 27-50.

Suciu, C. and Luca, S.A. IPCTE Radiocarbon Database. Romania and nearby areas from Neolithic to Eneolithic. Sibiu, Institutul pentru Cercetarea şi Valorificarea Patrimoniului Cultural Transilvănean în Context European, http://arheologie .ulbsibiu.ro /radiocarbon /2007%2010%20ian%20data.htm.

Traore, F. 1993. Cercetări etnoarheologice asupra ceramicii şi olăritului tradiţional din satul Mante (Republica Mali). *Acta Musei Napocensis* 26-30/1, (1989-1993), 335-350.

Ursulescu, N. 1977. Exploatarea sării din saramură în neoliticul timpuriu în lumina descoperirilor de la Solca (jud. Suceava). *Studii şi Cercetări de Istorie Veche şi Arheologie* 28/3, 307-317.

Ursulescu, N. 1995. L'utilisation des sources salées dans le Néolithique de la Moldavie (Roumanie). In M. Otte (ed.), *Nature et Culture, Colloque de Liège (13-17 décembre 1993)*, 487-495. Liège, E.R.A.U.L. 68.

Vitcu, D. 1987. *Salinele Moldovei în epoca modernă.* Iaşi, Universitatea "Al. I. Cuza" Iaşi.

Vlassa, N. 1970. Aşezarea neolitică de la Cluj. *Acta Musei Napocensis* VII, 3-39.

Vlassa, N. 1972. Cea mai veche fază a complexului cultural Starčevo-Criş în România. *Acta Musei Napocensis* IX, 7-38.

Vlassa, N. 1972a. Eine frühneolithischen Kultur mit bemalter Keramik der Vor- Starčevo-Körös - Zeit in Cluj-Gura Baciului, Siebenbürgen. *Prähistorische Zeitschrift* 47, 2, 174-197.

Vlassa, N. 1976. *Neoliticul Transilvaniei. Studii, articole, note.* Bibliotheca Musei Napocensis III, Cluj-Napoca.

Vlassa, N. 1980. Din nou despre poziţia stratigrafică şi cronologică a orizontului Gura Baciului I. *Marisia* X, 691-697.

Voinea, V. 2005. *Ceramica complexului cultural Gumelniţa-Karanovo VI. Fazele A1 şi A2.* Constanţa, Ex Ponto.

Weller, O. 2000. *Les premières formes d'exploitation du sel durant le Néolithique et le Chalcolithique européens: de la reconnaissance des techniques à l'analyse des dimensions socio-économiques.* Thèse de doctorat. Université de Paris I Pantheon-Sorbonne.

Weller, O. 2000a. Produire du sel par le feu: techniques et enjeux socio-économique dans le Néolithique européen, in P. Pétrequin, P. Fluzin, J. Thiriot, P. Benoit (eds.), *Arts du feu et productions artisanales, XXe Rencontre Internationales d'Archeologie et d'Histoire d'Antibes*, 565-584. Antibes, APDCA.

Weller, O. and Dumitroaia, Gh. 2005. The earliest salt production in the world: an early Neolithic exploitation in Lunca-Poiana Slatinii, Romania. *Antiquity* 79 (306), 11-18. www.Antiquity. ac. uk/projsall/weller/index.html

Weller, O., Brigand, R. and Alexianu, M. 2007. Cercetări sistematice asupra izvoarelor de apă sărată din Moldova. Bilanţul explorărilor din anii 2004-2007 efectuate în special în judeţul Neamţ. *Memoria Antiquitatis* XXIV, 121-190.

Weller, O., Brigand, R., Nuninger, L., Dumitroaia, Gh., Monah, D. 2007a. Analyses et modélisation spatiale autour des sources salées de Moldavie précarpatique durant la Préhistoire. In N. Morère Molinero (ed.), *Las salinas y la sal de interior en la historia: economia, medio ambiente y sociedad. Inland saltworks and salt history: economy, environment and society*, 165-184. Madrid, Dykinson.

Weller, O., Dumitroaia, Gh., Monah, D., Nuninger, L. 2007b. L'exploitation des sources salées de Moldavie: un exemple de resource structurante du territoire depuis le Néolithique, in D. Monah, Gh. Dumitroaia, O. Weller, J. Chapman (eds.), *L'exploitation du sel à travers le temps,* 99-113. Bibliotheca Memoriae Antiquitatis XVIII. Piatra Neamţ, Constantin Matasă.

Weller, O., Brigand, R., Nuninger, L., Dumitroaia, Gh. 2011. Spatial Analysis of Prehistoric Salt Exploitation in Eastern Carpathians (Romania), in M. Alexianu, O. Weller, R.-G. Curcă (eds.), *Archaeology and Anthropology of Salt: A Diachronic Approach, Proceedings of the International Colloquium, 1-5 October 2008, Al. I. Cuza University (Iaşi, Romania),* 69-80, BAR International Series 2198. Oxford, BAR Publishing.

Wollmann, V. 1996. *Mineritul metalifer, extragerea sării şi carierele de piatră din Dacia romană.* Bibliotheca Musei Napocensis XIII, Cluj-Napoca.

Wollmann, V. and Ciugudean, H. 2005. Noi cercetări privind mineritul antic în Transilvania (I). *Apulum* XLII, 96-116.

Online sources

Ghassulian Culture. pd.farli.org/the-southern-levant/The-Chalcolithic-Period/ghassulian-pottery.

http://bgu.academia.edu/IsaacGilead/Papers/1124900/The_Besorian_a_PreGhassulian_cultural_entity.

New Archaeological Research Relating to the Exploitation of Salt in Bucovina

Mugur Andronic
Muzeul Bucovinei, Suceava

Bogdan-Petru Niculică
Muzeul Bucovinei, Suceava

Abstract
Field archaeological research undertaken during the last years in Bukovina, both in Suceava County (Romania) and Chernivtsi Oblast (Ukraine), resulted into new important specifications regarding the capitalisation by prehistoric communities of the salt water springs from the area. First of all, within the county of Suceava, the locations of salted water sources were checked, some of them, mentioned in documents, were already in use in the Middle Ages. Due natural causes, most of them do not exist anymore; this is the case of the one of Putna, not far from the monastery, Gura Putnei, or Ostra. The salted water spring of Putna is connected to the presence in the neighbourhood of an Eneolithic settlement belonging to the Cucuteni culture. Due to the information received from a Ukrainian amateur archaeologist, we are now able to introduce in the scientific circuit unpublished valuable scientific information, on salt exploitation in the current region of Chernivtsi. It refers to the archaeological site of the village Solonetz-Cereşenka (district of Vyzhnytsia), where the salted water well, which is in use even nowadays, is connected to a Early Hallstatt habitation level (Gáva-Holihrady, the Grăniceşti cultural group), and also to the one of hamlet Slatina (Crasna village – district of Hlyboka), which produced ceramic materials belonging to the Cucuteni-Trypillia, Komariv, Noua and Gáva-Holihrady cultures. Finally, the location of a former salted water well of Vyzhnytsia, on the valley of the Ceremuş, known from as resulting from documents already in the period of the Austrian administration of Bucovina, well which does not exist anymore, did not provide archaeological materials, except a flake made of local rock, produced by knapping, discovered at a larger distance from the place of the former well. All these data unpublished so far complete the repertory of prehistoric salted water sources in the north-eastern Carpathian area; the ceramic and lithic material discovered in the neighbourhoods, the occurrence in the areas neighbouring the mountain frame of the Eastern Carpathians of the human habitat must be studied though systematically in the future.

Keywords
Bukovina, salt exploitation, Chalcolithic, Bronze Age, Iron Age

Archaeological research during the past few years in Bucovina, today Suceava County (Romania) and the region of Chernivtsi (Ukraine), has provided new, and more accurate, information regarding the history of salt exploitation in this historically well-defined area (Figure 1).

These data complement what is already known about the area east of the Carpathians (Andronic 2008, 12-13, 45, 68; Andronic, 2009). It is now the case, thanks to a series of truly surprising prehistoric discoveries, that even the higher mountain landscape of Bucovina is known to have witnessed a human presence in antiquity (Andronic, 2009a).

First of all, the locations of saline springs in the county of Suceava were checked, some of which were known from documentary sources to have been in use during the Middle Ages. Most of the settlements in question have ceased to exist as a result of acts of nature. Examples of such are Putna (not far from the monastery of the same name), on the Valley of Glod, which was blocked up following flooding in 2008, or Gura Putnei and Ostra. The first two are connected with the existence not far away of a Cucuteni Eneolithic settlement (Dealul Ţurana, Gura Putnei) (Andronic *et al.* 2004, 119), and also with the ceramic material discovered when excavations were carried out in the southwest part of the monastic compound at Putna (Cantacuzino, Rădulescu 1974, 537, 539).

The spring of Gura Putnei, in the middle of a forest near the right bank of a brook named scarcely by coincidence "Slatina" (salted water), has also now disappeared over time, in fact during the last two decades (Figure 2).

Figure 1. Map of Bucovina (Austrian epoch), with the positioning of the main salted water sources (the sites of the Chernivtsi region are marked with green, the numbers correspond to those in the text).

Figure 2. The area of the former salted water source of Gura Putnei, Putna commune, Suceava County.

Finally, much better known has been the saline spring at Ostra, first mentioned in a document dated 25 April 1475 as belonging to the feudal domain of the monastery of Humor (DRH II, 304, doc. no. 201). Our field survey which included information from local inhabitants, led us to conclude that it ceased to exist several decades ago, the local inhabitants even forgetting its location, in the valley of the Slatina brook, as a small left tributary of the Băişescu brook (Figure 3).

Figure 3. The valley of brook Slatina of Ostra, Suceava County.

Figure 4. The location of the former salted water source of Pripor – Ostra, Suceava County.

A few kilometres upriver in this forested microzone there are more saline springs that are still active, two at Fundul Glodeţelor and one at Piciorul Neamţului. The latter is much further from Ostra and we consider that this, and the more accessible above-mentioned spring on the Slatina brook, only witnessed the presence of humans sporadically, when Paleolithic people would search for herds of animals in the primeval forests of those days. We also recorded a saline spring that is now dry at "Pripor" (Figure 4) in the Păltiniş valley at Băişescu brook, but it lacked any archaeological remains. The octogenarian Vasile Flocea, who showed it to us, has extracted salt water from this saline spring ever since he was a child, as have many other inhabitants of the Ostra microzone.

Today, local inhabitants still use, as they did before, a saline spring in a place known as "Pârâul Florii" in the neighboring commune, Stulpicani, beyond Poiana Iepei hill in the Negrileasa Valley.

We were less fortunate in obtaining any real results—at least for now—from our researches at the saline well near Slătioara village (com. Râşca), nowadays filled in (Figure 5), but where in the inter-war period, a small resort known as the "Crăescu spa/ Băile Crăescu" was created; nor from a saline well in the forest of Slatina brook (south of the same commune), nor from another near the monastery of Slatina (Slatina commune).

Important archaeological remains associated with the exploitation of salt in prehistory are also to be found in the northern part of historical Bucovina (nowadays in Ukraine), all connected with saline water sources.

1. There is documentary evidence for a saline spring at Vijniţa, in the district of the same name, resulting from investigations undertaken by the Austrian authorities shortly after Bucovina was annexed by the Habsburg Empire in 1775.

Figure 5. The former salted water source Slătioara, Râşca commune, Suceava County.

There is no evidence for a saline spring surviving today, but there certainly was until ca. 1947-1948 such a well protected by a stone wall at 2 Beregometska Street (Figure 6)[1], on the way into the small town from Berhomete (information from Elena Mikitiuk, a local inhabitant born in 1930). The Soviet authorities decided that the consumption of this salt water should cease in favour of salt water sold by the State, and so the well was filled in. Our enthusiastic Ukrainian informant also provided the following information:

It is difficult to make any observations around the site of the former saline spring since there little land left free, and that of limited dimensions, between the many older and newer buildings;

There are numerous pieces of modern brick in these small allotments, evidence for much modern building construction;

Figure 6. The location of the former salted water well of Vijniţa, Vijniţa district, Chernivtsi region, Ukraine.

Archaeological remains are lacking on the plateau beyond the road above the former saline well. Most of this plateau, at the foot of a wooded hill, is occupied by the local cemetery. Nevertheless, Mr Covalschi found there a small fragment of a core made of local rock, dark in colour, with a hardness that allows for cleavage, and which still bears traces of splitting by means of two microliths. It is clear from this evidence that we may deduce that prehistoric man also used this source of salt water (although the fact is in any case inherently likely).

2. Another important settlement for the history of salt in Bucovina is to be found in the microzone of Vijniţa, situated at the foot of the pre-Carpathians. This is the saline well near the village named—not by chance—"Soloneţ" which belongs administratively to the neighbouring village of Cereşenca (Rom. Cireşoaia; the name before the Soviet occupation of the northern part of Bucovina).

It can be seen from the image provided by our Ukrainian informant, the well is occasionally used by locals even today. It is important to note here that a good deal of ceramic material was found on the land that slopes gently downwards on the left bank of a small brook. We can also distinguish several places with a considerable residue of chunks of fired clay, all of which suggest the presence of a former dwelling. The settlement is only 30 or 40 m from the well (Figure 7). These archaeological deposits could be clearly seen in the side of a newly dug ditch by the side of a road leading to a group of newly constructed houses.

Figure 7. The salted water well and the archeological site of Soloneţ – Cereşenca, Vijniţa district, Chernivtsi region, Ukraine.

These archaeological features lie at the foot of a wooded hill called Chicera, and in prehistory the settlement—probably seasonal and used only in summer—must have been situated in the middle of a forest, fact also demonstrated by the pedological characteristics of this type of soil. Under a vegetal layer at a depth of 30 cm, there is a dense mass of ceramic fragments and fired clay (Figure 8). This suggests that settlement had a long period of use, perhaps over a century or more, or—very likely—that there was intensive ceramic production, necessary for the production of solid salt through the famous method of re-crystallizing salt from a saline water source. There is in fact a considerable number of potsherds from vessels used to boil salt water; among the ceramic types identified here we should note not only large vessels (Figure 8), but fragments with straight rims, probably from a type of flat-based tray specially designed for such purpose (Figure 9). The boiling of the salt water could have taken place in numerous large bowls which may have been spared from breaking of salt was obtained in a moistened condition, like porridge.

[1] We should like to thank the amateur archaeologist Igor Covalschi for images and other information relating to the archaeology of Chernivtsi region.

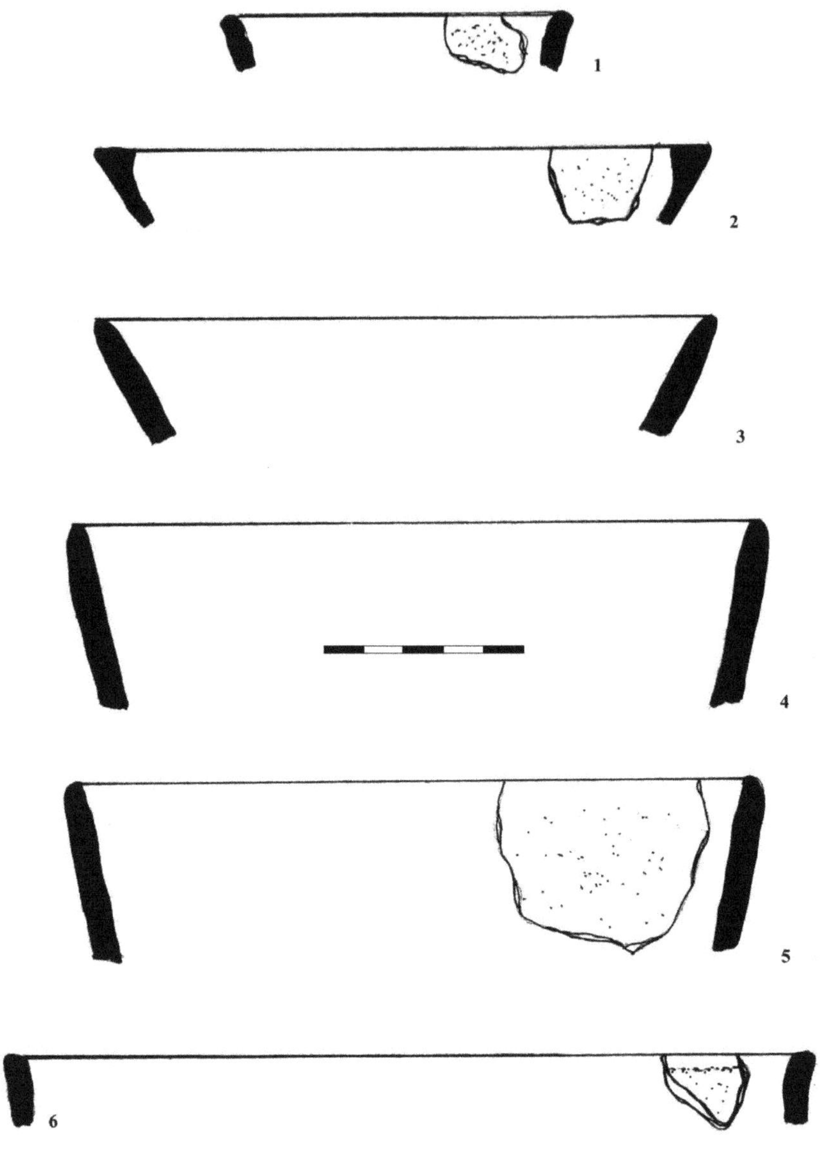

Figure 8. Ceramicware discovered at Soloneț – Cereșenca.

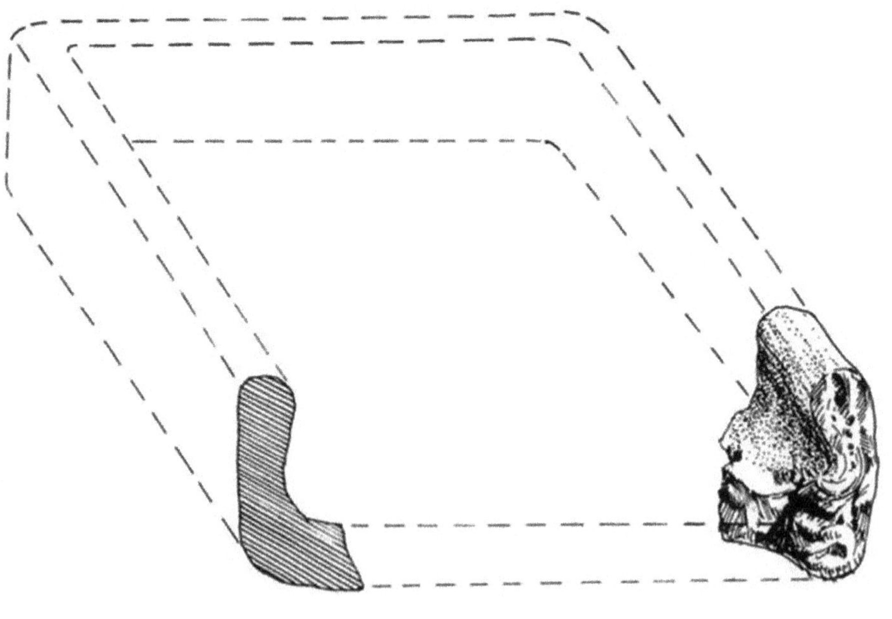

Figure 9. Clay tray discovered at Soloneț – Cereșenca.

To date, no lithic inventory is known, and the ceramic material, exclusively hand-made, with its generally brick-red hue, seems to belong to a single habitation level, dating to the Early Iron Age.

The ceramic wares discovered at Soloneţ contain fine and semi-fine sand as ingredients in the clay from which they were made. In general, the pottery was not decorated, only few fragments displaying traces of ornamentation with vertical, horizontal or arched motifs. Some fragments of sack-shaped vessel have oval projections. Since the material was exceptionally fragmentary it was possible only to reconstruct a part of the typological repertory.

We can thus take note of: vessels with a high cylindrical or slightly curved neck; large vessels with thick walls; fragments of amphorae; pots with flared rims; pots with rounded rims; bowls. Typologically and ornamentally the ceramic material belongs to the cultural group Grăniceşti of the Gáva-Holihrady culture, as defined by A. László (László, 1994), and is has clear characteristics in common with the pottery of the open settlement of Grăniceşti, and the fortified settlement at Siret: "Ruina Hill" (both in Suceava county, Romania).

The presence of a potsherd ornamented with an alveolar belt, with inclusions of broken potsherds, leaves open the possibility of later La Tène period occupation.

3. The third archaeological point in the northern part of historical Bucovina is located at only a few kilometres away from the modern Romanian border, to the southwest of Crasna, Hliboca district, in the hamlet of Slatina (Figure 10). Since no large expanses of land are worked here, our surveys could only be done with difficultly. On the slightly terraced, unploughed, south facing slopes, we observed several potsherds, flint flakes and calcareous marl, which can be best associated with the Cucuteni-Tripolie culture, together with ceramic material of the Bronze and Early Iron Age. More information will surely be forthcoming should Ukrainian colleagues undertake systematic excavations.

Figure 10. The location of the salted water well of Crasna, Hliboca district, Chernivtsi region, Ukraine.

References

Andronic, M. 2008. *Istoria Bucovinei, De la începuturi până la epoca cuceririi romane a Daciei.* Suceava.

Andronic, M. 2009. L'exploitation du sel à l'est de Carpates Orientaux dans la préhistoire et dans la protohistoire. *SCIVA* 59-60, 17-32.

Andronic, M. 2009a. Prezenţe ale omului preistoric în Carpaţii Bucovinei. *Arheologia Moldovei* XXXII, 209-220.

Andronic, M., Batariuc, P.V., Hău, Fl., Gogu, M., Niculică, B.P., Mareş, I. 2002-2003. Noi cercetări arheologice de teren în judeţul Suceava. *Suceava* XXIX-XXX, 117-226.

Cantacuzino, Gh.I., Rădulescu, Al. 1974. Consideraţii asupra ceramicii descoperite la Mănăstirea Putna, în anul 1971. *Studii şi Cercetări de Istorie Veche* 25, 4, 527-544.

László, A. 1994. *Începuturile epocii fierului la est de Carpaţi,* Bucureşti.

Sources of Salt and the Territorial Dynamics of Late Bronze Age Communities in the Northern Moldavian Subcarpathian Region

Vasile Diaconu
Muzeul de Istorie şi Etnografie Târgu-Neamţ

Abstract
The author proposes an analysis of the relation between the saline springs and the characteristic settlements of the Late Bronze period (Noua culture) of the northern Moldavian sub-Carpathian area. There is a special focus on the sources of salt, many of which were known and exploited by ancient communities. Through mapping the saline springs and the settlements, it is possible to define theoretically three micro-zones containing a high density of sites. A potential explanation might lie in the abundance of sources of salt that offered not only an essential mineral for human and animal sustenance, but also a trade commodity and a prestige good. At the present stage of research, only one of more than 16 known saline springs, that of Lunca-*Poiana Slatinei*, has produced archaeological evidence for the exploitation of salt water in the Late Bronze Age. There must, however, have been other similar sources, but in those cases crystallization was probably carried out within the settlements, after the salt water was transported, rather than in the vicinity of the springs.

Keywords
territorial dynamic, salt sources, Late Bronze Age, Noua culture, sub-Carpathian Moldavia

Introduction

The study of the territorial dynamics of human communities at different stages of prehistory requires an analysis based on multiple parameters connected with geographical factors: habitat, economic specificity and social organization. For an overview, these parameters should be situated within a given area inhabited by a given cultural entity, which is also useful when applied to a micro-zone. Such research can reveal the way in which habitation elements influence the dynamics of human communities, the different resources that were exploited, population density or possible contacts with neighbouring cultural entities (Champion 1980, 116).

Here we study the territorial dynamics between the Late Bronze communities of the northern Moldavian sub-Carpathian area and the principal available sources of salt. We have concentrated on a limited part of the sub-Carpathian region because there is a higher concentration of settlements there, in contrast with the central and southern parts of the region. Furthermore, the territory is located at the crossroads of different geographical units (a mountainous area, the sub-Carpathian area, the Moldavian Plateau), which had a certain impact on communities regarding access to certain resources, to mobility and to territorial control.

We are well informed with regard to the spatial analysis of prehistoric salt exploitation in the Neolithic and Eneolithic periods (Weller *et al.*, 2007a; 2011), but this is not the case for later historical epochs which are only treated in a general fashion (Cavruc *et al.*, 2006a; 2006b) although it is reasonable to suppose that salt was exploited more intensively than in earlier chronological stages (Cavruc *et al.*, 2006a, 41).

In chronological terms, we only consider here the final stage of the Bronze Age, approximately the second half of the 2nd millenium BC. At this time, the Noua culture extended over a vast territory, from the Middle Dniester to the Apuseni Mountains (Florescu 1964, 1991; Florescu and Florescu 2000; Petrescu-Dîmboviţa 2001; Sava 1994).

Natural background

The Moldavian sub-Carpathians are bounded to the north by the Moldova Valley and to the south the Trotuş Valley. Made up of a range of depressions (Neamţ, Cracău-Bistriţa, Tazlău-Caşin), the sub-Carpathian region has a simple, large, yet well-defined structure, being bounded the west by the Oriental Carpathians, to the east by hills and foothills connecting with the Moldavian Plateau and the Siret Corridor (Tufescu 1966, 81-116).

The area treated in our study matches the northern part of the Moldavian sub-Carpathians and approximates to the Neamţ Depression, and includes a few micro-zones that also lie in depressions (Figure 1). The limits are represented by the Stânişoarei Mountains to the west, the Moldova Valley to the north and east, while the southern limit consists of the Bălţăteşti-Ghindăoani saddle and Corni Hill (Davidescu 2000; Dimitriu 2007).

Without insisting on living conditions, we must stress that one of the most abundant resources of the area is salt. In the northern sub-Carpathians, numerous saline springs are attested (Alexianu *et al.*, 2007; Weller *et al.*, 2007b; 2010a) (Figure 3), and in many cases there is substantial evidence for human activity close by, bearing witness to their exploitation in different historical periods. The salt springs at Lunca-*Poiana Slatinei* (Dumitroaia 1987; 1994; 2000; Weller and Dumitroaia 2005; Munteanu 2006; Weller *et al.*, 2010b), Oglinzi-*Băi* (Dumitroaia 1994) and Ţolici-*Hălăbutoaia* (Dumitroaia *et al.*, 2008a; 2008b; 2009) are just a few examples. It is also likely that similar places that at present lack any archaeological confirmation were exploited in antiquity. In any case, the high percentage of salinity in the water from these (Figure 2) is an essential reason for their exploitation and capitalization, for human consumption, animal diet, or for trade.

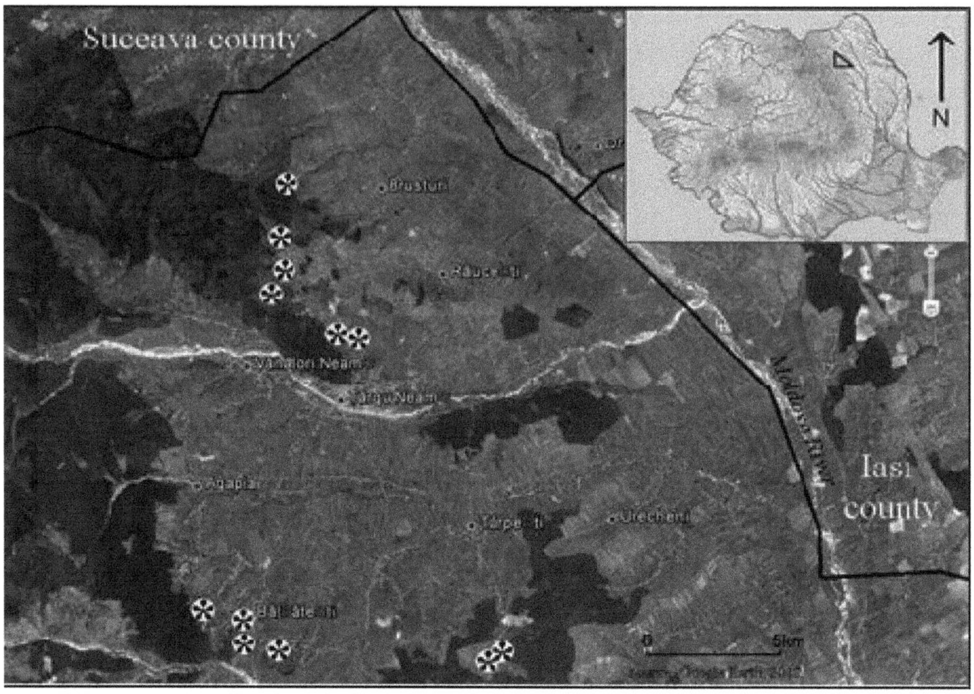

Figure 1. The area studied and the principal saline springs (*).

Saline spring	Anions					Cations			
	F (mg/l)	Br (mg/l)	Cl (mg/l)	NO3 (mg/l)	SO4 (mg/l)	Ca (mg/l)	Na (mg/l)	Mg (mg/l)	K (mg/l)
Bălţăteşti-*Slatina B*	0	50	164	0	35.96	0.18	89	6.11	16.8
Filioara-*Slatina II*	0	0	8.73	0	3.38	0.17	4.1	0.02	0.002
Ghindăoani-*Slatina*	0	307.3	136.5	0	8.29	1.43	76	2.51	3.6
Lunca-*Poiana Slatinei*	40	0	158	0	9.4	0.76	94	0.21	0.08
Oglinzi-*Băi*	0	0	21	0	4.47	0.14	9	0.04	0.03
Oglinzi-*Fântâna Corugea*	20	0	152	0	9.55	1.27	79	0.26	0.06
Oglinzi-*Slătior*	10	30	106	0	6.29	1.06	66	0.64	0.12
Poiana-*Fântâna Culeşa*	10	0	89	0	7.6	1.12	75	0.33	0.07
Ţolici-*Hălăbutoaia*	0	0	134.05	0	6.09	0.81	85	0.13	0.046

Figure 2. The chemical components of samples from saline springs located in the northern part of the Moldavian sub-Carpathians (after Weller *et al.*, 2007b).

Saline springs and the territorial dynamics of Late Bronze communities

In a recent study we analyzed the spatial distribution of Noua culture communities from the sub-Carpathian area (Diaconu 2011) and observed that settlement density was uneven. This could well correspond to historical reality, although there were major discrepancies in the degree to which sites had been investigated archaeologically. Generally speaking, the highest density of Noua culture settlements is to be found in the northern part of the Moldavian sub-Carpathians, with over 30 sites known to date (Cucoş 1992; Dumitroaia 1992a; 1992b; Diaconu 2007; 2010; 2012). The preference of Late Bronze communities for this area was, in our opinion, determined not only by general living conditions, but also by other, more specific factors, such as the existence of communication networks with the inner Carpathian region, and also the abundance and accessibility of salt sources.

Figure 3. Saline springs (Lunca-*Poiana Slatinei*, Oglinzi-*Slătior*, Ghindăoani-*Slatina*, Bălţăteşti-*Slatina B*) (Photos: V. D.).

It is clear that Late Bronze populations had an economy based on animal husbandry, which in turn involved the consumption of salt (Sava 2004). Nevertheless, direct evidence for the actual exploitation of saline springs is scarce for the period in question. For the east Carpathian region there is only one site, at Lunca-*Poiana Slatinei*, where Noua culture finds have been made (Dumitroaia 2000, 146; 2001, 33-34). This does not mean that other, similar, sources were not used, since it was quite possible for the water to be carried substantial distances in various kinds of vessel, whether of pottery, wood, sor leather. Unlike the Eneolithic period, when special containers (*briquetage*) were used to obtain crystallized salt and salt blocks, no such containers have been identified for the Bronze Age; most likely common pottery vessels were used for such activities. It is difficult to establish whether certain pottery vessels found in the settlements were used only incidentally for obtaining salt. There is archaeological evidence for the extraction of common salt through boiling salty water, and it is therefore possible to envisage the harvesting of natural crystals formed in the vicinity of springs (Alexianu *et al.*, 2011, 17-18).

In analyzing the correspondence between saline springs and Noua culture settlements in the northern Moldavian sub-Carpathian region, a few concentrated micro-zones can be distinguished. Even though the establishment is theoretical, it could well reflect certain historical realities, triggered by the need to use such an essential resource for humans and animals.

In mapping saline springs and Noua culture settlements, we were able to pinpoint three main areas (A, B, C).

A. The Lunca-Oglinzi-Răuceşti micro-zone (Figure 4).

Here, within an arbitrarily selected area of 12 km^2, six saline springs and 10 Late Bronze settlements are documented. It is important to note that finds attesting exploitation by Noua culture groups have only been made in the vicinity of one salt water spring, of Lunca-*Poiana Slatinei* (Dumitroaia 2000, 146; 2001, 33-34). In the light of this, we can speculate that the salt was used not only for local consumption, but was also destined for long distance trade. Another interesting situation is the settlement from Oglinzi-*Faţa Slatinei* (Dumitroaia 1992a, 86-87), located approximately 300 metres from the springs at Oglinzi-*Băi*. This site shows signs of intense habitation, which encourages the view that a stable community controlled the nearby saline springs.

B. The Agapia-Bălţăteşti-Ghindăoani micro-zone (Figure 4). Here, seven springs and 11 Noua culture settlements have been found. No clear archaeological evidence exists for the use made by these communities of local saline springs. Most of the settlements are located away from the springs, which may be the result of topographical conditions. A group of three settlements stood close to a saline spring at Filioara-*Slatina I*, but since it has disappeared it means that it is difficult for us to establish its degree of salinity or to speculate on its importance for communities in the area. As is also clear from Figure 4, some of the settlements included here may have used salt sources belonging to micro-zone C.

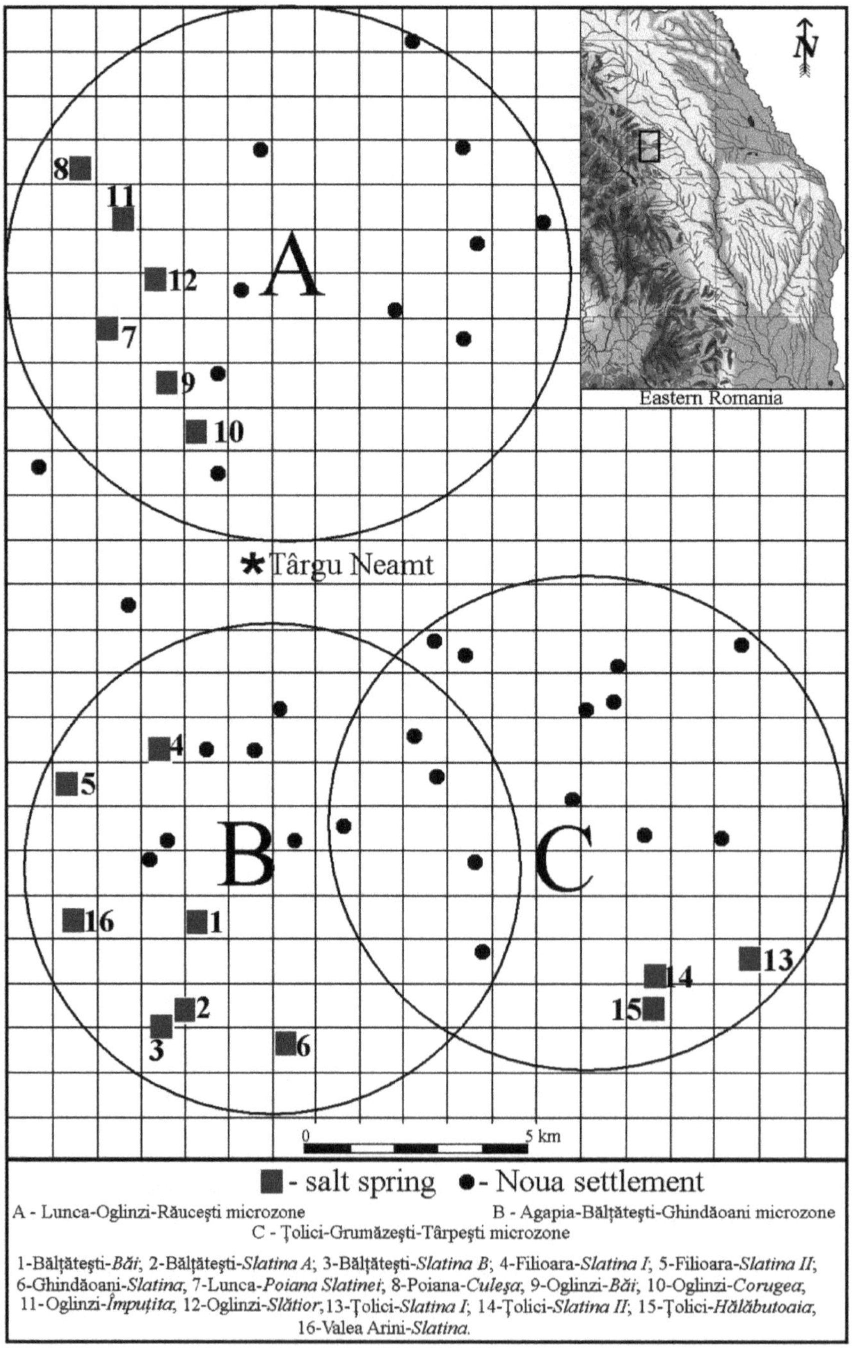

Figure 4. Sources of salt and the territorial dynamics of Late Bronze communities.

C. The Ţolici-Grumăzeşti-Târpeşti micro-zone (Figure 4). Unlike the other situations, only three saline springs are known here, but the number of settlements, at 14, is high.

The springs are clustered together to the west of Ţolici village. The nearest settlements are located at a distance of 2km, while the farthest is 7km away. Here too there is no archaeological evidence for the use of the springs by the Noua culture communities. As mentioned earlier, some of the settlements could have been using salt resources belonging to the micro-zone B.

In the light of all this, we can state the following: First, our analysis is strictly theoretical and is based on the spatial distribution of saline water sources and of Noua culture settlements. Nor is it entirely certain that Late Bronze sites were contemporary with one another, which perhaps puts a different light on things. Nor do we know precisely how many of those settlements were in use over longer period of time or were just seasonal habitations, determined by pastoral activities and the increased mobility of the communities. These unknowns render the conclusions presented here somewhat speculative, but we also believe that by extending the research, the theoretical pattern can still be applied.

Local consumption and trade

It goes without saying that Late Bronze populations used salt resources principally for their own consumption. Both humans and animals require a certain amount of salt in their diet, in order to preserve bodily health. Even if supporting archaeological evidence is absent, we believe that all (or almost all) saline springs in the northern part of the Moldavian sub-Carpathians were used by Noua culture groups. Our view receives tentative support from the fact that animal husbandry was a large scale occupation necessarily requiring a large amount of salt. In this context, it has been observed in different historical periods that salt played an essential role in animal diet, assisting growth and increasing milk production (Monah 1991, 388-389; Cavruc 2006, 33).

Nevertheless, part of the salt production was destined for trade, being carried to those areas where it was lacking (Cavruc 2010, 8). If we consider the Noua culture zone, we can see that that salt from the sub-Carpathian region was destined for communities living at great distance. We can assume that some saline springs, due to their concentration and abundance, were meeting places of macro-regional importance. For the Moldavian sub-Carpathians such a place might have been the spring at Lunca-*Poiana Slatinei*. A similar potential meeting place is known in Ukraine, at Loeva (Kruše'lnic'ka 2002; 2006, 68-75), and the rich salt resources of the Transylvanian area are noteworthy in this context (Wollmann and Ciugudean 2005, 97-102; Cavruc 2006, 34; Cavruc and Harding 2008; 2011; 2012, 175-176).

Returning to the Moldavian sub-Carpathians, suppose that some resources (minerals, metals; Cavruc 2011) and goods from Noua culture settlements were brought here as a result of inter-community exchange. The bronze daggers discovered in the settlements at Vânători-Neamţ (Dumitroaia 1986, 18-19, fig. 7/1) and Târpeşti (Marinescu-Bîlcu 1981, fig. 215/6) were thus traded for salt. Our hypothesis may receive support from the fact that the two artefacts are of a specifically eastern type and are the only examples from the sub-Carpathian area and the farthest to the west in Moldavia.

Conclusions

The high number of Late Bronze settlements in the northern part of the Moldavian sub-Carpathians illustrates the fact that their territorial dynamics were closely related to habitation factors, among which saline springs played an important role. Although few traces of exploitation are known, it is possible to envisage some springs having served as meeting places for different groups of people. In this context, we have proposed that there are within the area we have studied a few micro-zones in which there is a high density of sites whose existence can best be accounted for by presence of rich salt resources. The fact that the Noua culture had an economy based on animal husbandry allows us to conclude that the exploitation of salt sources was a well-coordinated and intensive activity.

At the present stage of research we can conclude that in the northern part of the Moldavian sub-Carpathians, saline springs can be divided into three categories based on archaeological evidence and on their distribution in relation to settlements of the Late Bronze Age. There were saline springs of macro-regional importance (Lunca-*Poiana Slatinei*), springs controlled by a nearby community (Oglinzi-*Băi*), and saline sources of local importance (the majority).

The precise extent to which salt was used in exchange for goods that were unavailable locally can be debated, and is subject to further, more complex, investigation in the future, not only in the Subcarpathian area of Moldavia but elsewhere.

References

Alexianu, M., Weller, O. and Brigand, R. 2007. *Izvoarele de apă sărată din Moldova Subcarpatică. Cercetări etnoarheologice.* Iaşi, Casa Editorială Demiurg.

Alexianu, M., Weller, O., Brigand, R., Curcă, R.-G., Cotiugă, V., Moga, I. 2011. Salt springs in today's rural world. An ethnoarchaeological approach in Moldavia (Romania), in M. Alexianu, O. Weller and R.-G. Curcă (eds.), *Archaeology and Anthropology of Salt: A Diachronic Approach. Proceedings of the International Colloquium, 1-5 October 2008, Al. I. Cuza University (Iaşi, Romania)*, 7-23. British Archaeological Reports, International Series 2198. Oxford, BAR Publishing.

Cavruc, V. 2006. Sarea în societăţile arhaice. Consideraţii generale, in V. Cavruc and A. Chiricescu (eds.), *Sarea, timpul şi omul*, 33-36. Sfântu Gheorghe, Angustia.

Cavruc, V. 2010. Consideraţii privind schimbul cu sare în mileniile VI-II î.Hr. în spaţiul carpato-dunărean. *Tyragetia* S.N., 4 (XIX), 1, 7-34.

Cavruc, V. 2011. Olovo i soly v karpatskom basseine v bronzovom veke. *Revista Arheologică* VII, 1-2, 5-46.

Cavruc, V., Ciugudean, H. and Harding, A.F. 2006a. Vestigiile arheologice privind exploatarea sării pe teritoriul României în epoca bronzului, in V. Cavruc and A. Chiricescu (eds.), *Sarea, timpul şi omul*, 41-49. Sfântu Gheorghe, Angustia.

Cavruc, V., Munteanu, R.E. and Ciugudean, H. 2006b. Vestigiile arheologice privind exploatarea sării pe teritoriul României în epoca timpurie a fierului, in V. Cavruc and A. Chiricescu (eds.), *Sarea, timpul şi omul*, 50-52. Sfântu Gheorghe, Angustia.

Cavruc, V. and Harding, A.F. 2008. Noi cercetări arheologice privind exploatarea sării în nord-estul Transilvaniei. Raport preliminar, in D. Monah, Gh. Dumitroaia and D. Garvăn (eds.), *Sarea, de la prezent la trecut*. Bibliotheca Memoriae Antiquitatis XX, 149-178. Piatra Neamţ, Constantin Matasă.

Cavruc, V. and Harding, A.F. 2011. New archaeological researches concerning saltworking in Transylvania. Preliminary report, in M. Alexianu, O. Weller and R.-G. Curcă (eds.), *Archaeology and Anthropology of Salt: A Diachronic Approach. Proceedings of the International Colloquium, 1-5 October 2008, Al. I. Cuza University (Iaşi, Romania)*, 111-122. British Archaeological Reports, International Series 2198. Oxford, BAR Publishing.

Cavruc, V. and Harding, A.F. 2012. Prehistoric production and exchange of salt in the Carpathian-Danube Region, in V. Nikolov and K. Bacvarov (eds.), *Salz und Gold: die Rolle des Salzes im prähistorischen Europa / Salt and Gold: The Role of Salt in Prehistoric Europe*, 173-200. Provadia & Veliko Tarnovo.

Champion, S. 1980. *Dictionary of Terms and Techniques in Archaeology*. New York, Everest House.

Cucoş, Şt. 1992. Contribuţii la repertoriul arheologic al judeţului Neamţ. *Memoria Antiquitatis* XVIII, 5-61.

Davidescu, G. 2000. *Depresiunea subcarpatică Ozana-Topoliţa. Studiu de geografie fizică*. Iaşi, Editura Cugetarea.

Diaconu, V. 2007. Recunoaşteri arheologice de suprafaţă în zona oraşului Târgu Neamţ. *Memoria Antiquitatis* XXIV, 87-119.

Diaconu, V. 2010. Noi situri arheologice în zona de nord-est a judeţului Neamţ. *Memoria Antiquitatis* XXV-XXVI, (2008-2009), 417-435.

Diaconu, V. 2011. Comunităţi umane din bronzul târziu în Subcarpaţii Moldovei. Observaţii privitoare la dinamica teritorială. *Cercetări Istorice* XXVII-XXIX, (2008-2010), 45-66.

Diaconu, V. 2012. *Depresiunea Neamţ. Contribuţii arheologice*. Bibliotheca Memoriae Antiquitatis XXVIII. Piatra Neamţ, Constantin Matasă.

Dimitriu, R. 2007. *Depresiunea Neamţului. Studiu de geografie umană*. Iaşi, Editura Universităţii "Alexandru Ioan Cuza".

Dumitroaia, Gh. 1986. Sondajul arheologic de la Vânători-Neamţ. *Memoria Antiquitatis* XII-XIV, (1980-1982), 15-29.

Dumitroaia, Gh. 1987. La station archéologique de Lunca-Poiana Slatinii, in M. Petrescu-Dîmboviţa *et al.* (eds.), *La civilisation de Cucuteni en contexte européen*, 253-258. Iaşi, Bibliotheca Archaeologica Iassiensis I.

Dumitroaia, Gh. 1992a. Materiale şi cercetări arheologice din nord-estul judeţului Neamţ. *Memoria Antiquitatis* XVIII, 63-143.

Dumitroaia, Gh. 1992b. Cultura Noua pe teritoriul judeţului Neamţ. *Carpica* XXIII, 119-141.

Dumitroaia, Gh. 1994. Depunerile neo-eneolitice de la Lunca şi Oglinzi, judeţul Neamţ. *Memoria Antiquitatis* XIX, 7-82.

Dumitroaia, Gh. 2000. *Comunităţi preistorice din nord-estul României. De la cultura Cucuteni până în bronzul mijlociu*. Bibliotheca Memoriae Antiquitatis VII. Piatra Neamţ, Constantin Matasă.

Dumitroaia, Gh. 2001. Sat Lunca, com. Vânători-Neamţ, jud. Neamţ, in V. Cavruc and Gh. Dumitroaia (eds.), *Cultura Costişa in contextul epocii bronzului din România*, 31-40. Piatra Neamţ, Constantin Matasă.

Dumitroaia, Gh., Munteanu, R., Garvăn, D., Weller, O., Brigand, R. 2008a. Ţolici, com. Petricani, jud. Neamţ. Punct: Hălăbutoaia. *Cronica Cercetărilor Arheologice din România. Campania 2007*, 325-326.

Dumitroaia, Gh., Munteanu, R., Garvăn, D., Weller, O., Diaconu, V., Brigand, R. 2008b. Un nou punct de exploatare a apei sărate în preistorie: Ţolici-Hălăbutoaia, in D. Monah, Gh. Dumitroaia and D. Garvăn (eds.), *Sarea, de la prezent la trecut*. Bibliotheca Memoriae Antiquitatis XX, 203-224. Piatra Neamţ, Constantin Matasă.

Dumitroaia, Gh., Munteanu, R., Garvăn, D., Diaconu, V., Weller, O., Brigand, R. 2009. Ţolici, com. Petricani, jud. Neamţ, punct Hălăbutoaia. *Cronica Cercetărilor Arheologice din România. Campania Campania 2008*, 227-228.

Florescu, A.C. 1964. Contribuţii la cunoaşterea culturii Noua. *Arheologia Moldovei* II-III, 143-193.

Florescu, A.C. 1991. Repertoriul culturii Noua-Coslogeni din România. Aşezări şi necropole. *Cultură şi Civilizaţie la Dunărea de Jos* IX. Călăraşi.

Florescu, M. and Florescu, A.C. 2000. Noua. In *Enciclopedia Arheologiei şi Istoriei Vechi a României*, M-Q, 199-204. Bucureşti, Editura Enciclopedică.

Kruše'lnic'ka, L. 2002. Un nou sit din epoca bronzului timpuriu în Ucraina Subcarpatică. *Angustia* 7, 99-114.

Kruše'lnic'ka, L. 2006. *Kultura Noua na zemleah Ukrajini*. Lviv.

Marinescu-Bîlcu, S. 1981. *Târpeşti. From Prehistory to History in Eastern Romania*, British Archaeological Reports 107. Oxford, BAR Publishing.

Monah, D. 1991. L'exploitation du sel dans les Carpates Orientales et ses rapports avec la culture de Cucuteni-Tripolye, in V. Chirica and D. Monah (eds.), *Le Paléolithique et le Néolithique de la Roumanie en contexte européen*, 387-400. Iaşi, Bibliotheca Archaeologica Iassiensis IV.

Munteanu, R.E. 2006. Lunca, com. Vânători-Neamţ, jud. Neamţ. Punct: Poiana Slatinei, in V. Cavruc and A. Chiricescu (eds.), *Sarea, timpul şi omul*, 64-66. Sfântu Gheorghe, Angustia.

Petrescu-Dîmboviţa, M. 2001. Perioada târzie a epocii bronzului, in M. Petrescu-Dîmboviţa and Al. Vulpe (eds.), *Istoria Românilor*, Volume I, 272-287. Bucureşti, Editura Enciclopedică.

Sava, E. 1994. Epoca bronzului - Perioada mijlocie şi târzie (sec. XVII-XII î.e.n.). *Thraco-Dacica* XV, 1-2, 141-158.

Sava, E. 2004. Unele aspecte economice din perioada târzie a epocii bronzului (complexul cultural Noua-Sabatinovka). In *Studii de istorie veche şi medievală. Omagiu profesorului Gheorghe Postică*, 68-75. Chişinău, Pontos.

Tufescu, V. 1966. *Subcarpaţii şi depresiunile marginale ale Transilvaniei*. Bucureşti, Editura Ştiinţifică.

Weller, O. and Dumitroaia, Gh. 2005. The earliest salt production in the world. A Neolithic exploitation in Lunca-Poiana Slatinei, Romania. *Antiquity* 79 (306). Online: antiquity.ac.uk/projgall/weller [last accessed: 10.09.2014].

Weller, O., Dumitroaia, Gh., Monah, D., Nuninger, L. 2007a. L'exploitation des sources salées en Moldavie: un exemple de ressource structurante du territoire depuis le Néolithique, in D. Monah, Gh. Dumitroaia, O. Weller, J. Chapman (eds.), *L'exploitation du sel à travers le temps*. Bibliotheca Memoriae Antiquitatis XVIII, 99-113. Piatra Neamţ, Constantin Matasă.

Weller, O., Brigand, R. and Alexianu, M. 2007b. Cercetări sistematice asupra izvoarelor de apă sărată din Moldova. Bilanţul explorărilor din anii 2004-2007 efectuate în special în judeţul Neamţ. *Memoria Antiquitatis* XXIV, 121-190.

Weller, O., Brigand, R. and Alexianu, M. 2010a. Recherches systématiques autour des sources salées de Moldavie. Bilan des prospections 2008-2010. *Memoria Antiquitatis* XXIV-XXVII, (2008-2010), 437-504.

Weller, O., Dumitroaia, Gh., Sordoillet, D., Dufraisse, A., Gauthier, E., Munteanu, R., 2010b. Lunca-Poiana Slatinei (jud. Neamţ): cel mai vechi sit de exploatare a sării din preistoria europeană. *Arheologia Moldovei* XXXII (2009), 21-39.

Weller, O., Brigand, R., Nuninger, L., Dumitroaia, Gh. 2011. Spatial analysis of prehistoric salt exploitation in Eastern Carpathians (Romania), in M. Alexianu, O. Weller, R.-G. Curcă (eds.), *Archaeology and Anthropology of Salt: A Diachronic Approach. Proceedings of the International Colloquium, 1-5 October 2008, Al. I. Cuza University (Iaşi, Romania)*, 69-80. Oxford, British Archaeological Reports 2198, BAR Publishing.

Wollmann, V. and Ciugudean H. 2005. Noi cercetări privind mineritul antic în Transilvania (I). *Apulum* 42, 95-116.

Sources of Salt in Dâmboviţa County: Geological, Ethnographical, Historical and Archaeological Perspectives

Dragomir Nicolae Popovici
Muzeul Naţional de Istorie a României, Bucureşti

Ovidiu Cîrstina
Complexul Muzeal Naţional „Curtea Domnească", Târgovişte

Ana Ilie
Complexul Muzeal Naţional „Curtea Domnească", Târgovişte

Gheorghe Olteanu
Complexul Muzeal Naţional „Curtea Domnească", Târgovişte

Mihai Năstase
Complexul Muzeal Naţional „Curtea Domnească", Târgovişte

Florin Petrică
Complexul Muzeal Naţional „Curtea Domnească", Târgovişte

Abstract
Salt is prominent among the natural resources found in Dâmboviţa County, which administratively and topographically, geographically and geomorphologically, lies within the sub-Carpathian Curvature. Information about human exploitation of salt in this region is somewhat disparate and discontinuous, scattered among a variety of more or less accessible publications of a geomorphological, historical, archaeological and ethnographical nature. The remarkable importance that this mineral has had for human communities over time has encouraged a multidisciplinary approach providing an assessment and evaluation of the ways in which the exploitation of various resources (rock salt and brine) and their use evolved through time. The first stage of the research programme has been to assess the current state of knowledge on the subject, in order to allow the development of a future research strategy. Data resulting from this first phase is presented here.

Keywords
Curvature Subcarpathians, Dâmboviţa county, salt sources, ethnography, archaeology

The intensification of research into the mining and use of salt in various regions of the country during the past two decades have highlighted not only its significance over time for various human communities, but also its economic, social and symbolic value (Alexianu *et al.*, 1992; 2007; Cavruc, Chiricescu 2006; Monah 1991; 2002; Monah *et al.*, 2007; 2008).

The existence of important salt deposits on the territory of the county of Dâmboviţa, as yet insufficiently studied, lies behind this research programme.

Geological studies (Velcea 1982) report the existence of salt massifs in places such as Gura Ocniţei-Moreni-Ocniţa, Glodeni and even Bezdead, have encouraged our team to concentrate on these three areas of the county (Figure 1).

The most important historical information we have today concerns the mining of rock salt during the middle Ages. (Ilieş 1956; Toma *et al.*, 2008). The sources mainly refer to the results of the exploitation of salt, without providing any significant information about the extent of mining, or the technology employed and its possible social implications.

Further information came from publications about villages or communes (regional units), written by enthusiastic amateurs.

The commune of Ocniţa
The commune of Ocniţa, located on the incidence line of the Subcarpathians with the Romanian Plain, is set in a depression bounded by hills with heights varying from 300 to 500m. Salt, oil, natural gas and sulphur resources of the Mio-Pliocene age are to be found in this area. The village is situated on a salt massif 6 km in diameter. The salt deposit crosses the depression from north to south, and it begins to rise from a depth of approximately 150 m, in Ochiuri village, reaching the surface in *Mătău* (Veseliu 2007, 190-191). Archaeological data to date from the commune is sporadic (Olteanu *et al.*, 2003, 13-17) (Figure 2). There are two Bronze Age sites (Tei culture), at *Scoarţei-Buzduga Hill* and *Cornişor.* At the confluence of the Grecea Mare and the Grecea Mică springs, in the region of Grecea Mare, there are a tumuli cemetery and a Daco-Roman settlement of the 2nd-3rd centuries AD, and the settlements of the 17th-19th century at Dealul Cornis, Piscul Viei and *Biserica.*

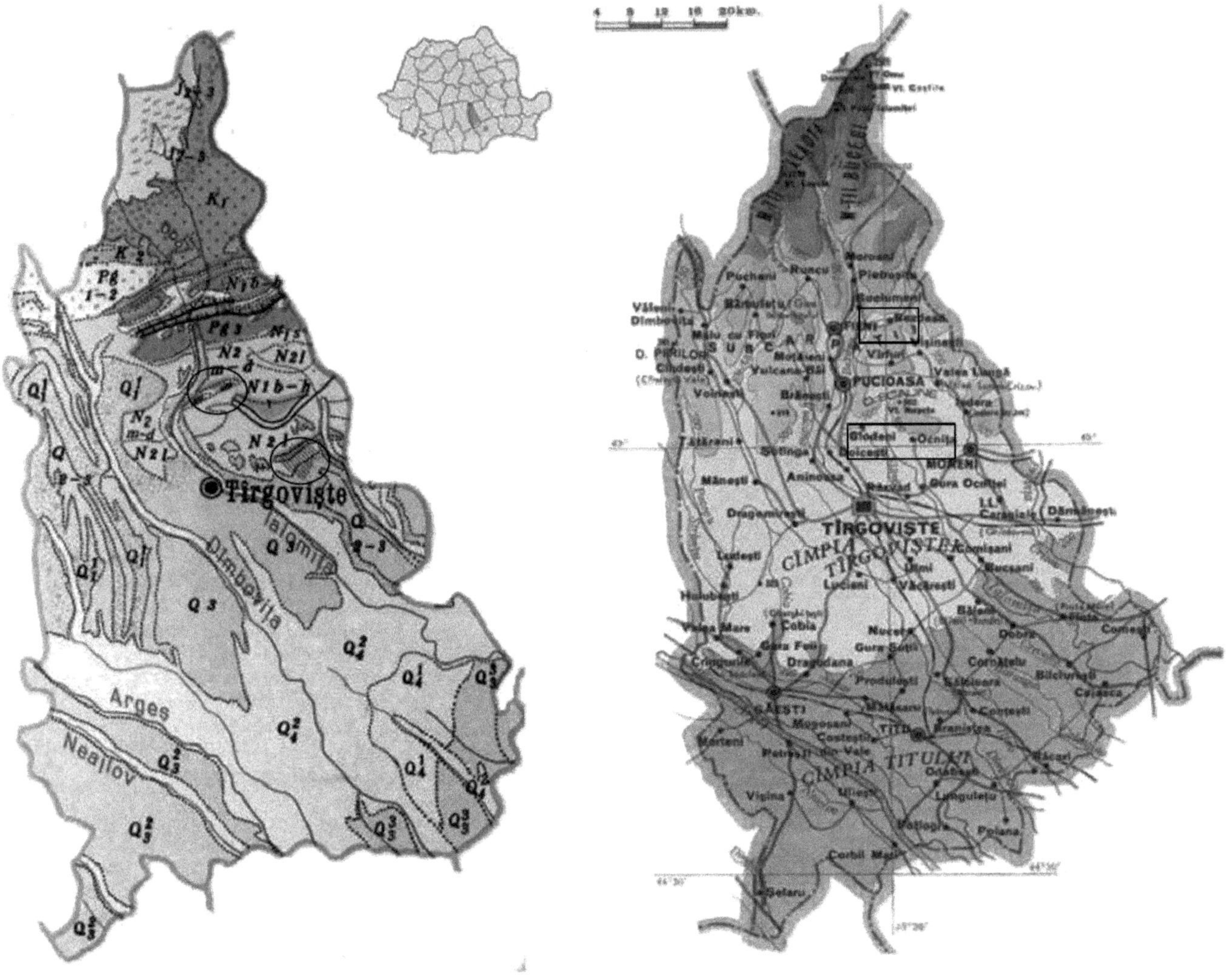

Figure 1. Geological map of the district of Dâmboviţa with the sources of salt (after Velcea 1982, 422) and a geographical map of the county with the areas prospected.

The only archaeological research to have taken place in this area was at *Hoagă*, situated on both banks of the Slănic/Ocniţa creek and 300 m south-east of Lacul Sărat, where there was a small Dacian settlement during the 2nd-1st centuries BC (Diaconescu and Drob 1980). Regulation of the creek in the 1980s led to the destruction of the archaeological site (information from M. Diaconescu; Veseliu 2007, 18).

Although currently available archaeological evidence relating to the exploitation of salt and/or brine in the vicinity of this settlement is sparse and relatively insignificant, the question can be approached by other means.

A rock salt mining tradition is reflected in the place-names of the area (Oproiu and Samoilă 2009, 255; Stoicescu and Oproiu 1983, 168-170; Veseliu 2006). They indicate the existence of sources of salt or brine, *Malul cu Sare, Slănicul, Islazu Slănic, Apa Sărată, Slănicul Sărat, Sărătura, Valea Sării* ("Bank with Salt", "Salted Water", "Salted Slanic", "Salt Valley" and the like). Place-names such as *Plaiul Salinei Vechi, Valea Ocnei, Fundul Ocnei,* most likely indicate areas where rock salt was exploited, and *Drumul Sării* (Salt Road) relates to the transportation of salt.

Mining galleries/salt mines are listed in places such as *Mătău, Ruşeţu-Valea lui Neag* and in the Cuţa Elena garden, where the subsidence of the land is blamed on old mining galleries (Veseliu 2007, 16, 53).

Historical information on salt mining in the commune of Ocniţa

From a historical point of view the most important surface source of rock salt in the county comes from Ocniţa. In the Middle Ages it was princely property, mentioned in documents of the 16th and 17th centuries. The name of the village comes from the salt mine exploited in the Middle Ages, which came to be known as *Ocna Mică de la Târgovişte* ("The Small Salt Mine of Târgovişte"). This name was given to the salt mine in the 16th century to differentiate it from *Ocna Mare* ("Large Salt Mine") at Râmnic (*Ilieş 1956*).

Surviving medieval documents record the exploitation of the mine and the privileges given to Mânăstirea Dealu ("Monastery on the Hill") as part of the princely revenue from Ocniţa, starting in the reign of Radu cel Mare (1495-1508), the founder of the monastery.

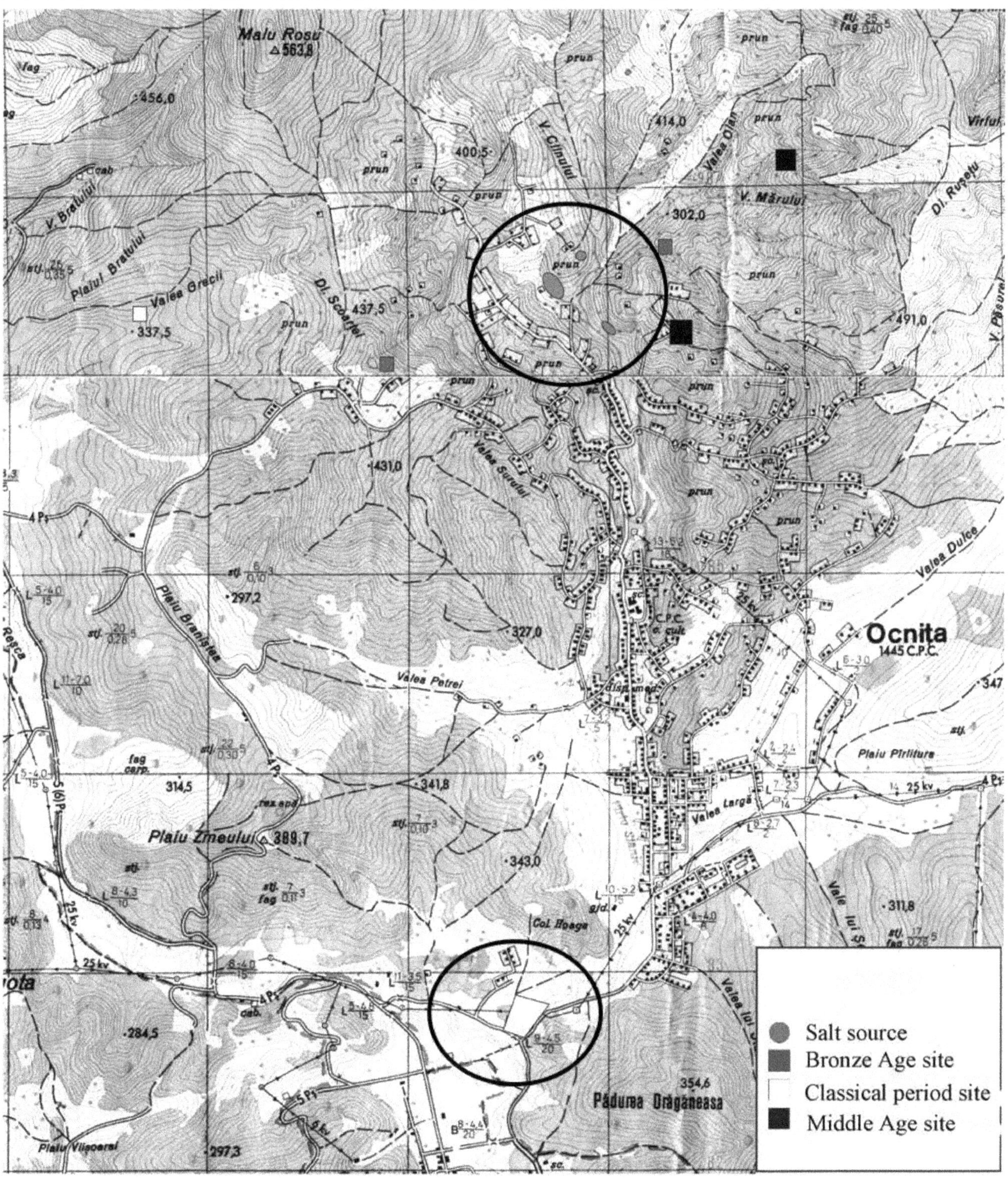

Figure 2. Topographical map (1: 25,000) with archeological sites and surveyed areas in the commune of Ocniţa-Lacul Sărat (below) and *Malu' cu Sare* (above).

Ten years later, this privilege was shared with Argeş Monastery. This situation generated conflict between the two monasteries, conflict which lasted for more than a century. In this context, the age of the systematic exploitation of the salt resources from the area[1] has been mentioned, but also challenged (*Documenta Romaniae Historica*. B. *Ţara Românească* II, no. 148, 290, no. 215, 411, no. 221, 422; Toma *et al.*, 2008, 45-51).

Beyond this duality, 16th century testimony records that the income obtained by the monarchy from the exploitation of salt mines was almost 40,000 scudi a year (*Călători străini despre Tărîle Române* III, 14*)*. But in 1583, the date of the testimony in question, mines at Telega and Ghitioara, Prahova county and *Ocna Mare*, at Valcea, were also exploited (Ilieş 1956, 158-159).

[1] Oral tradition records the beginning of salt mining in the reign of Mircea cel Mare and recall the fact that "here were the large salt mines of the country" (Lahovari and Tocilescu 1901, 536).

It is difficult to ascertain how much money came specifically from the exploitation of the mines in the area under study, but the disagreements between the monasteries suggests that the sums were significant. The name of *Ocna Mică de la Târgovişte* ("the Small Salt Mine from Târgovişte") itself, found in documents, could be meaningful in this regard.

In 1715, in the reign of Ştefan Cantacuzino, documents record a change in the way the taxes were calculated and the way in which the monasteries obtained their income was different from how it had been at the beginning of the 16th century (Toma *et al.*, 2008, 46-47). This change is interpreted as reflecting impoverishment of the salt dike at Ocniţa mine (Toma *et al.*, 2008, 51).

The mine is generally thought to have been exhausted by the end of the 18th century, despite attempts to reopen it in 1795 and 1919 (Toma *et al.*, 2008, 51). Another theory has it that the mine was abandoned due to landslides caused by earthquakes in 1802, 1803 and 1830 (Veseliu 2007, 17, 191).

Franz Joseph Sultzer, however, mentions in his journal (after a journey through Romanian lands in1781-82) that he saw near Târgovişte a Salina cea Veche ("Old Salt Mine"), where traces of salt mining could still be seen (Broşteanu 1901, 140-144). This information raises the possibility that the mines may have been abandoned before Sultzer's travels.

Also, the stipulations of the Organic Law (the first law system of Wallachia) concerning salt mines and their exploitation (1831) mentions only three working salt mines in Romania: Ocnele Mari, Telega and Slănicul (Broşteanu 1901, 383). This suggests that the Ocniţa salt mine was not officially reopened after the end of the 17th century.

Data obtained during fieldwork mention the fact that salt mining could have been restarted during the First World War. Afterwards, unregulated exploitation was carried on by individuals for themselves until the 1950s (Veseliu 2007, 41, 48, 54, 142-143; *pers. comm.* Ilie Toader, 2011).

We have no direct information about how rock salt was extracted in the Middle Ages, but it is assumed that the technology survived until modern times. For the same reason, it is also assumed that teams of workers extracted salt from wells of approximately 2 m in diameter, which were strengthened with rods and wooden slats (Ilieş 1956; Veseliu 2007, 142, 23-32).

The current exploitation of salt sources

In the commune of Ocniţa, there are two places of interest for the presence of salt sources (Figure 2).

1. The area around Lacul Sărat ("Salted Lake"), situated at the confluence of Văii Reşca with Pârâul Sărat/Ocniţa brook (Figure 3).

Our field survey showed that the area was profoundly modified by human activity, thanks to attempts that began in the late 19th century to install and maintain oil exploitation. This activity continues today. The lake suffers badly from oil pollution.

Figure 3. View to the Lacul Sărat (Photo A. Ilie).

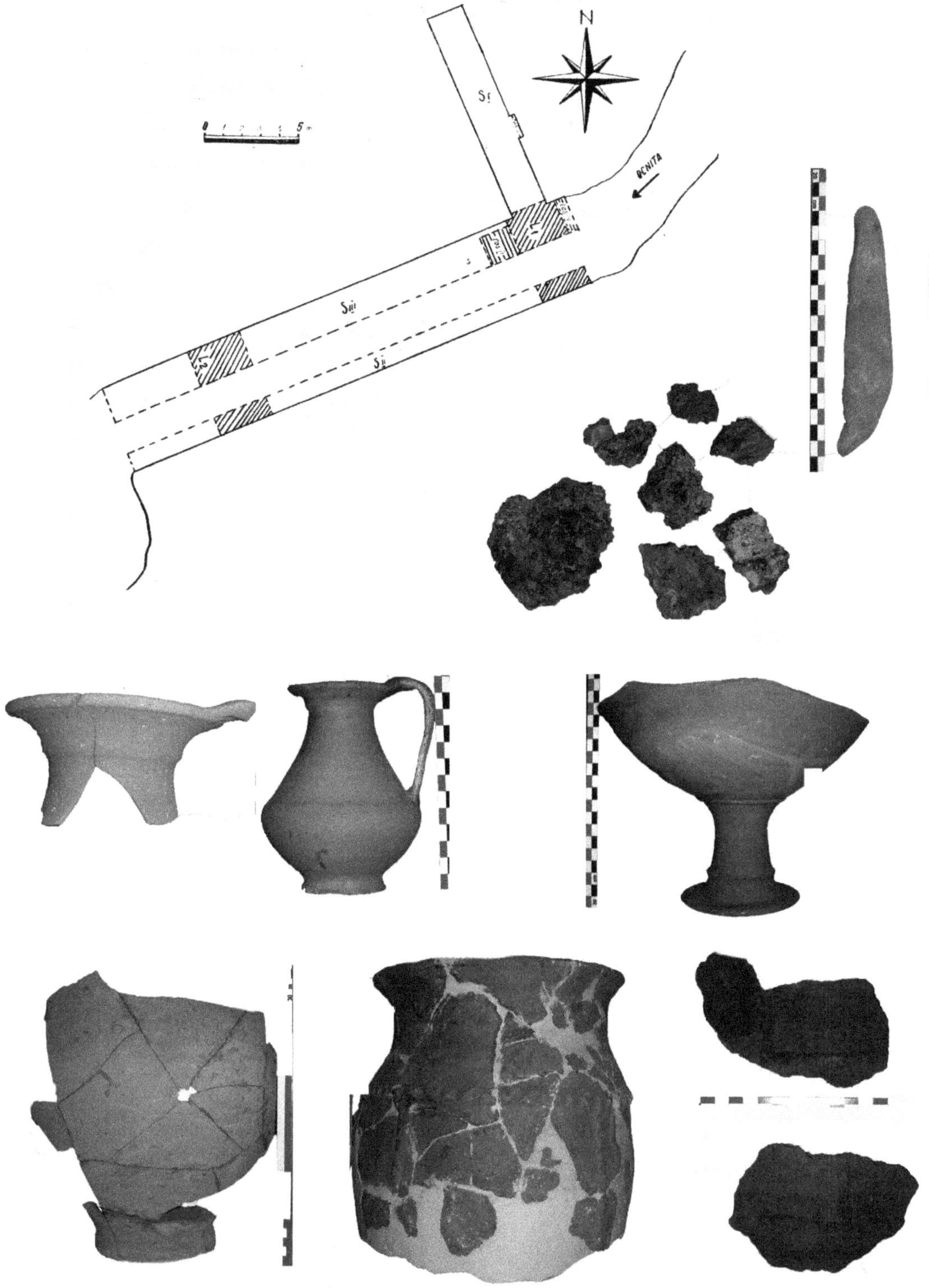

Figure 4. Plan of the excavation of the Dacian site at *Hoagă* (after Diaconescu, Drob, 1980, fig. 3) and associated archaeological material (Photos A. Ilie).

Local residents were unable to provide any conclusive information concerning the origin of the lake, but we believe it to be the result of oil exploitation dating back to the 19th century. Most of those questioned settled in the area after oil exploitation had begun.

Local tradition suggests the possibility that the lakes were formed on the sites of earlier salt exploitation, dating back to the time of the Dacians in view of the presence of an archaeological site nearby (Veseliu 2007, 68, 72). This, however, is unlikely and supported by no evidence, even if the nearby archaeological site was supposed to be associated with local salt resources (Diaconescu and Drob 1980, 204) (Figure 4).

What is known is that at some point in the 1930s local oil workers built a public bath for the treatment of rheumatic conditions in the settlement of *Hoagă*. Elegant huts were built on the banks of the salty "lakes", which were in fact three saline marshes, or the present Lacul Sărat (Veseliu 2007, 16).

2. The second place is situated in the northern part of the commune, where, in the valley of Slănic brook, outcrops of rock salt can be seen and numerous areas are exploited by local residents. The terrain of the entire area is uneven. The valley of the brook is shallow, the slopes are steep, and the small plateaux are used as meadows or orchards.

This exploitation is visible at ground level because of the salt which sometimes expands over 100 to 200 square metres. The area is known by locals as *Malu' cu Sare* ("Bank with Salt") or *Beznă* (Figure 5). The name of the upper course of Slănic/Ocniţa brook is Benţa (Ujvári 1972, 463, fig. 136). The toponym *Bezna* (pitch dark) may be seen as a local interpretation of the brook's name or as a memory of salt mining in the galleries.

Nowadays, salt is mostly exploited by the villagers by extracting salt water through small holes (Figure 6). They say that spring water comes from these small holes.

In the summer the salt water is not intensively exploited. At the beginning of autumn, however, a large number of holes are made to collect the brine in order to prepare pickles.

Our local informants told us that brine is used to make pickles, to pickle cabbage, and to preserve cheese or pork. The villagers mentioned the old custom of re-crystallization of salt water using sun or fire.

The same informants also mentioned the use without restriction of brine or rock salt by anyone who needed it. Shepherds used to arrive "from all over the county" in spring in order to obtain supplies of salt water necessary for the preservation of cheese and to make *pastramă* (mutton preserved in salt). In the autumn the exploitation of the saline spring would increase, mainly for the benefit of the local inhabitants.

According to some villagers, the local community would also exploit this resource by extracting blocks of rock salt. In the area surveyed in the summer of 2011, however, no traces of such exploitation could be seen. A local informant mentioned that somewhere at the foot of the bank, at *Floarea lui Bărbosu* (higher up on the valley of the brook), there could be a place where a there was exploitation of rock salt on a larger scale. The data were inconsistent, which suggests that exploitation is no longer practiced today.

During another campaign of field research in the autumn of 2011, small surface exploitation was observed and the villagers mentioned that just before the cold season people who still bred animals would put salt rock in shelters for animals. This method of salt mining consists of digging holes and extracting small a quantity of salt rocks (Figure 5, 7).

The commune of Glodeni

The village of Lăculeţe[2] is located in the river bed of Lăculeţe brook, which is bounded by high hills between 300 and 500 m in height.

There are four salt springs within the village, but only two are of interest from an ethnographic point of view, namely those located on the bed of Lăculeţe brook (Spring No. 1 and 2) (Figure 8).

The other two springs located in the Plopului Valley have a lower sodium chloride concentration; on the other hand, these springs contain iodine and bromine (A. Istrate, 2008), making them unsuitable for food preservation. Locals know about them, but do not use them on account of their inaccessibility.

The salt sources and the saline springs in the Glodeni area are first mentioned at the beginning of the 20th century (Lahovari and Tocilescu 1901, 574).

Toponyms also refer to salt sources in the area: *Saramură* ("brine"), a street in the village of Lăculeţe; *Slănicul Banului* and the *Sărata* ("salty") brook that flow from the village of Schela (Spiridon and Ion, 1998, 37, 40; Oproiu and Samoilă 2009, 188-189).

The only archaeological evidence is a reference to a cist tomb, but with no precise dating or location. The medieval village known as Solomon's Pasture should be added to the archaeological evidence (Olteanu 2002, 137-138). We should add that a Neolithic axe may have come from the village (Oproiu and Samoilă 2009, 189).

From an archaeological point of view most interest lay in an area near the exit of the village of Livezile, in Glodeni commune, towards the forest of Aninoasa, after the discovery of a mound which could well be man-made. It was surmounted by what was probably a triptych but which has now been replaced by a massive wooden cross. This mound is approximately 8m high and 30-40m in diameter. It is now disturbed on its eastern and northern sides by cart tracks. From this area some atypical ceramic fragments have been collected, attributed to the Bronze Age, but without further information. Although scattered, they probably belong to a nearby Bronze Age settlement (Figure 9a).

The entire plateau has been highly disturbed, probably through oil exploitation. At the end of the 19th century and the beginning of the 20th, numerous oil extraction companies are known to have been active in Glodeni (Toma *et al.*, 2008, 81-88). The depressions are mostly oval or circular pits with a diameter of 6-8m, lower by 2-3m than the surrounding area (Figure 9b).

[2] The explanations given by the villagers after being asked where in the area sources of salt could be found were because even the name of the village refers to small lakes/lake and even mentioned a toponym *Lacul Sărat* ("Salted Lake"), but in fact it is the one situated on the territory of Ocnita commune.

Figure 5. The northern area of the commune of Ocniţa, where salt deposits are visible: *Bezna* and *Malu' cu Sare* (Photo A. Ilie, D.N. Popovici).

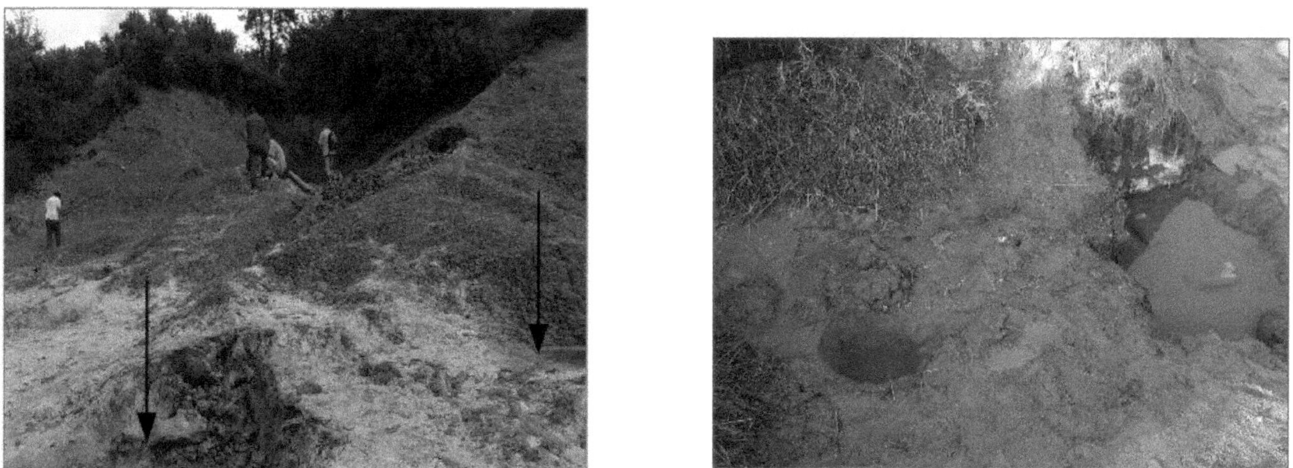

Figure 6. Present day exploitation of salt water along the river Benţa/Pârâul Sărat/Ocniţa (Photos A. Ilie).

Figure 7. Present day exploitation of salt along the river Benţa/Pârâul Sărat/Ocniţa (Photo A. Ilie).

We can thus start from the hypothesis that these earthworks could mark the traces of such exploitation. The few modern ceramic fragments found (dating from the 19th and 20th centuries) are probably the result of contemporary settlement. On the left bank of the valley of Reşca, near a fresh water pumping station, an area of approximately 500 square metres has been identified in which rare prehistoric pottery and stone, as well as modern artefacts have been found (Figure 9c).

Installations and salt use

1. On the right sight of the road which links the village of Priboiu with Lăculeţe, at the entrance of the village of Lăculeţe, in the place named *La pod or La fântâna sărată* or *Lacul sărat,* (Spring no. 1) there is now a saline spring (Figure 10).

The spring is contained by means of a concrete pipe and covered with a metal grid. No other area of exploitation (contemporary or earlier) has been observed, which suggests that the topography of the area has not been affected by major changes, such as landslides, during the last century.

The residents did not know if there were better times in the year for collecting salt water, which suggests that the flow and concentration are constant. Geomorphological studies have shown that the flow is 0.0015 l/s, but in terms of salinity, it has a lower concentration than spring No. 2 (A. Istrate, 2008).

2. The second place, *La Ţarc* or *Podul Poacov* (Spring No. 2), is situated in the village, at the confluence of two brooks near the road which crosses the village. A small path leads to the spring and people have laid bricks around the source to facilitate access.

The exploitation is done here via spring catchments through a series of circular holes, no deeper than about 0.70m, now protected with plastic barrels that allow the retention of water by filtration and decanting owing to the low flow of the source, measured at 0.001 l/day (Figure 11). The good mineralisation of the source is explained by the fact that the Moetian level is in direct contact with the surface.

The area is not clean, which indicates a certain lack of interest on the part of the local inhabitants. Indeed some locals do not use the springs, although they have transport facilities. The villagers remembered that until the Second World War they only used salt from these sources. Nowadays, they buy it from elsewhere.

Informants in the area (people grazing animals, mainly cattle) declare that they use brine (from both sources) for pickles, cheese, and for both animal and human food. Sometimes it was used to make traditional soap (with fat, ash and salt).

They mentioned the extraction of blocks of rock salt without being able to provide any clear or reliable information; this is probably the result of the reduction of animal husbandry over the last twenty years which has also reduced the need for salt.

Access to the sources of salt water is free, with no restrictions for local inhabitants or those from other nearby localities. They noted that brine is used both by locals and by some who come from more distant places, even coming with cars to transport larger quantities of brine. Their knowledge probably does not go back more than a century.

The commune of Bezdead

The village sits along the valley of the Bizdidel in the Slănic-Costesti-Bezdead syncline depression in the Ialomita Subcarpathians. The hills surrounding the city are 700-1000m in height.

Two sources of salt water were mapped in Bezdead, but only the spring of Valea Sărată ("Salted Valley") is known to have been exploited from the late 19th century. Chemical analyses made in the 1970s found a salinity concentration of 263g/l (Stoicescu and Oproiu 1983, 62).

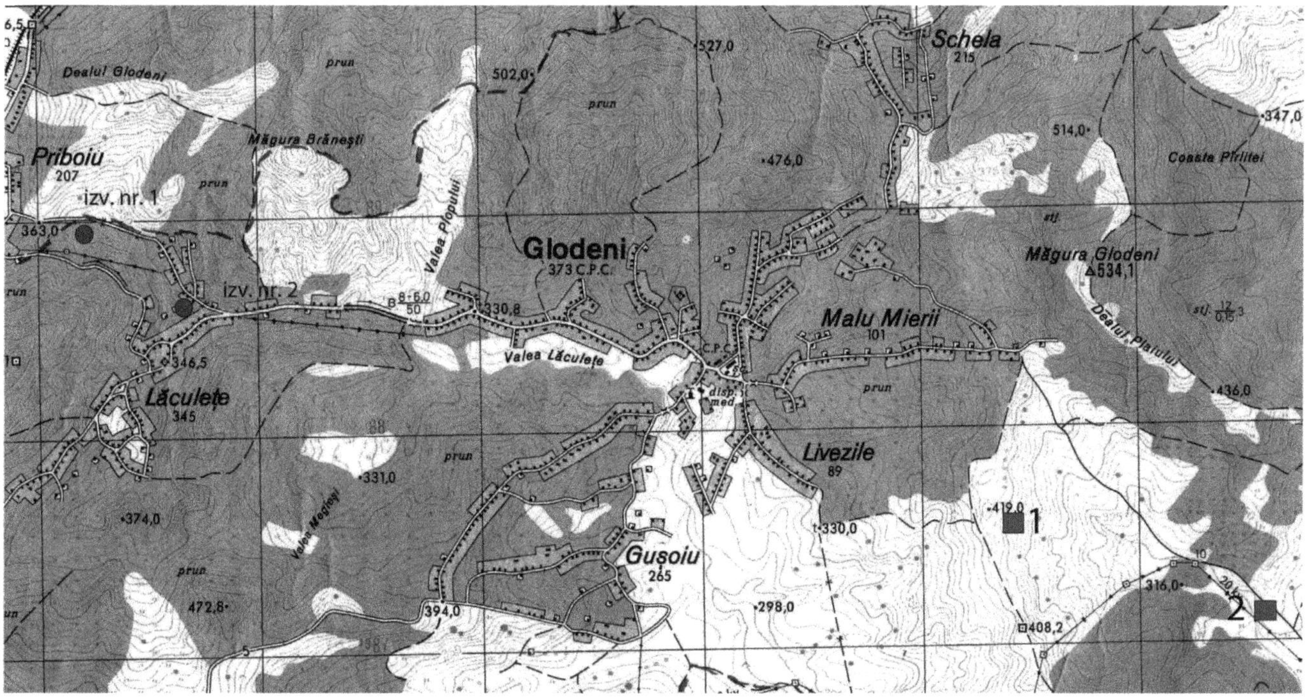

Figure 8. Topographical map (1:25,000) of the commune of Glodeni showing two saline springs (izv. nr. 1 şi nr. 2) and two archaeological sites of the Bronze Age (Nos 1 and 2).

Figure 9. a. Hill in Livezile village and fragments of Bronze Age pottery; b. Landscape towards the Aninoasa forest with traces of 19th century oil exploitation; c. Fragments of Bronze Age pottery, and modern pottery from Valea Reşca (Photos: D.N. Popovici, A. Ilie).

Figure 10. Glodeni-Lăculeţe. Saline spring No. 1: landscape and installation (Photos A. Ilie).

One of the monographs written about the village indicates the existence of salty soils along the Sărata valley (Davidoiu 2001, 20), which is now the common pasture of the village. Brine is thought to have played a minor role in the Middle Ages, being used only for preserving vegetables (Brezeanu 2008, 76).

A cist tomb attributed to the Monteoru culture was once found in Bezdead, but its precise location is unknown (Olteanu 2002, 51-52). Enquiries among local residents provided archaeological information about the original medieval village, but it position remains unknown (Figure 12).

Installation and salt use

1. In the valley Sărată or Sărăţelului, as it is called in a monograph written in 1940s (Arsenoiu 1943), in the place known as *Sarea Mare* ("Great Salt") is one of the two salt springs of the village of Bezdead (Figure 13).

It lies outside the present village, approximately 1km east of the northern end, and has a moderate flow rate.
In another monograph on the village we find the statement: "Also on the valley of Sărată, appear springs, poorly mineralized, many of them have been covered over by landslides, with flow rates of 0.05 l/hour" and "locals use the curative properties of the water of these springs in a rough and ready manner" (Dumitrescu 1980, 28).

There is no easy access to the spring, and the source can only be found by those who know where it is. The spring is equipped with a barrel in which salted water is collected.

The area is subject to landslides. The present point of exploitation displays the remains of equipment such a steps and a wooden supporting structure which has been affected by a landslide and can only be used nowadays to a limited extent (Figure 13).

Upstream from this point, traces of two other exploitation points are still visible. They are older, and have been more affected by those processes which probably forced the villagers to abandon them. They are in the form of oval or circular grooves, partly filled with mud, located on the path of a small spring, and which have been altered by landslides. At these points were gathered a few small ceramic fragments which, judging by the fabric, could be attributed to the Bronze Age.

2. The second point is located on Valea Morii ("Valley of the Mills") brook at *Izvorul Sărat* ("Salt Spring"): Spring No. 2 (Figure 14). It is built within the village, towards its southern end. There is a road that follows the course of the stream and comes close to the spring, but during summer the road is flooded and is unusable, and so access to the salt spring is made through people's orchards.

3. This spring is arranged in the same way as the other one from Bezdead.

It has a weaker flow and lower salinity compared to the other spring, observed in the field and confirmed by villagers, who say that only the elderly or feeble go there.

The springs of Bezdead are common property and people use it freely whenever they need to; the residents of the surrounding villages can also use it freely.

Other monographs (Arsenoiu, 1943, Rădulescu 1938) provide some useful data concerning the use of saline springs:
The springs never fail, although the water appears to decrease during the cold season, as we observed during our field survey of February 2012.

Re-crystallization was used by some villagers in order to obtain "pure, common, fine, white salt, about 1/3 of a litre", which was necessary for the inhabitants, as "they did not need to spend great sums of money on luxury salt".

Salt water was used in a domestic context by the villagers and the inhabitants of nearby villages, by bakers, and for the preparation of various kinds of cheese, pickles, etc. (Figure 15).

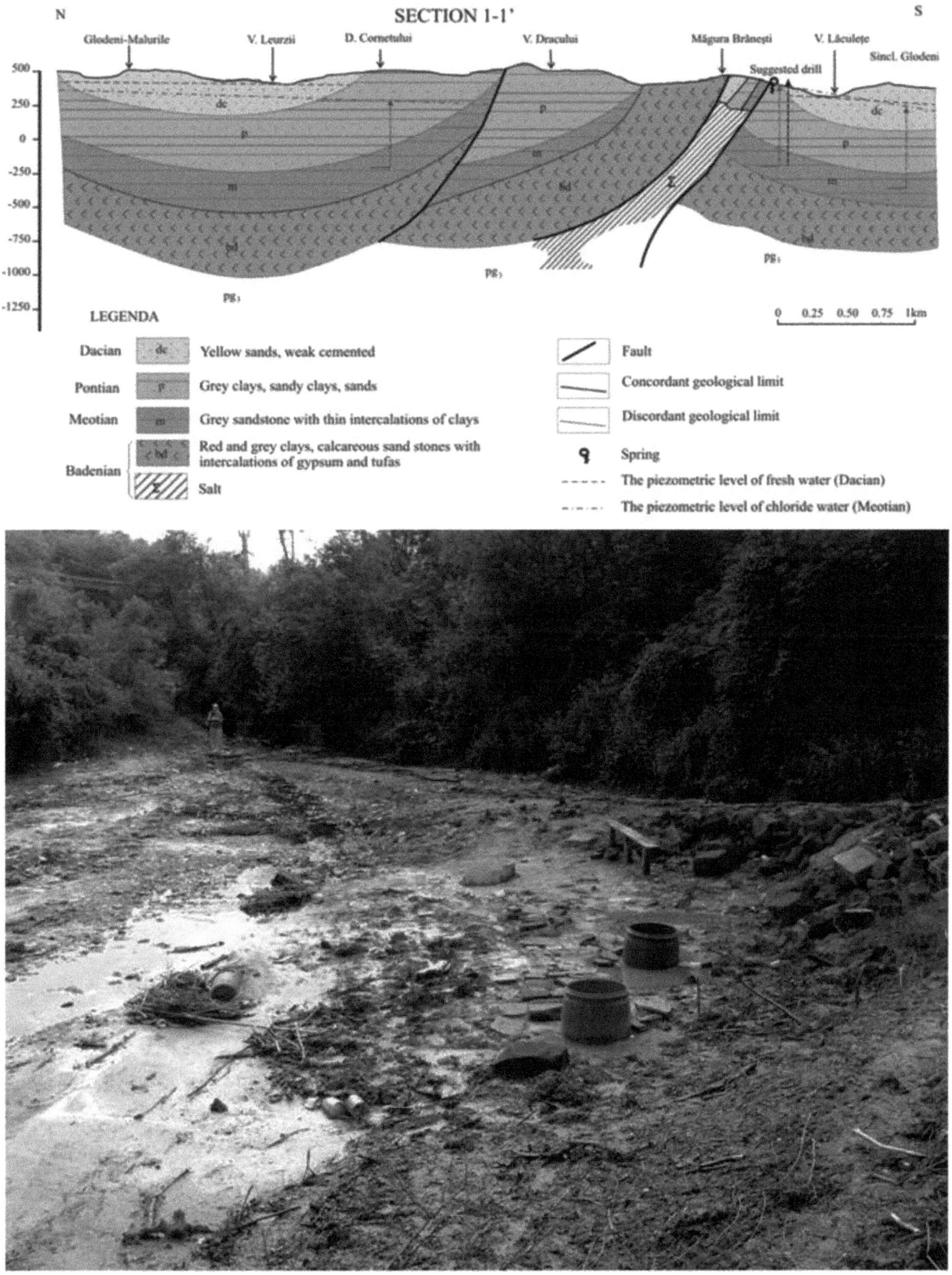

Figure 11. Geological profile of saline spring No. 2 (from A. Istrate, 2008) and installations of spring No. 2 (Photo D.N. Popovici).

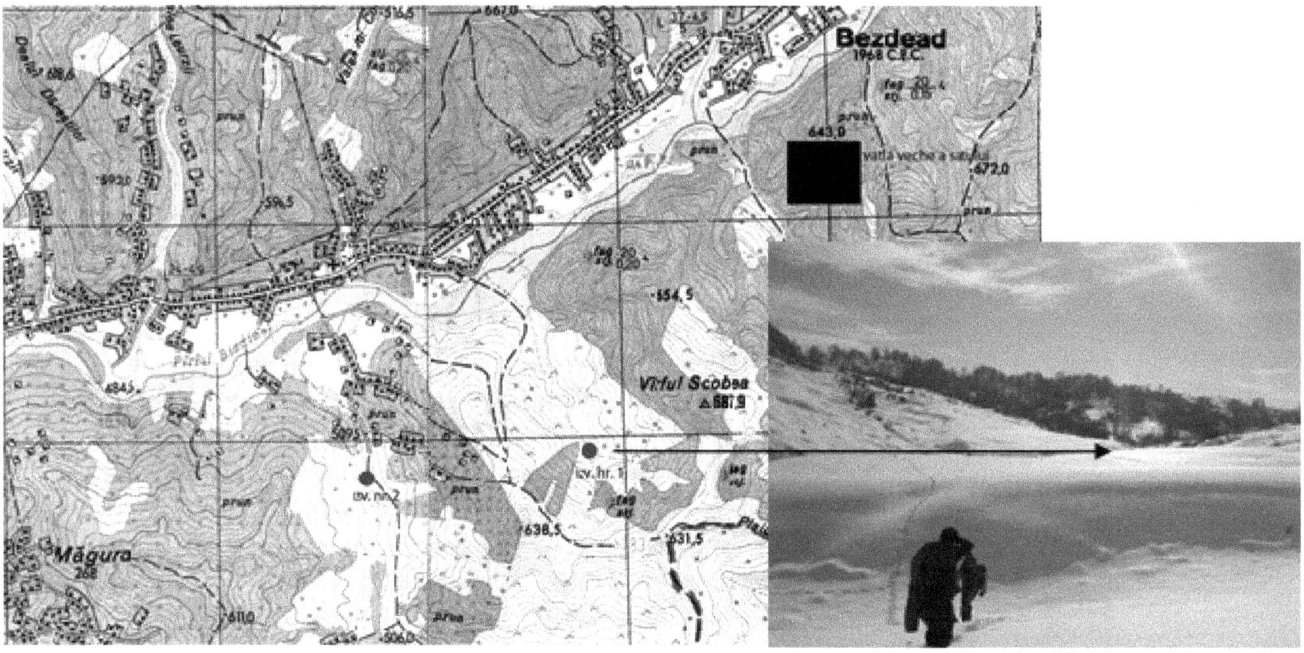

Figure 12. Topographical map (1:25,000) of the commune of Bezdead with areas of ethnographic and archaeological interest, and a photograph of the spring of salt water *Sarea Mare*.

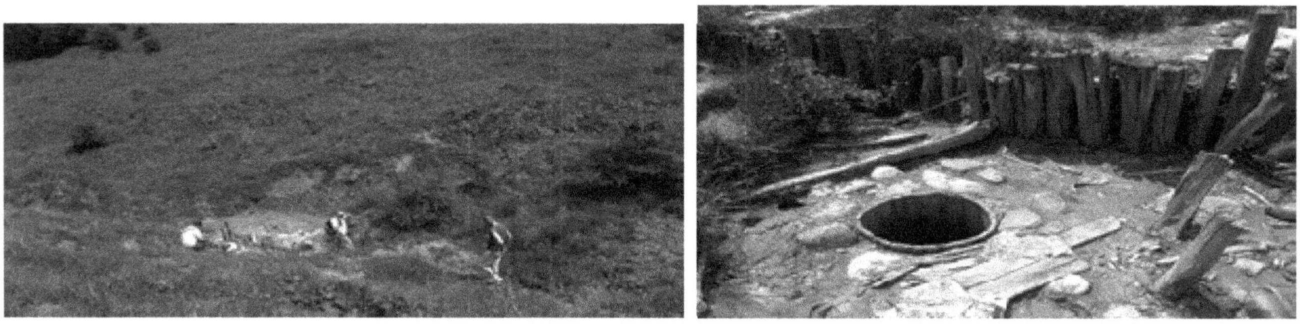

Figure 13. Saline spring *Sarea Mare* (Photos: A. Ilie, D. N. Popovici).

Figure 14. Bezdead. *Izvorul sărat*/Spring No. 2 – landscape and installation.

The sources are also used by the inhabitants of other villages, from where "women come everyday from Varfuri, Magura, Buciumeni with bottles on their back" while in another monograph further villages are mentioned: Urseiu, Valea Leurzii, Fieni, Țața, Moroieni, Teslea and Ocina in Prahova county.

Ethnographic research conducted in the village showed in addition that brine was used in the region for:

- the preservation of different kinds of vegetables as pickles;
- preserving cheese;
- the preservation of pork and lard, these being placed in the pickling brine in a wooden trough for relatively

short periods of time. In this case, salting was a preparatory step to smoking;
- the pickling brine was and is still used by many families for the daily seasoning/salting of meals;
- the pickling brine is used in the winter food of animals;
- it is also used with therapeutic aims, especially in anti-rheumatic baths, in this case the brine is heated and sometimes diluted, the natural concentration being considered too high.

Preliminary conclusions

At this stage of our research some observations can be made. We must recognise that the subject has not been studied in this area before, which made it necessary to evaluate from the very beginning the existing data and their significance. The study of different kinds of publication from various sources allowed for some preliminary statements.

The geomorphology studies, where they exist, are mainly prompted by economic factors, which explain how they were conducted and also their relevance to our subject.

The most consistent historical data, though incomplete, cover the Middle Ages, and mainly give economic data on the exploitation of salt and (principally) halite.
The archaeological data are also defective and were not correlated with our subject.

Figure 15. Bezdead. Every day use of salt water by the inhabitants of the village: transportation and storage.

The ethnographical data are sporadic and have not been synthesized. Thus, extremely important aspects of the exploitation and of the domestic or common consumption of salt, the modes of transportation, its therapeutic and symbolic reputation, or its consumption and/or its use in pickling brine have not yet been properly analysed.

Another matter was raised during the initial stage of this regional research: when the industrial exploitation of petrol increased, there were major changes in the composition and the social structure of the population of the area. Many workers came from other regions (especially in the second half of the 19th century), attracted by the possibility of better pay and longer periods of employment. From our point of view, this situation brought about the loss of part of the collective memory. In these conditions it was considerably more difficult to obtain ethnographic or archaeological information. The recording of what remains of this memory now has the character of an emergency rescue.

On the other hand, the economic evolution of the area we studied over the last twenty years or so, has had some far-reaching effects. Animal husbandry has dramatically decreased, with major implications for the consumption of halite, as well as of pickling brine. In addition, the availability of cheap salt in the shops, as well as the pollution of local sources of salt, has contributed to the reduction in the traditional consumption of salt and of pickling brine.

There is little archaeological evidence of salt exploitation, and installations have been found in the vicinity of the sources, at least nothing comparable to those which are well known to scholars elsewhere in Romania.

In general, none of the inhabitants we interviewed were able to provide significant data of an archaeological nature. The oral tradition in the area seems to preserve only general information concerning the last century. The local elite gave us some information partially derived from documentary sources, but without a sufficiently strong basis to enable us to draw any firm general conclusions.

The ethnographic information we obtained enabled us to establish the existence of small-scale local exploitation of halite and also of pickling brine. This exploitation is seasonal. These elements were used both for immediate consumption and for making and preserving certain products (such as vegetables, cheese or meat). The data obtained from our informants did not accentuate any specifically local features. Salt water is thus used in all its aspects for the preservation commodities such as meat, lard, bones, cheese, or vegetables, and more rarely nowadays for therapeutic purposes. The only difference seems to be in Bezdead, where salt water is frequently employed in the daily activities of the inhabitants, possibly due to its remote location isolated from urban centres.

The correlation of the existing data leads to the preliminary conclusion that there may have been sources of salt supplies for local communities limited to the lowland, at least in the last century.

The documentary evidence for the existence of salt mines in the Middle Ages allows for the hypothesis that there were also such establishments in preceding periods. The opacity of the archeological data, at least at present, shows we must continue to conduct fuller and more systematic research over wider areas.

References

Alexianu, M., Dumitroaia, G. and Monah, D. 1992. Exploatarea surselor de apă sărată din Moldova: o abordare etnoarheologică, *Thraco-Dacica* XIII/1-2, 159-167.

Alexianu, M., Weller, O. and Brigand, R. 2007. *Izvoarele de apă sărată din Moldova subcarpatică. Cercetări etnoarheologice*. Iaşi, Demiurg.

Arsenoiu, I. C. 1943. *Bezdead.* Monographs collection CNMCD Târgovişte, inventory no. 6400/VII.

Brezeanu, S. 2008. *Moşnenii din Bezdead-Dâmboviţa (1620-1820)*. Bucureşti, Meronia.

Broşteanu, C. 1901. *Salinele nostre. Studiu istoric, juridic şi economic asupra exploatărei salinelor şi monopolului sărei la romani şi români (753 a. Crist. – 1900 d. Crist.) după documente literare, epigrafice, legislative şi economice*. Bucureşti.

Cavruc, V. and Chiricescu, A. (eds.) 2006. *Sarea, Timpul şi Omul*. Sfântu Gheorghe, Angustia.

Davidoiu, V. 2001. *Vatră şi oameni în freamătul vremii. Bezdead – CDXXVIII*. Târgovişte, Bibliotheca.

Diaconescu, M. and Drob, V. 1980. Săpăturile de la Ocniţa-Dâmboviţa, *Materiale şi Cercetări Arheologice* XIV, 200-209.

Dumitrescu, D.I. 1980. *Comuna Bezdead. Studiu monografic complex*. www.noidacii.ro/...17/Monografia%20prof[1].Dumitrescu.pdf

Holban, M., Alexandrescu-Dersca Bulgaru, M.M., Cernavodeanu, P., Totoiu, I. 1971. *Călători străini despre Tările Române III*. Bucureşti, Editura Ştiinţifică.

Ilieş, A. 1956. Ştiri în legătură cu exploatarea sării în tara Românească până în veacul al XVIII-lea. *Studii şi Materiale de Istorie Medie* I, 155-197.

Istrate, A. 2008. The chloride mineral waters from the Glodeni locality, Dâmboviţa, *The Annals of Valahia University of Târgovişte. Geoghaphical Series* 8, 109-118.

Lahovari, G.I. and Tocilescu, G.G. 1901. *Marele Dicţionar Geografic*. Bucureşti.

Monah, D. 1991. L'exploitation du sel dans les Carpates orientales et ses rapports avec la culture de Cucuteni-Tripolye, in V. Chirica and D. Monah, (eds.), *Le Paléolithique et le Néolithique de la Roumanie en contexte européen*, 387-400. Bibliotheca Archaeologica Iassiensis IV, Iaşi.

Monah, D. 2002. L'exploitation du sel dans les Carpates orientales. In O. Weller (ed.), *Archéologie du sel. Techniques et sociétés dans la Pré- et Protohistoire européenne. Actes du colloque 12.2 du XIVe Congrès de l'UISPP, 2001, Liège et de la Table Ronde du Comité des Salines de France, 1998, Paris. Internationale Archäologie, ASTK* 3, 135-146. Rahden/Westfalie, Verlag Marie Leidorf GmbH.

Monah, D., Dumitroaia, G., Weller, O., Chapman, J. (eds.) 2007. *L'exploitation du sel à travers le temps*. Bibliotheca Memoriae Antiquitatis XVIII. Piatra Neamţ, Constantin Matasă.

Monah, D., Dumitroaia, G. and Garvăn, D. (eds.) 2008. *Sarea de la prezent la trecut*. Bibliotheca Memoriae Antiquitatis XX, Piatra Neamţ, Constantin Matasă.

Olteanu, G. 2002. *Repertoriul arheologic al judeţului Dâmboviţa (I)*. Târgovişte, Cetatea de Scaun.

Olteanu, G., Cârstina, O. and Căprăroiu, D. 2003. *Repertoriul arheologic al judeţului Dâmboviţa (II)*. Târgovişte, Cetatea de Scaun.

Oproiu, M. and Samoilă, E. 2009. *Înfruntând veacurile. Aşezări şi monumente dâmboviţene*. Târgovişte, Transversal.

Pascu, Ş. (ed.).1966. *Documenta Romaniae Historica. II. B. Ţara Românească*. Bucureşti, Editura Academiei Române.

Rădulescu, C.E. 1938. *Bezdead.* Monographs collection CNMCD Târgovişte, inventory no. 6617/VII.

Spiridon, S.N. and Ion, I.Ş. 1998. *Glodeni, file de monografie*. Glodeni.

Stoicescu, N. and Oproiu, M. 1983. *Dicţionar istoric al judeţului Dîmboviţa*. Unpublished, CNMCD Târgovişte Archives.

Toma, G., Voiculescu, C. and Oproiu, M. 2008. *O lume în mişcare. Dâmboviţa. Iniţiative şi capital*. Târgovişte, Transversal.

Velcea, V. 1982. Judeţul Dâmboviţa. Geologia. Resursele subsolului. In G. Posea (ed.) *Enciclopedia geografică a României*, 417-418. Bucureşti, Editura Ştiinţifică şi Enciclopedică.

Veseliu, G.T. 2007. *Monografia comunei Ocniţa. Ocniţa-Scaun Domnesc*. Târgovişte, Bibliotheca.

Ujvári, I. 1972. *Geografia apelor României*. Bucureşti, Editura Ştiinţifică.

A New Technique for Salt Block Preparation at Coza (Tulnici, Vrancea County, Romania)

Robin Brigand
Maison de l'Archéologie et de l'Ethnologie UMR 8215 *Trajectoires*, Nanterre

Olivier Weller
Maison de l'Archéologie et de l'Ethnologie UMR 8215 *Trajectoires*, Nanterre

Marius Alexianu
Universitatea „Alexandru Ioan Cuza" din Iaşi

Abstract
Recent surveys conducted from 2009 to 2011 in the Vrancea County (Romania) have allowed to get more information on the original practices concerning the exploitation of rock salt outcrops. This contribution focuses on the extraction sites situated in the Coza Valley and on the way to prepare salt lick blocks for animal nutrition.

Keywords
Vrancea county, rock salt extraction, pastoralism, animal consumption

Considering this is the era of freezers and of the generalization of nutritional habits, salt has gradually lost its utilitarian and symbolical meaning, characteristic to preindustrial societies. Nonetheless, in certain regions little explored yet, salt has still an undeniable social role which involves a variety of behaviours, from the plain salt collecting to a whole industry whose target is to separate salt from its natural supports. The diversity of salt production systems known throughout the world, its complexity and its degree of technical investment should be related to the socio-economic and cultural stakes of this resource for human societies.

In Romania, the recent surveys conducted from 2009 to 2011 by our Romanian–French collective in the Vrancea County have allowed us to get more information on the original practices concerning the exploitation of rock salt outcrops, their uses, and their related redistribution forms. The present contribution is based on three phases. The first is to present the scientific context of our discoveries. The second is to provide details on the extraction sites situated in the Coza Valley (*Grochile de Sare* and *La Siga*), with a focus on their specificities regarding the preparation of salt lick blocks for animal nutrition. The third is to include this local study within the regional exploitation of salt rock, considering that this resource is rich in this area and the exploitation is still traditional.

Romania has the most abundant and most accessible salt resources on the European level and it represents one of the richest countries in the world. Related to the Carpathian orogeny, these resources are distributed under various forms along the external arch of the mountain range (in Moldavia, Wallachia, and Oltenia) and in the Transylvanian basin. Today it is generally accepted that the first forms of liquid salt exploitation (salt springs) should be related Furthermore, the Moldavian slopes of the Eastern Carpathians preserve the traces of the oldest salt production evidence, dating to the Starčevo-Criş culture, around the middle of the sixth millennium BC (Dumitroaia 1987; Weller and Dumitroaia 2005; Weller *et al.*, this volume). Hence, it was only natural for the first ethnographic research to have been dominated by archaeological issues underpinned by the remarkable unity of the place. Here, the still very much alive traditional practices and archaeological vestiges attesting the ancient and perennial character of salt production sites have been converging for over 7000 years.

More than twenty years after the first Romanian ethnoarchaeological study (Alexianu *et al.* 1992), a close collaboration between a French programme of the Ministry of Foreign Affairs and another Romanian programme financed by the CNCS enabled the application of a systematic approach on the Moldavian salt resources. The purpose was to identify them through a study comprising, on one hand, the understanding of technical procedures, traditional uses, and current knowledge and, on the other, the archaeological study.

Since 2003, when the first French–Romanian programme conducted by one of us (O.W) began, we have improved considerably our knowledge on the socio-economic place of salt in the Moldavian rural societies. This improvement is due to the systematic field prospection of almost 240 points where the resource is available in a liquid or solid form (Weller *et al.* 2007; 2010) and to more than 180 joint ethnographic surveys conducted, with 230 informants. We have already included the main results in several publications, which should be consulted for a better understanding of our ongoing research (Alexianu *et al.* 2007; 2008; 2011).

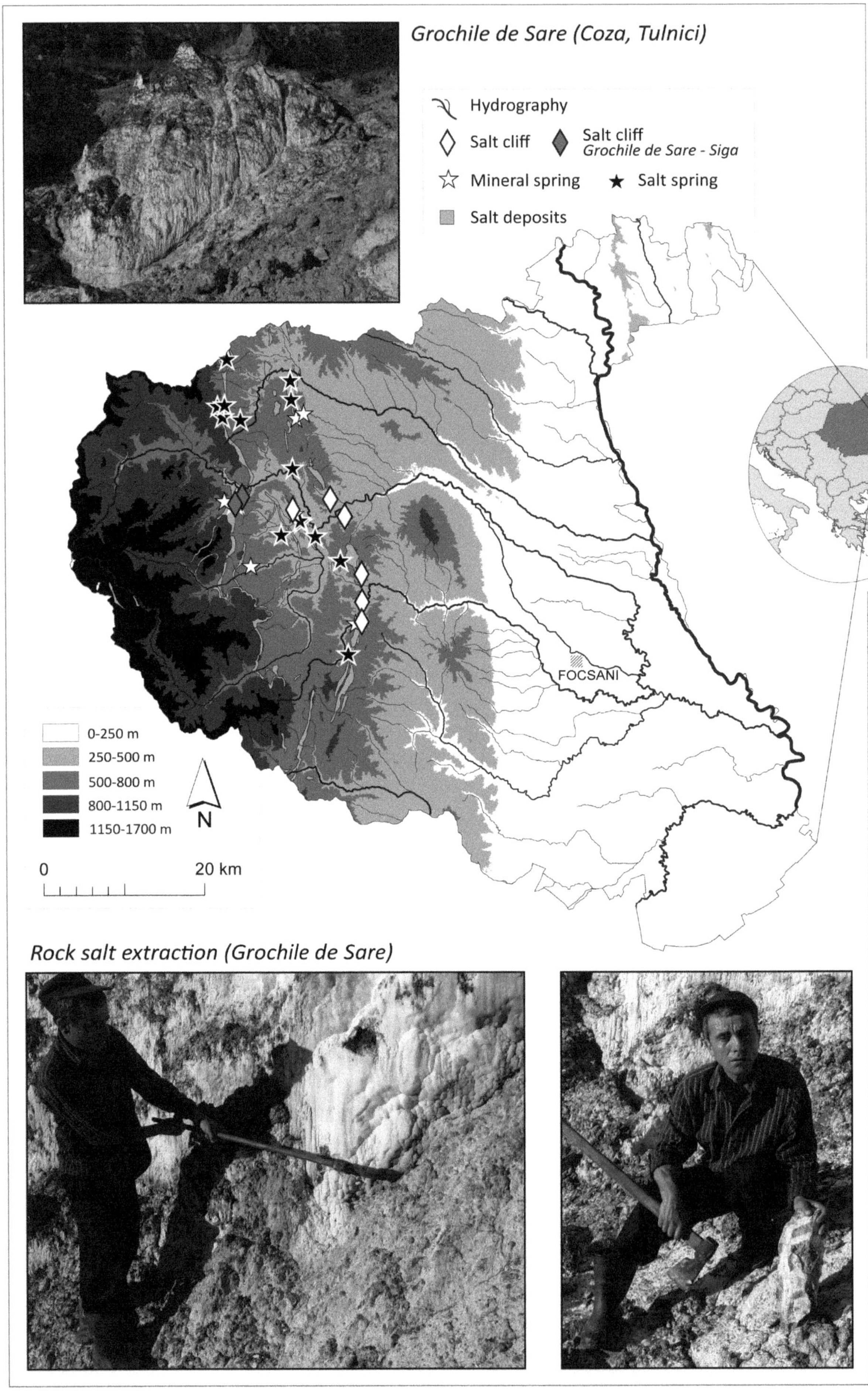

Figure 1. Localisation of the deposits studied (photos R.B.).

Starting with 2009, the systematic investigations have been extended to the southern county of Vrancea (Figure 1). Since then and considering the amount of rock salt outcrops in this region, we have been interested in the forms of exploitation and preparation concerning salt blocks. The most surprising results were those obtained in Coza Valley, upstream from its confluence with the Putna River[1]. They regard two exploitations: the one of *Grochile de Sare* situated on a right tributary of the Coza River, and the one of *La Siga*, on the right bank of Coza.

The rock salt of Coza: extraction, preparation, and use

The region we are interested in belongs to the Curvature sub-Carpathians, characterized by a structural type of landform, more complicated than the one of the Moldavian sub-Carpathians. More precisely, it has important tectonic dynamics which, associated with the process of linear erosion and of the slopes, contribute to the outcropping of the folds with salt core of Lower Miocene (Aquitanian) and Middle Miocene (Helvetian).

The site of *La Siga* is situated at the edge of the Coza village, on the right bank. It is made of a grey salt outcrop, locally covered by landslides, with more than 30 m in height and almost 100 m in width. As for *Grochile de Sare*, it presents two sectors of white salt outcrop, 5–6 m high and 25 m wide, situated in the forest, in a collapse-sensitive, little accessible area, 15–20 m above and to the right of Sării Brook (or Bogdan Brook), and at least 500 m from its confluence with Coza River. This extraction site is directly related to two captured salt springs (*Piciorul Babei Trife* and *La Cabana*), situated 200 and 300 m away, respectively. Villagers have been using these springs to conserve and prepare food for people or animals, as well as for the production of crystallized salt.

Most probably because the two high quality and accessible sodium chloride springs are so close, the two rock salt outcrops are almost exclusively destined to animal consumption, especially to bovine and ovine. The extraction is done with axes or spades on vertical cliff faces, by removing blocks with various sizes (from 2–3 kg to several dozens, depending on the thickness of the salt level). The extraction is all the easier to make, as the two deposits are not very compact, considering the insertion of silty clay lenses. Where the salt is pure (as for instance in the valley of the Alghianu brook, near the village of Poiana, com. of Vrâncioaia), the use of chisels and iron coins is attested. At *Grochile de Sare*, there is also a mention of the extraction method through chambers with 2–2.50 m in depth and 1–1.50 m in width, without stay baulks. This situation is different from the mining exploitation at Valea Sării, where a horizontal extraction with 5 to 10 m in depth (sometimes down to 50 m) involves baulks made of pinewood. The *grochile de sare* (eng. salt holes) attested at Coza (regionalism of the word *gropile*) have formed because of successive "withdrawals": more or less deep holes emerged and their origin had nothing to do with any planned organization of the rock face. It is also interesting mentioning that Coza Valley has numerous microtoponyms deriving from *Groapa*, which could be an indication of these ancient forms of extraction.

Depending on the needs of the users and on the importance of the herd, the shepherds refill their stocks once or twice a year with 200–300 kg, even up to 500 kg per cart. However, the originality of these two deposits compared to other outcrops in the county is the preparation of the blocks before bringing them to the animals. Indeed, the blocks extracted present an alternation of salt and silty clay deposits, which makes them unfit to consumption (Figure 2). The issue is that they have to be removed, and people found an original methodology to this purpose. More specifically, after a rapid shaping on the extraction site and the transportation to the village or to the sheepfold, they are placed on a beechwood or alder fire and then covered with wood charcoal or with white-hot coniferous leafy branches. This way, the clay dries up and falls off the salt gangue, which turned slightly burnished/purple–blue. This task is performed until the salt block is suitable for animals as dietary supplement. If, after a certain period of use, other silty clay lenses become visible in the block, the process is resumed until pure salt is obtained.

Figure 2. Salt rock sample from the site of Coza – *Grochile de Sare.*

In parallel with the use of salt blocks for animal consumption, it is worth mentioning usages that are more classical. The salt rock is turned into pieces and crushed with a hammer in a metal container or in a thick piece of cloth. This type of salt is used to install haystacks in gardens, in the hayloft, but also to prepare brines (for cheese and bacon) or food portion for swine. If rock salt is exclusively destined to human consumption, the extraction is done with extreme care, in order to identify the least impure and the whitest sectors.

Surface extraction and use: a first assessment

In the Vrancea region, except for the use of salt springs, the exploitation of rock salt deposits could become a precious witness of the oldest forms of extraction. Nonetheless, the systematic absence of archaeological vestiges directly associated at the surface to the outcrops

[1] The surveys were conducted on the 2nd of July 2010 (M.A. and O.W.) and on the 22nd of June 2011 (R.B. and M.A.).

does not allow a determination of the origin, unlike the situations in Transylvania, where the mining exploitation of the rock salt deposits is attested since the Bronze Age (Cavruc *et al.* 2006). The extraction techniques observed in Moldavia attest specific iron tools with an influence on the nature and complexity of exploitation (vertical or horizontal).

The tools and the operating chain. Beyond the situation encountered in the valley of the Alghianu River (villages of Bodeşti and of Poiana) where the use of iron coins and chisels is attested, the classic tools are the hatchets, the spades, and the pickaxes. There is also evidence for the existence of wooden coins, but they appear to have been used for carpentry exclusively. The horizontal exploitation upstream from the Alghianu River seems to be possible following the excavation of a small-depth groove where the iron coin is inserted; with the help of an iron object, it is possible to extract blocks varying from 1 to 5 kg. At Valea Sării, these blocks can reach 30 kg. Their sizes vary depending on the accessibility of the deposit and, of course, on whether they have to be carried on foot. In the last case, a sack and sometimes a shaft are used. At Coza, the extraction areas are selected after a series of axe-based tests meant to measure the thickness and quality of salt. The extracted blocks are shaped up before being arranged in the cart, and then transported to the village.

The uses. In most cases, the salt blocks are destined to animal consumption, especially for ovine. Indeed, the shepherds sometimes have to regroup in order to exploit a deposit before the summering on mountain pastures (at the beginning of spring), but also in order to prepare for winter housing. The users associate with each other and the repartition depends on the amount necessary for each livestock. Considering that sheep need 2 to 3 kg of salt, the estimate amount necessary is 60 kg for a herd comprising 300 sheep during the summer season (6 months). As regards the human consumption, the salt is used after being crushed in an appropriate wooden or iron container of after being diluted in freshwater (usually tepid), in order to prepare the brine.

Conclusion

Though they are only preliminary, the first researches on the rock salt deposits in the Vrancea County have allowed us to outline a first ethnographic frame of reference and they have opened interesting opportunities to reflect on in order to measure better the variability of the use forms, which depend largely on its nature and accessibility. Other studies (even more recent), conducted in regions lacking liquid or solid sources of salt, for instance, have shown the importance of salt soils within pastoral activities. In these different cases, the point is to illustrate the remarkable capacity to adapt to a resource both polymorphic and fundamental for the development of human societies. These researches are a challenge to the archaeological community because, by underlining certain current social and economic mechanisms, we will be able to set light on ancient forms of use and of salt exploitation. In Moldavia, the unity of place, the permanence of traditional salt exploitation and, mostly, its attractive power are the strong points that allow us to relate the past and the present, the ethnography and the archaeology. This is why we believe it is indispensable to continue these researches in two complementary directions. On one side, we wish to complete and finish off the elaboration of a frame of reference on salt and its uses. On the other side, we wish to measure the nature of the spatial relations between the ancient implantations and the resource *per se*. It is at this price that the socio–economic dimension of salt will be valorised in its entire complexity.

Acknowledgement

This work was supported by a grant of the Romanian National Authority for Scientific Research, CNCS-UEFISCDI, project number PN-II-ID-PCE-2011-3-0825, *The Ethno-archaeology of Salt Springs and Salt Mountains from the extra-Carpathian areas of Romania.* We would like to thank for the support of the University Panthéon-Sorbonne in Paris, the Commission for Foreign Excavations (French Ministry of Foreign Affairs, *Exploitations pré- et protohistoriques du sel en Roumanie orientale*, O. Weller), the National Centre of Scientific Research (CNRS, UMR 8215 *Trajectoires*).

References

Alexianu, A., Dumitroaia, Gh. and Monah, D. 1992. Exploatarea surselor de apă sărată din Moldova: o abordare etnoarheologică. *Thraco-Dacica* XIII/1-2, 159-167.

Alexianu, M., Weller, O. and Brigand, R. 2007. *Izvoarele de apă sărată din Moldova subcarpatică. Cercetări ethnoarheologice.* Iaşi, Demiurg.

Alexianu, M., Weller, O. and Brigand, R. 2008. Usages et enjeux autour des sources salées de Moldavie précarpatique, Romania, in O. Weller, A. Dufraisse and P. Pétrequin (eds.), *Sel, eau et forêt. D'hier à aujourd'hui.* Cahiers de la MSH Ledoux XII. Besançon, 49-72.

Alexianu, M., Weller, O., Brigand, R., Curcă, R., Cotiugă, V., Moga, I. 2011. Salt springs in today's rural world. An ethnoarchaeological approach in Moldavia (Romania), in M. Alexianu, O. Weller and R. Curcă (eds.), *Archaeology and anthropology of salt: a diachronic approach,* 7-23. BAR International Series 2198. Oxford, BAR Publishing.

Cavruc, V., Ciugudean, H. and Harding, A.F. 2006. Vestigii arheologice privind exploatarea sării pe teritoriul României în epoca bronzului, in V. Cavruc and A. Chiricescu (eds.), *Sarea, timpul şi omul,* 41-48. Sfântu Gheorghe, Angustia.

Dumitroaia, Gh. 1987. La Station archéologique de Lunca-Poiana Slatinei, in M. Petrescu-Dimbovita, N. Ursulescu, D. Monah and V. Chirica (eds.), *La civilisation Cucuteni en contexte européen*, 253-258. Bibliotheca Archaeologica Iassiensis I, Iaşi.

Weller, O. and Dumitroaia, Gh. 2005. The earliest salt production in the World. An early Neolithic exploitation in Poiana Slatinei-Lunca, Romania. *Antiquity*, 79, 306.

Weller, O., Brigand, R. and Alexianu, M. 2007. Cercetări sistematice asupra izvoarelor de apă sărată din Moldova. Bilanţul explorărilor din anii 2004-2007 efectuate în special în judeţul Neamţ. *Memoria Antiquitatis* XXIV, 121-190.

Weller, O., Brigand, R. and Alexianu, M. 2010. Recherches systématiques autour des sources salées de Moldavie. Bilan des prospections 2008-2010. *Memoria Antiquitatis* XXV, 437-504.

The Radial Model of Salt Supplying. Preliminary Remarks

Marius Alexianu
Universitatea „Alexandru Ioan Cuza" din Iaşi

Abstract
Based on over 300 ethnological investigations conducted in resilient areas of the extra-Carpathian area of Romania, for the most part in the framework of two ethnoarchaeological research projects, the author proposes a new spatial model of salt supplying. Diametrically opposed to site catchment analysis, this model shifts the emphasis on the role of a natural resource of the utmost importance for the subsistence and development of the human communities settled concentrically around the salt springs and salt outcrops. The applicability in the field of archaeology of the proposed ethnological model presents a high potential.

Keywords
radial model, spatial analysis, salt supplying, salt springs, salt outcrop

The publication in 1977 of the results of much earlier archaeological investigations[1], following which the use of briquetage was first identified in Romanian literature, introduced a new theme in Romanian archaeology — the prehistoric exploitation of salt springs (Ursulescu 1977). The impact of this discovery was quite limited in the Romanian scientific literature until the discovery in 1983 of another salt water spring exploitation point (Dumitroaia 1987). What heightened interest in these findings was the fact that the first traces of human occupation (and thus exploitation) of both of the two salt springs dated to the Early Neolithic (Starčevo-Criş culture). Echoes of these two discoveries have attracted the attention of specialists both in the country (Monah 1980; 1987) and abroad (Ellis 1984; Nandriş 1987) which highlighted in general terms the significance that exploitable salt reserves had in the development of Neolithic and Eneolithic communities, at a multiscalar level. Starting from the observations of field archaeologists D. Monah and Gh. Dumitroaia who, in personal communications, pointed out to me the continuity of some traditional practices of salt water supplying from several sources in Neamţ and Bacău counties, and also as a result of the contact with ethnoarchaeologist John Nandris (London), in the context of field research in Neamţ county (1986), the undersigned took the initiative of carrying out ethnoarchaeological research on the exploitation of salt springs. By checking the work of Romanian ethnographers, I found that, except for a few general references in synthesis works, salt exploitation issues in rural areas had received practically no consideration. In this situation, I, along with archaeologist Dan Monah and ethnographer Elena Florescu (all three from Piatra Neamţ), conducted in 1986 an ethnographic research in Bacău county, the area surrounding the salt springs from Berzunţi. I also held long and fruitful discussions with my colleague from the Piatra Neamţ History Museum, Gheorghe Dumitroaia, director of the research team from Poiana Slatinei and who, as his birthplace was near the salt springs from Oglinzi, possessed extremely detailed information on their traditional exploitation. Likewise, research focused on the issue of the salt springs were carried out in the areas of Poduri, Lunca-Oglinzi-Răuceşti, Târgu Neamţ, Piatra Neamţ, Vânători-Lunca-Răuceşti, Balţăteşti, Ţolici, and Piatra Şoimului (Monah *et al.* 2008b, 84-86).

This is the backdrop of the first ethnoarchaeological study of salt springs carried out in Romania, which took into account over 30 salt springs from Neamţ and Bacău counties, among others, and which incorporated the idea of checking whether there were traces of exploitation from early archaeological times in their proximity (Alexianu *et al.* 1992, 160-161). We want to emphasize that this ethnological research was not conceived *ab initio* in terms of the site catchment area theory (Roper 1979), but in the completely opposite perspective, namely the importance of a salt spring for human communities; This perspective is embodied in a clear indication of the research objectives, namely the identification and mapping of salt springs (including abandoned ones), conducting archaeological surveys in the close proximity of the salts springs, and conducting systematic excavations in the detected archaeological sites (Alexianu *et al.* 1992, 160). Relevant in this regard is an important conclusion I reached after this pioneering research, even if the field information was empirical and fragmented, having not been obtained from systematic ethnographic research: "We believe that the distribution scheme (of brine from the salt springs, A/N), of a radial nature, is likely, of course on other spatial proportions, to have also worked for habitations detected archaeologically" (Alexianu *et al.* 1992, 162). I cautiously assumed at that moment that the scheme was currently available to a distance of about 10 km.

More than a decade after the publication of this study, the involvement in a Franco-Romanian project of the French archaeologist and ethnologist Olivier Weller marked a particularly important turning point, by instilling a rigorous character into the project and by adopting a systematic method regarding the ethnoarchaeological approach. Of particular importance

[1] In 1968, archaeologist Mircea Ignat from the Suceava Museum dug a trial pit near the Slatina Mare salt spring, Solca, Suceava County, the results of which have not been published (Monah 2008, 16-17).

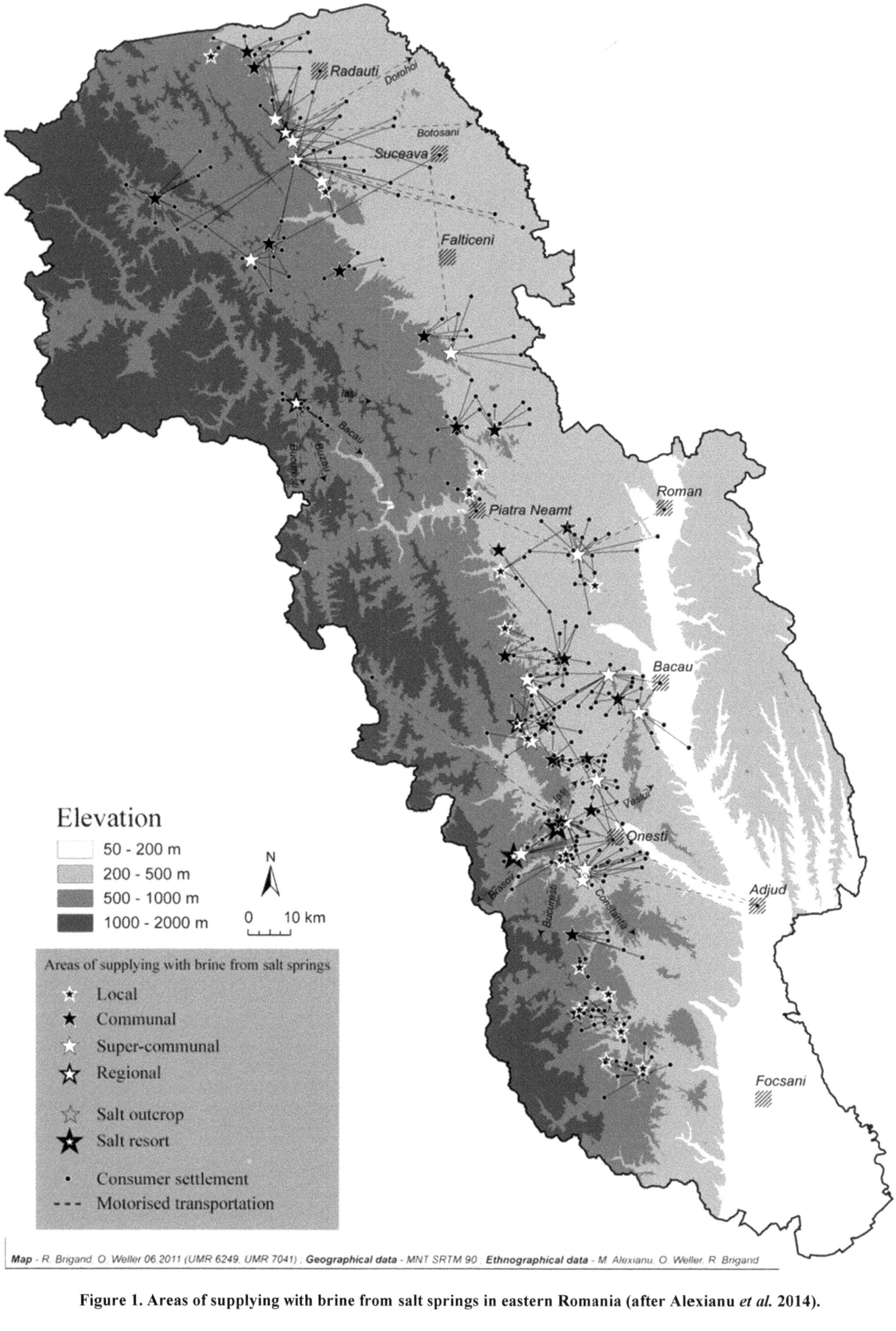

Figure 1. Areas of supplying with brine from salt springs in eastern Romania (after Alexianu *et al.* 2014).

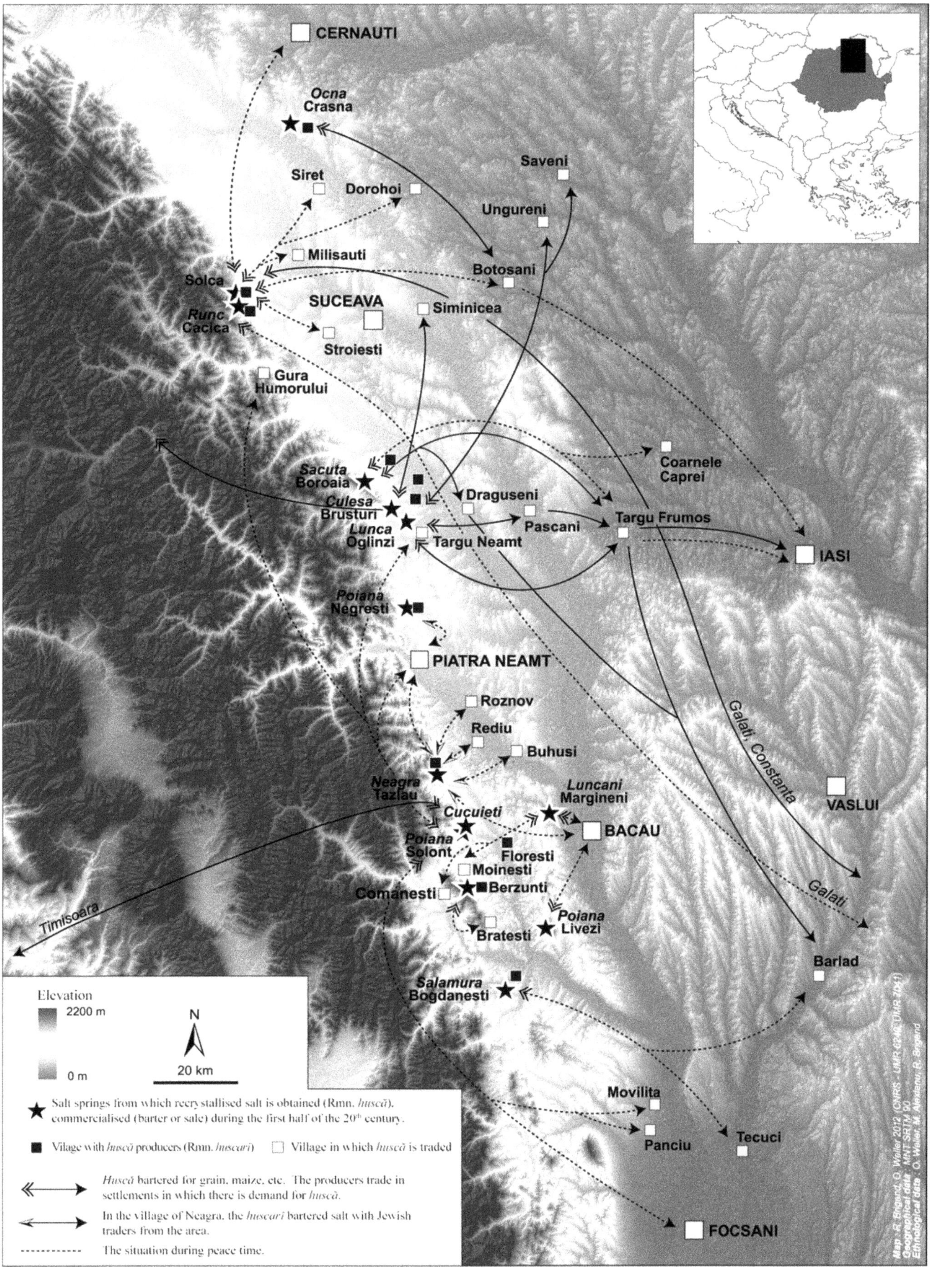

Figure 2. Areas of supplying with recrystallized salt from salt-spring brine in eastern Romania (after Alexianu *et al.* 2014).

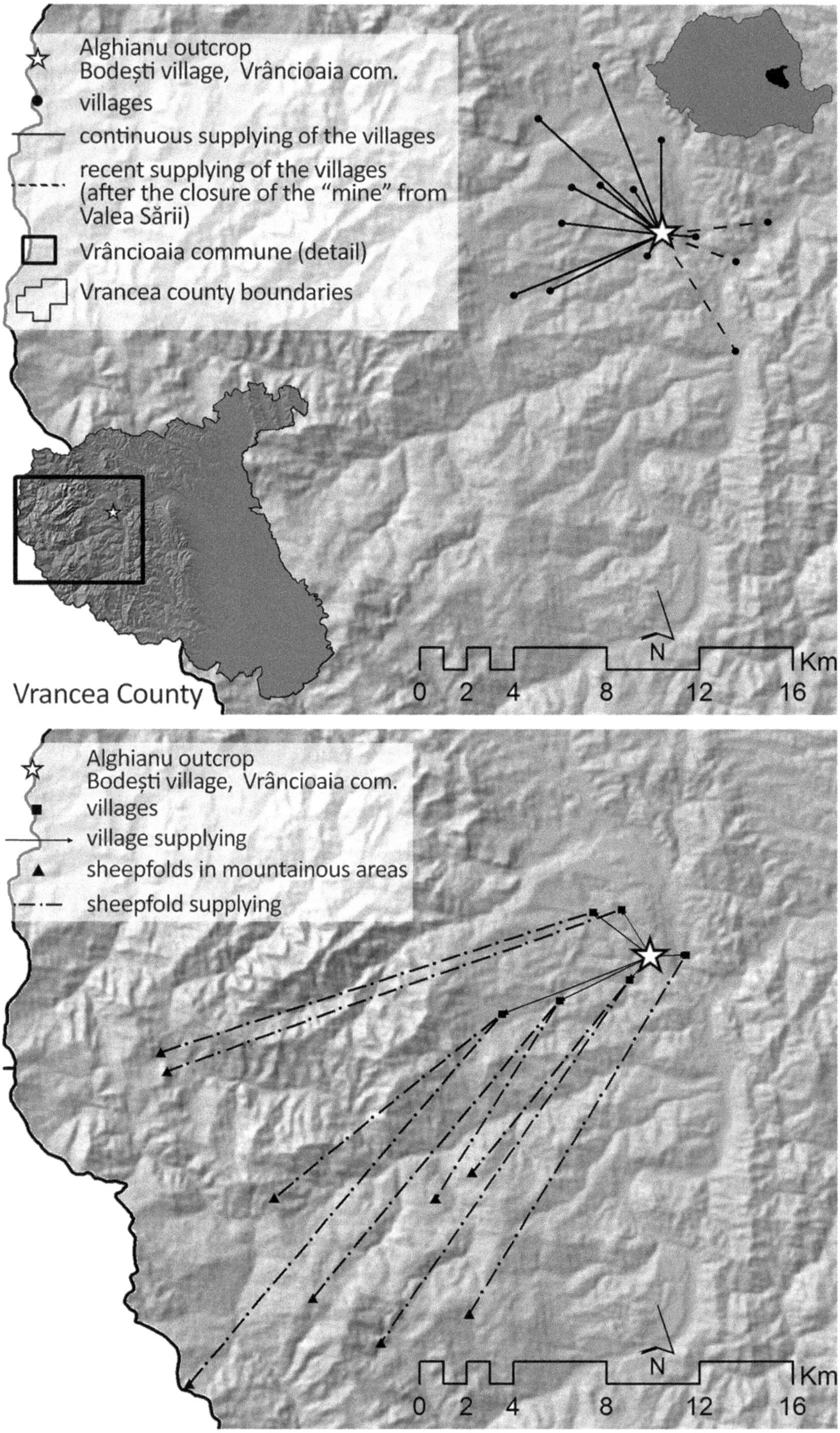

Figure 3. Supplying of villages with salt boulders from the Alghianu outcrop (top) and supplying of sheepfolds from mountainous areas from the Alghianu outcrop (bottom) (after Alexianu *et al.* 2015).

in this regard was designing three types of original questionnaires in late 2003 by Olivier Weller, Marius Alexianu and Laure Nuninger, respectively one for the salt spring, one for the sheepfolds, and one for the user settlements (villages, but also cities). We emphasize that the first important question of all three types of questionnaires referred to the sheepfolds or settlements that used a certain salts spring. The use of this questionnaire technique on the field demonstrated from the onset of the campaign (2004) that it is an extremely effective investigation tool (Alexianu and Weller 2007). A grant for ethnoarchaeological research on the salts springs in the sub-Carpathian area of Eastern Romania in 2007 (Alexianu and Weller 2009), and another in 2011, this time concerning the salt springs and salt deposits from the entire extra-Carpathian area of Romania (Alexianu *et al.* 2012), led to the accumulation of large amounts of information able to meet the requirements of a saturated model (Alexianu *et.al.* 2013). At the moment of speaking, the areas of distribution of salt springs and salt deposits were studied in the framework of these two projects.

Regarding the distribution areas of the salt springs, we operated a distinction between salt springs' brine supplying areas and the ignigenous salt supplying areas.

The study of maps that have already been published (Alexianu *et al.* 2011, fig. 2) reveals a variety of situations; thus, the area of distribution of salt springs occupies fractions of a circle, semicircle or more, while the area of distribution of other springs (less numerous) shows a completely radial aspect (Figure 1). In fact, the complete radial aspect represents the reference model; other sub-models are due to the absence of human communities in certain areas (usually mountain areas, not adequate for human settlement).

The absence of a full radial pattern (with some rare exceptions) in the case of areas of distribution of ignigenous salt obtained from salt water springs is due to the particularities of the environment in the area between the Eastern Carpathians and the eastern border of Romania. Therefore, the distribution mainly from the West to the East of this ignigenous salt at the end of the Second World War and during the drought of 1946–1947 is due to the easy access to the resources of grain, corn, potatoes located in the plains of the mentioned area, and to the ports of Constanța and Galați from which grains were exported (Figure 2).

The EthnosalRo ethnological research project on the distribution of salt boulders from salt outcrops point to the same complete radial distribution model (Alexianu *et al.* 2015).

Very interesting is the situation of the Alghianu deposit which, in a first phase, presented a semi-radial distribution model, and in a second phase, determined by the ceasing of the exploitation of another nearby salt deposit, the distribution model of salt boulders from Alghianu too once again become complete (Figure 3/top).

Another situation when the distribution model is always semi-radial is that when the distribution of salt boulders accompanies the pendulum-like shepherding movements towards sheepfolds situated far away, in which case the distribution is necessarily from the villages to the mountainous areas, i.e. from East to West (Figure 3/bottom).

Regarding the distances for which the radial model of brine from salt springs distribution is observable, ethnological research revealed that, in recent years, they usually vary from a few kilometres to a maximum of 30 km, while, at the end of and right after the Second World War, they could reach up to 100 km (Figure 4). In the case of the distribution of ignigenous salt obtained from salt spring brine, the distances reach up to 300 km and more.

Depending on several important factors cumulatively — discharge, salinity, taste — the area of attractiveness of salt water springs varies from a strictly local level (1–3 villages) to a micro-zonal level (villages located up to 30 km away). The area of direct attractiveness of salt water springs has to be differentiated from the area of distribution of salt water, which, in exceptional circumstances (war, drought), can reach approx. 100 km, and, in the case of ignigenous salt, up to 300 km and more (regional level).

Regarding the direction of movement for the supply of salt water, at the local level, it is done from the human settlements to the spring, while at a regional level, we can speak of a two-way direction, the initial movement from the settlement (located a few kilometres from the spring) to the source, the subsequent movement from the spring to the carrier's village, and from here to the supplied localities situated at distances up to 100 km.

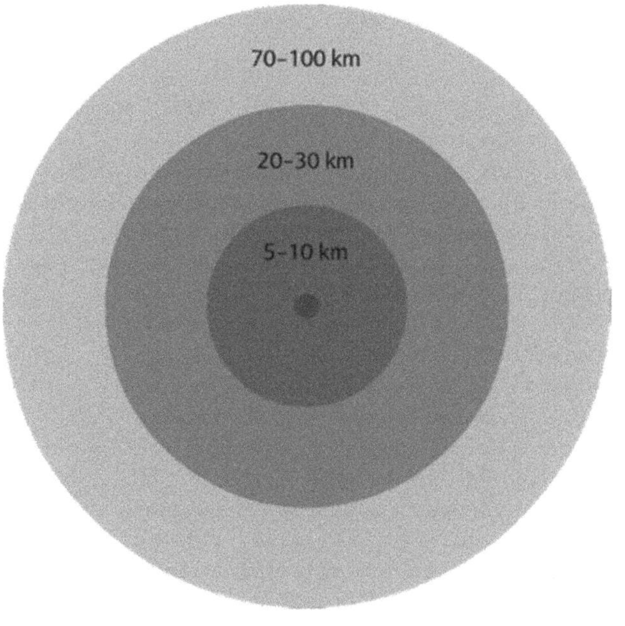

Figure 4. Areas of supplying with brine from one salt spring: direct supplying up to 20–30km, and indirect supplying up to 70–100km. Supplying frequency: frequent (darker shades) to once or twice a year (lighter shades).

The issue of salt boulder supplying is more complicated. So, if in the case of supplying to localities and sheepfolds located in their territory, the radial model obviously applies. In the case of mountain sheepfolds, located at considerable distances, the radial model does not apply.

In conclusion, based on the approx. 300 ethnological investigations carried out so far within the framework of two extensive projects on the salt springs and salt deposits from Romania, we were able to build a radial supply model for brine from salt springs, ignigenous salt obtained from salt spring brine, and salt boulders from salt outcrops. The complete radial model is available when all around the salt spring or the salt deposit there are human settlements. We can speak of an incomplete radial model where human settlements (permanent and seasonal) are located only in certain areas compared to a salt spring or a salt deposit.

Acknowledgement
This work was supported by a grant of the Romanian National Authority for Scientific Research, CNCS–UEFISCDI, project number PN-II-ID-PCE-2011-3-0825, 219/5.10.2011, *The ethno-archaeology of the salt springs and salt mountains from the extra-Carpathian areas of Romania* – www.ethnosalro.uaic.ro

References

Alexianu, M. 2013. Saturated model. A first application in world and Romanian ethno-archaeology, in A. Marciniak and N. Yalman (eds.), *Contesting Ethnoarchaeologies*, 211-225. Series: One World Archaeology, Vol. 7. New York, Springer.

Alexianu, M., Dumitroaia, Gh. and Monah, D. 1992. Exploatarea surselor de apă sărată din Moldova: o abordare etnoarheologică, *Thraco-Dacica* 13/1-2, 159-167.

Alexianu, M. and Weller, O. 2007. Recherches ethnoarchéologiques sur le sel: les enquêtes de 2004 et les premiers résultats obtenus dans la zone de Poiana Slatinei à Lunca (dép. Neamţ), Roumanie, in D. Monah, Gh. Dumitroaia, O. Weller, J. Chapman (eds.) *L'exploitation du sel à travers le temps*, 299-318. Piatra Neamţ, Constantin Matasă.

Alexianu, M., Weller, O. and Brigand, R. 2007. *Izvoarele de apă sărată din Moldova subcarpatică. Cercetări etnoarheologice*. Iaşi, Demiurg.

Alexianu, M. and Weller, O. 2009. The Ethnosal project. Ethnoarchaeological investigation at the Moldavian salt springs. *Antiquity*, vol. 83, issue 321. Online: antiquity.ac.uk/projgall/weller321 [last accessed: 15.03.2015].

Alexianu, M., Weller, O., Brigand, R., Curcă, R.-G., Cotiugă, V., Moga, I. 2011. Salt Springs in Today's Rural World. An Ethnoarchaeological Approach in Moldavia (Romania), in M. Alexianu, O. Weller and R.-G. Curcă (eds.) *Archaeology and Anthropology of Salt: a Diachronic Approach Proceedings of the International Colloquium, 1-5 October 2008 "Al. I. Cuza" University (Iaşi, Romania)*, 7-23, BAR 2198. Oxford, BAR Publishing.

Alexianu, M., Weller, O. and Brigand, R. 2012. EthnosalRo: an ethnoarchaeological project on Romanian salt. *The European Archaeologist* 38, 17-22.

Alexianu, M., Weller, O., Brigand, R. Curcă, R.-G. 2014. Annäherungen an eine Unsichtbare Vergangenheit: Ethnoarchäologische Forschungen zu den Salzwasserquellen der Moldauischen Vorkarpaten (Rumänien). *Studia Antiqua et Archaeologica* 20, 7-35.

Alexianu, M., Tencariu, F.A., Asăndulesei, A., Weller, O., Brigand, R., Sandu, I., Romanescu, Gh., Curcă, R.-G., Caliniuc, Şt., Asăndulesei, M. 2015. The salt from Alghianu beck (Vrancea County, Romania): a multifaceted ethnoarchaeological approach, in R. Brigand and O. Weller (eds.), *Archaeology of salt. Approaching an invisible past*, 47-63. Leiden, Sidestone Press.

Dumitroaia, Gh. 1987. La station archéologique de Lunca-Poiana Slatinei, in M. Petrescu-Dîmboviţa, N. Ursulescu, D. Monah, V. Chirica (eds.), *La Civilisation de Cucuteni en contexte européen*, 253-258, Iaşi, Bibliotheca Archaeologica Iassiensis, I.

Ellis, L. 1984. *The Cucuteni-Tripolye Culture. A Study in Technology and the Origins of Complex Society*, BAR 217. Oxford, BAR Publishing.

Monah, D., Antonescu, S. and Bujor, A. 1980. Raport preliminar asupra cercetărilor arheologice din comuna Poduri, jud. Bacău. *Materiale şi cercetări arheologice*, 86-89.

Monah, D., Popovici, D., Dumitroaia, Gh., Cucoş, Ş., Bujor, A. 1987. Raport preliminar asupra săpăturilor arheologice de la Poduri-Dealul Ghindaru (1984-1985). *Memoria Antiquitatis* XV-XVII (1983-1985), 9-20.

Monah, D. 2008a. Arheologia preistorică a sării în România. Scurt istoric, in D. Monah, Gh. Dumitroaia and D. Garvăn (eds.), *Sarea, de la prezent la trecut*, 13-39. Piatra-Neamţ, Constantin Matasă.

Monah, D. 2008b. Noi investigaţii etnoarheologice asupra izvoarelor sărate de pe Valea Muntelui, in D. Monah, Gh. Dumitroaia and D. Garvăn (eds.), *Sarea, de la prezent la trecut*, 81-105. Piatra-Neamţ, Constantin Matasă.

Nandriş, J.G. 1987. Romanian ethnoarchaeology and the emergence and development of Cucuteni in the European context, in M. Petrescu-Dîmboviţa, V. Chirica and D. Monah (eds.), *La civilisation de Cucuteni en contexte international*, 201-222. Iaşi, Bibliotheca Archaeologica Iassiensis, I.

Roper, D.C. 1979. *The method and theory of site catchment analysis: a review*. Advances in archaeological method and theory, vol. 2, 119-140. New York, Springer.

Ursulescu, N. 1977. Exploatarea sării din saramură în neoliticul timpuriu, în lumina descoperirilor de la Solca (jud. Suceava). *Studii şi Cercetări de Istorie Veche* 28/3, 307-317.

Weller, O. and Dumitroaia, Gh. 2005. The earliest salt production in the World. An Early Neolithic exploitation in Poiana Slatinei-Lunca, Romania. *Antiquity* 79, 306. Available: antiquity. ac.uk/ ProjGall/ weller [accessed: 10.11.2014].

Saline Springs on the Moldavian Plateau. Dates in Unpublished 19th Century Archival Sources

Mădălin-Cornel Văleanu

Complexul Naţional Muzeal „Moldova" Iaşi

Abstract

Taking as a starting point some information about a saltwater spring in the region of Băiceni in a document dated August 1884, this article presents some unpublished archival sources of the 19th century preserved in the National Archives, Iaşi County, in which there are references to records of mineral resources in Moldavia. We discuss the question of the genesis of some types of "saltwater" and mineral springs, their appearance in the Moldavian Plateau, and the possibility that these resources were exploited in prehistory.

Keywords

Moldavian plateau, Băiceni, Cucuteni, salt springs, archival sources, habitat, prehistory, Neolithic

Over the past 15 years the question of the use of mineral springs by the prehistoric communities of the eastern Carpathians has been a constant preoccupation. In our doctoral thesis, and also in a published volume, we included references to the ways by which our prehistoric ancestors could be sure of acquiring necessary mineral salts (Văleanu 2003, 203-204). This subject is also of interest for us in the context of a Romanian Academy/Iaşi Institute of Archaeology post-doctoral research project[1].

The information we discuss in the present study came to light during research on another research project relating to an earlier interest, namely the history of archaeological research on Cetăţuia de la Cucuteni (Petrescu-Dîmboviţa and Văleanu 2004)). This project involved a thorough search through the documentary reserves of the National Archives of Iaşi, and we were able to identify valuable unpublished material as a result. This included hitherto unrecognised references to saline springs in the area of Cucuteni-Băiceni, near the Eneolithic settlement on Cetăţuiei Hill.

The documents where we found this are in Iaşi City Archives and are related to the land holdings of the city, dating from the second half of the 19th century. We thus discover that at the beginning of 1884, Iaşi City Hall took steps to establish the limits of the city's land holdings. Among these lands there was also the domain of Băiceni, acquired through the "Law of the Ten Million" of 1882.

The Tribunal of Iaşi authorized the engineer Iulius Dobias to establish the limits of the domain of Băiceni. He was known in Iaşi fashionable society at the end of the 19th century, being a member of the Masonic Lodge "Steaua României", together with important members of the "Junimea" Society and of the Society of Physicians and Naturalists of Iaşi.

On 15 March 1884, Iulius Dobias asked Iaşi City Hall to provide all the documentation relating to Băiceni, and he received the documents in May. The Tribunal of Iaşi laid down 15 September 1884 as the date by which any litigation relating to the boundaries of properties in Băiceni should be resolved.

For the preparation of the action of September 1884, Iulius Dobias elaborated the *Mémoire of the Domain of Băiceni* (Figure 1) on the basis of "scrupulous observations" made on the spot acting on the basis of a document dated August 1884 (National Archives, Iaşi, Iaşi City Archives, File 174, vol. II, sheets 41-45), as the author of the report himself notes.

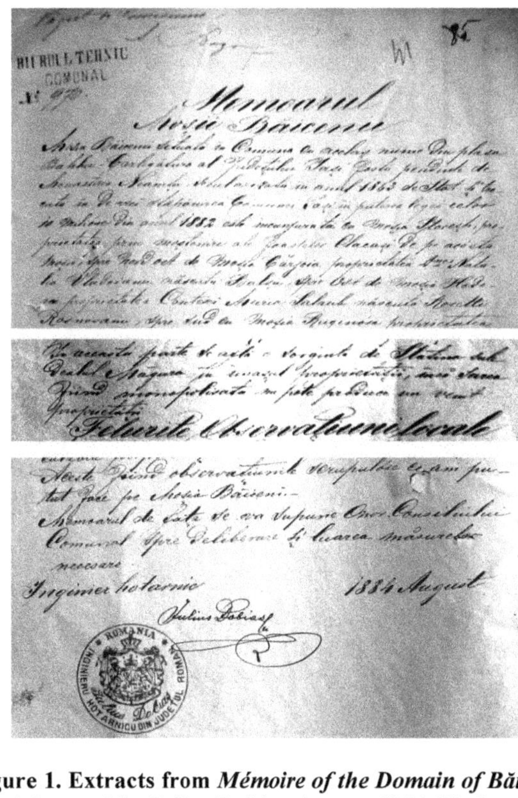

Figure 1. Extracts from *Mémoire of the Domain of Băiceni* elaborated by Iulius Dobias, August 1884 (National Archives, Iaşi, Iaşi City Archives, File 174, vol. II, sheets 41-45).

[1] Work within the Project *Social sciences and humanities in the context of global development—development and implementation of postdoctoral research*, project number POSDRU/89/1.5/S/61104, with the financial support of the Sectoral Operational Programme for Human Resources Development 2007-2013, co-financed by the European Social Fund.

An analysis of this document provides important data about this domain, which was the property of the Monastery of Neamţ that became State property in 1863 following secularization. The observations in the *Mémoire* made on the spot are very detailed. Even details that at a first sight appear insignificant are mentioned, such as fruit trees that are too old, or that in the centre of the village there is a small plot said to belong to Probota. The list could go on, with information about a series of buildings that existed on the property, many of them derelict, including houses, enclosures, a pub and the village mill. This report is full of interest and value, since the acute attention to detail displayed in recording what was there allow for exact reconstruction.

Once the property limits were established in the document of 15 September 1884, another deed was drawn up, called a deed of land limits, to which a plan of the domain was later added, dated 12 April 1885.

Our initial document, the *Mémoire of the Domain of Băiceni* contains an interesting remark: Iulius Dobias relates that within the perimeter of the larger domain, to the north, there was a saline spring beneath Măgura Hill (National Archives, Iaşi, Iaşi City Archives, File 174, vol. II, sheet 42 recto. The place is located on the edge of the Cârjoaia domain, the property of Natalia Balş; cf. the *Deed of land limits of the commune Băiceni* of 18 September 1884 (National Archives, Iaşi, Iaşi City Archives File 174, vol. II, sheet 12 recto) in the meadow of the property, but that it could bring no income to the property in the light of the state salt monopoly.

The site of the saline spring is not marked on the plan of the domain, but if we consult a military map of the area of 1894, we can easily identify the area (Figure 2). It is interesting to note that southeast of Măgura Hill (in fact, on its southeast slope) is Slandola hill, a place name that does not appear on later maps of the area.

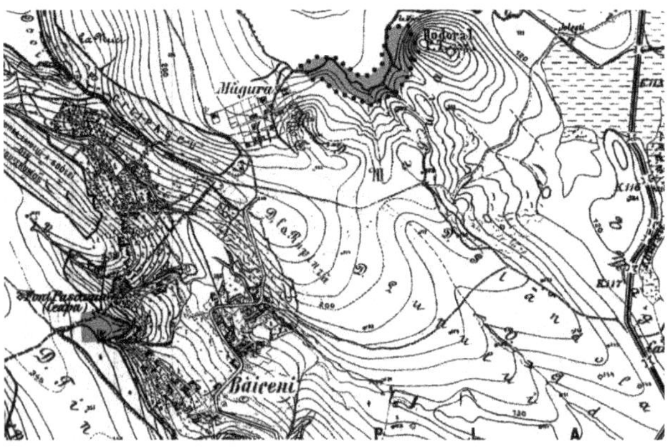

Figure 2. The position (in grey) of the salt spring (top) and Cucuteni-Cetăţuie Eneolithic settlement (bottom) on a military map of the area of 1894.

It is not altogether unexpected to find saline springs on Moldovei Plateau, specifically in the area between the River Siret (to the west) and the River Prut (to the east), for they are generated by the geology of the area, and springs with high mineralization (0.5 - 4.5 g/dm^3), of what is termed brackish water, occur frequently in the area.

Springs with salt or brackish water on Moldovei Plateau came about as a result of the mineral salts in the geological layers being dissolved by rain water which seeps into the ground or into underground aquifers. In fact, the salinity of such springs varies, especially during dry seasons, when spectacular increases in the quantity of salt can occur. To call these waters salt waters is something of a misnomer, since in reality they do not just contain salt (i.e. sodium chloride), but a multiplicity of other minerals including chlorides, fluorides, iodides, sulphides, or carbonates that are leeched from the geological layers.

Springs with salt and brackish water occur in just the same way as other mineral springs in the area (e.g sulphurous or ferruginous springs), as the result of the action of water on soluble substances in the geological layers. The occurrence of these springs depends on the micro-geology of the area. Since Neogene geological deposits display a multitude of lithofacial variations, being generated by sedimentary deposits in a salt-withdrawal marine basin, springs with higher mineralization are only to be expected in this area.

Some of the Neogene formations present on the Moldovei Plateau attest the undoubted existence of a littoral area in the region. Geological formations related to the former littoral areas can indeed be the genetic cause of the presence here of springs whose water tastes salty. We should add though that there have been no specialized studies of the valorization potential of these types of springs on the Moldovei Plateau, understandable in that they have little economic potential, and in fact mostly create water supply problems for local inhabitants.

This was probably also the situation described by Iulius Dobias at Băiceni, a situation that can be relatively frequently encountered in this area. It should be repeated that the occurrence of a saline spring in this part of the Moldovei Plateau, between the Siret and the Prut, is not related to the presence of underground salt beds, as in the case of the sub-Carpathian region to the west.

Another phenomenon relatively frequently encountered in this area is that of salts. Relevant place names are often mentioned in 19th century and earlier documents, or on military plans and maps already at the end of the 19th century (Figure 3a and b). The genesis of these formations is related to springs with a richer mineralization overflowing, springs whose origins were explained above.

Through the repeated evaporation of water, mineral salts are concentrated on the surface, and this also determines what will grow in the vicinity. Zones with salts are easier to observe in spring, in the shape of areas marked by white stains (Figure 4). This substance (similar to fine salt, but akin to the somewhat erroneous expression of "salt water" discussed above) usually has a marked salty taste, the result of a high sodium chloride content, but sometimes it may only be slightly bitter or even tasteless. In reality, this fine powder contains a variety salts, and not only sodium chloride, reflecting the local geological situation.

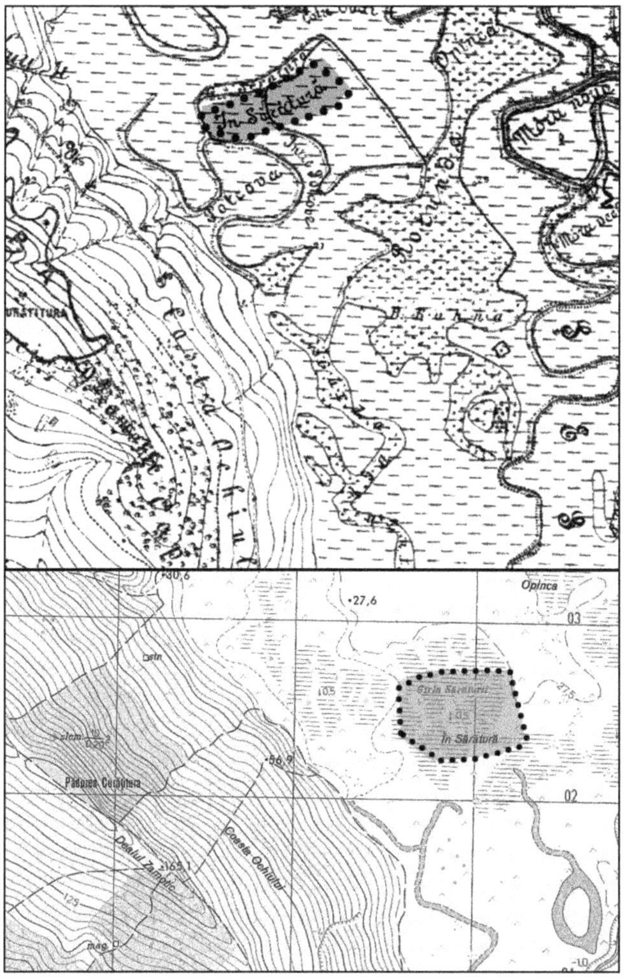

Figure 3. Sărătura, the place names, near the Isaiia-Balta Popii Neolithic settlement, on military maps of the area of 1894 (a) and 1984 (b).

Figure 4. A zone bearing salts (sărătură) at Valea Oii (Iaşi County) in April 2012.

In today's context these springs, and we refer to the saline springs and waters in the area of the Moldovei Plateau, clearly possess little or no potential. Nevertheless, in a crisis, they can provide the salt necessary for survival for limited communities. The question is whether prehistoric communities exploited these springs. Theoretically, the answer would be yes, considering the ways in which they are known to have acted, but actual proof is lacking.

In conclusion, we might note that already in the first half of the 19th century, the potential of the mineral springs of Moldova attracted attention, especially for their curative role. The authorities in fact gave this matter careful consideration, and we identified more than 20 relevant files in the archives of the Prefecture of Iaşi County preserved in the National Archives in Iaşi.

Also worthy of note is a series of works published then, such as the volume published in the year 1851 by Atanasie Fătu with the title *Description and usage of plain and mineral waters in Moldova*, and printed in Iaşi (Figure 5), which provided an inventory of mineral springs known at the time. Among these mineral springs, the author mentions the existence of an important source with sulphurous water at Băiceni, at the foot of Cetăţuiei Hill (Fătu 1851, 188), relatively close to the saline water source mentioned by Iulius Dobias in 1884, and also another mineral source, named Recea (Fătu 1851, 188).

Figure 5. The cover of Atanasie Fătu, *Description and usage of plain and mineral waters in Moldova* with the title *Description and usage of simple water and of mineral waters in Moldova*, Iaşi, 1851.

References

Fătu, A. 1851. *Descrierea şi întrebuinţarea apei simple şi a apelor minerale din Moldova*, Iaşi.

Petrescu-Dîmboviţa, M. and Văleanu, M.C. 2004. *Cucuteni-Cetăţuia. Monografie arheologică*. Piatra Neamţ, Constantin Matasă.

Văleanu, M.C. 2003. *Omul şi mediul natural în neo-eneoliticul din Moldova*, Iaşi, Helios.

The Symbolism of Salt in Holidays as Expressed in the Pioneering Works of Romanian Ethnographers Simion Florea Marian and Tudor Pamfile

Mihaela Asăndulesei
Universitatea „Alexandru Ioan Cuza" din Iaşi

Abstract
This paper studies the works of two of the most important Romanian ethnographers, Simeon Florea Marian and Tudor Pamfile, who brought to light both religious and traditional holidays. The article outlines the Romanian holidays that place salt at the centre of these customs. The author believes that though the Romanian people ascribed supernatural powers to numerous elements, mostly within nature, salt has a special position, possibly because its practical use was discovered since prehistory. This opinion is also supported by the table mentioning holidays, the customs focusing on salt and the meaning of these customs.

Keywords
symbolism of salt, Romanian holidays, Romanian folk traditions

Starting with the last decades of the 19th century and in the first decades of the following century, scientists began to notice worryingly that Romanian folk traditions were losing ground to the advance of civilization. They realized that they should focus on intensive research within an organized institutional framework, to prevent irrecoverable losses.

One of those who contributed fundamentally to the research in this field was Simeon Florea Marian (1847–1907). Titular member of the Romanian Academy, he was attracted from the beginning by folk creations and he was the first to study all its compartments (monographically, analytically, descriptively and morphologically) (Rusu 2003, 507). He is considered one of the most influential Romanian folklorists and ethnographers; his written inheritance has no comparison in the scientific literature, and it brought him the admiration and praise of many scholars and the public of his time. During his sixty years of life, he managed to combine the vocations of priest and teacher with that of active folklore researcher. He wrote several synthesis works about the most diverse components of folk culture (Datcu I., Introducere in vol. I, S. Fl. Marian 1994, VI; Rusu 2003, 507), with the most notable being *Ornitologia poporană română* (1883), *Descântece poporane române* (1886), studies on Romanian weddings, funerals and births (1890, 1892), *Vrăji, farmece şi desfaceri* (1893) and *Serbătorile la români* (I-1898, II-1899, III-1901). Through his work, he contributed to laying the foundation of Romanian ethnography and he stimulated other researchers to collect numerous folk creations.

Among those who continued his work, who studied folk customs and presented them to society, it is worth mentioning Tudor Pamfile (1883–1821), who was said to be the spiritual son of Simeon Florea Marian. Though he was an active army officer, Pamfile's works are those of a remarkable ethnographer. During the 38 years of his life, he published 40 volumes comprising a great variety of themes. Starting in 1906, when he published a volume on children games, he wrote books on the beliefs of Romanian villagers on the sky and stars, and on riddles. He also wrote several volumes of folk tales, a volume on Romanians' household industry, one on diseases and cures for people, cattle and birds, a collection of Christmas carols, songs for New Year's Eve, New Year and the day after the New Year, a volume of lyrical songs, another on Romanian agriculture, a short story about the Romanian people, a volume on the chromatic of Romanian people, three volumes of Romanian mythology, etc. (Pamfile 2006, 5).

Tudor Pamfile also printed three volumes on Romanian holidays, as a continuation of the ethnographic studies of his spiritual father, whom he had known and appreciated; the two had a great collaboration in the field. If we compare the trilogy of Simeon Florea Marian to the one by Tudor Pamfile, we realize that the first paid equal attention to traditional holidays, not related to the Church, which people often celebrate more solemnly than the religious ones, while Tudor Pamfile focused on religious holidays. It is well known that deeply religious holidays are preceded, accompanied and followed by a series of profane festive customs; there is no separation of the two, but a fusion between the sacred and the profane (Datcu I., Introducere in vol. I, S. Fl. Marian 1994, XII).

The monumental work written by Simion Florea Marian, *Serbătorile la români* (I – *Cîrnilegile,* 1898, II – *Păresimile,* 1899, III – *Cincizecimea,* 1901), and the studies elaborated by Tudor Pamfile, printed as a trilogy (*Sărbătorile la români.* Studiu etnografic – I - *Sărbătorile de vară la români.* Studiu etnografic, 1910; II, III – *Sărbătorile de toamnă şi postul Crăciunului.* Studiu etnografic; *Crăciunul.* Studiu etnografic, 1914) are still works of reference, a century after their publication. They proposed a complex image of the holidays and customs they studied and of the people who practiced them. The importance of these works is all the greater as they can be considered the first works comprising very old testimonies on Romanian holidays, including mentions of salt and its symbolic use.

Another reason for which I chose the works of Simeon Florea Marian and Tudor Pamfile as the main subject of my present investigation, is that in the current stage of research, the surveys of scientists involved in the study of salt symbolism have reported few mentions (Chiricescu 2013) within Romanian holidays of this crucial mineral in the life of human communities.

It is acknowledged that salt has played an essential role in the evolution of human communities. Beyond the practical utility of this element, I also underline that in traditional societies, both in the past and present, it is ascribed a series of imaginary characteristics; the symbolism of salt is extremely diverse. Since time immemorial until the present, in all regions of the world, salt has been considered a primordial, purifying element, a symbol of vitality, health, wealth, hospitality and unity. This symbolism has generated a multitude of magical-religious manifestations, focusing on purification, fortune telling, sanctification, finding love, protection against curses and charms, and disenchantment (Cavruc and Chiricescu 2006, 7).

This indispensable daily mineral is present in various customs related to the sequence of seasons or in those overlapping important Christian holidays. Generally, through their symbolism, the elements within the practice of these customs - in our case, salt - have supernatural powers; they are used exclusively to cast away evil, to protect children, people in general, the household and livestock (Dăncuţ 2006, 147).

In the works I relied on for this investigation, salt is present within several holidays, but it is correlated mostly with profane customs accompanying the religious celebration. Events involving salt in the work of Simeon Florea Marian include Saint Basil or the New Year, Epiphany, the Meeting of the Lord, Saint Theodore (Rmn. Toader), Forty Saints, *Alexiile* (Saint Alexios (Rmn. Alexie)), the Annunciation, Easter and Saint Basil. In the works of Tudor Pamfile, the few holidays where the use of salt is mentioned include Pentecost (*Căluşarii* dance), *Filipii* (Saint Philip), Christmas Eve and the *Ignat (Saint Ignatius)* (pig sacrificing).

In order to analyse concisely the existence and meaning of salt within each custom, we decided to introduce all the information in Table 1, as to provide a complete overview of all the essential data for the study.

Table 1. Habits based on the use of salt and their significance in the celebrations mentioned in the works of Simion Florea Marian and Tudor Pamfile.

HOLIDAY	TRADITION	MEANING	REFERENCES
Saint Basil or the New Year Saint Basil Holiday or New Year's Eve is celebrated during the night of 31 December. It is a holiday with rich traditions, with both laic and religious valences.	Mothers put salt in the same number of onion layers, as children in the respective family. The onion layers are left by the window until the next morning, when a fortune can be predicted.	Foretelling good fortune and its size	Marian 1898, 7
	Marriageable young girls fill one onion layer with salt and leave it by the window until the next morning, when they find out the result.	Foretelling good fortune and its size	Marian 1898, 7
	Marriageable young girls, who want to find out the qualities of their - still unknown - future husband, go to the house of a woman who has the role of spirit of fate. She hides various household items (including salt) under pots, and each girl gets to choose one.	Discovering the features of the destined one	Marian 1898, 63 Marian 1898, 64
	Marriageable young girls put four elements (charcoal, clay, bread and salt) - in the shape of a cross - on an upside down sieve; they pick one of the four with a threaded needle.	Discovering the features of the destined one	Marian 1898, 75-76
	Marriageable young girls put four elements (a coin, earth, bread and salt) - in the shape of a cross - and they choose one blindfolded, at midnight.	Discovering the features of the destined one	Marian 1898, 75-76
	Men put salt in twelve onion layers, which they name after the months of the year, and they leave them until the next morning, when they find out the result.	Weather forecasting	Marian 1898, 90
	If salt starts to melt, then the weather will be rainy next year.	Weather forecasting	Marian 1898, 119

	Marriageable young girls often cast spells, the most frequent ones involving the evocation of nature. Alongside specific incantations and religious expressions, they also make the well-known gesture of rewarding nature with bread and salt.	Discovering the features of the destined one	Marian 1898, 130-131
	Marriageable young girls often cast spells, through which - alongside specific incantations, some ritual gestures and physical manifestations - they also make the well-known gesture of rewarding nature with bread and salt.	Discovering the features of the destined one	Marian 1898, 137
	Marriageable young girls prepare a loaf of round bread from equal amounts of flour, water and salt and they eat it.	Discovering the features of the destined one	Marian 1898, 142
	Marriageable young girls make a loaf of round bread from equal amounts of flour, water and salt; they eat half of it and they put the other half under the pillow until the next morning.	Discovering the features of the destined one	Marian 1898, 143
	Men give water to the animals; in it, they put silver coins, hay sprinkled with saltwater and smoked with the compost collected within a certain period.	Protecting the animals from troubles/unfortunate events	Marian 1898, 148
Epiphany Epiphany is celebrated on 6 January and it represents Christ's Baptism; it is also a very important holiday comprising a variety of traditions.	Before putting the food on the table, women put hay, salt and bran under the tablecloth.	Ensures the well-being of the household	Marian 1898, 164-165
	Men put on the table - alongside the rest of the food - two loaves of bread, salt and a glass of water, thus waiting for the spirits of the deceased persons.	Showing respect to the deceased persons	Marian 1898, 165
	The boys who want to know details of their future wives gather and unveil an item from a group of several covered items (among which is salt).	Discovering the features of the destined one	Marian 1898, 189
Candlemas Saint Tryphon (Rmn. Trifon) is celebrated on 1 February; it is the eve of the holiday of Meeting of the Lord; folk belief mentions this saint as the protector of orchards from caterpillars and locusts; it also protects people from madness.	Men sprinkle trees and cattle with holy water especially prepared for this holiday, and they pour some of it in the animals' water before giving it to them.	Protecting the trees from pests; Protecting the animals from trouble/ unfortunate events	Marian 1898, 246
***Sân-Toader* (Saint Theodore)** This holiday is not celebrated on a certain date, but in the first Saturday of the Easter Fast. This holiday is a combination of the Christian representation of Saint Theodore and the mundane elements ascribed to him,	Men avoid giving salt to the cattle around this holiday, except for the animals used for work, which receive charmed salt.	Protecting the animals from trouble/ unfortunate events	Marian 1899, 44
	The head of the household offers corn, oats and salt to the neighbours' horses to ensure a symbolic meal for the Horses of Saint Theodore.	Protecting the animals from trouble/ unfortunate events	Marian 1899, 51
	Within a ritual, young girls collect certain herbs; they make an infusion which they use	Thicker and longer hair;	Marian 1899, 58

materialized especially in accompanying horses.	to wash their hair; the infusion also includes the bread and salt used in the ceremony.	Curing people of illness	
	Within a ritual rewarding nature with bread and salt, young men collect healing herbs.	Thicker and longer hair; Curing people of illness	Marian 1899, 63
	When they wish to become blood brothers, most individuals perform a ceremony based on bread and salt.	Consolidating an oath; salt – an element indispensable to life	Marian 1899, 85
	Young girls make a loaf of round bread from honey, flour and some salt, which they eat while reciting some incantation.	Ensuring the individual's success	Marian 1899, 88
Forty Saints The holiday of Forty Saints, also known as the Forty Martyrs, celebrates - on 9 March - the memory of forty martyrs who sacrificed their lives for the truthful teaching and salvation. Concerning the tradition of setting in motion the plough, it is related to the saints - mentioned by certain legends - who had the power to make crops richer, mostly considering that this holiday is celebrated when the weather becomes warmer.	The woman incenses and sprinkles holy water on the ploughmen and the ploughs of the household before their first day in the field, and she offers them bread and salt.	Protecting the individuals; Protecting the household; Protecting the crops; Protecting the animals from trouble/unfortuna te events	Marian 1899, 173-174
Alexiile **(Saint Alexios)** This holiday is the folk tradition that celebrates Saint Alexios on 17 March; he is also known as the "man of God" or the patron of all living things.	Men perform a ceremony through which they "feed" bread and salt to an elderberry.	Protecting the households from pests	Marian 1899, 206
Annunciation The Annunciation is celebrated on 25 March to remember the moment when Archangel Gabriel announced that the Virgin Mary would give birth to Jesus Christ.	The women of the house make a fire in the courtyard for the angels, alongside which they put bread, water and salt; later, they give them as gifts.	Rewarding the angels for announcing Christ's upcoming birth[1]	Marian 1899, 225
Easter The Easter holiday celebrates Christ's resurrection. Alongside	Easter pie - according to another legend, also from Bukovina - is made because Jesus Christ, before being caught and crucified, told his disciples that they had been eating non-	Providing a new meaning for this food after Easter changed its	Marian 1901, 5 Marian 1901, 7

[1] The volume *Sărbători şi obiceiuri* (Ghinoiu 2004, vol. 4, 335) mentions the custom according to which during the Easter Week, people used pine branches taken from staves, for fire (at night), which announced the Birth of Christ somehow. Hence, for the practice of the Annunciation eve, reminded by Simion Florea Marian, I extended the meaning stated by Ghinoiu, because it shows clearly the essence of the 25 March holiday.

the Birth of Christ, it is one of the most important religious holidays for Christians; specific foods are prepared in that period: red eggs and Easter pie.	salted, non-leavened baked bread (*turte* or *azime*), but that they would eat, from then on, salted and leavened baked bread, called *paşti*. This custom has deep roots: the *Exodus* of the Old Testament mentions that, when Jews escaped the Egyptian slavery, they left in a hurry and they did not have the time to put salt in the dough or to let it leaven. This is why, as long as the Jewish Easter celebrated the liberation from Egyptian slavery, those involved ate the *azime*. After Easter changed its meaning, by celebrating Christ's resurrection during this holiday, all those who observe tradition eat salted and leavened baked bread. It is also well known that bread and salt play a very important role in Christian religion and in the Romanian culture.	meaning; it symbolizes the Body of Christ	
	Men sprinkle the Easter pie and other items (including salt) with holy water, which they use throughout the year for various things.	Curing people; Protecting the individuals; Ensuring the well-being of the household	Marian 1901, 80
	The men of the house put Eucharist bread in the animals' salt.	Multiple gestations for household animals	Marian 1901, 83
	Individuals use salt in various activities (taste, purification of wells, in the animals' food, as a cure).	Purifying the wells; Curing people; Spicing food; Protecting the animals from trouble/unfortunate events	Marian 1901, 86
Small Easter Small Easter - Easter of the *Blajini*, Easter of *Rohmani*, Easter of the Dead or *Mătcălăul* - is celebrated on Monday inthe second week after Easter; the belief is that the *Blajini* - fabulous beings that live on the Netherworld - celebrate Easter that day. People give things for charity and priests say prayers in cemeteries. *Mătcălăul* is also considered the patron of young women who become *mătcuţe*, meaning blood sisters.	When they wish to become blood brothers, most individuals perform a ceremony based on bread and salt.	Consolidating an oath; salt – an element indispensable to life	Marian 1901, 191 Marian 1901, 193
Saint George Saint George is celebrated on 23 April; he was a martyr who chose to state publicly his Christian faith,	Some women who cast spells also perform a complex ceremony that involves collecting dew, with which they sprinkle on the bran and salt of their own animals. The cattle that eat grass from where the dew was collected will run out of milk.	Taking/Stealing milk from cattle	Marian 1901, 205

despite the decree issued in 303 by Diocletian against Christians. In folk legends, Saint George is the "guardian of time"; he holds the keys of time and is the patron of shepherds.	Some women hide a salt rock under a stone on the road; after all the cattle of the village have passed, they give the salt to a cow they wish to make more productive.	Taking/Stealing milk from cattle	Marian 1901, 207 Marian 1901, 208
	The women who believe their cows are not productive enough recite some incantation while performing a ritual that involves collecting certain herbs accompanied by bread and salt, which they feed to their cattle.	Giving the milk back to the cattle	Marian 1901, 217
	Some women perform a ceremony that involves ants which they reward with bread and salt.	Giving the milk back to the cattle	Marian 1901, 223
	On the *Ales* (chosen) day through a complex ritual, the mistress of the herd makes a "cream" *using special elements*, among which is grease from a black pig, sacrificed on a certain day and then salted.	Protecting the animals from trouble/unfortunate events; Protecting the animals from spells	Marian 1901, 231
	The women who milk the sheep make a loaf of braided bread using only flour, water and salt, covered with an egg, which they later use in the milking process (Obiceiuri de primăvară…).	Protecting the animals from trouble/unfortunate events; Ensuring their productivity	Marian 1901, 233
	After finishing the milking process, the shepherds make a symbolic offer to the sheep: fine salt mixed with maize flour or bran. The remaining salt is given to the lambs; afterwards, the head of the shepherds eats from each braided bread and he symbolically cuts off three pieces, which he throws in three buckets. The women who milk the sheep take these pieces home, they let them dry and they use them for the sheep's food throughout the year.	Protecting the animals from spells; Ensuring their productivity	Marian 1901, 235
	If the Chosen day is a fast day, then all must fast, and they regularly eat beans and salted fish with corn bread.	In this case, we are not dealing with salt as an element *per se*, but with a food based on salt. However, we failed to find a meaning for this tradition in the literature we have studied thus far.	Marian 1901, 237
	Shepherds overturn a bucket with "pure" water, in which they also put salt and fresh grass.	Protecting the animals from spells	Marian 1901, 238
	Young girls collect dew, which they keep and use throughout the year. Young girls also wash themselves in calm waters, performing a ritual that involves rewarding the waters with bread and salt.	Ensuring people's health	Marian 1901, 277
Pentecost Pentecost is also called - according to the Orthodox-Christian calendar - the Fiftieth Day or the Descent of the	The head of the household receives the *călușari* with salt, sagebrush and garlic.	Protecting the household; Protecting the animals from trouble/unfortunate events;	Pamfile 2006, 49

Holy Spirit. It is celebrated fifty days after Easter and it symbolizes the Descent of the Holy Spirit on the Holy Disciples, who disseminated Christian faith. It is also considered the "birthday" of the Christian Church. In folk tradition, this holiday is the day when evil spirits (also called fairies, *iele* and *rusalii*) become stronger than usual, reason for which they must be defeated and cast away. The *Călușari* dance is closely connected to this holiday because it is performed from the Ascension to the Pentecost and it illustrates the symbolic fight of good male forces against evil female forces. This dance should not be interpreted only from the perspective of protagonists - members under oath and under the authority of a *vătaf* -, but also from that of beneficiaries - the villagers who accept the dance for wealth and health, but also because it animates the atmosphere during the holiday.		Protecting the individuals	
	The head of the household puts a rock salt in the centre of the courtyard where the *călușari* dance and the animals will lick the rock.	Protecting the household; Protecting the animals from trouble/unfortunate events; Protecting the individuals	Pamfile 2006, 52 Pamfile 2006, 53
***Filipii* (Saint Philip)** *Filipii* (Saint Philip) represents the holiday of Philip the Disciple and it is celebrated on 14 November. The Orthodox calendar does not mention it as a religious holiday per se, but mainly like a folk celebration, which the Romanian people know and cherish dearly (Pamfile 2006, 211). The *filipii* are gods of the house, in the honour of	Women are prohibited from working or offering gifts.	Protecting the animals from trouble/unfortunate events	Pamfile 2006, 212

which only the woman observes the holiday, by not doing any kind of chores. In other areas, it is believed that *filipii* are greater saints, governing wild beasts or wolves. *Filipii* defend the house from evil and from various dangers - from fire to wolves (Pamfile 2006, 211). This holiday is some kind of household celebration and it can take from one to six or seven days. It is related to the Christmas Fast and it is celebrated around the same time.			
Christmas Eve Christmas Eve is not a religious holiday per se, but it precedes the great celebration of Christmas. However, it is very important because this day includes numerous rituals more or less related to the holiday celebrated the following day.	People who own sheep hide a rock salt under the threshold, until the eve of Saint Basil's holiday; then they smash it and feed it to the animals	Protecting the animals from troubles/unfortun ate events;	Pamfile 2006, 282
	The heads of the household are prohibited from throwing away the ashes on this day, before offering onion and salt.	We have found no meaning for this celebration in the literature studied thus far[2]	Pamfile 2006, 282
	Marriageable young girls make a salted round loaf of bread, using equal amounts of flour and salt, and they avoid drinking water throughout the day.	Finding out the identity of one's future lover	Pamfile 2006, 287
Pig sacrificing Pig sacrificing is not a holiday per se, but a tradition that occurs before Christmas, usually on Christmas Eve.	After the pig is sacrificed, the head of the household cuts a cross on its forehead or back of the neck, and puts salt on it.	Purifying the animals; Conserving food	Pamfile 2006, 390

[2] However, to emit an opinion, even without strong arguments, I have to mention that the *Dictionary of symbols* (Chevalier, Gheerbrant 2009, 284) provides the following explanations for *ash*: bodily residue after extinguishing the fire of life. The definition also states that we must not forget that all things related to death are also related to the symbolism of *eternal return.* Hence, considering these meanings and the prohibition of throwing away the ash on Christmas Eve, I suggest an analogy, according to which people throw the remains of what "died" to favour the "birth" of a new phase. As for the gesture of giving to others onion and salt in the aforementioned context, I found no explanation, as already pinpointed, reason for which I cannot state an opinion on this matter.

Hence, in close and direct connection with the previous table, I pinpoint that salt is symbolically present within many Romanian holidays. In addition, in order to make the reading easier, I will outline a concise conclusion of the information presented in the table. Therefore, this paper underlines that the core element of the study is included within 13 holidays, comprising 29 significances, corresponding to the meanings expressed during those days.

The most common traditions including salt are those which meaning is related to protecting livestock, crops, households, fruit trees and individuals from spells or various pests - 15 traditions, within the following holidays: Saint Basil, Meeting of the Lord, Saint Theodor, Forty Saints, Easter, Saint Basil, Pentecost, *Filipii* (Saint Philip), Christmas Eve, *Alexiile* (Saint Alexios). Next in line, we find the customs invoking salt for foretelling of fate, its characteristics or the future profession - nine traditions, within the following holidays: Saint Basil, Epiphany and Christmas Eve. Taking the milk from or giving the milk back to the cattle represents a folk tradition, encountered four times in the works used for this paper; they are all included in the feast of Saint Basil. Another meaning for salt is that of blood brother/sisters present during the holidays of Saint Theodor and Small Easter. It is known even presently that folk weather forecast is based on the use of salt; within this investigation, I encountered the practice in two cases, both traditions during the Saint Basil holiday. The significance of salt for foretelling of fate is featured within two traditions, during the Saint Basil holiday. Two other practices performed during Easter and Saint Theodor holiday have the meaning of curing people. For hair growth, there are various traditions related to the Saint Theodor holiday, among which I underlined two in this paper. During Epiphany and Easter, two traditions have deep roots, and their significance is ensuring the wealth of the household. Present in daily life and one of the meanings that will never disappear, food preservation using salt was pinpointed in two traditions related to Easter and to pig sacrificing. The Saint George holiday is closely connected to sheep breeding; hence, I found two customs in which the meaning is to ensure sheep productivity. Respect for the deceased is a meaning comprised within a tradition associated to the Epiphany holiday. Another meaning of salt use - ensuring the success of the individual - is featured within a tradition related to Saint Theodor. During Easter, a certain tradition involves feeding salt to the livestock in a certain manner, to ensure multiple gestations. Salt use for purifying wells is present in daily life; for this paper, I discovered this meaning within a custom associated to Easter. Another custom comprising salt use belongs to Saint George holiday, for ensuring people's health. Livestock purification is a meaning encountered within the custom of pig sacrificing.

Therefore, I pinpoint the various symbolic uses of salt within Romanian holidays; once again, they confirm the importance of the role played by this indispensable element in the life of human communities, in both a practical and a figurative way. This study can be a starting point for a new research direction, permanently based on complex enquiries concerning the use of salt. In the opinion of renowned researchers (Alexianu *et al.*, 2007, 15), these enquiries will multiply in time, in various cultural spaces, often with great developments significant for the traditions and mindsets of various communities. This is not an exhaustive study and I intend to continue the research on various approaches to salt symbolism during holidays, as I continue reading on the matter; however, this investigation can be considered a pioneer study and an interpretative reading pattern. It is also worth highlighting the words of Simeon Florea Marian, according to whom certain customs fade in time and others take their place, but historians, anthropologists and ethnographers need to study them (Datcu I., Introducere in vol. I, S. Fl. Marian 1994, IX).

Acknowledgement
This research was financed by the project „MINERVA – Cooperare pentru cariera de elită în cercetarea doctorală şi post-doctorală", POSDRU/159 /1.5/S/ 137832, co-financed by the European Social Fund, Sectorial Operational Programme Human Resources Development Human Resources 2007-2013.

References

Alexianu, M., Weller, O. and Brigand, R. 2007. *Izvoarele de apă sărată din Moldova subcarpatică. Cercetări etnoarheologice*. Iaşi, Demiurg.

Avram, S. 2005. *Sarea – tradiție, simbol şi ritual.* Sfântu Gheorghe, Angustia.

Cavruc, V. and Chiricescu, A. (eds.) 2006. *Sarea, Timpul şi Omul*. Sfântu Gheorghe, Angustia.

Chevalier, J. and Gheerbrant, A. (eds.) 2009. *Dicționar de Simboluri. Mituri, vise, obiceiuri, gesturi, forme, figuri, culori, numere.* Iaşi, Polirom.

Chiricescu, A. 2013. *Civilizația tradițională a sării în estul Transilvaniei: raport de cercetare.* Sfântu Gheorghe, Angustia.

Dăncuţ, M. 2006. *Sarea – fragmentarium etnologicum*, in V. Cavruc and A. Chiricescu (eds.), *Sarea, Timpul şi Omul,* 147-148. Sfântu Gheorghe, Angustia.

Ghinoiu, I. (ed.) 2004. *Sărbători şi obiceiuri: Răspunsuri la chestionarele Atlasului Etnografic Român, Moldova*, Vol. IV, Romanian Academy, "C. Brăiloiu" Institute of Ethnography and Folklore, Corpus of Ethnographic Documents. Bucureşti, Editura Enciclopedică.

Hulubaş, A. 2012. *Obiceiuri de naştere din Moldova. Tipologie şi corpus de texte.* Iaşi, Editura "Alexandru Ioan Cuza".

Marian, S.Fl. 1898. *Serbătorile la români. Studiu etnografic. Cârnilegile,* vol. I. Bucureşti, Editura Academiei Române.

Marian, S.Fl. 1899. *Serbătorile la români. Studiu etnografic. Păresimile,* vol. II. Bucureşti, Editura Academiei Române.

Marian, S.Fl. 1901. *Serbătorile la români. Studiu etnografic. Cinci-decimea,* vol. III. Bucureşti, Editura Academiei Române.

Marian, S.Fl. 1994. *Sărbătorile la români. Studiu etnografic,* vol. I. Bucureşti, Editura Fundaţiei Culturale Române.

Pamfile, T. 2006. *Sărbătorile la români.* Bucureşti, Saeculum I.O.

Rusu, N.D. 2003. *Membrii Academiei Române (1866-2003). Dicţionar. Ediţia a III-a, revăzută şi adăugită cu un cuvânt înainte de Academician Eugen Simion, Preşedintele Academiei* Române. Bucureşti, Editura Enciclopedică.

Online source

Obiceiuri de primăvară – Sfântul Gheorghe http://www.cimec.ro/Etnografie/sf-gheorghe/Obiceiuri-de-primavara-Sf-Gheorghe.html

Salt: Beliefs and Practices. Some Ethnographical Case Studies from Moldavia

Roxana-Elena Diaconu
Muzeul de Istorie şi Etnografie, Târgu-Neamţ

Vasile Diaconu
Muzeul de Istorie şi Etnografie, Târgu-Neamţ

Abstract
The authors present some of the uses of salt and/or of salt water from an ethnographic point of view. The starting point is some data from villages in Neamţ County, which find parallels in other regions of Moldavia. Different ways of using salt derived from ethnographic research provide the framework of the article; these include: the symbolic value of salt especially in rites of passage (such as birth) or weather prediction, as well as more practical uses, in halotherapy, and in the workshop context. The significance of this research is that it lays emphasis on beliefs and practices and that are unique, or less well known to the public of today. Ethnographic information concerning salt and salt water usage is extremely varied and requires careful recording and subtle analysis.

Keywords
salt, beliefs, symbolism, rites of passage, halotherapy, ethnography, Moldavia

During the past few decades a multi-disciplinary Romanian research programme has concentrated on issues arising from the phenomenon of salt. The first results have been encouraging. The archaeological investigation of sites near sources of salt has produced a variety of fascinating results; we might even begin to speak of a sub-discipline, namely the prehistoric archaeology of salt (Monah 2008a, 14).

The desire to identify likely ancient techniques of salt exploitation, whether of rock salt, salt crystals or brine, and the ways in which it was transported and used, its properties and symbolism has meant that the ethno-archaeology of this mineral has become a very attractive area of study for many Romanian and foreign scholars (Alexianu *et al.* 1992; 2007; 2011; Alexianu and Weller 2007; Avram 2005; Monah and Dumitroaia 2007; Monah *et al.* 2008).

Because salt was used so extensively, it is difficult to identify all the effects that it had on communities in different times and places.

Although the question of salt requires an interdisciplinary approach, some directions of study lack a consistent documentary base.

Even if matters are relatively easy to understand from an archaeological perspective, given the existence of artefacts and stratigraphy, the ethnography aspect requires more detailed research.

We noticed that in the sub-Carpathian area of Moldavia (but also in other areas of Romania) there was a continuity of practice relating to the exploitation of saline springs that had their origins in pre-industrial times (Alexianu *et al.* 2007, 8). It is necessary therefore to bring together the rich informative material through the recording of different kinds of ethnographic evidence.

Two case studies arising from the ethnographic data recorded in different parts of Moldavia are treated here; in particular in the sub-mountainous area of Neamţ County (an area well known for its numerous sources of salt, especially saline springs; Alexianu *et al.* 2007; Weller *et al.* 2007; 2010).

Part of the information that follows comes from the ethnographic literature, but we also introduce unpublished or unfamiliar data that we obtained through interviews with number of persons from the villages in Neamţ, Bacău and Botoşani Counties (Figure 1).

The archaeological and ethnographical literature frequently refer to salt as a resource principally used for human and animal food. The chemical composition of salt and the numerous forms it takes indicate that usage varies considerably (Monah 1991; Alexianu and Weller 2007, 307-311; Alexianu *et al.* 1992; 2008, 166-174). Within the framework of the traditional village, sodium chloride was used as a preservative for food or various organic materials (such as animal skins), for curing disease, both human and animal, for making soap, as an exchange commodity, and also in rituals and customs related to health, welfare and protection from evil (Alexianu *et al.* 2007, 103-134; Alexianu *et al.* 2011, 17-19).

Two factors emerge when we examine the topic of salt or saline water usage, namely practical use and symbolic value (Bonte and Izard 1999; Cavruc 2010, 10). It is usually difficult to draw a line between the two because most people's daily activities were guided by a wide range of beliefs and superstitions.

We shall present two ethnographic examples related to salt or saline water usage without necessarily attempting to provide a comprehensive repertory of every known custom down to the present. We believe that it is useful just to stress unfamiliar aspects that generally involve the symbolic value of salt, although we have not overlooked certain aspects related to the practical use of salt among the traditional communities of Moldavia (Figure 2).

No.	Informant	Age/sex	Residence	Occupation	Information on salt usage
1	Buhăceanu Mihai	57/M	Hăneşti, Hăneşti commune Botoşani County	beekeeper	halotherapy
2	Ciocârlan Maria	62/F	Topoliţa, Grumăzeşti commune, Neamţ County	housewife	preservation weather prediction
3	Cozma Gheorghe	80/M	Humuleşti-Târgu Neamţ, Neamţ County	blacksmith	crafts
4	Cuciureanu Ortansa	72/F	Hăneşti, Hăneşti commune, Botoşani County	housewife	symbolic value
5	Diaconu Constantin	72/M	Humuleşti-Târgu Neamţ, Neamţ County	shepherd	alimentation crafts halotherapy
6	Mosor Neculai	67/M	Humuleşti-Târgu Neamţ, Neamţ County	shepherd	halotherapy alimentation crafts
7	Năvrăpescu Aurelia	70/F	Târgu Neamţ, Neamţ County	housewife	halotherapy weather prediction
8	Năvrăpescu Constantin	65/M	Târgu Neamţ, Neamţ County	craftsman	alimentation symbolic value
9	Păscăriţa Ileana	90/F	Prăjeşti, Traian commune, Bacău County	midwife	alimentation halotherapy
10	Tanasă Gheorghe	57/M	Oglinzi, Răuceşti commune, Neamţ County	worker	crafts

Figure 1. The individuals who supplied ethnographic information.

1. Birth, christening and salt

One of the most important rites of passage is birth. Since ancient times this moment was accompanied by numerous traditions, beliefs and practices that aimed to protect newborn children and their mothers, and also had the objective of preparing them for life and including them in the community to which they belonged.

Because the totality of symbolic action associated with birth rites is highly uniform (Burada 1978; Marian 1995; Ţârcomnicu 2009, 15, 75), we shall only focus attention on rites that involve the use of salt.

Dan Monah has recently introduced such a rite into the scholarly debate, namely the salting of the new born, a rite widely practiced since Antiquity by the Macedo-Romanian communities of Greece, Macedonia, Albania and Bulgaria (Monah 2008b, 117-131).

According to Monah, in the Macedo-Romanian culture babies were sprinkled with salt immediately after birth and kept like this for twelve hours; they were then washed with warm water and wine (Monah 2008a, 126). There were both medical and hygienic reasons for this, but the rite also possessed a quite complex symbolism that is difficult to enunciate at present. Although Monah states that the practice of salting the newborn is poorly attested north of the Danube (Monah 2008b, 127), there is some information that confirms the use of sodium chloride in beliefs and practices related to small children.

Simion Florea Marian indicates that in Valahia, salt was added to infants' first bath in order not the scald the new born children (Marian 1995, 59). Also in the period between the birth of a baby and its christening, visitors to a new born child's family brought to the mother an alcoholic drink (*rachiu* in Romanian) mixed with honey and a piece of salt (Marian 1995, 89). He also describes how on the day of the christening the midwife puts bread and salt in the baby's cot so that the infant would have bread and salt, the staples of a healthy diet, for the rest of its life (Marian 1995, 112). In the Bucovina region, also on the day of the christening, a child's godparents would give the priest a chicken, a bottle of alcohol (*rachiu* again), and salt (Marian 1995, 114).

Some 18th century sources indicate that in some regions in Moldavia newborn children were sprinkled with salt to prevent them from catching smallpox, but it was observed that the custom was ineffective (Samarian 1938, 351).

In the universe of the traditional Romanian village, there was also the belief that the baby should be bathed in salt water at birth in order for him or her to be protected from spells and the Evil Eye (Ciauşanu 2007, 230-231).

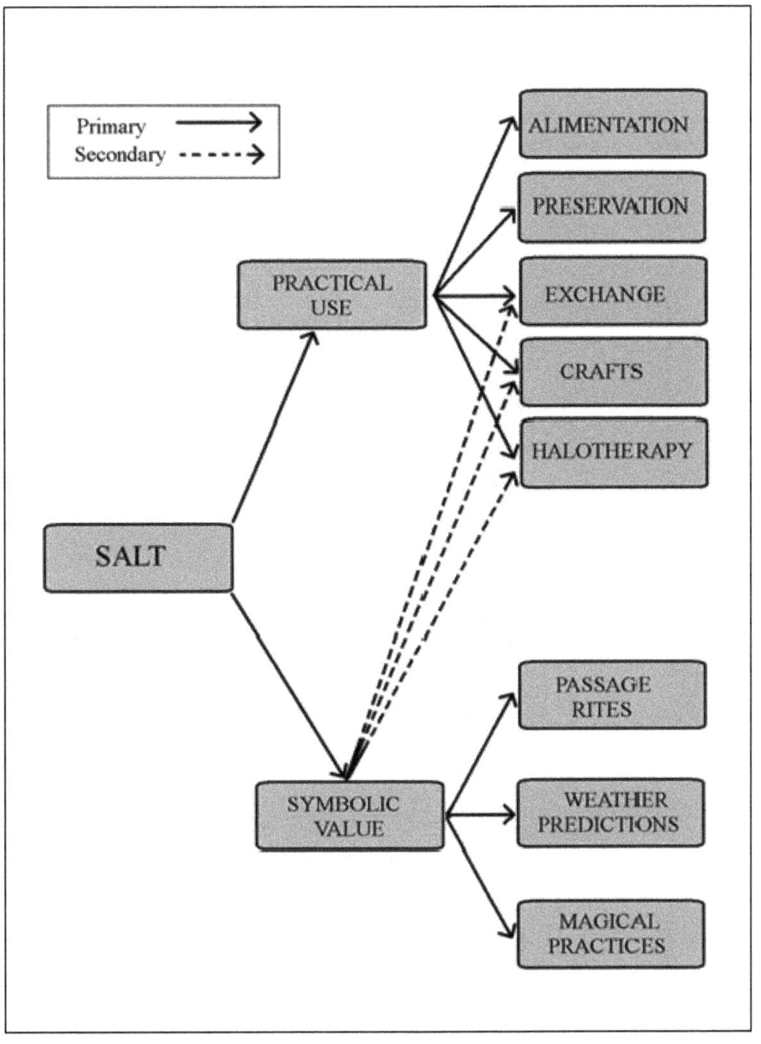

Figure 2. Salt usage based on ethnographic information.

Then there is another custom that should not be overlooked. When an infant could not sleep and it was considered that its "sleep was stolen", the mother would take it in her arms and turn towards the forest or a tree and would throw pieces of bread and salt while whispering an incantation aiming to reassure the child (Marian 1995, 230-231).

2. Halotherapy

It is well known that salt is a mineral with strong therapeutic properties, a fact that has long been known judging by observations of ancient writers (Curcă 2007, 259-269). Monah's study shows that medical practices involving salt were widespread and many have survived until the present. The recording of such information through ethnographic investigation constitutes solid evidence for the world of the Romanian traditional village having archaic customs with therapeutic purposes that would often involve salt or salt water.

There is no point in recounting here the extensive literature relating to halotherapy (Curcă 2007; Alexianu *et al.* 2008, 167-168; Sandu *et al.* 2010), but we should mention that salt and/or salt water were employed for rheumatic diseases, eye disorders, inflammation, wounds, headaches and toothache (Cantimir 1893, 7, 13; Chiricescu 2006, 163; Dăncuş 2006; Curcă 2007; Alexianu *et al.* 2007, 112-121; Sandu *et al.* 2010).

We should add a treatment known to some shepherds according to which melted butter is mixed with salt to produce an ointment good for treating cuts and burns in humans (Diaconu Constantin, *pers. comm.*).

We also recorded the information that a mixture of vegetable oil and salt was also recommended for the treatment of burns (Năvrăpescu Aurelia, Păscărița Ileana, *pers. comm.*).

3. Salt and weather

In the world of the traditional village, people had different ways of predicting changes in the weather that had a direct influence on agricultural activity. Some of these methods of prognostication had a saline element. For example, when salt in the salt cellar (*sărăriță* in Romanian) became damp, it was a sign that rainy weather was on the way.

An interesting custom known in several parts of Moldavia was the "onion calendar" (Figure 3). At the beginning of the year, on Saint Vasile's Eve (December 30 - January 1st) people would draw up such this calendar in order to predict the periods when it might rain during the coming year. The procedure was as follows: they wrote the name of every month on a piece of paper and then cut an onion into four. A teaspoon of salt was placed on every slice of onion and left until the next morning. The pieces of onion on which the salt had melted indicated the months when rain might be expected. These were written down and constituted landmarks for agricultural activity (Năvrăpescu Aurelia, Buhăceanu Mihai, Ciocârlan Maria, *pers. comm.*).

Our informants maintained that predictions made at the beginning of the year by means of the "onion calendar" were in many cases true.

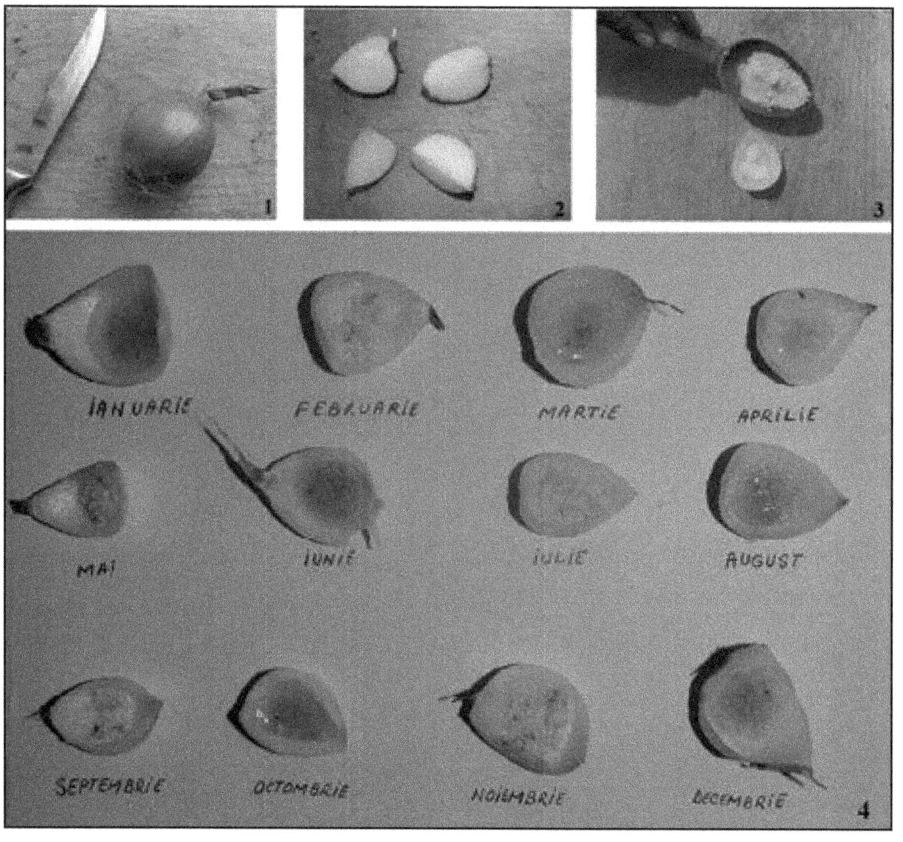

Figure 3. The 'onion calendar'.

4. Salt and domestic animals

Salt on the farm was mainly employed in animal feed (Cavruc 2006, 33), but we recorded cases when salt also had medical attributes. We discovered that ovine eye inflammation (*albeață*) might be treated by sprinkling a mixture of salt and finely crushed snail shell in an animal's eyes (Mosor Neculai, *pers. comm.*). Our informant told us that when he was in Dobruja almost 30 years ago, the mixture was made from finely crushed conch shells.

We know of the custom of salting recently born calves and lambs in the Bucovina region (Marian 1995, 60). Some shepherds in the north-eastern part of the Neamț County, unaware of the practice, believed that it was done because new born animals had to be licked by their mothers in order to be accepted, something that would ensure their survival and normal growth.

Moreover, when milk is given or sold in some villages from Botoșani County, "a dash" of salt is sprinkled into the vessel so that the "mana" (a popular term meaning the capacity of the animals to produce enough milk) does not leave the boundaries of the house (*hotarul* or *vatra casei*) (Cuciureanu Ortansa, *pers. comm.*). Ethnographic testimony indicates that in the past "mana" was made an incantation in order to protect the animal; people who uttered this incantation had to carry sharp objects (a spindle, a bodkin, or an iron fork), and a small bag containing bran, salt and a copper coin (Pamfile 1909, 19-20). We could not discover a similar practice in Neamț County.

In certain regions people say that it is not a good idea to steal salt from animals because they will then all die (Niculiţă-Voronca 1998, 152). There is also the belief that salt should be put on the back of a newborn calf to prevent it from being bewitched (Ciauşanu 2007, 230-231).

In the ethnographic literature there is also a reference to the custom of not offering salt, corn flour or sauerkraut on Saint Philip's day (14 November) because it was not good for the animals in the household (Pamfile 1997, 212). None of the people we interviewed knew of such a practice.

Another interesting, but less common custom, is the hiding of a lump of salt covered in cloth beneath the doorstep on Christmas Eve. The lump remained there until the spring (in fact until April 22 when the sheep were herded together for transhumance), when it was pulled out, broken up and mixed with the animal's food (Pamfile 1997, 282).

At Christmas, in the Roman region (Neamţ County), when pigs were sacrificed, it was common to incise a cross covered in salt on the animal's back so that the meat "would be received by God" (Pamfile 1997, 390). At present, in the sub-Carpathian area of the Neamţ County, the pig is covered in salt after the sacrifice, but this practice carries no symbolic significance; instead a cross is made on its head or on its back (Năvrăpescu Constantin, Diaconu Constantin, *pers. comm.*).

5. Salt and crafts

As a result of ethnographic inquiry, but also from the specialized literature, we discovered that salt was employed in various traditional crafts. For example, skinners used it at various stages of their activity. After skinning, the animal skins had to be salted to preserve them, and then they were dried.

Traditional tanning also involved the use of salt as part of the liquid (a mixture of salt, whey and bran) in which animal skins were soaked so that they could subsequently be washed and processed (Vulcănescu 1973, 459-462).

In 2011 we recorded one piece of information from a blacksmith (Cozma Gheorghe, *pers. comm.*) who told us that the clay that was applied to the hearth of a forge had to contain salt so that it would last without cracking. Also connected with iron processing was the information that in order to harden metal it was dipped in salt water after forging (Tanasă Gheorghe, *pers. comm.*).

Salt was also employed, with an obvious symbolic value, in the construction of stoves in peasants' houses. People say that when these heating installations were being built they placed inside them some Bird Cherry (*Padus avium Mill.*) and pigs' bristles, and put salt on the hearth, (*o huscă de sare*) (Pamfile 1909, 445).

Conclusions

Although we did not intend to analyze all the activities and customs involving the use of salt, the ethnographic information that we assembled attests to the complexity of the symbolic contexts in which salt was employed. Nevertheless we emphasize the fact that we need rather more information in order to permit a more general view. Even though our information comes from a small number of informants, the fact that it has parallels in other parts of the country indicates that some customs were diffused over a wide area.

Future ethnographic and ethno-archaeological research should analyse whether some customs and beliefs involving salt are stronger and better preserved in areas where salt is found or whether they are to be found over a wider area. With the information that we had at our disposal we were able to observe that the practical uses of salt were better known than any symbolic value it might have. This can be explained both by the weakening of village traditions and the disappearance of some ethnographic sources (such as spells, legends, or traditional songs).

Future researchers should record a larger assemblage of ethnographic information that should be classified and structured on well-defined criteria so that we might infer or confirm some activities related to the use of salt from the prehistoric and protohistoric periods.

References

Alexianu, M., Dumitroaia, Gh. and Monah, D. 1992. Exploatarea surselor de apă sărată din Moldova: o abordare etnoarheologică. *Thraco-Dacica* 13/1-2, 159-167.

Alexianu, M., and Weller, O. 2007. Recherches ethnoarchéologiques sur le sel: les enquêtes de 2004 et les premières résultats obtenus dans la zone de Poiana Slatinei à Lunca (dep. Neamţ, Roumanie), in D. Monah, Gh. Dumitroaia, O. Weller and J. Chapman (eds.), *L'exploitation du sel à travers le temps*. Bibliotheca Memoriae Antiquitatis, XVIII, 299-318. Piatra Neamţ, Constantin Matasă.

Alexianu, M., Weller, O. and Brigand, R. 2007. *Izvoarele de apă sărată din Moldova Subcarpatică. Cercetări etnoarheologice.* Iaşi, Casa Editorială Demiurg.

Alexianu, M., Weller, O., Brigand, R., Curcă, R.-G. 2008. Cercetări etnoarheologice asupra izvoarelor de apă sărată din Moldova Subcarpatică. *Tyragetia* S.N., 2 (17), 1, 155-177.

Alexianu, M., Weller, O., Brigand, R., Curcă, R.-G., Cotiugă, V., Moga, I., 2011. Salt springs in today's rural world. An ethnoarchaeological approach in Moldavia (Romania), in M. Alexianu, O. Weller and R.-G. Curcă (eds.), *Archaeology and Anthropology of Salt: A Diachronic Approach. Proceedings of the International Colloquium, 1-5 October 2008, Al. I.*

Cuza University (Iaşi, Romania), 7-23. British Archaeological Reports, International Series 2198. Oxford, BAR Publishing.

Avram, S. 2005. Sarea - tradiţie, simbol şi ritual. *Angustia* 9, 241-243.

Bonte, P. and Izard, M. 1999. *Dicţionar de etnologie şi antropologie*. Bucureşti, Polirom.

Burada, T.T. 1978. Obiceiuri la naşterea copiilor poporului român din Macedonia. In *Opere. Folclor şi Etnografie*, Volume III. Bucureşti, Editura Muzicală.

Cantimir, D. 1893. *Sărurile minerale de Bălţăteşti (Composiţia chimică şi întrebuinţarea lor)*. Bucureşti.

Cavruc, V. 2006. Sarea în societăţile arhaice. Consideraţii generale, in V. Cavruc and A. Chiricescu (eds.), *Sarea, timpul şi omul*, 33-36. Sfântu Gheorghe, Angustia.

Cavruc, V. 2010. Consideraţii privind schimbul cu sare în mileniile VI-II î. Hr. în spaţiul carpato-dunărean. *Tyragetia* S.N., 4 (XIX), 1, 7-34.

Ciauşanu, Gh. F. 2007. *Superstiţiile poporului român. În asemănare cu ale altor popoare vechi şi noi*. Bucureşti, Editura Saeculum I.O.

Chiricescu, A. 2006. Exploatarea tradiţională a sării în Transilvania, in V. Cavruc and A. Chiricescu (eds.), *Sarea, timpul şi omul*, 158-163. Sfântu Gheorghe, Angustia.

Curcă, R.-G. 2007. La halothérapie - textes antiques et pratiques actuelles dans la Roumanie Orientale, in D. Monah, Gh. Dumitroaia, O. Weller, J. Chapman (eds.), *L'exploitation du sel à travers le temps*. Bibliotheca Memoriae Antiquitatis, XVIII, 259-269. Piatra Neamţ, Constantin Matasă.

Dăncuş, M. 2006, Sarea - *fragmentarium etnologicum*, in V. Cavruc and A. Chiricescu (eds.), *Sarea, timpul şi omul*, 147-148. Sfântu Gheorghe, Angustia.

Marian, S.Fl. 1995. *Naşterea la români*. Bucureşti, Editura Grai şi Suflet - Cultura Naţională.

Monah, D. 1991. L'exploitation du sel dans les Carpates orientales et ses rapports avec la culture de Cucuteni-Tripolye, in V. Chirica and D. Monah (eds.), *Le Paléolithique et le Néolithique de la Roumanie en contexte européen*, 387-400. Iaşi, Bibliotheca Archaeologica Iassiensis, IV.

Monah, D. 2008a. Arheologia preistorică a sării în România. Scurt istoric, in D. Monah, Gh. Dumitroaia and D. Garvăn (eds.), *Sarea, de la prezent la trecut*. Bibliotheca Memoriae Antiquitatis, XX, 13-39. Piatra Neamţ, Constantin Matasă.

Monah, D. 2008b. O confirmare etnografică a versetului Iezecheil 16/4. Sărarea noilor născuţi, in D. Monah, Gh. Dumitroaia and D. Garvăn (eds.), *Sarea, de la prezent la trecut*. Bibliotheca Memoriae Antiquitatis, XX, 117-131. Piatra Neamţ, Constantin Matasă.

Monah, D. and Dumitroaia, Gh. 2007. Recherches sur l'exploitation préhistorique du sel en Roumanie, in D. Monah, Gh. Dumitroaia, O. Weller and J. Chapman (eds.), *L'exploitation du sel à travers le temps*. Bibliotheca Memoriae Antiquitatis, XVIII, 13-34. Piatra Neamţ, Constantin Matasă.

Monah, D., Dumitroaia, Gh. and Nicola, D. 2008. Noi investigaţii etnoarheologice asupra izvoarelor sărate de pe Valea Muntelui, in D. Monah, Gh. Dumitroaia and D. Garvăn (eds.), *Sarea, de la prezent la trecut*. Bibliotheca Memoriae Antiquitatis, XX, 80-101. Piatra Neamţ, Constantin Matasă.

Niculiţă-Voronca, E. 1998. *Datinele şi credinţele poporului român adunate şi aşezate în ordine mitologică*. Volume I. Iaşi, Polirom.

Pamfile, T. 1909. *Industria casnică la români. Trecutul şi starea ei de astăzi. Contribuţiuni de artă şi tehnică populară*. Bucureşti, Librăria Socec & Comp.

Pamfile, T. 1997. *Sărbătorile la români*. Bucureşti, Saeculum I. O.

Samarian, P.Gh. 1938. *Medicina şi farmacia în trecutul românesc*. Volume III. Bucureşti, Bucovina.

Sandu, I., Poruciuc, A., Alexianu, M., Curcă, R.-G., Weller, O. 2010. Salt and human health: science, archaeology, ancient texts and traditional practices of Eastern Romania. *The Mankind Quarterly* L, 3, 225-256.

Ţărcomnicu, E. 2009. *Obiceiuri şi credinţe macedoromâne. Dicţionar*. Bucureşti, Biblioteca Bucureştilor.

Vulcănescu, R. 1973. Cojocăria, in R. Vuia *et al.* (eds.), *Etnografia Văii Bistriţei*, 447-518. Piatra Neamţ.

Weller, O., Brigand, R. and Alexianu, M. 2007. Cercetări sistematice asupra izvoarelor de apă sărată din Moldova. Bilanţul explorărilor din anii 2004-2007 efectuate în special în judeţul Neamţ. *Memoria Antiquitatis* XXIV, 121-190.

Weller, O., Brigand, R. and Alexianu, M. 2010. Recherches systématiques autour des sources salées de Moldavie. Bilan des prospections 2008-2010. *Memoria Antiquitatis* XXV-XXVI (2008-2010), 437-504.

Paradigmatic Presentations of Salt in the Romanian Language and Folklore

Adrian Poruciuc
Universitatea „Alexandru Ioan Cuza" din Iaşi

Abstract
The first part of the present article contains the author's observations on the way the Romanian language reflects the fact that salt is not simply an important asset of Romania, but also a symbolic item. In the second part, the author's focus is on two meaningful presentations of salt in two pieces of Romanian folklore, namely a folktale and a riddle. The former presents salt as a vital nutriment, whereas the latter recalls an archaic way of obtaining salt.

Keywords
salt, Romanian phrases, Romanian folklore, folktale, riddle, ethnological aspects

Romanian and Aromanian salt-phrases
Romania is a country quite rich in salt, as one may conclude from the multitude of place names that mark salt mines and/or sources of natural brine – salt marshes, springs and wells – in various regions of Romania (cf. Poruciuc 2011). Under such circumstances it is not surprising that the Romanian noun *sare* 'salt' – from Vulgar Latin *sale(m)* – has many derivatives and occurs in many phrases, of which I will present only a concise selection below.

Romanian is certainly not the only language in the world that has phrases with conceptual-metaphorical references to salt – for instance, as an allusion to the *New Testament* ("Ye are the salt of the earth" – Matthew, 5, 13) speakers of English will apply the label *the salt of the earth* to ordinary-but-good-and-honest persons just as speakers of Romanian designate the same kind of persons as *sarea pământului* (with the same meaning as the English phrase). Nevertheless, Romanian has its specific usages of the term *sare*, as I will point out in this introductory part of my article. Since the two main dialectal branches of Romanian are Daco-Romanian (that is, the North-Danubian basis of today's Standard Romanian) and Aromanian (spoken in several South-Danubian countries, mainly in Greece, Albania and Macedonia)[1], I decided to select two series of illustrative examples of the ones given in the *sare* entries of a dictionary published by the Romanian Academy (MDA, IV, 2003) and of Papahagi's Aromanian dictionary (1974), respectively.

To begin with the latter dictionary (for my own reasons), I will observe that Papahagi's series of examples (1974, s.v. *sare*) is opened by an ethnographically relevant phrase that reflects the dominantly pastoral way of life of the Aromanians: *ca oile după sare alagă* ("they run like sheep after salt"). The second example refers to how true friendship could be built up within the context of Aromanian traditional culture: *are mîncată cu mine pîne cu sare* (literally, "he's eaten bread and salt with me" = "we are close friends"). The third example is actually based on what Papahagi gives as a specific secondary meaning of Aromanian *sare*, namely 'character': *aşi lă-i sarea a lor* (literally, "so is their salt" = "that's just like them"). Finally, Papahagi's fourth example shows salt-and-water as symbol of dissolution and oblivion: *hai tora, apă şi sare tute cîte dzîsim* (literally, "come now, water and salt [be] everything we've said" = "may everything we've said be forgotten").

In their turn, the authors of MDA included a comprehensive series of Daco-Romanian phrases in their *sare* entry. For instance, the rhyming formula *sarea şi marea* ("the salt and the sea") designates a big crowd. (I am surprised that the entry under discussion did not also mention the quite frequently used phrase *a promite marea cu sarea* – literally, "to promise the sea with the salt" – commonly said about wildly exaggerated promises.) Another phrase, *ca câinelui a linge sare* ("as a dog [likes] to lick salt") ironically indicates someone's extreme reluctance to do something. Further, the combination bread-and-salt is shown to have (as in Aromanian – see above) ritual-cultual implications, as manifest in the Daco-Romanian phrase *a primi cu pâine şi cu sare* (literally, "to receive with bread and salt" = "to receive with honors")[2]. The same entry of MDA gives two phrases that contain the combination *sarea-n bucate* ("salt in vittles")[3], with two quite different senses. The former is *a se amesteca în toate ca sarea-n bucate* (literally, "to get mixed up in everything like salt in vittles" = "to get involved in an unwanted way"). As for the latter phrase (much more "positive"), it is given as *a fi ca sarea-n bucate*

[1] There are people – including a number of specialists – who consider Romanian and Aromanian (also known as Macedo-Romanian, or Vlahica) as two distinct languages; but this is an issue that I will not dwell upon in the present article.

[2] That the East Slavs have similar customs is quite clearly indicated by the Russian adjective *khlebosol'nyj* 'hospitable' (< *khleb* 'bread' + *sol'* 'salt').

[3] I chose to translate the obsolete Romanian plural *bucate*, ultimately based on Latin *bucca* 'mouth' (cf. French *bouche* 'mouth') by nonstandard English *vittles*, rather than by *victuals* or *dishes*. I feel that my choice is sustained etymologically too, taking into consideration that English *vittles* comes from Old French, which also had two variants, *victuailles* and *vitailles* (« provisions de bouche, vivres »), both of which derived from Late Latin *victualia*, ultimately based on the verb *vivere* 'to live' (cf. Baumgartner and Ménard 1996, s.v. *victuaille*). Therefore, *vittles* rather transparently designates vital nutriments.

(literally, "to be like salt in vittles" = "to be perfectly suitable"); and such a phrase provides a perfect introduction to the central part of my article.

A Romanian folktale about salt

Romanian literature has nothing comparable to Shakespeare's *King Lear*. But a Romanian folktale with a King-Lear motif had the chance to be recorded more than a century ago. It was published exactly under the title *Sarea în bucate* (*Salt in Vittles*), by Petre Ispirescu, in 1887[4]. I will do my best to summarize that folktale, whose main characters are a nameless emperor[5] and his three nameless daughters. The youngest of them, we shall see, infuriated her father, unintentionally, by a statement whose sphere of reference was much more concrete that the one of Shakespeare's Cordelia.

So, once upon a time there was an emperor who had three daughters. One day the emperor felt like asking his eldest daughter how much she loved him; and she said that her love for him was "like honey" (*ca mierea,* in the original). Then he asked his second-born daughter the same question; and she said "like sugar" (*ca zaharul*). As the emperor was quite happy with such answers, he eventually looked at his youngest (rather bashful) daughter, and he repeated the question. Her answer was: *Ca sarea în bucate, tată!* – "Like salt in vittles, father!" He flew into a rage, and he burst out: "Get away from me, with your salt and all!" The banished daughter, disguised as a maidservant, had to flee to another country, where she had the chance to get a job at the court of the local emperor. It so happened that the emperor's son fell in love with the diligent maidservant, and he persuaded his imperial parents to allow him to marry her.

The final part of the tale is about a grand wedding party, to which the bride's own father was duly invited. The bride (because her face was veiled, I suppose) was not recognized by her father during the party. But, before that, she had been quite busy in the kitchen: she had herself cooked special dishes for her father, without any salt in them; and she had even arranged that the saltcellar on his table should contain sugar, not salt. It stands to reason that – while all the other guests obviously relished their food – the bride's father got angrier and angrier about the tasteless dishes served to him. Finally he shouted at the bridegroom's father: "Look here, Emperor, did you invite me to your son's wedding only to make a fool of me?" When the bridegroom's father heard about the cause of his guest's anger, he wanted to punish the cooks on the spot. But the bride stood up and said: "I am the one who cooked the dishes for the emperor who has got so angry…This emperor is my father…I wanted to show him that a man can live without honey and sugar, but not without salt; that is why for him I've cooked vittles without salt." And, as the happy ending of our tale has it, the bride's father had to admit that "he had failed to value his daughter's wisdom, and he begged for her forgiveness.[6]" I consider that the Romanian anonymous-illiterate maker(s) of the above-presented tale[7] managed to express, the best way they could, the very idea of salt as basic ingredient. It so happens that the Romanian riddle I am going to present next also shows emperors to be very fond of salt.

A singular Romanian salt-riddle

The second paradigmatic example of my choice is quite special, not only because the text represents another genre, but also because the kind of salt it refers to is not (or, rather, no longer) in common use in Romania. The text under discussion, included (under nr. 1599) in Gorovei's authoritative collection of 1898[8], actually is the only Romanian riddle given as an illustrative example in the chapter entitled *Sarea* (*Salt*) in that volume. Another remarkable aspect is the first-person voice of the riddle, which thus recalls some of the most archaic European texts[9]. Here is my translation of Gorovei's salt riddle: Water gives birth to me,
The sun brings me up,
Emperors and noblemen love me,
And, when I meet my mother,
I fall down and melt[10].

Above I have suggested that this riddle refers to an uncommon kind of salt. What I mean is that the typical varieties of salt known to most Romanians are rock-salt

[4] The folktale under discussion was included in the edition Ispirescu 1966, which I made use of for this article.

[5] For the sake of non-Romanian readers, I feel obliged to explain (as I did, in more detail, in endnote 18 of Poruciuc 2010, 120) that he dominant figures of most Romanian folktales are not kings, but emperors (*împăraţi*) – which certainly is a lingering echo of the Eastern Roman Empire.

[6] I translated after Ispirescu 1966, 399-400.

[7] I do not neglect the fact that nineteenth-century folklorists were commonly inclined to "improve" the language of tale-tellers, so I cannot claim that *Salt in Vittles* was published exactly in the form heard by Ispirescu. But I have no doubt that the basic motif and the "arguments" of the parable are the original ones.

[8] I made use of the 1972 edition of Gorovei's collection. Actually, the salt riddle I discuss in the present article does not represent Gorovei's field investigation, since – as indicated by him in a footnote (1972, 365) – the riddle had been included in a booklet published by G. T. Bulgărescu in Bucharest, in 1887.

[9] To confine myself only to Old English texts, I will first mention the Christian poem *The Dream of the Rood*, in which – quite "originally" one could now say – the Cross itself makes a first-person presentation of the Crucifixion: "I raised the powerful king,/ the lord of heavens – I durst not bow;/ they pierced me with dark nails" (my translation after the original given in Whitelock 1990, 155). The same convention is manifest in Old English riddles (some of which followed Latin models); in such texts "things" present their own distinctive features in first-person statements, and they expect to be "guessed" (see riddles of the Exeter Book in Whitelock 1990, 170-173). As for Gorovei's collection of riddles, I will mention that in many of them wooden objects (a stick, a club, a flute, an icon, etc.) conventionally make first-person presentations that start with the lines *În pădure născui,/ În pădure crescui* – literally, "In the forest I was born,/ In the forest I grew."

[10] In the original (Gorovei 1972, 365): *Apa mă naşte,/ Soarele mă creşte,/ Împăraţi şi domni mă iubesc,/ Şi cînd pe mama întîlnesc,/ Cad jos şi mă topesc.*

from mines and (regionally) salt obtained by boiling natural brine until a special kind of salt, known as *huscă*[11], remains on the bottom of the cauldron or pot that is used for the process. The riddle above seems to refer rather to the technique specific to the Aegean and Mediterranean *salinae*, where the salt-obtaining process can be defined as "solar evaporation of marine waters" (Carusi 2007, 228). In that respect, one could even presume that the original version of our riddle was actually Aromanian, because what Aromanian shepherds will sprinkle on the "lick-stones" placed close to their sheepfolds is "mostly pure sea-salt" (Kahl 2007, 240)[12]. However, Gorovei's riddle refers only to *water* and *sun*, not to any *sea*. Therefore, I am inclined to consider that the Romanian riddle under discussion makes reference to the primeval method of getting salt crust formed by solar evaporation around sources of natural salt water. It is the technique still practiced, for instance, around Tuz Gölü ("Salt Lake"), in Central Anatolia: what the inhabitants of that region do implies basically "breaking up of the salt crust [...] and processing it into powder with the aid of grinding stones" (Erdoğu and Özbaşaran 2007, 140).

To return to the Romanian singular salt-riddle, it is worth mentioning that Gorovei attached three comparable riddles recorded in other languages; and he meticulously indicated his sources. The first is a French one (or, rather, a Breton one, since it was taken from L. F. Sauvé's « Devinettes bretonnes », *Revue celtique*, IV, 1884, nr. 13); and it expressly refers to sea-salt:

Ma mère est la mer, et elle ou sa sœur est toujours mon bourreau.
[My mother is the sea, and she or her sister is always my hangman][13].

The second riddle is Italian (extracted by Gorovei from Eugène Rolland's *Devinettes ou énigmes populaires de la France*, Paris, 1877, 159); it somehow recalls the idea of salt-needed-at-table that is also visible in the Romanian folktale discussed above:

Quall'è quella cosa, che ogni persona ne vuole in tavola e sua madre lo disfà? – Il sale.
[Whichever is the thing that everybody wants at table and its mother undoes it? – Salt.]

The final and most interesting example is a Russian riddle, in French translation (from *Revue des traditions populaires*, VII, p. 756):
Né dans la terre, baptisé dans le feu, tombe-t-il dans l'eau, il disparaît entièrement.
[Born in the earth, baptized in the fire – if it falls into the water, it vanishes wholly.]

The answer to the last riddle should be quite easy for Romanian inhabitants of the northeastern sub-Carpathians, where the technique of obtaining *huscă* (that is, salt-by-fire) is still familiar. I would not be surprised to learn that the Russian riddle above was actually created by Slavs of the regions close to the Carpathians.

Brief conclusions

In a salt-rich country like Romania it is not surprising that *sare* is known to be not only as mere designation of a vital nutriment, but also as a symbolic "ingredient" in expressions of folk wisdom. The fixed phrases mentioned in the introductory part of the present article are proof of the fact that *sare* is a metaphorically strong term in the language of both Romanians and Aromanians. As for the two pieces of folklore I have discussed above, they are two notable paradigmatic examples, but certainly not the only ones that can illustrate the symbolic potential of salt within the system of Romanian traditional culture.

References

Baumgartner, E. and Ménard, P. 1996. *Dictionnaire étymologique et historique de la langue française.* Paris.

Carusi, C. 2007. Le commerce du sel dans l'Antiquité grecque, in D. Monah, Gh. Dumitroaia, O. Weller, J. Chapman (eds.), *L'exploitation du sel à travers le temps*, 221-233. Piatra Neamţ, Editura Constantin Matasă.

Erdoğu, B. and Özbaşaran, M. 2007. White, Pure, Immaculate and Incorruptible: Salt in Prehistoric Central Anatolia, in D. Monah, Gh. Dumitroaia, O. Weller, J. Chapman (eds.), *L'exploitation du sel à travers le temps*, 135-145. Piatra Neamţ, Editura Constantin Matasă.

Gorovei, A. 1972. *Cimiliturile românilor.* Bucureşti, Editura Eminescu.

Ispirescu, P. 1966. *Zîna zînelor – Basme, legende, snoave.* Bucureşti, Editura pentru Literatură.

Kahl, T. 2007. *Hirten in Kontakt – Sprach- und Kulturwandel ehemaliger Wanderhirten (Albanisch, Aromunisch, Griechisch).* Wien/Berlin.

Papahagi, T. 1974. *Dicţionarul dialectului aromân.* Bucureşti.

Poruciuc, A. 2007. Romanian *huscă* ('salt obtained by evaporation of salt water') explained as an Old Germanism in D. Monah et al. 2007, 319-324.
Poruciuc, A. 2010. *Prehistoric Roots of Romanian and Southeast European Traditions.* Sebastopol, CA.

Poruciuc, A. 2011. Etymological and Historical Implications of Romanian Place Names Referring to Salt, in: M. Alexianu, O. Weller and R.-G. Curcă, (eds.), *Archaeology and Anthropology of Salt. A Diachronic*

[11] In a previous article (Poruciuc 2007) I demonstrated that Romanian *huscă* 'salt obtained by evaporation of salt water' most probably is an Old Germanic loan.
[12] As for Daco-Romanian shepherds, in my childhood I saw them place big lumps of rock-salt around their sheepfolds, for a more direct kind of licking.
[13] I translated the three non-Romanian riddles as literally as I could.

Approach – Proceedings of the International Colloquium, 1-5 October, 2008, "Al. I. Cuza" University (Iaşi, Romania), 215-218. BAR International Series 2198. Oxford, BAR Publishing.

Whitelock, D. (ed.) 1990. *Sweet's Anglo-Saxon Reader.* Oxford.

Abbreviations

MDA – *Micul dicţionar academic.* 2003. Vol. IV (Pr-Z), eds. Marius Sala et al. Bucureşti.

The Potential Toponymic Field of Salt Springs in Romanian Microtoponymy

Marius Alexianu
Universitatea „Alexandru Ioan Cuza" din Iaşi

Abstract
This study is dedicated to the myriad and varied toponymy generated in Romanian by salt springs, by uniting the semantic and toponymic approaches. The recourse to the theory of toponymic fields led to the elaboration of a place-naming model that represents the potential toponymic field through polarization and differentiation of halocrenonyms. The connection established by the human mind between discrepant geographical entities (salt springs, mountains, hills, creeks, roads, etc.) evidences the attempt to anthropize the natural environment, by mentally organizing the landscape around a core element (the salt spring), meant to appropriate and ultimately dominate it. Testing this model in other languages can challenge or confirm (albeit partially) it.

Keywords
Romanian language, salt springs, toponymic field, micro-toponimy, place-naming model

The Romanian studies on salt slowly progressed from those of a geographical and geological nature to those treating historical, ethnographical and archaeological problematics. Very recently, ethnoarchaeological research has also started to be conducted here, a spearheading historical science that brought out a number of aspects that were overlooked up to that moment, sometimes even on the international stage. Thus, the ethnoarchaeological investigations conducted as part of two Romanian projects with French participation (ethnosal.uaic.ro and ethnosalro.uaic.ro) have evinced how the salt springs, even today of great importance in the daily life of the rural communities from the piedmont areas of Moldavia, generated an entire spiritual universe. Their importance is also reflected at the linguistic level, including through toponymic imprinting. The multitude and diversity of the toponymy generated by the salt water springs throughout Romania impelled us to elaborate a dedicated study of this subject. From the methodological point of view, we considered necessary for our enterprise to coalesce the semantic and toponymic approaches. Accordingly, the parameters of our research called for the coining of novel toponymic terms.

From the onset of our research, we discovered that the hydronymic terminology does not possess, as it is the case with other categories of waters, a specialised term from the name of springs, a situation which prompted us to advance specific terms. Ancient Greek had two generalising lexemes for designating springs, according to their stale or gushy flow. Thus, *krene* meant a "well spring fountain" (Liddel-Scott, *s.v.*), "spring, fountain, which does not spurt" (Bailly, *s.v.*), as opposed to *pege*, "running water, fount" (Liddel-Scott, *s.v.*), "spring, running water, waves, gushy source" (Bailly, *s.v.*). On account of this distinction, we find appropriate to propound for the names of springs the terms *crenonym* and *pegonym*.

The same differentiation logic operating in Ancient Greek for designating the two categories of springs is also present in Romanian, without generalising lexemes. Stale-water springs are designated with the appellatives *apă izvoroasă* (lit. 'springy water'), *crună*, *fântână* ('fountain'), *fântâniţă* ('little fountain'), izvor ('spring'), *mârvilă* ('plash'), *ochi* ('outlet), *răsuflătoare de apă* (lit. 'water vent'), *revenie*, *şiuşor*, *ţâţână de apă* (lit. 'water teat'), *zvorac* (derived from *izvor*) (Moldovanu 2010, 152). Conversely, Romanian uses a similarly wide array of appellatives for the springs with water that "gushes all of a sudden and with force: *bigăr*, *fierbătoare*, *fiertoare*, *gâlgoi*, *izbuc*, *izbucnitură*, *izvor puternic*, *şipot*, *izvor zbâcnitor*" (Moldovanu 2010, 153).

For toponyms that designate salt springs (stale or gushy), we propose the terms *halocrenonym* and, respectively, *halopegonym*.

In what concerns salt springs, Romanian has a relatively long series of lexemes only for the stale salt springs: "*gloduri*, *mlăştinău*, *moare*, *morugă*, *murătură*, *ocnă*, *ocnit*, *saramură*, *sărătură*, *slatină*, *slănic*, *slătioară*" (Moldovanu 2010, 153).

The notion of "stale salt spring" is represented in Romanian toponymy in an analytical or synthetic manner. The analytic mode is illustrated by:

- the lexeme *izvor* ('spring'), *fântână* ('fountain'), *apă* ('water'), followed by the qualificative *sărat* ('salty'): Izvoru Sărat (lit. 'the salty ~') (DTRM III, 2009, 254); Apa Sărată (lit. 'the salty ~') (Alessandrescu *et al.* 1897, 22; DTRO I 1993, 227). In the last case, even if the name refers to a settlement, it is obvious that its name derives from a microtoponym. From all these lexemes, only *fântână* has been treated in depth (Dumistrăcel 1983-1984; 1985).

- the lexeme *izvor* ('spring'), *fântână* ('fountain'), followed by a material determinative (noun in the accusative or genitive case): Fântâna Slatinii (lit. 'the salt spring ~') (Grămadă 1996, 97), Izvorul cu Slatină (lit. 'the ~ with brine') (Grămadă 1996, 97), Fântâna Slatinei (lit. 'the salt spring ~'), Fântâna de Slatină (lit. 'the ~ for brine'), Izvorul cu Slatină (lit. 'the ~ with brine') (Alexianu *et al.* 2007, 59), Fântâna de Slatină (*v.a.*) (Diacon 1989, 36), Fântâna Slatinii (*v.a.*) (Diacon 1989, 258), Izvorul Sării (lit. 'the salt's ~') (DTRM III 2009, 254).

- the lexeme *izvor* ('spring'), followed by determinative for a running salt water or a salt land (proper noun in the nominative or genitive case): Izvoru Sărata (lit. 'the Sărata ~'), Izvoru Sărățelu (lit. 'the Sărățelu ~') (DTRM III, 2009, 254), Izvorul Săraturilor (lit. 'the Săraturilor ~') (DTRM III, 2009, 254).

The synthetic mode is illustrated by:

- the lexeme *slatină*, defined as "wet salt place with mires" (Porucic 1931, 39), "wet salt land" (Porucic, 1931, 63) or "very salty water spring" (Porucic 1931, 78). The diminutive for Slatină is Slătinuță (Solca, Suceava County). The general rule for attestations is that of entopic syntagma, not as proper halocrenonym: La Slatină (lit. "at the ~").

(La) Slatină is the most frequently found halocrenonym in Romania. There is an entire series of toponyms composed starting from this entopic syntagma from which we mention, without claiming exhaustivity: Slatina Rea (lit. 'the bad ~'), Pârâul Slatinei (lit. 'the ~'s creek'), Slatina Voitinului [lit. 'the ~ of the Voitin (hill)'], Slatina Mare (lit. 'the large ~'), Măgura Slatinei (lit. 'the ~'s hillock'), Drumul Slatinei (lit. 'the ~'s road'), Podul Slatinei (lit. 'the ~'s bridge'), Slatina Mică (lit. 'the small ~') (Grămadă 1996, 97), Câmpia Slatinei (lit. 'the ~'s plain') (DTRM II 2007, 67), Dealu Slatinei (lit. 'the ~'s hill'), Dealu Slatinei de Jos (lit. 'the hill of the lower ~') (DTRM II 2007, 239), Dealu(l) Slatinei (*v.a.*) (Diacon 1989, 259; Alexianu *et al.* 2007, 60), Drumu Mare al Slatinei (lit. 'the large road of/to the ~') (DTRM II 2007, 334), Drumu(l) Slatinei (*v.a.*) (DTRM II 2007, 350; Alexianu *et al.* 2007, 59), Hotaru Slatinei (lit. 'the border of the ~') (DTRM III 2009, 215), Poiana Slatinei (lit. 'the ~'s glade') (Diacon 1989, 257; Alexianu *et al.* 2007, 204), Vârful Slatinei (lit. 'the ~'s peak'), Vârvu Slatinei (*idem*) (Diacon 1989, 257), Valea Slatinei (lit. 'the ~'s valley') (Iordan 1963, 126, 460), Calea Slatinei (lit. 'the way of the ~') (DTRO 1993, 342), Slatina de pe Islaz (lit. 'the ~ from the open field'), Slatina de pe Islazul Mihușca [lit. 'the ~ from the Mihușca common field'] (Alexianu *et al.* 2007, 204), Slatina Mică (lit. 'the small ~'), Slatina Mare (lit. 'the large ~') (Alexianu *et al.* 2007, 204), Slatina Veche (lit. 'the old ~') (Alexianu *et al.* 2007, 24, 137), Slatina Pănușița (lit. 'the Pănușița ~') (Alexianu *et al.* 2007, 122), Slatina Dulcea (lit. 'the Dulcea ~') (Alexianu *et al.* 2007, 131), Hegu Slatinii (Diacon 1989, 257), Bahna Slatinii (lit. 'the ~'s mire') (Diacon 1989, 258). Note should be made of the fact that the instances where *slatină* is in the genitive case, the lexeme is treated as a halocrenonym. The same occurs when the first element of the composed toponym (that is, when it is followed by a determinative) is *slatină*. Considering its countless attestations throughout Romania, we can state that *slatină* develops, in some cases, toponymic fields that generally comprise from one to three toponymic derivatives, according to the geographical features of the place where the salt spring is located.

- the lexeme *saramură*, defined as "very salty water, watery salt land" (Porucic 1931) or "very salty water spring" (Porucic 1931, 78), is also encountered in the form *salamură*. The variant for *saramură* is *saramură*, while the variant for *salamură* is *sălămură* (Porucic 1931, 78).

The composed toponyms derived from the halocrenonym *saramură/salamură* are much rarer: Dealul Saramurii (lit. 'the ~'s hill'), Pârâul Salamurii (lit. 'the ~'s creek') (Alexianu *et al.* 2007, 60), Salamura de la (Muntele) Stroiu [lit. 'the ~ from the Stroiu (mountain)'] (Alexianu *et al.* 2007, 86), Salamura de la Dealul Gălean (lit. 'the ~ from the Gălean hill') (Alexianu *et al.* 2007, 113).

- the lexeme *murătură/murătoare* (pl. *murători*) is defined as "a very salty water" (Porucic 1931, 38), "salt water spring with marshland around it", "marshland. A spring with very salty water so that it can be used for pickling cucumbers" (Porucic 1931, 44), "spring for very salted water" (Porucic 1931, 50). In this case, too, composed toponyms developed from this halocrenonym, such as Drumu spre Murători (lit. 'he road to the ~') (DTRM II 2007, 351).

Deciphering the mechanism through which a halocrenonym developed composed toponyms, on the face of it in a haphazard manner, is made much easier by the appeal to the theory of toponymic fields. A toponymic field is a "*sui generis* structure, through which associations of geographical objects (the designata) from a micro-area, performed by speakers, are reflected, at the level of expression, by either analytic or synthetic oppositions (equipollent, privative or gradual)" (Moldovanu 2010, 25).

The definition of the first type of toponymic field starts from the idea that the "toponymic structures are based [...] on relations of unilateral dependency (determination) between a constant (the base) and multiple variables (derivatives), on the one side, and of facultative dependency (constellation) between derivatives, on the other" (Moldovanu 2010, 18); the toponymic oppositions of this kind are called "polar oppositions", and the process of structuring is called polarisation, defined as "the process of creating new toponymic derivatives from a core-toponym, corresponding to the object most important in a continuous (unitary) geographical set". Through polarisation, "purely toponymic fields are generated, without any direct connection with the lexical fields"; "the result of the structuring process is a denominative assemblage" called "toponymic field, in which the oppositions between components are of another type than the lexical ones" (Moldovanu 2010, 18).

The second type of toponymic field is generated by toponymic differentiation, that is, "the process through which, with the help of lexical delimitators, the parts of named geographic object are designated. In this case, the toponymic oppositions are constituted on the basis of the lexical ones, since the differentiation opposes the corresponding denotants from the same hierarchized lexematic field" (Moldovanu 2010, 19).

The mixed toponymic fields result from the combination of the two types of fields, the polarising field and the differentiated field (Moldovanu 2010, 20).

Toponymic fields through polarisation

(La) Slatină, the most representative halocrenonymic constant, as it presents the most numerous microtoponymic occurrences in Romania, has generated via polarisation the following types of toponyms (Ungureanu, Boamfă 2006), classified:

a. according to the nature of the element designated, in:

- morphonyms (which designate landforms): Măgura Slatinei, Podul Slatinei, Câmpia Slatinei, Dealu Slatinei, Dealu Slatinei de Jos, Poiana Slatinei, Vârful Slatinei,Valea Slatinii, Hegu (from De. *Hügel*, 'hill') Slatinii.

- hydronyms:
    - potamonyms (names of running waters): Pârâul Slatinei;
    - limnonyms (names of lakes, ponds): Bahna Slatinei.

- hileonyms (names of plant associations): Pădurea Slatina, Pădurea Slatinei.

b. according to the nature of the anthropic element named in:

- hodonyms (names of communication routes): Drumu Slatinei, Calea Slatinei;

- limeonyms (names of limits, borders): Hotaru Slatinei.

We mention that the mentioned toponymic derivatives do not exist for each salt spring, but that they are a collection of all the attestations for all the salt springs known at the current stage of research. This situation allowed us to elaborate the following model, which represents in fact the potential toponymic field of the halocrenonym *(La) Slatină* (Figure 1).

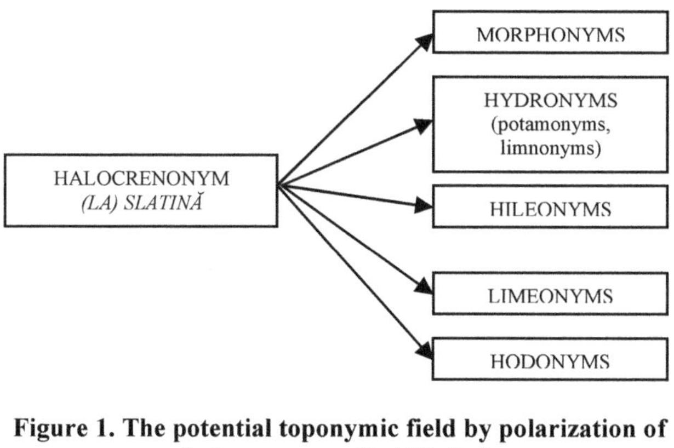

Figure 1. The potential toponymic field by polarization of halocrenonym *(La) Slatină*.

Toponymic fields through differentiation

The toponymic fields through differentiation of the halocrenonym *(La) Slatină* emerged thus (Figure 2):

- through gradual opposition in terms of discharge: Slatina Mare, Slatina Mică, Slătinuţa;

- through opposition in terms of use chronology: Slatina Veche;

- through opposition in terms of quality: Slatina Rea;

- through opposition in terms of elevation: Slatina de Jos;

- through differentiation of a geographical-spatial nature: Slatina Voitinului, Slatina de pe Islaz var. Slatina de pe Islazul Mihuşca, Slatina Pănuşiţa, Slatina (de la Pârâul) Dulcea;

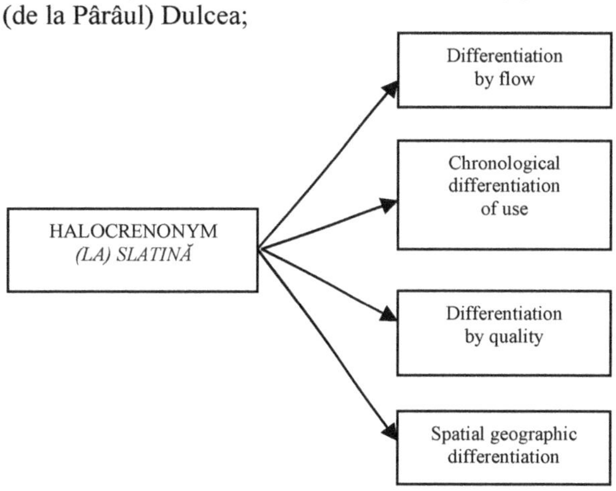

Figure 2. The potential toponymic field by differentiation of halocrenonym *(La) Slatină*.

Mixed toponymic fields

From among the mixed toponymic derivatives, the most illustrative are: Gura Văii Slatinei, Drumu Mare al Slatinei, Dealu Slatinei de Jos.

The characterisation from various angles of the synthetic halocrenonyms analysed above, the most frequent on Romanian territory, contributes to their better understanding.

(La) Slatină, (La) Salamură, Murătoarea are microtoponymic halocrenonyms formed from simple appellatives and belong to the category of spontaneous (or natural) toponyms (Ungureanu, Boamfă 2006). *Salamura and murătoarea* are formed on the basis of the Romanian lexis of Latin origin, while *slatina* is formed are formed on the basis of the Slavic superstratum in Romanian. With the exception of the inhabitants from the areas with salt springs, the lexeme *slatină* does not display semantic transparency for current speakers of Romanian. *Saramură* is a lexeme understood by all speakers of Romanian, while *murătoare* is perceived as concerning salt water, but with a certain difficulty, in a diffuse manner [associating it with *moare* ('pickle brine'), *murătură* ('pickle')]. We can state that the lexeme *slatină* has semantically fossilised for most Romanian speakers. But it combines with Romanian lexemes of Latin or Slavic origin that are semantically transparent for all Romanian speakers (Giuglea 1944, 16), in the great majority of cases of toponymic derivatives formed through polarisation.

From a syntactic point of view, we notice the assignment to the genitive case of the constant *Slatină* in most of the cases, but with different meanings. Thus, in the case of the toponyms Măgura Slatinei, Podul Slatinei, Câmpia Slatinei, Dealu Slatinei, Poiana Slatinei, Valea Slatinei, and Vârfu Slatinei we can speak of the determinative genitive, inherited from the syntax of Latin, in the case of the toponym Pârâul Slatinei by a genetic genitive, while in the case of the toponyms Drumul Slatinei and Calea Slatinei by a direction genitive, denoting too a final destination.

The study of the microtoponymy related to salt springs, a subject so far untouched by hydronymic research, reveals their importance to the local communities, open the way for analysing the original mechanism of toponymic domination, for understanding prehistoric man as deeply as possible. Further research of this sort for as many languages as possible could provide the foundations of a contrastive analysis that would eventually lead to the identification of universal structures of toponymic derivation. The research on the toponymic impact of the salt springs is tributary to a comprehensive approach to the relationships between Man and "the fifth element", a *sine qua non* of animal life. The main conclusion is that, in the terms of the toponymic fields theory, the most frequent halocrenonym in Romania, (*La*) *Slatină*, with hundreds of attestations, constitutes the core-toponym for a considerable series of toponymic derivatives. Taken together, they can be considered a potential toponymic field.

Acknowledgement
This work was supported by a grant of the Romanian National Authority for Scientific Research, CNCS-UEFISCDI, project number PN-II-ID-PCE-2011-3-0825, *The Ethno-archaeology of Salt Springs and Salt Mountains from the extra-Carpathian areas of Romania* – www.ethnosalro.uaic.ro.

References

Alessandrescu, C., Brătescu, P. and Moruzi, I. 1897. *Dicționar geografic al județului Prahova*. Târgoviște, Tipografia și legătoria de cărți „Viitorul", Elie Angelescu.

Alexianu, M., Weller, O. and Brigand, R. 2007. *Izvoarele sărate din Moldova subcarpatică. Cercetări etnoarheologice*. Iași, Editura Demiurg.

Ciorănescu, Al. 2001. *Dicționarul etimologic al limbii române* (Ediție îngrijită și traducere din limba spaniolă de Teodora Șandru Mehedinți și Magdalena Popescu Marin). București, Editura Saeculum I.O.

Dumistrăcel, S. 1983-1984. Rolul variabilei geografice: „lacul" de lîngă puț. *Anuar de lingvistică și istorie literară, XXIX, Iași*, 139-174.

Dumistrăcel, S. 1985. Sur le maintien de *Fîntînă* (<lat. *fontana*) „source, Quelle" dans les parlers roumains. *Revue Roumaine de Linguistique* T.XXX, 6, 539-542.

Giuglea, G. 1944. *Cheie pentru înțelegerea continuității noastre în Dacia, prin limbă și toponimie*, București.

Grămadă, N. 1996. *Toponimia minoră a Bucovinei* (Îngrijirea ediției, studiul introductiv, bibliografia, notele și indicele: Ion Popescu-Sireteanu), II. București, Editura Anima.

Iordan, I. 1963. *Toponimia romînească.* București, Editura Academiei.

Moldovanu, D. 2010. *Teoria câmpurilor toponimice (cu aplicație la câmpul hidronimului MOLDOVA).* Iași, Editura Universității „Alexandru Ioan Cuza" Iași.

Popescu-Voitești, I. 1943. *Sarea regiunilor carpatice românești.* București, Fundația Regală pentru Literatură și Artă.

Porucic, T. 1931. *Lexiconul termenilor entopici din limba Română în Basarabia.* Chișinău, Tipografia Eparhială Cartea Românească.

Stoica, C. and Gherasie, I. 1981. *Sarea și sărurile de potasiu și magneziu din România.* București, Editura Tehnică.

Ungureanu, Al. and Boamfă, I. 2006. *Toponomastica.* Iași, Editura Sedcom Libris.

Weller, O. and Dumitroaia, Gh. 2005. The earliest salt production in the World. A Neolithic exploitation in Lunca-Poiana Slatinei, Romania. *Antiquity*, 79 (306).

Abbreviations

DTRO - 1993. *Dicționarul toponimic al României. Oltenia, vol. I* (Sub redacția Prof. univ. dr. Gheorghe Bolocan). Craiova, Editura Universitaria.

DTRM – 2007. *Dicționar toponimic al României. Muntenia, vol. II* (Sub redacția Prof. univ. dr. Nicolae Saramandu). București, Editura Academiei Române.

DTRM – 2009. *Dicționarul toponimic al României. Muntenia, vol. III* (Sub redacția Prof. univ. dr. Nicolae Saramandu). București, Editura Academiei Române.

The Toponymy of Salt in Zamfir Arbore's *Geographical Dictionary of Bessarabia*

Mihaela Asăndulesei
Universitatea „Alexandru Ioan Cuza" din Iași

Abstract
The present paper aims to list the toponyms found on the territory of Bessarabia that derive from various terms referring to manifestations of salt, found in the geographical dictionary elaborated by Zamfir Arbore. The terms were analysed from several points of view, viz. geographical, lexical, semantic, morphological, and derivational, in order to highlight, on the one hand, the role and importance of the salt sources in the life of the communities, and, on the other hand, the toponymic fields—polarising, differentiating or mixed—generated by a toponymic core relating to the referential element 'salt'.

Keywords
salt, placenames, toponymic field, Bessarabia

Toponymy is one of the fields with a strong interdisciplinary undercurrent, demanding vast philological, geographical and historical knowledge for understanding the problems related to the origin, meaning and evolution of toponyms (Ungureanu, Boamfă 2006, 11). These designations of the different geographical elements, studied from various points of views, according to the goals and interests of the one investigating them, reveal the manner in which they emerged and the role they play in the life of the community, being likewise important for geographers, historians, ethnologists, or linguists (Iordan 1963, 1).

The toponymy of salt refers to those placenames that form through relation to the referential 'salt'. Salt-related toponymy is quite rich and widespread across the entire world. The designation of geographical elements and their subsequent extension to various forms of human settlement is a fully expected outcome of the fact that salt has played an important role in the life of past human communities. Salt has been used since prehistoric times to preserve food, and even as currency, for which reason it became since immemorial times an extremely sought good (Emilciuc 2011, 55). In support of this also comes the existence of numerous "salt routes", commercial roads along which for centuries or even millennia salt was transported to regions lacking this vital resource.

The present article tackles for the first time the issue of salt toponymy in Bessarabia, relying on the analysis of Zamfir C. Arbore's work *The Geographic Dictionary of Bessarabia* (in original *Dicționarul geografic al Basarabiei*), published in 1904 as the first instalment of a series entitled *The Geographical Dictionaries of the Romanian Provinces outside the Kingdom* (in original *Dicționarele geografice ale provinciilor române în afară de Regat*), commissioned by the Romanian Geographical Society. The idea of this endeavour was to study all the Romanian-inhabited regions found outside of the national state, in order to obtain an overall picture of the geopolitical realities of those times (Tabuncic 2011, 96). Alongside the work *Bessarabia in the 19th century* (in original, *Basarabia în secolul XIX*), published by the same author five years before, the *The Geographical Dictionary of Bessarabia* was also a statement of solidarity of the Romanians living within their own nation state with the inhabitants of the region beyond the Prut River, in their struggle against Russian rule (Tabuncic 2011, 96).

Bessarabia, the tract of land spanning between the Prut and Dniester rivers, is one of the regions inhabited by ethnic Romanians with a distinct history and political context, which is reflected by its variegated realities — including the toponymic one (Iorga 1912). In 1904, the year Arbore's work was published, Bessarabia was under Tsarist administration as a gubernyia also containing other nationalities (Jews, Bulgarians, Gagauzes, Lipovenians, Germans) that influenced the Romanian majority, including the language, an influence reflected in placenames too.

At the onset of my study, using the current works from the field of toponymy (Moldovanu 2010; Ungureanu, Boamfă 2006) and multiple editions of Arbore's *The Geographical Dictionary of Bessarabia* (Arbore 2001; Arbore 2012), I set to record all the salt-related toponyms found in Bessarabia, analysing them from the geographical, lexical, semantic, morphological, and derivational point of view. The aim was to appraise at the toponymic level the various influences incurred during the Russian rule or, conversely, the preservation of the indigenous Romanian specificity.

The following pages contain the alphabetically-ordered list of the toponyms related to the substance salt, according to the nature of the designated geographical element, as found in the geographical dictionary of Zamfir Arbore. For a clearer image of the classification of the placenames according to the aforementioned criterion, four categories have been employed: (1) *hydronyms*, which designate water bodies; (2) *oikonyms*, into which human settlements fall, expressed through territorial-administrative units; (3) *morphonyms*, which refer to landforms, whether positive or negative; and (4) *odonyms*, which denominate ways of communication (streets, ways, roads, etc.). The placenames, albethey microtoponyms or official toponyms, are expressed in two ways: as simple names, formed from common or proper nouns, or as complex designations, formed from syntagmas.

Hydronyms

Hydronyms formed from simple designations

Alcalia — Is a small river in Acherman County. It springs from the Armas Valley and Răpciuza (Eremia 2013, 92) and empties into the Lake Bazârian (Bicinoie) (Arbore 2012, 36) or Burnas, north-east of the village of Tuzla (Eremia 2013, 92). The river appears on the topographic maps from the 18th-19th century (Eremia 2013, 92). The term can be traced to the Turkic adjective *alkali* ('brackish', 'lye-like', 'bitter-salty'), often applied to steppe rivers or the salt lakes fed by the waters of the Black Sea (Eremia 2013, 92). In the edition published by Iurie Colesnic, the Alcalia rivulet also appears under the name Răpciuza (Răpşiuza) (Arbore 2012, 447; Arbore 2001, 6), after the name of the valley from which it originates. According to its meaning, the toponym refers to the particularities of the open natural landscape.

Sasâc — Is a large lake in Ismail County, also known as Cunduc, with a length of over 31 km and a width of almost 14 km. It is separated from the Black Sea by a stretch of land, narrow and 13-km long. In the North, into this lake empty the Cohâlnic or Cunduc and Sărata rivers (Arbore 2012, 181). The toponym's base is the entopic term *sasic* meaning 'swampy lake with saline water' (Porucic 1931, 39), 'swampy saltland with ponds and lakelets' (Porucic 1931, 50). This entopic is based on the term *sasık* found in Turkish and Gagauz, meaning 'foul swamp' (Eremia 2006, 166). According to its meaning, the toponym refers to the particularities and features of the designated natural landscape

Sărata — Is a lake, in Ismail County, at the mouth of the Sărata brook, a left-hand tributary of the Prut (Arbore 2011, 182). Its length is almost one km and its length of 50-800 m (Arbore 2012, 363). On the territory of Bessarabia there are more such geographical objects with this name or derived from it, being the most encountered toponym related to the element salt. The adjective *sărat*, of Latin origin, becomes a topic name through nominalisation, refers to a particularity of the soil, and not of the water — "land rich in soluble mineral salts", "salty land". In the saline areas, because of the abundant mineral salts, the soil turns white, resembling regular salt. From this feature, through visual comparison, the respective areas were designated *Sărata* (Eremia 2014). It is interesting how other toponyms polarise or derive through differentiation from it; in our case, the lake's designation polarises from the name of the brook that empties into it.

Sărata — Is a river in Acherman County. It emerges near the village Alexandrouca and, receiving the brooks from the Jidauca and Saiţu, runs along a South-East direction up to the village Negrova; from here it goes straight southwards, to Lake Cunduc, into which it empties at its northern edge, near the Bulgarian colony of Tropoclo. On the shores of the Sărata Lake are the villages: Alexandrouca on both shores, Manja, Petropaulouca, Goilovca, the German colony Sărata, on the right bank; and Moruzani, Parij Nou, Molodova, Naurotca, Leaşcova, Camcic, the stanitsa (Cossack village) Mihailouca, the stanitsa Acmanghit, the Bulgarian colony Tropoclo, and the farm Câşla, on the left bank of the river. During summer, the Sărata dries out (Arbore 2012, 363).

Sărata — A rivulet running in Bender and Ismail counties, formed by the brook that runs through the Sărata valley and the brook from the Nirişeni valley. Both brooks merge near the village Sărata Galbâna, and from here the river runs through the valley of the same name up to the Prut valley, into which it empties south of the village Hănăsenii Noi, forming the Sărata Lake. From the right, the river receives the waters of the Sărăţica brook, and from the left, those of the Orac brook. Besides these two brooks, the Sărata also gathers the waters of several creeks that run through abutting valleys opening in the Sărata valley (Arbore 2012, 363). In Iurie Colesnic's edition, the two affluents are mentioned in inverse locations with respect to the Sărata River (Arbore 2001, 182). In what concerns its left-hand tributary, its name is formed from a simple appellative, through suffixation, thus indicating the small size, with relation to the river into which it empties, being a toponym through differentiation. By analysing the geographical description of the Sărata rivulet, it is possible to generate a mixed toponymic field, extremely interesting.

Sărăţâca — Is a brook on the boundary separating Bender County from Ismail County. It describes a valley in which the Sărăţâca (Arbore 2012, 365) / Sărăţica (Arbore 2001, 184) village is found. Is a compound toponym, formed through the suffixation of the simple appellative *Sărata* with the diminutive particle *-ica*, in its variant *-âca* specific to the Moldavian speech.

Soloneţul — Is a rivulet in Soroca County. It springs from the potholes at the bottom of the Soloneţ valley in Bălţi County, running through this valley until it empties into the Răut River. The villages Drăgăneşti and Soloneţul Roşietici are found along its path (Arbore 2012, 378). The toponym is of Slavic origin, specific to the East Slavic branch, with the meaning of "saltworks, salt water, salty pond" (Poruciuc 2011, 215-216) and "white salty and hard land" (Eremia 2006, 173; Porucic 1931, 63). On account of the manner in which it was formed, it belongs to the category of toponyms rendered through simple appellatives, combined with the definite article, being a core of toponymic polarisation for the morphonym designating the valley of the same name.

Compounded hydronyms

Ocnele Buhnei — Is a lake in Ismail County, in the Prut Valley, north of Lămunda village (Arbore 2012, 311), according to the correction proposed by Iurie Colesnic, it means the village of Flămânda (Arbore 2001, 153). The toponym is formed by a single genitive syntagma represented by the plural form of term *ocnă* in the nominative case, with the definite article, and the element *buhnă*, in the genitive. This term belongs to the category of salt-related toponyms on account of the first term that forms it. Analysing from the lexical and semantic point of view the notions that form this hydronym, I emphasize the term *ocnă*, a borrowing from Old Slavonic, meaning "salt pit" (Poruciuc 2011, 2016) and "mine (particularly salt)", which meant "window" (DEX 2012, 733), and the term *buhnă*, the explanation of which is rather difficult to ascertain, through which the relation of the toponym with the denominated reality is clarified. In Anatol Eremia's dictionary of geographic terms, the word *buhnărie* is given as a "place with many *buhna*" (Eremia 2006, 29), but without providing a meaning for this word. The *Etymological Dictionary* (in original, *Dicționarul etimologic*; Ciorănescu 2001, 120) and the *Explanatory Dictionary* (in original, *Dicționarul Explicativ al Limbii Române*; DEX 2009, 123) state that *buhnă* is a form of the term *bufnă*, *bufniță* ('owl') Two directions can be drawn from these observations: either the evolution of the hydronym from a patronym, or the rendering of a particularity of the natural landscape, more precisely an element of the local fauna. With respect to the first variant, it is possible that the toponym indicates its vicinity to saltworks belonging to an individual with the name or moniker Buhna. With respect to the latter variant, the determinant that composes the toponym can render the presence of an owl near or inside the saltworks.

Răcăria Slănic — Is a lake, in Ismail County, located in the valley of the Prut between Lake Răcăria Mare and the Balaş pond (Arbore 2012, 340). The toponym belongs to the compounded ones, having as constituting elements the terms *răcăria* and *slănic*. As it refers to a lake, the first term refers, most probably, to its abundance in crayfish (Rmn. *rac*). The toponym belongs to the category of suffixed terms, and, according to its meaning, it refers to the nature of the local fauna. The determinant from the appositive toponymic syntagma is represented by a proper noun that designates the particularities and features of the denominated element. The element is of Slavic origin, derived from the Old Slavonic *slanŭ* ('salty') (Poruciuc 2011, 215), from which the entopic term *slănic* evolved, meaning "salty water" (Ungureanu, Boamfă 2006, 136). In support of this observation comes Moldovanu's demonstration that geographical objects are particularised by suffixing the abovementioned lexical base. Thus, *slanikŭ* refers to the saltworks, while *Slanĭcĭ* designates a salty brook; Romanian has the form derived from *slanikŭ*: "the primary suffixal opposition was neutralised in the use of the non-Slavic speakers that took the toponyms, so that the Romanians generalised this designation, while the variant *slanĭcĭ* is used by the Szekelys" (Moldovanu 2010, 19).

Oikonyms

Oikonyms formed from simple appellatives

Ocnița — Is a village, founded around 1807 in Acherman County, Alexandrouca volost[1], located along the Ciligider brook, also called Frumuşica Veche (Arbore 2012, 212). The toponym is formed by suffixing the term *ocnă* with the diminutive particle – *ița*, indicating either the small size of the denominated element, either simply fondness to it. In the *The Geographical Dictionary*, the terms listed as the main designation is Frumuşica Veche, and the first variant is indicated without other details. With respect to the dictionary entry *Ocnița*, found at letter *O*, I should mention that an error occurred, also carried to the newer editions, whereby the correspondence of the settlement is not Frumuşica Veche, but Frumuşica Nouă.

Ocnița — Is a village in Hotin County, Secureni volost, found at the spring of the Ciuhur, at 5 km from the Ocnița station of the Mohilău–Noua Suliță railway (Arbore 2012, 311). The toponym can be assigned to the category referring to the particularities of the economy, more specifically to the exploitation of the underground resources and the activities of the inhabitants (Ungureanu, Boamfă 2006, 193).

Ocnița — Is a village in Orhei County, Ciocolteni [Ciocâlteni] volost, composed of two hamlets, found in the valley of the Cohâlnic (Arbore 2012, 312). In the edition curated by Iurie Colesnic, the names of the two hamlets are specified, the names of which derive from the toponymic core under scrutiny, highlighting the social difference between their inhabitants: Ocnița-Răzeşi ('yeomen') and Ocnița-Țărani ('peasants').

Ocnița — Is a small market town in Soroca County, Otaci volost, founded near the station of the Bălți-Ocnița-Noua Suliță railway (Arbore 2012, 312).

Sărata — Is a German colony, in Acherman County, the centre of the former volost with the same name, found on the Sărata brook (Arbore 2012, 363). It was founded in 1825 by a number of immigrants from Baden, Würtemberg, Bavaria, and Austria, on the place of the former Romanian village of Gura Curudere, destroyed by the Tatars. The toponym of Latin origin is derived through polarisation from the name of the brook along which shores the settlement is located, indicating the particularities of the designated natural landscape.

Sărățâca — Is a German colony also known as **Matildendorf**, in Bender County, in the Ceaha valley, on the right side, Iosefdorf volost (Arbore 2012, 365), The polarisation of the Romanian toponym from the name of the brook is obvious.

Sărățeni — Is a village in Orhei County, Ciocâlteni volost, along the Răut, near the road going to Bălțiu (Arbore 2012, 365). Iurie Colesnic's edition provides the present-day name of village, namely Sărătenii Vechi (Arbore 2001, 183). The toponym is of Latin origin, formed through suffixation with the specific particle -*eni*. Its formation indicates the geographic origin of the inhabitants, revealing the migration of the population from one settlement to the other (Ungureanu, Boamfă 2006, 96), with the suffix -*eni* indicating the origin from which the incomers originate (Iordan 1963, 159). Thus, it can be stated that it probably formed following the migration of a community coming from a settlement

[1] An administrative subdivision in Bessarabia during Russian rule.

called *Sărata* into the denominated area, whose members were called by the surrounding population *sărățeni*, a designation extended to the settlement they founded.

Sărățica — Is a village in Ismail County, found in the Sărâțâca (Arbore 2012, 365) / Sărățica (Arbore 2001, 185) valley. The polarisation of the designation from the valley of the same name is obvious; according to the manner in which it formed, the term belongs to the category of toponyms formed by suffixation, with the diminutive particle *-ica*.

Slănina — Is a village in Soroca County, Târnova volost, on the left bank of the Cubolta rivulet, near the village of Mandâc. This is why it is also known as Noul Mandâc ("the new Mandâc") or after its Russian form Novîi-Mandîc (Arbore 2012, 374). Since 1965 the village is called Lozeni, and in 1970 is integrated into the settlement of Livădeni, under the same name. The name Mândâc was later restored (Arbore 2001, 189). Even though the collective mind doesn't make a connection between the term *slănină* ('bacon', 'fatback') to *sare* ('salt'), from the lexical point of view the toponym is based on the Old Slavonic term *slanŭ*, meaning "salted" (Poruciuc 2011, 215). It is difficult to analyse the toponym according to its meaning, considering the fact that its notion is limited to fatback cuts preserved (with added salt, paprika, and/or smoked) or melted for lard (DEX 2009, 1026).

Solonceni — Is a village in Orhei County, Răzâna volost, on the shore of the Dniester, between the villages Tarasova and Cerna, with the Russian variant given as **Solonceanî** (Arbore 2012, 378). Based on a Russian etymon, *solonceac*, referring to soil containing soluble salts, of very limited natural fertility, *sărătură*, *soloneț* (Eremia 2006, 173), this toponym is formed by suffixing with the particle *-eni*, in Romanian, and an adaptation specific to the Russian language. Its formation reveals the migration of a population from a settlement to a new one (Ungureanu, Boamfă 2006, 96), the original possibly being Solonet, from which the toponym under consideration polarised. In the analysis of the stage just before, the derivation from the designation of the inhabitants that moved from one settlement to the other is obvious.

Solonețul — Is a village in Soroca County, Văscăuți volost, in the Sloboziei valley (Arbore 2012, 378). Iurie Colesnic's edition specifies that the current orthography is Soloneț (Arbore 2001, 191). The toponym polarizes from the name of the rivulet near which it is found.

Tuzlî — Is a small market town in Ismail county, on the western shore of Lake Hadji Ibrahim (Arbore 2012, 415). Zamfir Arbore mentions that the denizens work in agriculture and fishing, with the small town housing three factories for packaging salted fish and two for salt. The correct form of the designation, namely Tuzla, is mentioned in Iurie Colesnic's edition (Arbore 2001, 213). The toponym comes from the Turkish term *tuzla*, meaning saltworks; it is encountered as an entopic term in this form, *tuzla*, as a place where salt is extracted from sea water (Eremia 2006, 198). This entopic is based on the Turkish term *tuz* ('salt'). The toponym is part of the category of designations referring to the particularities of the economy, namely the exploitation of underground resources. From the phonetic point of view, it is recorded with the form specific to the Moldavian speech, or Russian influence, with a final *î*.

Compounded oikonyms

Gura Sărată — a rural commune, in Ismail County, located on the valley of the Prut, near the mouth of the Lăpușna rivulet, a kilometre from the Prut, officially also known as **Sărata Răzeși** (Arbore 2012, 364). The designation is formed from an adjectival toponymic syntagma represented by an appellative and a determinant, related to the term *sare* ('salt'). As it's located near the mouth (Rmn. *gură*) of the Lăpușna rivulet, the name of the territorial-administrative unit probably derived from this appellative, also receiving the influence of the better known and more used toponym.

Ocnița-Răzeși — Is a hamlet that constitutes the village of Ocnița, in Orhei County (Arbore 2001, 154). The toponym derives through differentiation from the toponymic core represented by the village to which it belongs from the territorial-administrative point of view, concurrently underlining the social difference between the inhabitants of the two component hamlets. From the point of view of the manner in which it formed, it can be assigned to the category of appositive toponymic syntagmas, given that the determinant is in the nominative case.

Ocnița-Țărani — Is a hamlet part of the Ocnița village, Orhei County (Arbore 2001, 154). It formed in the same manner as the toponym analysed above, by adjoining the name of the village to which it belongs to, and the predominant social class of the area it describes.

Sărata Galbână — Is a village, in Bender County, located in the Sărata valley, at the mouth of the Nirișenilor valley, also known as Slobozia (Arbore 364). Is part of the Gura Galbână volost, from which it derives its name through differentiation. The road from Gura Galbână runs south of the village, going to the valley of the Prut. The second edition of the dictionary specifies the current written form of the toponym, namely Sărata Galbenă, and the new administrative status, inside Lăpușna County (Arbore 2001, 183). The toponym is composed from a adjectival syntagma that derives through an appellative from the name of the valley in which it is located, and through a determinative from the volost to which it belongs. The differentiating oikonym is chromatic (*galben* var. *galbân* (m.), *galbână* (f.) = 'yellow') (Alexianu 2011, 580), making both terms of Latin origin.

Sărata Nouă — Is a village, in Bălți County, found in the Gârla Mare valley, along the Folești–Cornești road, with a Russian variant **Sărata-Novaia** (Arbore 2012, 364) / **Sarata-Novaia** (Arbore 2001, 183). The toponym is formed by an appellative and a determinant, expressed through an adjective indicating the "age" of the settlement (*nou* (m.), *nouă* (f.) = 'new'). With respect to the Russian version, the variant listed in Iurie Colesnic's edition, without diacritics, is more plausible, since Russian doesn't use the letter *ă*, and when it is borrowed from Romanian it is replaced with *a* or *e*. Therefore, the appellative, being a proper noun and, more critically, of Latin origin, was borrowed and adapted by Russian speakers, and the determinative, translated and updated as to agree in number and gender with the appellative.

Sărata Nouă — Is a German colony, in Acherman County, Sărata volost, also known as Gnadenthal (Arobre 2012, 364). This settlement was founded in 1833 by colonists from Würtemberg (Arbore 1898, 103) and named Gnadenthal. Throughout time, several German colonies were founded in Bessarabia, which were organised by the Russian administration into three ranges (Rmn. *ocol*; Rus. *okrug*). One of them was called Sărata, encompassing the colonies Sărata, Gnadenthal and Lichenthal (Arbore 1898, 223). Probably the name of the settlement evolved into Sărata Nouă from the range to which it belonged from the territorial and administrative point of view, with the determinative from the adjectival syntagma advertising the "age" of the settlement. The edition curated by Iurie Colensnic mentions that the current name of the settlement is Dolinovka, in Artsyz raion, Odessa Oblast, Ukraine. This name is fully Russian, justified by the new territorial-administrative status. It can be assigned to the category of compounded toponyms, obtained from the merger of a determinative with a generic element, according to the specificities of the Russian language (Ungureanu, Boamfă 2006, 170). Thus, it is possible that the particle *doli* is a transformation of the Russian *soly*, соль ('salt'), and the particle *novka* a translation, adaptation and agreement of the original Romanian adjective *nou*.

Sărata Răzeşi — Is a rural commune, in Ismail County, located in the Prut valley, near the mouth of the Lăpuşna rivulet, a kilometre away from the Prut River, also known as **Gura Sărată** (Arbore 2012, 364). The toponym is formed by an appositive syntagma, represented by the appellative Sărata and the determinative Răzeşi, acting as proper noun, reflecting the social status of the inhabitants.

Sărata Veche — Is a large village, found inside a pothole, to the right of the Gârla Mare valley, called by Russian speakers **Sărata Staraia** (Arbore 2012, 364) / **Sarata-Staraia** (Arbore 2001, 183). The designation is formed by an appellative, as a proper noun, and an adjective indicating the "age" of the settlement (*vechi* (m.), *veche* (f.) = 'old') in comparison with Sărata Nouă (cf. supra), located nearby; as such, it is a toponym formed by differentiation. The name used by Russian speakers is a borrowing of the proper name, whereby the letter *ă* was replaced by *a*, and the adjective that determines it was translated.

Sărata de Mijloc — Is a village mentioned in the description of the village Sărata Nouă, in Bălţi County (Arbore 2012, 364), without a dedicated entry in the dictionary. The toponym is formed by a toponymic syntagma composed of a noun, as a proper name, a preposition and an adverb of place (*mijloc* = 'middle'), by means of which the geographical position of the denominated object is indicated in relation to other elements in its vicinity, from which it derives through differentiation.

Sărătenii Vechi — Is the current name of the village Sărăţeni, in Orhei County (Arbore 2001, 183). The toponym is formed from an adjectival syntagma, derived from the previous name of the settlement, stressing its "age" in the context of the current reality. The appellative is given by a noun with a definite article that underwent a graphical change, namely the replacement of *ţ* with *t*.

Sărăţica Veche — Is the current name of the village Sărăţica, which belongs to Lăpuşna County (Arbore 2001, 183), formed by an adjectival syntagma by means of which the "age" of the settlement is underlined, deriving through differentiation from the toponymic core expressed by the old name of the settlement.

Sărăţâca Nouă — Is a village in Bender County, also recorded under the name Tartaul. It belongs to the Gura Galbână volost, located in the valley near Sărâţâca (Arbore 2012, 365). The second edition of the dictionary specifies the literary variant of the appellative, and the present-day territorial-administrative status of the settlement inside Lăpuşna County (Arbore 2001, 184). The determinant is an adjective through which the "age" of the settlement and its affinity to the differentiatory toponymic field of the hydronym Sărata.

Sărăţenii Noi — Is a village in Ismail County, in the Sărata valley, at the mouth of the Ciadâr, also known under its Russian variant **Novîe-Sărăţeni** (Arbore 2012, 365)/ **Novâe Saraţenî** (Arbore 2001, 183). Iurie Colesnic lists the current name and territorial-administrative status of the settlement as Sărăţeni, Lăpuşna County (Arbore 2001, 183). The Romanian variant requires few comments, being most definitely an adjectival toponymic syntagma composed of a proper noun articulated with the definite article, and the adjective *nou* agreeing in gender and number with the appellative that determines it. With respect to the Russian variants listed in the two editions of the dictionary, the influence of the current grammar rules is visible, as is the absence of a direct contact with this language in the case of the 2012 edition. The Russian form of the Latin toponym, which derives through differentiation from the Sărata valley, is a Russian translation and adaptation of the Romanian name.

Soloneţul Roşietici — Is a village in Soroca County, Floreşti volost, located on the right bank of the Răut River, at the mouth of the Soloneţ (Arbore 2012, 378-379). Iurie Colesnic mentions that the current name of the village is **Roşietici** (Arbore 2001, 191). The toponym derives through differentiation from the name of the river at the mouth of which it is located. It is composed of an adjectival syntagma, represented by an appellative of Slavic origin articulated with the definite article, and an unarticulated adjective acting as a proper name. The determinative that composes the toponym is based on the adjective *roşu* ('red'), nuanced by means of the particle *-ietici*, indicating the masculine plural. Noteworthy is the association of two terms of different origins (Slavic and Latin) and their officialisation in a region where the Russian language and alphabet entrenched itself along all levels.

Morphonyms

Morphonyms formed from simple appellatives

Salomeţul — Is a valley in Hotin County, south of the village Glina, with a Russian variant **Soloneţ**. It opens in the Larga valley, in the area of the mouth of the Hosman valley (Arbore 2012, 359). This toponym should be assigned to the category of salt-related designations since it appears to be a variant of the term *soloneţ*, which underwent graphical changes. It is formed from a simple appellative represented by a noun articulated with the definite article.

Sărata — Is a valley in Chişinău and Ismail counties. It starts in Chişinău County, between the market town of Hânceşti and the village of Lăpuşna, and goes southwards, entering, near the village of Nirişeni, into Bender County; east of the village Orac, it enters into Ismail County. From the area of Trajan's Wall, it changes its direction towards SW, going this way up to near the Prut valley, into which it opens, south of the village of Hănăseni. The length of the valley is 63 km. From the left it receives the Nirişeni and Sărăţica valleys; from the right, the Orac, Ciadâr, Ţâghira, Cugurlui, Cazanciu, and Copcui valleys. At the mouth of the valley there is Lake Sărata. At the mouth of the Sărăţica, the Sărata valley is crossed by Trajan's Wall. The Sărata valley contains the villages Sărata Galbână, Sărăţenii Noi and Cazanciuc, while a pothole at its left houses the village of Caracui (Arbore 2012, 363-364). The toponym is a simple appellative of Latin origin, often encountered in the Bessarabian toponymic landscape and analysed in detail by other dictionaries listing it. A vast mixed toponymic field can be easily drawn from the description of this valley, which provides the manner by which designations for geographical elements are formed in this area.

Sărătura — Is a valley in Bălţi County, called in Russian **Sarutura** (Arbore 2012, 365) / **Saratura** (Arbore 2001, 183). It starts west of Mihăileni village, runs on a S-E direction up to near the Ochiul Alb valley, near the village of the same name and Nicoreni. The toponym is formed from the appellative *sărătură*, meaning "floodplain with soil rich in efflorescences" (Porucic 1931, 37), "region with land rich in soluble soils" (Porucic 1931, 50) or "salty land" (Porucic 1931, 63), articulated with the definite article. In what concerns the Russian variant of the toponym of Latin origin, it is a case of borrowing and adaptation to the rules of the Russian language; I have already mentioned the absence of the vocal *ă*, replaced in this case with *a* or *u*.

Sărăţâca — Is a valley on the border separating Bender County from Ismail County. The length of the valley is 15 km, running along a S-S-W direction. The valley receives, from the left, the Ciauru, Musa Murza and Haidarzui valleys. The valley contains the village of Sărăţâca. The mouth of the Ciauru valley is crossed by the Upper Trajan's Wall (Arbore 2012, 365). The toponym has already been analysed in the categories dedicated to the hydronyms and oikonyms, developing together a polarising toponymic field.

Soloneţul (Valea) — Is a valley in Hotin County, Noua Suliţă volost. It starts from the potholes near the village of Rângaciuş, turning southwards and going in parallel with the Malica valley. At the end of the valley there are the springs of the Rângaciul brook. The valley is 7 km in length, according to the map of the Russian General Staff (Arbore 2012, 378). It is interesting that this valley is not mentioned in the edition curated by Iurie Colesnic. Considering how the toponym is mentioned, it can be assumed that the official designation in Soloneţul, while the appellative *vale* ('valley'), articulated, even though written with initial capital letter, is only used in order to apply a cartographic rule of recording geographic nomenclature. The observation can be supported by empirical reality, which proves that the populations doesn't use the construction *Valea Soloneţul*, but only *Valea Soloneţului* or *Soloneţul*.

Soloneţul — Is a valley in Soroca County. It starts in Bălţi County, on the domains of Osmolovski; up to Drăgăneşti, it goes eastwards; from here it goes north-eastwards up to the Răut valley, which opens into the Soloneţul Roşieşti village. The length of the valley: 37 km, according to the map of the Russian General Staff. From the village of Rădoaia, it enters Soroca village. From the left it receives the mouths of the Rădoaia and Hurienii valleys. The rivulet Soloneţul runs along its course (Arbore 2012, 378). The designation is part of the mixed toponymic field of the hydronym Soloneţul.

Soloneţul — Is a valley in Soroca County. It starts east of the village Dinjeni, and goes eastwards, then south-eastwards until it opens into the Cubolta valley, into which the village Moşeni opens to the right. From the left it receives the Taraga valley. It is 14 km long, according to the map of the Russian General Staff (Arbore 2012, 378).

Soloneţului (Valea) — Is a valley in Soroca County, with a length of 3 km according to the map of the Russian General Staff. It starts from the village of Soloneţul and ends at Trajan's Wall, whence it is called Sloboziei valley (Arbore 2012, 378). The toponym derives, through differentiation from the village Soloneţ, which is found in the space described by it. Even though in the available editions the appellative is indicated in parentheses, the form of the toponym is obviously Valea Soloneţului. According to the manner of formation, it is composed of a genitival toponymic syntagma that emphasizes the proximity of the village from which it evolved. It is very interesting the association of Latin-origin appellative and a Slavic-origin determinant.

Odonyms

Odonyms formed from simple appellatives

Ocnița — Is the railway station in Soroca County, on the Ocnița-Bălți line, with a length of 86 versts, linked: from one side with Jmerinca, and from the other side with Noua Suliță (Arbore 2012, 312). The toponym polarizes from the oikonym that designates the settlement served by the railway.

After analysing the array of toponyms related to the element salt from across the territory of Bessarabia, identified in Zamfir Arbore's geographical dictionary, it can be stated that the salt-related toponyms are quite frequent in the region. The majority are based on the Latin adjective *sărat*, in the nominalised form. From the collection of designations encompassing variations of the toponym *Sărata*, the most prominent are those that underline a chronological differentiation (Alexianu 2011, 580), e.g. *Sărata Nouă, Sărata Veche, Sărătenii Vechi, Sărătenii Noi, Sărățica Veche* sau *Sărățâca Nouă*, or of location, e.g. *Sărata de Mijloc*. Another differentiation is found among the oykonymic determinations, in the nominative case, of designations of a social status: *Sărata Răzeși*. In the southern part of the region, the toponym *Tuzlî*, of Turkish origin, was recorded. In what concerns the toponyms of Slavic origin, there is a wider variety, namely designations derived from the notions *solonț, slănic* and *ocnă*. The toponyms based of these lexemes formed, for the most part, through suffixation (*Ocnița, Solonceni*) and differentiation, creating toponymic syntagmas (*Solonețul Roșietici, Valea Solonețului, Ocnele Buhnei, Ocnița-Țărani, Ocnița-Răzeși, Răcăria Slănic*).

Analysing the placenames listed above from the perspective of the theory of toponymic fields (Moldovanu 2010), several toponymic cores become apparent, from which other designations derive through polarisation or differentiation, as follows (Figure 1-7):

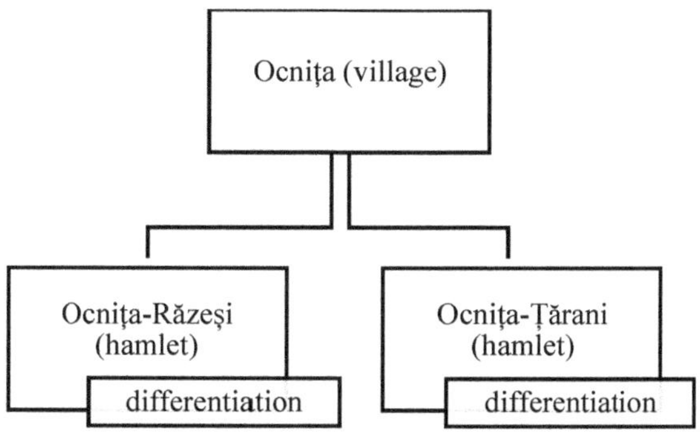

Figure 1. The toponymic field through differentiation developed around the oikonymic core *Ocnița*.

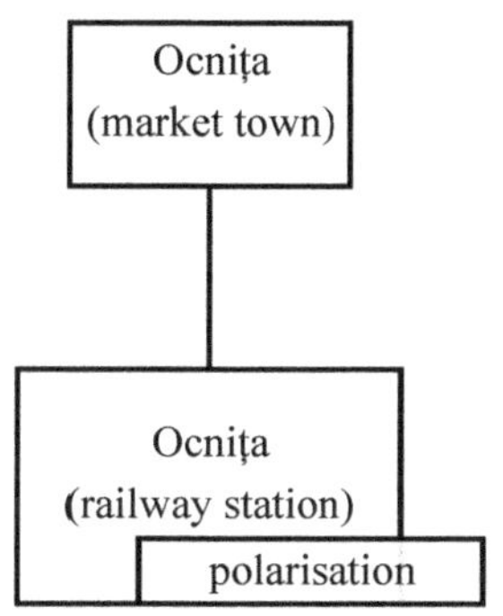

Figure 2. The toponymic field through polarisation developed around the oikonymic core *Ocnița*.

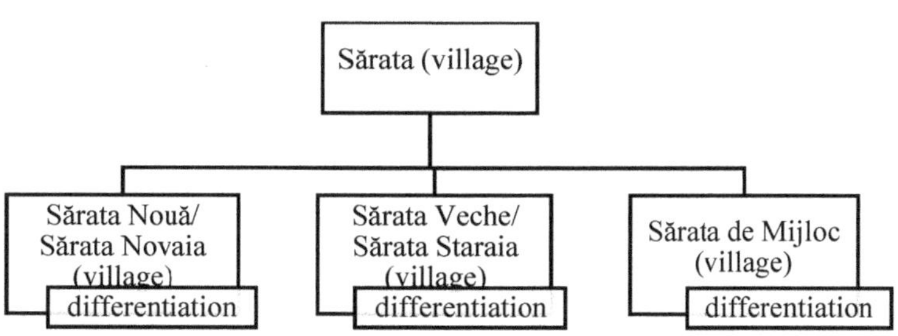

Figure 3. The toponymic field through differentiation developed around the toponymic core *Sărata*.

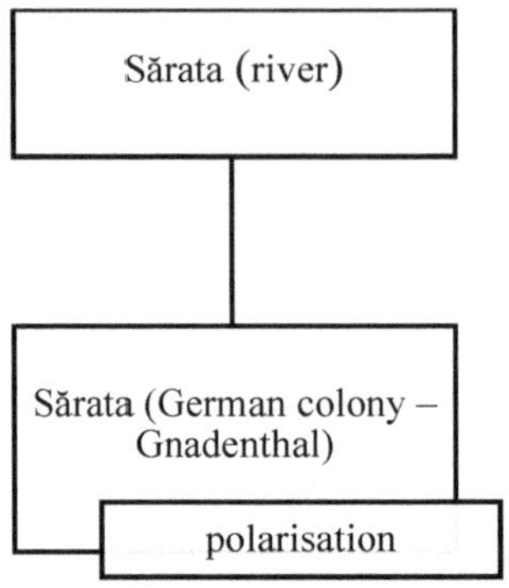

Figure 4. The toponymic field through polarisation developed around the hydronymic core *Sărata*.

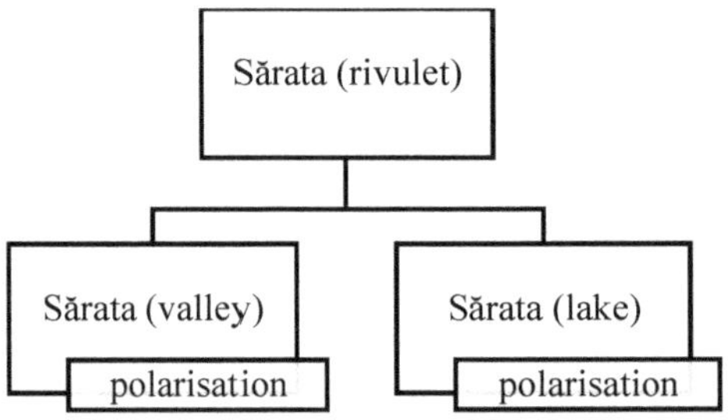

Figure 5. The toponymic field through polarisation developed around the hydronymic core *Sărata*.

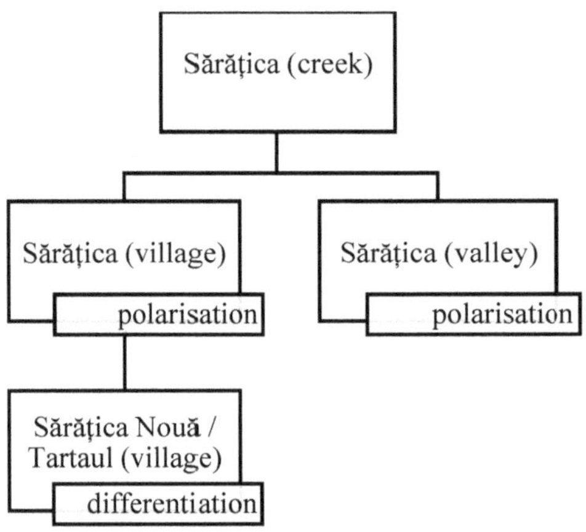

Figure 6. The mixed toponymic field developed around the hydronymic core *Sărăţica*.

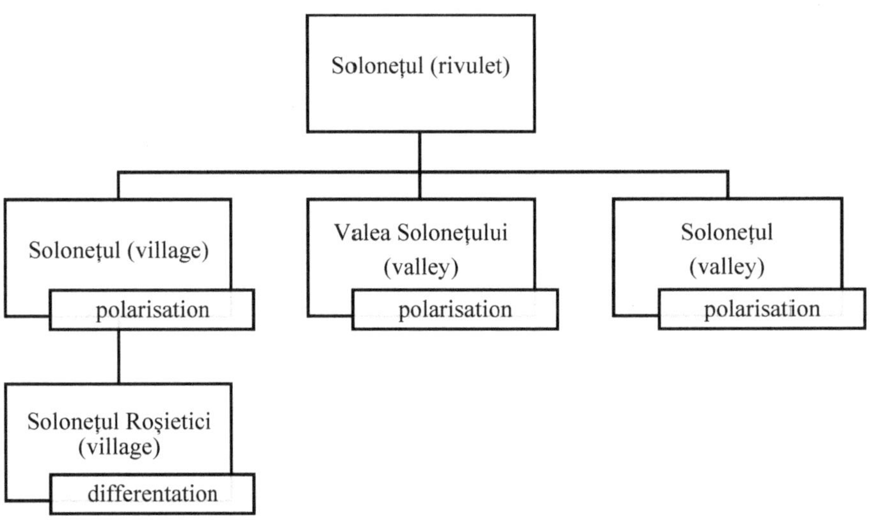

Figure 7. The mixed toponymic field developed around the hydronymic core *Soloneţul*.

Compared with the area west of the Prut, Bessarabia lacks toponyms based on the notions both of Latin origin *saramură*, *murătoare*, and of Slavic origin, *salină* or *slatină*, perhaps due to the absence of the saline manifestations that gave rise to these designations.

The strong Slavic (and not only) influence in the area east of the Prut is conspicuous in the common speech, with the speakers combining words of different origins and even from different languages, but also in the official language, and even in the toponymy. Thus, if in some cases of placenames identified in Arbore's dictionary it was possible to ascertain the radical related to the element *salt*, even if it underwent transformations, in the case of other toponyms it is not possible to establish, at the current state of research, if they are or aren't halotoponyms. This is valid for the following placenames: Dormindere (Colonceac) (Bulgarian colony), Dolineni (village), Morovaia (hamlet), Muravlevca (village), Murtaza (lake), Sariar (brook and valley), Sarlâc (village), Selemet (village), and Tuzoara (market town).

In conclusion, I must mention that the toponymy of salt, both at the level of microtoponymy and of official toponymy in Bessarabia is concurrent with that from all Romanian territories (Alexianu 2011), being extremely similar.

Acknowledgement
This work was supported by a grant of the Romanian National Authority for Scientific Research, CNCS-UEFISCDI, project number PN-II-ID-PCE-2011-3-0825, *The Ethno-archaeology of Salt Springs and Salt Mountains from the extra-Carpathian areas of Romania* – www.ethnosalro.uaic.ro

References

Alexianu, M. 2011. Sarea: reflexe onomastice în limba română. In O. Felecan (ed.), *Numele şi numirea. Actele conferinţei Internaţionale de Onomastică. Ediţia I: Interferenţe multietnice în antroponimie, Baia Mare, 19-21 septembrie 2011*, 579-583. Cluj-Napoca, Mega.

Arbore, Z. 1898. *Basarabia în secolul XIX* (operă premiată şi tipărită de Academia Română). Bucureşti, Institutul de Arte Grafice Carol Gobl, Furnisor al Curţii Regale, 16, Strada Doamnei, 16.

Arbore, Z. 2001. *Dicţionarul geografic al Basarabiei* (Editor Iurie Colesnic). Colecţia Testament, Museum. Chişinău, Fundaţia Culturală Română.

Arbore, Z. 2012. *Dicţionarul geografic al Basarabiei* (Ediţie îngrijită şi prefaţă de I. Oprişan). Bucureşti, Saeculum I.O.

Emilciuc, A. 2011. Comerţul cu sare în Basarabia în contextul includerii provinciei în sistemul economic al Imperiului Rus (1812-1850), *Tyragetia* serie nouă, vol. V [XX], 2, Istorie. Muzeologie, 55-70.

Eremia, A. 2012. Identitate şi spiritualitate românească în toponimia Basarabiei. *Akademos* 4, decembrie, 140-143.

Eremia, A. 2013. Hidronimia bazinului hidrografic al Mării Negre. *Buletin de lingvistică* 14, 89-93.

Eremia, A. 2014. Hidronimia Basarabiei. Originea şi semnificaţia numelor de râuri. *Limba Română* 1, XXIV.

Iordan, I. 1963. *Toponimia românească.* Bucureşti, Editura Academiei Republicii Populare Române.

Moldovanu, D. 2010. *Teoria câmpurilor toponimice (cu aplicaţie la câmpul hidronimului Moldova).* Iaşi, Editura Universităţii „Alexandru Ioan Cuza" din Iaşi.

Ungureanu, A. and Boamfă, I. 2006. *Toponomastică.* Iaşi, Sedcom Libris.

Porucic, T. 1931. *Lexiconul termenilor entopici din limba română în Basarabia.* Arhiv. Bas. An. II. Nr. 1, 2, 3, 4 şi An. III nr. 2. Chişinău, Tipografia Eparhială Cartea Românească.

Poruciuc, A. 2011. Etymological and Historical Implications of Romanian Place-Names Referring to Salt, in M. Alexianu, O. Weller, R.-G. Curcă (eds.), *Archaeology and Anthropology of Salt: a Diachronic Approach. Proceedings of the International Colloquium, 1-5 October 2008 "Al. I. Cuza"University (Iaşi, Romania)*, 215-217. BAR International Series 2198. Oxford, BAR Publishing.

Tabuncic, S. 2011. Vechi cărţi bisericeşti semnalate de Zamfir Arbore în *Dicţionarul Geografic al Basarabiei* la începutul veacului al XX-lea. In Departamentul pentru Românii de Pretutindeni, Universitatea „Dunărea de Jos" Galaţi and Asociaţia Profesorilor de Istorie de la Dunărea de Jos (Coord.), *La Frontierele civilizaţiilor — Basarabia în context geopolitic, economic, cultural şi religios: conferinţă internaţională: Galaţi, 21-23 octombrie 2010*, 95-112. Galaţi, Galaţi University Press.

Online source

http://limbaromana.md/index.php?go=articole &n=2370

Use of Saline Aerosol Devices in School Gymnasiums: a Romanian Experiment

Ion Sandu
Universitatea „Alexandru Ioan Cuza" din Iaşi

Maria Canache
Şcoala „Ştefan cel Mare" Zemeş

Andrei-Victor Sandu
Universitatea Tehnică „Gheorghe Asachi" Iaşi

Viorica Vasilache
Universitatea „Alexandru Ioan Cuza" din Iaşi

Ioan Gabriel Sandu
Universitatea Tehnică „Gheorghe Asachi" Iaşi

Abstract
The paper presents the current aspects related to installing, in gymnasiums, certain devices generating saline aerosols, in order to have "fresh air" environments, adequate for an optimal unfolding of physical education and sports classes. The correct and systematic practice of motion games within an environment purified with saline aerosols may have a positive impact, under the current circumstances of intense environment pollution. It may also have a positive impact upon students' school and competitive activity, and upon human performances, implicitly. It contributes substantially to developing students' resistance to illness, to a better body functioning, and to higher sports and school performances.

Keywords
saline aerosols, SALIN devices, halochamber, physical education

Our research team elaborated a series of superficial artificial halochamber systems (Sandu *et al.* 2010a; Sandu *et al.* 2010b; Sandu *et al.* 2011a; Sandu *et al.* 2011b; Sandu *et al.* 2011c; Sandu *et al.* 2011d; Sandu *et al.* 2011e; Sandu *et al.* 2013) for multiple uses, such as practicing physical exercises or training practices for increasing sports performances, increasing the mitigation rate for certain cardiorespiratory and psycho-neuromotor diseases, or for prophylactic antivirus purposes (Canache 2013).

In order to improve the parameters of the cardiorespiratory and the psycho-neuromotor system for students involved in intense physical activities and in order to treat or mitigate certain diseases, it is possible to install saline aerosol-generating systems in gymnasiums, too. Depending on the size of the gymnasium and on its other structural-functional characteristics, these saline aerosols-generating systems must allow certain concentrations of saline aeroions, with a predetermined composition and a constant variation in time. It all depends on its use; there are multiple applications, in the therapy of respiratory conditions and even for improving the parameters of the cardiorespiratory and the psycho-neuromotor system of the human subjects involved in intense physical activities (Sandu *et al.* 2010a; Sandu *et al.* 2010b; Sandu *et al.* 2011a; Sandu *et al.* 2011b; Sandu *et al.* 2011c; Sandu *et al.* 2011d; Sandu *et al.* 2011e; Sandu *et al.* 2013).

The gymnasium is a sealed, dry room, with double pane glass and ionized glass, with UV filters, (Figure 1) whose unique entry is through an antechamber (from which one can also access the other rooms with destinations related to the activity of the halochamber), with a capacity of 588 m^3 air and 14(L)x7(l)x6(h), respectively.

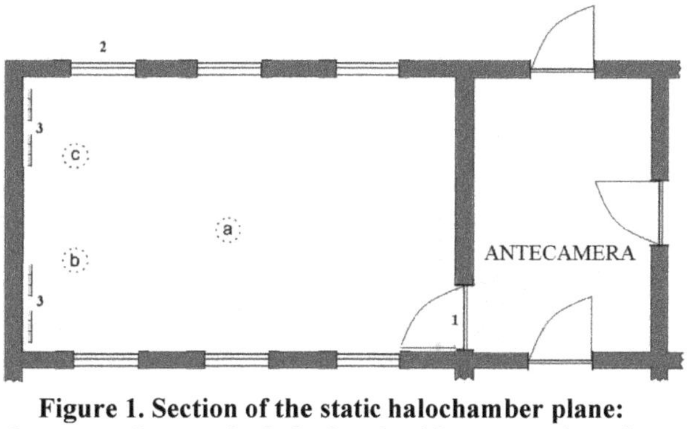

Figure 1. Section of the static halochamber plane: 1 – access door to the halochamber/the gymnasium, 2 – double pane glass with UV filter, 3 – aerosol-generating devices; a, b, and c are the areas from where we take saline aerosol samples for analysis.

In the room, in the upper corners of the wall opposed to the entrance, (with no windows) four devices generating semi-dry aeroions were installed; they are called SALIN Plus (produced by Tehnobionic Buzău). These devices comprise a fan and a diaphragm that contains extruded porous grains of recrystallized halite from supersaturated solutions. The four devices continuously recycle the air within the gymnasium.

The fans allow the circulation of 28-32m^3/h, thus of 112-128m^3/h together. In order to drop to the minimum level in the saline aerosols, it is necessary to turn on the four devices at least 3-6 ore before using the halochamber for sports activities. Through erosion, the four devices take the salt nanoparticles from the labile crystallites of the surface efflorescences and then they disperse them into the halochamber's atmosphere where, under the influence of air humidity, they reformulate structurally as oligomers of salt nanocrystallites and pentahydrol, taking the shape of a spherical glomerule, which behaves as a negative aeroion. At the surface, glomerules feature a monomolecular layer of water dipoples with negative charge on the outside.

In the gymnasium, practices comprise one hour, twice a week, with groups of students (by age or form). Certain groups practice on Monday and Thursday, while others on Tuesday and Friday, from 10.00 AM to 6.00 PM, with at least one-hour break after each working session for determining the parameters required by the experiment. The gymnasium can be ventilated periodically and it does not have to be perfectly sealed. It is relatively easy to obtain and maintain for a long period optimal concentration level with constant activity in aerosols; it allows co-assisting the dry NaCl aerosols with other salts necessary for mitigating diseases or for improving human performance. They are highly viable and they provide the possibility of self-regulating negative aerosols, which is ideal for various therapeutic and sporting activities. Initially, the air in the chamber was climatized for humidity (RH) of 60...65% and a temperature of 20...22°C.

Determining the level of saline aerosols and the microclimatic parameters

Circa six hours before the beginning of sports training, the saline aerosol-generating devices were switched on in the gymnasium. Half an hour before beginning the practice with students, the environment and functional characteristics of the room are determined; the same characteristics are measured after the physical education class. For measuring and monitoring the atmosphere composition in the halochamber, we used both methods for establishing the climatic parameters of a room, and a series of specific methods, such as the conductometric determination of saline aerosols concentration, amount of particles, and lifespan by using the laser optical particle counter and the aeroions dosimeter (Sandu and Chirazi 2010; Pascu, Sandu, Ciobanu, Sandu, Vasile, Ciobanu, Sandu, Pascu 2009).

Because the dynamic of saline aerosols emissions within the gymnasium is influenced by environmental conditions, all measurements must consider constant values for the parameters of pumped air. The microclimate created in the gymnasium must be monitored permanently; through climatization, we create the artificial conditions of temperature and humidity required by the training routine. The measurements were conducted in three areas of the halochamber: in the central point of the room, in an upper corner opposed to the door wall, around 400 mm away from the walls and the ceiling and in the middle, near the floor, 400 mm away from the window and the floor (Figure 1- a, b, and c). Relative temperature and humidity (ET and RH) were measured by means of a digital hygro-thermometer; pressure using a barometer, and illumination using an electronic lux meter. The climatic characteristics were maintained, according to Table1.

Table 1. Climatic characteristics of the gymnasium (school halochamber).

System	Parameter	Variations of climatic characteristics
Gymnasium	C, mg/m³, saline aerosol	0.649-0.705
	T°C	20-21
	P,atm	740-750
	UR, %	60-65
	I, lx	80-85

In the gymnasium, both before the working sessions and at the end, we measured the concentration, granulometry, volume, and lifespan of saline aerosols in the halochamber by means of a differential conductivity meter and of a particle counter based on an optical laser system (SIBATA GT 321).

The action of aerosols on the body

Halotherapy is considered a very simple therapeutic procedure; it is attractive even for children because it does not involve drugs or supplements, movement restriction, or diets. It consists of both *inhaling* saline aerosols through the airways, and *absorbing* saline aerosols through the skin (Canache *et al.* 2011; Canache *et al.* 2012).

It is known that saline aerosols dissipate in a relatively uniform manner within the halotherapeutic chamber due to diffusion. They are permanently combined with water oligomers as vapours that make up atmospheric humidity. Exposing the body (and thus the skin) to this atmosphere determines the direct contact with aerosols, which are absorbed by skin capillaries and then carried to the tissues (to the cell level). Salt ions are absorbed by both the lymphatic system and the cells, through diffusion and osmosis. The living cells of the body get their supply of minerals, lipids, carbohydrates, proteins, water, etc, through these processes (Crişan 2011).

The beneficial effects of halotherapy are as follows:

-increase in expectoration efficiency, by increasing the volume of sputum and secretions in the airways, which leads to an improvement of breathing and to the patient's feeling of liberation, of pulmonary release (because coughing is more productive);
-long-term bacteriostatic effect, proven by the mitigation of current respiratory infections or by the complete elimination of seasonal episodes (it is beneficial for patients with COPD, for whom an acute episode of respiratory infection can be fatal);
-improvement of sleep quality, by reducing the oedema of the upper aerodigestive tract (palatal vault, tongue root, tonsil crypts), often present in chronic snorers (Chiruţă, Moscaliuc and Pascu 2008).

Salt also comprises other chemical elements, a multitude of minerals and particles that help balancing the most important physiological processes of the body, mineralizing and hydrating it. Therefore, halotherapy has a wide spectrum of beneficial effects, as follows:
-iodine is necessary for the normal functioning of the thyroid and the digestive processes;
-magnesium has an important role in the prevention of allergic reactions; it is necessary for vitamin intake, for the normal functioning of muscle and nerves, for the stabilization of heart rate and of blood sugar levels;
-calcium contributes to the formation of bone and muscle tissue and it is responsible for blood clogging;
-iron helps eliminating toxins from the body and transporting oxygen, (it is included in the haemoglobin) it prevents anaemia, and it increases immunity;
-potassium improves the metabolic processes and contributes to the regulation of water balance within the cells;
-sulphur has a main role in immunological processes and it is compulsory for the synthesis of albumins;
-sodium controls blood pressure and it has an important role in maintaining the acid-base balance of the body;
-selenium prevents free radical damage and it helps regulating the thyroid;
-lithium reduces the levels of body chemicals, which cause behaviour issues and irritability;
-manganese contributes to healthy skin, bones, and cartilages.

Conclusions

By constantly using halotherapy in gymnasiums during school classes, we can achieve effective prophylaxis for both children who live and activate in urban agglomerations, and for those who cannot benefit from natural salt mine effects (they lack either time or money).

Long-term treatments ensure an effective remedy against stress, a remineralisation of all tissues down to cell level, and the longevity of the modern man, so much affected by the current environment, overwhelmed by permanent pollution.

In order to influence positively the body functioning and the balance and stability between segments and the body, and between the body and the environment, physical education classes may take place in a therapeutic setting, with natural or artificial saline aerosols. The program of sports activities, adapted to the age particularities of students, in a setting that includes saline aerosols, may be beneficial for the organism. In time, it may increase the ventilation power of the breathing system, the functional capacity of the body, the sanguine level of hormones (accelerating metabolism, stimulating the growth process in children, increasing muscle mass, reducing osteoporosis risks). In addition, it may delay aging, increase stress resistance, and general comfort.

Acknowledgement

This work was supported by a grant of the Romanian National Authority for Scientific Research, CNCS-UEFISCDI, project number PN-II-ID-PCE-2011-3-0825, *The Ethno-archaeology of Salt Springs and Salt Mountains from the extra-Carpathian areas of Romania* – www.ethnosalro.uaic.ro.

References

Canache, M. 2013. *Cercetări privind influenţa aerosolilor salini asupra îmbunătăţirii activităţii şcolare a elevilor,* Doctoral dissertation, Iaşi.

Canache, M., Sandu, I., Drochioiu, G., Lupaşcu, T., Sandu, I.G. 2011. Therapeutic role of saline solions in apolluted environment, in *Romanian Creativity in European Context*, 233-240. Iaşi, Tehnopress.

Canache, M., Sandu, I., Chirazi, M., Lupaşcu, T., Sandu, I.G. 2012. Saline aerosols influence on growth staturo-weight children. *Present Environment and Sustainable Development* 6(2), 221-234.

Chiruţă, R., Moscaliuc, Gh.V., Pascu, C. 2008. *Considerente fiziologice asupra aerosoloterapiei saline*, 1-2. Iaşi.

Crişan, R. 2011. *Folosirea haloterapiei în tratamentul patologiei obstructive pulmonare.* Doctoral dissertation. Iaşi.

Pascu, C., Sandu, I., Ciobanu, G., Sandu, I.G., Vasile, V., Ciobanu, O., Sandu, A.V., Pascu, A. 2009. *Patent RO122232 (B1).*

Sandu, I. and Chirazi, M. 2010. *Ecologia sistemelor sportive.* Iaşi, Performantica.

Sandu, I., Canache, M., Lupaşcu, T., Sandu, I.G., Sandu, A.V., Vasilache, V. 2010a. *Patent MD4040 (B1).*

Sandu, I., Ştirbu, C., Lupaşcu, T., Ştirbu, C., Sandu, A.V. 2010b. *Patent MD4039 (B1).*

Sandu, I., Canache, M., Sandu, I.G., Sandu, A.V., Vasilache, V. 2011a. *Patent RO126283 (A2).*

Sandu, I., Ştirbu, C., Chirazi, M., Ştirbu, C., Sandu, A.V. 2011b. *Patent RO126285 (A2).*

Sandu, I., Ştirbu, C., Sandu, A.V. 2011c. *Patent RO126284 (A2).*

Sandu, I., Ştirbu, C., Lupaşcu, T., Chirazi, M., Ştirbu, C., Sandu, A.V. 2011d. *Patent MD4089 (B1).*

Sandu, I., Canache, M., Vasilache, V., Sandu, I.G. 2011e. The effects of salt solions on the health of human subjects. *Present Environment and Sustainable Development* 5, 67-88.

Sandu, I., Canache, M., Lupaşcu, T., Chirazi, M., Sandu, I.G., Pascu, C. 2013. The influence of physical doping with other salts on aerosols Generations. *Aerosol and Air Quality Research* 13(6), 1731-1740.

Index of Authors

Marius ALEXIANU
(Universitatea „Al. I. Cuza" din Iaşi, România)
alexianumarius@yahoo.com

Mugur ANDRONIC
(Muzeul Bucovinei, Suceava, România)
andronicmugur@yahoo.com

Mihaela ASĂNDULESEI
(Universitatea „Al. I. Cuza" din Iaşi, România)
mihaela.asandulesei@yahoo.com

Ludmila BEJENARU
(Universitatea „Al. I. Cuza" din Iaşi, România)
ludbej2059@yahoo.com

Robin BRIGAND
(Maison de l'Archéologie et de l'Ethnologie UMR 8215 *Trajectoires*, Nanterre, France)
robinbrigand@yahoo.fr

Maria CANACHE
(„Ştefan cel Mare" Middle School, Zemeş, România)
mariacanache@yahoo.com

Ovidiu CÎRSTINA
(Complexul Muzeal Naţional „Curtea Domnească" Târgovişte, România)
ovidiu_cristina@yahoo.com

Alexandra COMŞA
(Institutul de Arheologie "Vasile Pârvan" Bucureşti, România)
alexcomsa63@yahoo.com

Roxana-Gabriela CURCĂ
(Universitatea „Al. I. Cuza" din Iaşi, România)
roxanigabriela@yahoo.com

Roxana-Elena DIACONU
(Muzeul de Istorie şi Etnografie Târgu-Neamţ, România)
roxana_b82@yahoo.com

Vasile DIACONU
(Muzeul de Istorie şi Etnografie Târgu-Neamţ, România)
diavas_n82@yahoo.com

Tilman B. DRÜECKE
(INSERM Unit-1018, Team 5, CESP, Villejuif, France)
tilman.drueke@inserm.fr

Gheorghe DUMITROAIA
(Complexul Muzeal Judeţean Neamţ, Piatra-Neamţ, România)
muzeupn@yahoo.com

Sebastian FINK
(Universität Innsbruck, Österreich)
sebastian.fink@uibk.ac.at

Daniel GARVĂN
(Muzeul Judeţean Buzău, România)
daniel_garvan@yahoo.com

Ana ILIE
(Complexul Muzeal Naţional „Curtea Domnească" Târgovişte, România)
ana_arheo@yahoo.com

Cornelia-Magda LAZAROVICI
(Institutul de Arheologie, Iaşi, România)
magdamantu@yahoo.com

Gheorghe LAZAROVICI
(Universitatea „Lucian Blaga" Sibiu, România)
ghlazarovici@yahoo.com

Ileana Oana MACARI
(Universitatea „Al. I. Cuza" din Iaşi, România)
oanam@mail.uaic.ro

Bernard MOINIER
(Consultant, Paris, France)
bernard.moinier@tele2.fr

Roxana MUNTEANU
(Muzeul Judeţean Buzău, România)
e_r_mro@yahoo.com

Mihai NĂSTASE
(Complexul Muzeal Naţional „Curtea Domnească" Târgovişte, România)
nastase_mc@yahoo.com

Bogdan NICULICĂ
(Muzeul Bucovinei, Suceava, România)
niculicab@yahoo.com

Lăcrămioara OCHIUZ
(Universitatea de Medicină şi Farmacie „Grigore T. Popa" Iaşi, România)
ochiuzd@yahoo.com

Gheorghe OLTEANU
(Complexul Muzeal Naţional „Curtea Domnească" Târgovişte, România)
olteanu_muzeu@yahoo.com

Răzvan Victor PANTELIMON
(Universitatea Ovidius din Constanţa, România / Universidad Católica de Valparaíso, Chile)
pantelimonr@yahoo.com

Mihaela PARASCHIV
(Universitatea „Al. I. Cuza" din Iaşi, România)
mihaelaparaschiv_clasice@yahoo.com

Florin PETRICĂ
(Complexul Muzeal Naţional „Curtea Domnească" Târgovişte, România)
floringabrielpetrică@yahoo.com

Dragomir Nicolae POPOVICI
(Muzeul Naţional de Istorie a României, Bucureşti, România)
mirel_d_n_p@yahoo.com

Adrian PORUCIUC
(Universitatea „Al. I. Cuza" din Iaşi, România)
apor@uaic.ro

Constantin PREOTEASA
(Complexul Muzeal Judeţean Neamţ, Piatra-Neamţ, România)
constantin.preoteasa@yahoo.com

Gheorghe ROMANESCU
(Universitatea „Al. I. Cuza" din Iaşi, România)
romanescugheorghe@gmail.com

Andrei-Victor SANDU
(Universitatea Tehnică „Gheorghe Asachi" Iaşi, România)
sav@tuiasi.ro

Ion SANDU
(Universitatea „Al. I. Cuza" din Iaşi, România)
sandu_i03@yahoo.com

Ioan Gabriel SANDU
(Universitatea Tehnică "Gheorghe Asachi" Iaşi, România)
gisandu@yahoo.com

Nicolae URSULESCU
(Universitatea „Al. I. Cuza" din Iaşi, România)
n.ursulescu@gmail.com

Viorica VASILACHE
(Universitatea „Al. I. Cuza" din Iaşi, România)
viorica_18v@yahoo.com

Mădălin-Cornel VĂLEANU
(Complexul Naţional Muzeal „Moldova" Iaşi, România)
madalinvaleanu@yahoo.com

Olivier WELLER
(CNRS-UMR 7041 Protohistoire européenne, Maison de l'Archéologie et de l'Ethnologie, Paris, France)
olivier.weller@mae.cnrs.fr